Statistics for Industry and Technology

In Honor of
Shanti S. Gupta

SHANTI S. GUPTA

Advances in Statistical Decision Theory and Applications

S. Panchapakesan
N. Balakrishnan

Editors

1997

Birkhäuser
Boston • Basel • Berlin

S. Panchapakesan
Southern Illinois University
Department of Math
Carbondale, IL 62901-4408

N. Balakrishnan
Department of Mathematics
and Statistics
McMaster University
Hamilton, Ontario L8S 4K1
Canada

Library of Congress Cataloging-in-Publication Data

Advances in statistical decision theory and applications / S. Panchapakesan and N.
  Balakrishnan, editors.
      p.   cm. -- (Statistics for industry and technology)
      Includes bibliographical references and index.
      ISBN 0-8176-3965-9 (alk. paper). -- ISBN 3-7643-3965-9 (Basel :
alk. paper)
      1. Statistical decision.   I. Panchapakesan, S.
   II. Balakrishnan, N., 1956-   .   II. Series.
   QA279.4.A38   1997
   519.5'42--dc21   97-26991
                                                                          CIP

Printed on acid-free paper
© 1997 Birkhäuser Boston

Birkhäuser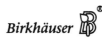

ISBN 0-8176-3965-9
ISBN 3-7643-3965-9
Typeset by the Editors in L^{A}T$_E$X.
Cover design by Vernon Press, Boston, MA.
Printed and bound by Hamilton Printing, Rensselaer, NY.
Printed in the U.S.A.

9 8 7 6 5 4 3 2 1

Contents

Preface

Shanti S. Gupta has made pioneering contributions to ranking and selection theory; in particular, to subset selection theory. His list of publications and the numerous citations his publications have received over the last forty years will amply testify to this fact. Besides ranking and selection, his interests include order statistics and reliability theory.

The first editor's association with Shanti Gupta goes back to 1965 when he came to Purdue to do his Ph.D. He has the good fortune of being a student, a colleague and a long-standing collaborator of Shanti Gupta. The second editor's association with Shanti Gupta began in 1978 when he started his research in the area of order statistics. During the past twenty years, he has collaborated with Shanti Gupta on several publications. We both feel that our lives have been enriched by our association with him. He has indeed been a friend, philosopher and guide to us.

The importance of statistical theory and applications to practical problems is not a matter of opinion anymore but an accepted truth. Decision theory provides valuable tools to choose from competing strategies in business and industry. Statistical problems are formulated in different ways so as to allow the user to make the type of decisions he or she is interested in. This allows and, in a way, necessitates a variety of goals such as estimation, testing of hypotheses, selecting the best of several options using classical methods as well as decision-theoretic and Bayesian approaches. This volume has been put together in order to (i) review some of the recent developments in statistical decision theory and applications, (ii) focus and highlight some new results and discuss their applications, and (iii) indicate interesting possible directions for further research.

With these goals in mind, a number of authors were invited to write an article for this volume. These authors were not only experts in different aspects of statistical decision theory and applications, but also form a representative group from former students, colleagues and other close professional associates of Shanti Gupta. The articles we received have been peer reviewed and carefully organized into 28 chapters. For the convenience of the readers with different interests, the volume is divided into the following seven parts:

- BAYESIAN INFERENCE

- DECISION THEORY

- POINT AND INTERVAL ESTIMATION—CLASSICAL APPROACH

- TESTS OF HYPOTHESES

- RANKING AND SELECTION

- DISTRIBUTIONS AND APPLICATIONS

- INDUSTRIAL APPLICATIONS

The above list has taken into account various types of inferential problems of interest. It should also be stressed here that this volume is **not** a proceedings, but a carefully planned volume consisting of articles consistent with the clearly set editorial goals mentioned earlier.

Our sincere thanks go to all the authors who have contributed to this volume. They share our admiration for the contributions and accomplishments of Shanti Gupta and have given us their hearty cooperation in bringing this volume out. Special thanks are due to Mrs. Debbie Iscoe for the excellent typesetting of the entire volume. Finally, we thank Mr. Wayne Yuhasz (Editor, Birkhäuser) for the invitation to undertake this project and Mrs. Lauren Lavery for her assistance in the production of the volume.

With great pleasure, we dedicate this volume to Shanti Gupta.

S. PANCHAPAKESAN
Carbondale, Illinois, U.S.A.

N. BALAKRISHNAN
Hamilton, Ontario, Canada

APRIL 1997

Shanti S. Gupta—Career and Accomplishments

Shanti Gupta received his B.A. Honours and M.A. degrees, both in mathematics, from the Delhi University in 1946 and 1949, respectively. Soon after getting his M.A., he joined the faculty of mathematics at Delhi College, Delhi, where he spent the next four years as Lecturer in Mathematics. Early in this period, influenced by Professor P. L. Bhatnagar, he was interested in doing research in astrophysics. However, he was gradually attracted to the then emerging field of statistics which was fast developing in India under the pioneering leadership of P. C. Mahalanobis and C. R. Rao. While teaching at Delhi College, he joined the one-year diploma course in applied statistics conducted by the Indian Council of Agricultural Research at New Delhi. This became the launching pad for a long, fruitful and still active career in statistics.

In 1953, Shanti Gupta came as a Fullbright student to the University of North Carolina at Chapel Hill for graduate study in statistics. He received his Ph.D. in mathematical statistics in 1956 for his thesis entitled *On A Decision Rule for A Problem in Ranking Means* written under the guidance of Raj Chandra Bose. With this, he started his research career in Ranking and Selection Theory as one of the three reputed pioneers in the area.

From 1956 to 1962, Shanti Gupta held positions as Research Statistician (1956–1957) and Member of Technical Staff (1958–1961) at Bell Telephone Laboratories; Associate Professor of Mathematics (1957–1958) at University of Alberta, Canada; Adjunct Associate Professor (1959–1961) at Courant Institute of Mathematical Science, New York University; and Visiting Associate Professor (1961–1962) at Stanford University. In 1962, he joined the Department of Mathematics at Purdue University, West Lafayette, Indiana, as Professor of Mathematics and Statistics. He became the Head of the newly created Department of Statistics in 1968 and successfully continued in that capacity until July 1995 when he stepped down to devote full time to teaching and research.

Under the leadership of Shanti Gupta, the Statistics Department at Purdue became one of the topmost in the U.S. He founded the Purdue International Symposium Series on *Statistical Decision Theory and Related Topics* with five conferences to date. He has guided 28 Ph.D. dissertations (list on page xxxix). In recognition of his services to the University and the statistical profession, Purdue University created in 1996 the Shanti S. Gupta Distinguished Professorship for excellence in teaching which is currently held by Prof. David S. Moore.

Shanti Gupta has provided distinguished service to the statistical community at large by his activities in various capacities for professional societies. These include: Director of Visiting Lecturers Programs in Statistics of the Committee of Professional Statistical Societies (COPSS) (1967–1970), Member of COPSS (1988–1991), Founding Editor of the IMS Lecture Notes–Monograph Series in Statistics and Probability (1979–1988), Member of IMS Committee on Mathematical Tables (1960–1973), Member (elected) of IMS Council (1977–1980, 1983–1987), Member of IMS Executive Committee (1983–1987, 1988–1991), Chairman of the Joint Management Committee of the ASA and IMS for *Current Index to Statistics* (1981–1988), and Member of NRC Advisory Committee on U.S. Army Basic Scientific Research (1983–1988). He served as President of the Institute of Mathematical Statistics during 1989–1990.

He has provided valuable service to many research journals in various capacities. Included in this list are: Associate Editor of *Annals of Statistics*, Coordinating Editor and Editor-in-Chief of *Journal of Statistical Planning and Inference*, and Editorial Board Member of *Communications in Statistics*. Currently, he is an Associate Editor of *Sequential Analysis*, Editorial Board Member of *Statistics and Decisions*, and Governing Board Member of *Journal of Statistical Planning and Inference*.

Shanti Gupta has been elected a Fellow of the American Statistical Association, the American Association for the Advancement of Sciences, and the Institute of Mathematical Statistics, and an elected member of the International Statistical Institute. He has also held special short-term visiting positions such as Special Chair, Institute of Mathematics, Academia Sinica, Taipei, Taiwan; and Erskine Fellow, University of Canterbury, Christchurch, New Zealand. He was one of the Special Invitees at the 1995 Taipei International Symposium in Statistics and was presented with the key to the city by the Mayor of Taipei.

Shanti Gupta, through his phenomenal research in the area of ranking and selection over the last forty years, has influenced the trend of research in the area and has provided inspiration and encouragement to many young researchers. Now that he is freed from the administrative burden, we are sure that he will continue his contributions to the field with added vigor and energy.

Publications

Books:

1. *Statistical Decision Theory and Related Topics* (co-edited with J. Yackel), New York: Academic Press, 1971.

2. *Statistical Decision Theory and Related Topics II* (co-edited with D. S. Moore), New York: Academic Press, 1977.

3. *Multiple Decision Procedures: Theory and Methodology of Selecting and Ranking Populations* (co-authored with S. Panchapakesan), New York: John Wiley & Sons, 1979.

4. *Multiple Statistical Theory: Recent Developments*, Lecture Notes in Statistics, Volume 6 (co-authored with D.-Y. Huang), New York: Springer-Verlag, 1981.

5. *Statistical Decision Theory and Related Topics III (Volumes I and II)* (co-edited with J. O. Berger), New York: Academic Press, 1982.

6. *Statistical Decision Theory and Related Topics IV (Volumes I and II)* (co-edited with J. O. Berger), New York: Academic Press, 1988.

7. *Statistical Decision Theory and Related Topics V* (co-edited with J. O. Berger), New York: Springer-Verlag, 1994.

Articles:

1957

1. On a statistic which arises in selection and ranking problems (with M. Sobel), *Annals of Mathematical Statistics*, **28**, 957–967.

2. Selection and ranking problems with binomial populations (with M. J. Huyett and M. Sobel), *Transactions of the American Society for Quality Control*, 635–644.

1958

3. On selecting a subset which contains all populations better than a standard (with M. Sobel), *Annals of Mathematical Statistics*, **29**, 235–244.

4. On the distribution of a statistic based on ordered uniform chance variables (with M. Sobel), *Annals of Mathematical Statistics*, **29**, 274–281.

1959

5. Moments of order statistics from a normal population (with R. C. Bose), *Biometrika*, **46**, 433–440.

1960

6. Order statistics from the gamma distribution, *Technometrics*, **2**, 243–262.

7. Selecting a subset containing the best of several binomial populations (with M. Sobel), In *Contributions to Probability and Statistics*, Harold Hotelling Volume (Eds., I. Olkin *et al.*), Chapter 20, pp. 224–248, Stanford: Stanford University Press.

1961

8. Percentage points and modes of order statistics from the normal distribution, *Annals of Mathematical Statistics*, **32**, 888–893.

9. Gamma distribution in acceptance sampling based on life tests (with Phyllis A. Groll), *Journal of the American Statistical Association*, **56**, 942–970.

1962

10. Gamma distribution, In *Contributions to Order Statistics* (Eds., A. E. Sarhan and B. G. Greenberg), Chapter 12D, pp. 431–450, New York: John Wiley & Sons.

11. Life tests sampling plans for normal and lognormal distributions, *Technometrics*, **4**, 151–175.

12. On selecting a subset containing the population with the smallest variance (with M. Sobel), *Biometrika*, **49**, 495–507.

13. On the smallest of several correlated F statistics (with M. Sobel), *Biometrika*, **49**, 509–523.

1963

14. Probability integrals of multivariate normal and multivariate t, *Annals of Mathematical Statistics*, **34**, 792–828.

15. Bibliography on the multivariate normal integrals and related topics, *Annals of Mathematical Statistics*, **34**, 829–838.

16. On a selection and ranking procedure for gamma populations, *Annals of the Institute of Statistical Mathematics*, **14**, 199–216.

1964

17. On the distribution of linear function of ordered, correlated normal random variables with emphasis on range (with K. C. S. Pillai and G. P. Steck), *Biometrika*, **51**, 143–151.

1965

18. On some multiple decision (selection and ranking) rules, *Technometrics*, **7**, 225–245.

19. On linear functions of ordered correlated normal random variables (with K. C. S. Pillai), *Biometrika*, **52**, 367–379.

20. Exact moments and percentage points of the order statistics and the distribution of the range from the logistic distribution (with B. K. Shah), *Annals of Mathematical Statistics*, **36**, 907–920.

21. A system of inequalities for the incomplete gamma function and the normal integral (with M. N. Waknis), *Annals of Mathematical Statistics*, **36**, 139–149.

1966

22. Selection and ranking procedures and order statistics for the binomial distributions, In *Classical and Contagious Discrete Distributions* (Ed., G. P. Patil), pp. 219–230, Oxford, England: Pergamon Press.

23. Estimation of the parameters of the logistic distribution (with M. Gnanadesikan), *Biometrika*, **53**, 565–570.

24. On some selection and ranking procedures for multivariate normal populations using distance functions, In *Multivariate Analysis* (Ed., P. R. Krishnaiah), pp. 457–475, New York: Academic Press.

25. Distribution-free life tests sampling plans (with R. E. Barlow), *Technometrics*, **8**, 591–613.

1967

26. On selection and ranking procedures, *Transactions of the American Society for Quality Control*, 151–155.

27. On selection and ranking procedures and order statistics from the multinomial distribution (with K. Nagel), *Sankhyā, Series B*, **29**, 1–34.

28. Best linear unbiased estimators of the parameters of the logistic distribution using order statistics (with A. S. Qureishi and B. K. Shah), *Technometrics*, **9**, 43–56.

1968

29. On the properties of subset selection procedures (with J. Deely), *Sankhyā, Series A*, **30**, 37–50.

1969

30. Selection procedures for restricted families of probability distributions (with R. E. Barlow), *Annals of Mathematical Statistics*, **40**, 905–917.

31. On the distribution of the maximum and minimum of ratios of order statistics (with R. E. Barlow and S. Panchapakesan), *Annals of Mathematical Statistics*, **40**, 918–934.

32. Selection and ranking procedures (with S. Panchapakesan), In *Design of Computer Simulation Experiments* (Ed., T. H. Naylor), pp. 132–160, Durham: Duke University Press.

33. Some selection and ranking procedures for multivariate normal populations (with S. Panchapakesan), In *Multivariate Analysis-II* (Ed., P. R. Krishnaiah), pp. 475–505, New York: Academic Press.

1970

34. Selection procedures for multivariate normal distributions in terms of measures of dispersion (with M. Gnanadesikan), *Technometrics*, **12**, 103–117.

35. On some classes of selection procedures based on ranks (with G. C. McDonald), In *Non-parametric Techniques in Statistical Inference* (Ed., M. L. Puri), pp. 491–514, Cambridge, England: Cambridge University Press.

36. On a ranking and selection procedure for multivariate populations (with W. J. Studden), In *Essays in Probability and Statistics*, S. N. Roy Memorial Volume (Eds., R. C. Bose *et al.*), Chapter 16, pp. 327–338, Chapel Hill: University of North Carolina Press.

1971

37. On some contributions to multiple decision theory (with K. Nagel), In *Statistical Decision Theory and Related Topics* (Eds., S. S. Gupta and J. Yackel), pp. 79–102, New York: Academic Press.

38. Some selection procedures with applications to reliability problems (with G. C. McDonald), In *Operations Research and Reliability*, Proceedings of NATO Conference, pp. 421–439, New York: Gordon and Breach Publishers.

1972

39. Review of "Sequential Identification and Ranking Procedures (with Special Reference to Koopman-Darmois Populations)", by Bechhofer, Kiefer and Sobel, *Annals of Mathematical Statistics*, **43**, 697–700.

40. A class of non-eliminating sequential multiple decision procedures (with A. Barron), In *Operations Research-Verfahren* (Eds., Henn, Künzi and Schubert), pp. 11–37, Meisenheim am Glan, Germany: Verlag Anton Haim.

41. On a class of subset selection procedures (with S. Panchapakesan), *Annals of Mathematical Statistics*, **43**, 814–822.

42. On multiple decision (subset selection) procedures (with S. Panchapakesan), *Journal of Mathematical and Physical Sciences*, **6**, 1–72.

1973

43. On the order statistics from equally correlated normal random variables (with K. Nagel and S. Panchapakesan), *Biometrika*, **60**, 403–413.

44. On order statistics and some applications of combinatorial methods in statistics (with S. Panchapakesan), In *A Survey of Combinatorial Theory* (Eds., J. N. Srivastava *et al.*), Chapter 19, pp. 217–250, Amsterdam: North-Holland Publishing Company.

45. On selection and ranking procedures—a restricted subset selection rule (with T. J. Santner), In *Proceedings of the 39th Session of the International Statistical Institute*, Volume 45, Book I, pp. 478–486, Vienna, Austria.

1974

46. Selection and ranking procedures for multivariate normal populations, In *Proceedings of the Nineteenth Army Conference on Design of Experiments*, ARO74-1, pp. 13–24.

47. On subset selection procedures for Poisson populations and some applications to the multinomial distribution (with D.-Y. Huang), In *Applied Statistics* (Ed., R. P. Gupta), pp. 97–109, New York: Academic Press.

48. Nonparametric subset selection procedures for the t best populations (with D.-Y. Huang), *Bulletin of the Institute of Mathematics, Academia Sinica*, **2**, 377–386.

49. On a maximin strategy for sampling based on selection procedures from several populations (with W.-T. Huang), *Communications in Statistics*, **3**, 325–342.

50. A note on selecting a subset of normal populations with unequal sample sizes (with W.-T. Huang), *Sankhyā, Series A*, **36**, 389–396.

51. Moments, product moments and percentage points of the order statistics from the lognormal distribution for samples of size twenty and less (with G. C. McDonald and D. I. Galarneau), *Sankhyā, Series B*, **36**, 230–250.

52. On moments of order statistics from independent binomial populations (with S. Panchapakesan), *Annals of the Institute of Statistical Mathematics Supplement*, **8**, 95–113.

53. Inference for restricted families: (A) multiple decision procedures: (B) order statistics inequalities (with S. Panchapakesan), In *Reliability and Biometry; Statistical Analysis of Lifelength* (Eds., F. Proschan and R. J. Serfling), pp. 503–596, Philadelphia: SIAM.

1975

54. On selection procedures for the t best populations and some related problems (with R. J. Carroll and D.-Y. Huang), *Communications in Statistics*, **4**, 987–1008.

55. On gamma-minimax classification procedure (with D.-Y. Huang), In *Proceedings of the 40th Session of the International Statistical Institute*, Volume 46, Book 3, Contributed Papers, pp. 330–335, Warsaw, Poland.

56. On some parametric and nonparametric sequential subset selection procedures (with D.-Y. Huang), In *Statistical Inference and Related Topics*, Volume 2 (Ed., M. L. Puri), pp. 101–128, New York: Academic Press.

57. Some selection procedures based on the Robbins-Monro stochastic approximation (with W.-T. Huang), *Tamkang Journal of Mathematics*, **6**, 301–314.

58. On a quantile selection procedure and associated distribution of ratios of order statistics from a restricted family of probability distributions (with S. Panchapakesan), In *Reliability and Fault Tree Analysis* (Eds., R. E. Barlow, J. B. Fussell and N. D. Singpurwalla), pp. 557–576, Philadelphia: SIAM.

1976

59. On subset selection procedures for the entropy function associated with binomial populations (with D.-Y. Huang), *Sankhyā, Series A*, **33**, 153–173.

60. Subset selection procedures for the means and variances of normal populations: unequal sample sizes case (with D.-Y. Huang), *Sankhyā, Series B*, **38**, 112–128.

61. On ranking and selection procedures and tests of homogeneity for binomial populations (with D.-Y. Huang and W.-T. Huang), In *Essays in Probability and Statistics*, Ogawa Volume (Eds., S. Ikeda *et al.*), Chapter 33, pp. 501–533, Tokyo: Shinko Tsusho Co., Ltd.

62. Subset selection procedures for finite schemes in information theory (with W.-Y. Wong), In *Colloquia Mathematica Societatis János Bolyai*, 16: *Topics in Information Theory* (Eds., I. Csiszar and P. Elias), pp. 279–291, Amsterdam: North-Holland Publishing Company.

63. Some subset selection procedures for Poisson processes and some applications to the binomial and multinomial problems (with W.-Y. Wong), In *Methods of Operations Research* (Eds., R. Henn *et al.*), pp. 49–70, Meisenheim am Glan, Germany: Verlag Anton Hain.

1977

64. On the probabilities of rankings of populations with applications (with R. J. Carroll), *Journal of Statistical Computation and Simulation*, **5**, 145–157.

65. On some optimal sampling procedures for selection problems (with D.-Y. Huang), In *The Theory and Applications of Reliability* (Eds., C. P. Tsokos and I. N. Shimi), pp. 495–505, New York: Academic Press.

66. Some multiple decision problems in analysis of variance (with D.-Y. Huang), *Communications in Statistics—Theory and Methods*, **6**, 1035–1054.

67. Selection and ranking procedures: A brief introduction, *Communications in Statistics—Theory and Methods*, **6**, 993–1001.

68. On the monotonicity of Bayes subset selection procedures (with J. C. Hsu), In *Proceedings of the 41st Session of the International Statistical Institute*, Volume 47, Book 4, Contributed papers, pp. 208–211, New Delhi, India.

69. On some gamma-minimax selection and multiple comparison procedures (with D.-Y. Huang), In *Statistical Decision Theory and Related Topics II* (Eds., S. S. Gupta and D. S. Moore), pp. 139–155, New York: Academic Press.

70. Nonparametric selection procedures for selecting the regression equations with larger slope (with D.-Y. Huang), *Journal of Chinese Statistical Association*, **15**, 5717–5724.

1978

71. On the performance of some subset selection procedures (with J. C. Hsu), *Communications in Statistics—Simulation and Computation*, **7**, 561–591.

1979

72. Locally optimal subset selection procedures based on ranks (with D.-Y. Huang and K. Nagel), In *Optimizing Methods in Statistics* (Ed., J. S. Rustagi), pp. 251–260, New York: Academic Press.

73. Some results on subset selection procedures for double exponential populations (with Y.-K. Leong), In *Decision Information* (Eds., C. P. Tsokos and R. M. Thrall), pp. 277–305, New York: Academic Press.

74. On subset selection procedures for Poisson populations (with Y.-K. Leong and W.-Y. Wong), *Bulletin of the Malaysian Mathematical Society*, **2**, 89–110.

75. Subset selection procedures for restricted families of probability distributions (with M.-W. Lu), *Annals of the Institute of Statistical Mathematics*, **31**, 235–252.

76. On selection rules for treatments versus control problems (with A. K. Singh), In *Proceedings of the 42nd Session of the International Statistical Institute*, pp. 229–232, Manila, Philippines.

1980

77. Minimax subset selection with applications to unequal variance (unequal sample size) problems (with R. L. Berger), *Scandinavian Journal of Statistics*, **7**, 21–26.

78. Subset selection procedures with special reference to the analysis of two-way layout: Applications to motor-vehicle fatality data (with J. C. Hsu), *Technometrics*, **22**, 543–546.

79. Some recent developments in multiple decision theory: A brief overview (with D.-Y. Huang), In *Proceedings of the Conference on Recent Developments in Statistical Methods and Applications*, pp. 209–216, Taipei: Institute of Mathematics, Academia Sinica.

80. A note on optimal subset selection (with D.-Y. Huang), *Annals of Statistics*, **8**, 1164–1167.

81. An essentially complete class of multiple decision procedures (with D.-Y. Huang), *Journal of Statistical Planning and Inference*, **4**, 115–121.

82. Gamma-Minimax and minimax decision rules for comparison of treatments with a control (with W.-C. Kim), In *Recent Developments in Statistical Inference and Data Analysis* (Ed., K. Matusita), pp. 55–71, Amsterdam: North-Holland Publishing Company.

83. Some statistical techniques for climatological data (with S. Panchapakesan), In *Statistical Climatology* (Ed., S. Ikeda), pp. 35–48, Amsterdam: North-Holland Publishing Company.

84. On rules based on sample medians for selection of the largest location parameter (with A. K. Singh), *Communications in Statistics—Theory and Methods*, **9**, 1277–1298.

1981

85. On mixtures of distributions: a survey and some new results on ranking and selection (with W.-T. Huang), *Sankhyā, Series B*, **43**, 245–290.

86. On gamma-minimax, minimax, and Bayes procedures for selecting populations close to a control (with P. Hsiao), *Sankhyā, Series B*, **43**, 291–318.

87. On the problem of selecting good populations (with W.-C. Kim), *Communications in Statistics—Theory and Methods*, **10**, 1043–1077.

88. Optimality of subset selection procedures for ranking means of three normal populations (with K. J. Miescke), *Sankhyā, Series B*, **43**, 1–17.

89. ASA and Statistical Education: An active role for evaluation (with S. Panchapakesan), In *Proceedings of the American Statistical Association, Section on Statistical Education*, pp. 27–29.

1982

90. Comments on the paper "Quality of Statistical Education: Should ASA Assist or Assess?" by Judith M. Tanur, *The American Statistician*, **36**, 94.

91. Selection procedures for a problem in analysis of variance (with D.-Y. Huang), In *Statistical Decision Theory and Related Topics III*, Volume 1 (Eds., S. S. Gupta and J. O. Berger), pp. 457–472, New York: Academic Press.

92. On the least favorable configurations in certain two-stage selection procedures (with K. J. Miescke), In *Statistics and Probability: Essays in Honor of C. R. Rao* (Eds., G. Kallianpur, P. R. Krishnaiah and J. K. Ghosh), pp. 295–305, Amsterdam: North-Holland Publishing Company.

93. On the problem of finding a best population with respect to a control in two stages (with K. J. Miescke), In *Statistical Decision Theory and Related Topics III*, Volume 1 (Eds., S. S. Gupta and J. O. Berger), pp. 473–496, New York: Academic Press.

94. Nonparametric procedures in multiple decisions (ranking and selection procedures) (with G. C. McDonald), In *Colloquia Mathematica Societatis János Bolyai, 32: Nonparametric Statistical Inference*, Volume I (Eds., B. V. Gnedenko, M. L. Puri, and I. Vincze), pp. 361–389, Amsterdam: North-Holland Publishing Company.

95. Subset selection procedures for the means of normal populations with unequal variances: Unequal sample sizes case (with W.-Y. Wong), *Selecta Statistica Canadiana*, **6**, 109–149.

1983

96. Empirical Bayes rules for selecting good populations (with P. Hsiao), *Journal of Statistical Planning and Inference*, **8**, 87–101.

97. An essentially complete class of two-stage selection procedures with screening at the first stage (with K. J. Miescke), *Statistics and Decisions*, **1**, 427–439.

98. On subset selection procedures for means of inverse Gaussian populations (with H.-M. Yang), In *Proceedings of the 44th Session of the International Statistical Institute*, Volume 2, Contributed papers, pp. 591–594, Madrid, Spain.

1984

99. Selection procedures for optimal subsets of regression variables (with D.-Y. Huang and C.-L. Chang), In *Design of Experiments: Ranking and Selection*, Bechhofer Volume (Eds., T. J. Santner and A. C. Tamhane), pp. 67–75, New York: Marcel Dekker.

100. On some inequalities and monotonicity results in selection and ranking theory (with D.-Y. Huang and S. Panchapakesan), In *Inequalities in Statistics and Probability* (Ed., Y. L. Tong), IMS Lecture Notes—Monograph Series, Volume 5, pp. 211–227, Hayward, California: Institute of Mathematical Statistics.

101. A two-stage elimination type procedure for selecting the largest of several normal means with a common unknown variance (with W.-C. Kim), In *Design of Experiments: Ranking and Selection*, Bechhofer Volume (Eds., T. J. Santner and A. C. Tamhane), pp. 77–93, New York: Marcel Dekker.

102. Sequential selection procedures—A decision-theoretic approach (with K. J. Miescke), *Annals of Statistics*, **12**, 336–350.

103. On two-stage Bayes selection procedures (with K. J. Miescke), *Sankhyā, Series B*, **46**, 123–134.

104. On isotonic selection rules for binomial populations better than a standard (with W.-T. Huang), In *Proceedings of the First Saudi Symposium on Statistics and Its Application* (Eds., A. M. Abouammoh *et al.*), pp. 89–111, Riyadh: King Saud University Press.

105. Edgeworth expansions in statistics: Some recent developments (with S. Panchapakesan), In *Colloquia Mathematica Societatis János Bolyai: Limit Theorems in Probability and Statistics*, Volume I (Ed., P. Révész), pp. 519–565, Amsterdam: North-Holland Publishing Company.

106. Isotonic procedures for selecting populations better than a control under ordering prior (with H.-M. Yang), In *Statistics: Applications and New Directions* (Eds., J. K. Ghosh and J. Roy), pp. 279–312, Calcutta: Indian Statistical Institute.

1985

107. Optimal sampling in selection problems, In *Encyclopedia of Statistical Sciences*, Volume 6 (Eds., S. Kotz and N. L. Johnson), pp. 449–452, New York: John Wiley & Sons.

108. Subset selection procedures: Review and assessment (with S. Panchapakesan), *American Journal of Mathematical and Management Sciences*, **5**, 235–311.

109. On the distribution of the studentized maximum of equally correlated normal random variables (with S. Panchapakesan and J. K. Sohn), *Communications in Statistics—Simulation and Computation*, **14**, 103–135.

110. Minimax multiple *t*-tests for comparing *k* normal populations with a control (with K. J. Miescke), *Journal of Statistical Planning and Inference*, **12**, 161–169.

111. Bayes-P^* subset selection procedures (with H.-M. Yang), *Journal of Statistical Planning and Inference*, **12**, 213–233.

112. A new Bayes procedure for selecting the best normal population in two stages (with J. K. Sohn), In *Proceedings of the International Statistical Institute, 45th Session*, Book 1, Contributed papers, pp. 287–288, Amsterdam, The Netherlands.

1986

113. *User's Guide to RS–MCB: A Computer Package for Ranking, Selection, and Multiple Comparisons with the Best* (with J. C. Hsu), Columbus, Ohio: The Ohio State University Research Foundation. Also as Chapter 38: The RSMCB Procedure, In *SUGI Supplemental Library User's Guide, Version 5*, pp. 497–502, Cary, North Carolina: SAS Institute Inc.

114. A statistical selection approach to binomial models (with G. C. McDonald), *Journal of Quality Technology*, **18**, 103–115.

115. Isotonic procedures for selecting populations better than a standard: Two-parameter exponential distributions (with L.-Y. Leu), In *Reliability and Quality Control* (Ed., A. P. Basu), pp. 167–183, Amsterdam: Elsevier Science Publishers B. V..

116. Empirical Bayes rules for selecting good binomial populations (with T. Liang), In *Adaptive Procedures and Related Topics* (Ed., J. Van Ryzin), IMS Lecture Notes-Monograph Series, Volume 8, pp. 110–128, Hayward, California: Institute of Mathematical Statistics.

1987

117. Locally optimal subset selection rules based on ranks under joint type II censoring (with T. Liang), *Statistics and Decisions*, **5**, 1–13.

118. Optimum two-stage selection procedures for Weibull populations (with K. J. Miescke), *Journal of Statistical Planning and Inference*, **15**, 147–156.

119. Asymptotic distribution-free selection procedures for a two-way layout problem (with L.-Y. Leu), *Communications in Statistics—Theory and Methods*, **16**, 2313–2325.

120. Selecting the best unknown mean from normal populations having a common unknown coefficient of variation (with T. Liang), In *Mathematical Statistics and Probability Theory*, Volume B: *Statistical Inference and Methods* (Eds., P. Bauer *et al.*), pp. 97–112, Dordrecht: D. Reidel Publishing Company.

121. Statistical selection procedures in multivariate models (with S. Panchapakesan), In *Advances in Multivariate Statistical Analysis* (Ed., A. K. Gupta), pp. 141–160, Dordrecht: D. Reidel Publishing Company.

122. On Bayes selection rules for the best exponential population with type-I censored data (with T. Liang), In *Proceedings of The International Institute of Statistics, 46th Session*, Contributed papers, pp. 153–154, Tokyo, Japan.

123. On some Bayes and empirical Bayes selection procedures (with T. Liang), In *Probability and Bayesian Statistics* (Ed., R. Viertl), pp. 233–246, New York: Plenum Publishing Corporation.

1988

124. Empirical Bayes rules for selecting the best binomial population (with T. Liang), In *Statistical Decision Theory and Related Topics* IV, Volume 1 (Eds., S. S. Gupta and J. O. Berger), pp. 213–224, New York: Springer-Verlag.

125. On the problem of finding the largest normal mean under heteroscedasticity (with K. J. Miescke), In *Statistical Decision Theory and Related Topics* IV, Volume 2 (Eds., S. S. Gupta and J. O. Berger), pp. 37–50, New York: Springer-Verlag.

126. Selection and ranking procedures in reliability models (with S. Panchapakesan), In *Handbook of Statistics 7: Quality Control and Reliability* (Eds., P. R. Krishnaiah and C. R. Rao), pp. 131–156, Amsterdam: North-Holland.

127. Selecting important independent variables in linear regression models (with D.-Y. Huang), *Journal of Statistical Planning and Inference*, **20**, 155–167.

128. Comment on "The Teaching of Statistics" and "The Place of Statistics in the University" by Harold Hotelling, *Statistical Science*, **3**, 104–106.

1989

129. Isotonic selection procedures for Tukey's lambda distributions (with J. K. Sohn), In *Recent Developments in Statistics and Their Applications* (Eds., J.P. Klein and J.C. Lee), pp. 283–298, Seoul: Freedom Academy Publishing Co..

130. On selecting the best of k lognormal distributions (with K. J. Miescke), *Metrika*, **36**, 233–247.

131. On Bayes and empirical Bayes two–stage allocation procedures for selection problems (with T. Liang), In *Statistical Data Analysis and Inference* (Ed., Y. Dodge), pp. 61–70, Amsterdam: Elsevier Science Publishers B.V..

132. Parametric empirical Bayes rules for selecting the most probable multinomial event (with T. Liang), In *Contributions to Probability and Statistics: Essays in Honor of Ingram Olkin* (Eds., L. J. Gleser *et al.*), Chapter 18, pp. 318–328, New York: Springer-Verlag.

133. Selecting the best binomial population: Parametric empirical Bayes approach (with T. Liang), *Journal of Statistical Planning and Inference*, **23**, 21–31.

134. On sequential ranking and selection procedures (with S. Panchapakesan), In *Handbook of Sequential Analysis* (Eds., B. K. Ghosh and P. K. Sen), pp. 363-380, New York: Marcel Dekker.

1990

135. On detecting influential data and selecting regression variables (with D.-Y. Huang), *Journal of Statistical Planning and Inference*, **53**, 421-435.

136. On a sequential subset selection procedure (with T. Liang), In *Proceedings of the R. C. Bose Symposium on Probability, Statistics and Design of Experiments*, pp. 348–355, New Delhi: Wiley Eastern.

137. On lower confidence bounds for PCS in truncated location parameter models (with L.-Y. Leu and T. Liang), *Communications in Statistics— Theory and Methods*, **19**, 527–546.

138. Selecting the fairest of $k(\geq 2)$ m–sided dice (with L.-Y. Leu), *Communications in Statistics—Theory and Methods*, **19**, 2159–2177.

139. On finding the largest mean and estimating the selected mean (with K. J. Miescke), *Sankhyā, Series B*, **52**, 144–157.

140. Recent advances in statistical ranking and selection: Theory and methods. In *Proceedings of the 1990 Taipei Symposium in Statistics* (Eds., M.-T. Chao and P. E. Cheng), pp. 133–166, Taipei: Institute of Statistical Science, Academia Sinica.

1991

141. An elimination type two-stage procedure for selecting the population with the largest mean from k logistic populations (with S. Han), *American Journal of Mathematical and Management Sciences*, **11**, 351–370.

142. On a lower confidence bound for the probability of a correct selection: Analytical and simulation studies (with T. Liang), In *The Frontiers of Statistical Scientific Theory and Industrial Applications* (Volume II of the Proceedings of ICOSCO-1) (Eds., A. Öztürk and E. C. Van der Meulen), pp. 77–95, Columbus, Ohio: American Sciences Press.

143. Selection of the best with a preliminary test for location–scale models (with L.-Y. Leu and T. Liang), *Journal of Combinatorics, Information and System Sciences*, **16**, 221–233.

144. On the asymptotic optimality of certain empirical Bayes simultaneous testing procedures (with T. Liang), *Statistics and Decisions*, **9**, 263–283.

145. Logistic order statistics and their properties (with N. Balakrishnan), In *Handbook of the Logistic Distribution* (Ed., N. Balakrishnan), pp. 17–48, New York: Marcel Dekker.

1992

146. On empirical Bayes selection rules for negative binomial populations (with T. Liang), In *Bayesian Analysis in Statistics and Econometrics* (Eds., P. K. Goel and N. S. Iyengar), pp. 127–146, Lecture Notes in Statistics, No. 75, New York: Springer-Verlag.

147. On some nonparametric selection procedures (with S. N. Hande), In *Nonparametric Statistics and Related Topics* (Ed., A. K. Md. E. Saleh), pp. 33–49, Amsterdam: Elsevier Science Publishers B. V..

148. Selection and ranking procedures for logistic populations (with S. Han), In *Order Statistics and Nonparametrics* (Eds., I. A. Salama and P. K. Sen), pp. 377–404, Amsterdam: Elsevier Science Publishers B. V..

1993

149. Selection and ranking procedures for Tukey's generalized lambda distributions (with J. K. Sohn), In *The Frontiers of Modern Statistical Inference Procedures*, Volume 2 (Ed., E. Bofinger *et al.*), pp. 153–175, Columbus, Ohio: American Sciences Press.

150. On combining selection and estimation in the search for the largest binomial parameters (with K. J. Miescke), *Journal of Statistical Planning and Inference*, **36**, 129–140.

151. Subset selection procedures for binomial models based on a class of priors (with Y. Liao), In *Multiple Comparisons, Selection and Applications in Biometry: A Festschrift in Honor of C. W. Dunnett* (Ed., F. M. Hoppe), pp. 331–351, New York: Marcel Dekker.

152. Bayes and empirical Bayes rules for selecting multinomial populations (with T. Liang), In *Statistics and Probability: Bahadur Festschrift* (Eds., J. K. Ghosh, S. K. Mitra, K. R. Parthasarathy and B. L. S. Prakasa Rao), pp. 265–278, Calcutta: Indian Statistical Institute.

153. Single–stage Bayes and empirical Bayes rules for ranking multinomial events (with S. N. Hande), *Journal of Statistical Planning and Inference*, **35**, 367–382.

154. Selection and screening procedures in multivariate analysis (with S. Panchapakesan), In *Multivariate Analysis: Future Directions* (Ed., C. R. Rao), Chapter 12, pp. 233–262, Amsterdam: Elsevier Science Publishers B.V..

155. Selecting the best exponential population based on type-I censored data (with T. Liang), In *Advances in Reliability* (Ed., A. Basu), pp. 171–180, Amsterdam: Elsevier Science Publishers B.V..

1994

156. On empirical Bayes selection rules for sampling inspection (with T. Liang), *Journal of Statistical Planning and Inference*, **38**, 43–64.

157. An improved lower confidence bound for the probability of a correct selection in location parameter models (with Y. Liao, C. Qiu and J. Wang), *Statistica Sinica*, **4**, 715–727.

158. Bayesian look ahead one stage sampling allocations for selecting the largest normal mean (with K. J. Miescke), *Statistical Papers*, **35**, 169–177.

159. Empirical Bayes two-stage procedures for selecting the best Bernoulli population compared with a control (with T. Liang and R.-B. Rau), In *Statistical Decision Theory and Related Topics V* (Eds., S. S. Gupta and J. O. Berger), pp. 277–292, New York: Springer-Verlag.

160. A minimax type procedure for nonparametric selection of the best population with partially classified data (with T. Gastaldi), *Communications in Statistics—Theory and Methods*, **23**, 2503–2532.

161. Bayes and Empirical Bayes rules for selecting the best normal population compared with a control (with T. Liang and R.-B. Rau), *Statistics and Decisions*, **12**, 125–147.

1995

162. Estimation of the mean and standard deviation of the logistic distribution based on multiply type–II censored samples (with N. Balakrishnan and S. Panchapakesan), *Statistics*, **27**, 127–142.

163. Estimation of the location and scale parameters of the extreme value distribution based on multiply type–II censored samples (with N. Balakrishnan and S. Panchapakesan), *Communications in Statistics—Theory and Methods*, **24**, 2105–2125.

164. A two-stage procedure for selecting the population with the largest mean when the common variance is unknown (with B. Miao and D. Sun), *Chinese Journal of Applied Probability and Statistics*, **11**, 113–127.

165. Empirical Bayes two-stage procedures for selecting the best normal population compared with a control (with T. Liang and R.-B. Rau), In *Symposium Gaussiana: Proceedings of the 2nd Gauss Symposium Conference, B: Statistical Science* (Eds., V. Mammitzsch and H. Schneeweiss), pp. 171–196, Berlin: Walter de Gruyter & Co..

1996

166. Bayesian look-ahead sampling allocations for selecting the best Bernoulli population (with K. J. Miescke), In *Research Developments in Probability & Statistics—M. L. Puri Festschrift* (Eds., E. Brunner and M. Denker), pp. 353-369, Utrecht, The Netherlands: VSP International Science Publishers.

167. Selecting good normal regression models: An empirical Bayes approach (with T. Liang), *Journal of the Indian Society of Agricultural Statistics*, **49** (Golden Jubilee No. 1996–1997), 335–352.

168. Life testing for multi-component systems with incomplete information on the cause of failure: A study on some inspection strategies (with T. Gastaldi), *Computational Statistics & Data Analysis*, **22**, 373–393.

169. On detecting influential data and selecting regression variables (with D.-Y. Huang), *Journal of Statistical Planning and Inference*, **53**, 421–435.

170. Bayesian look ahead one-stage sampling allocations for selection of the best population (with K. J. Miescke), *Journal of Statistical Planning and Inference*, **54**, 229–244.

171. Simultaneous selection for homogeneous multinomial populations based on entropy function: An empirical Bayes approach (with L.-Y. Leu and T. Liang), *Journal of Statistical Planning and Inference*, **54**, 145–157.

172. Design of experiments with selection and ranking goals (with S. Panchapakesan), In *Handbook of Statistics 13: Design and Analysis of Experiments* (Eds., S. Ghosh and C. R. Rao), pp. 555–585, Amsterdam: Elsevier Science Publishers B.V..

To Appear

173. Estimation of the mean and standard deviation of the normal distribution based on multiply type–II censored samples (with N. Balakrishnan and S. Panchapakesan), *Journal of the Italian Statistical Society.*

174. Higher order moments of order statistics from exponential and right-truncated exponential distributions and applications to life-testing problems (with N. Balakrishnan), In *Handbook of Statistics 16: Order Statistics and Their Applications* (Eds., N. Balakrishnan and C. R. Rao), Amsterdam: Elsevier Science Publishers B.V..

175. Simultaneous lower confidence bounds for probabilities of correct selections (with T. Liang), *Journal of Statistical Planning and Inference.*

176. On simultaneous selection of good populations (with T. Liang), *Statistics and Decisions.*

Ph.D. Students

1. John J. Deely, *Multiple Decision Procedures from an Empirical Bayes Approach*, 1965.

2. Mrudulla Gnanadesikan, *Some Selection and Ranking Procedures for Multivariate Normal Populations*, 1966.

3. Austin N. Barron, *A Class of Sequential Multiple Decision Procedures*, 1968.

4. Gary C. McDonald, *On Some Distribution-Free Ranking and Selection Procedures*, 1969.

5. Jerone Deverman, *A General Selection Procedure Relative to the t Best Populations*, 1969.

6. S. Panchapakesan, *Some Contributions to Multiple Decision (Selection and Ranking) Procedures*, 1969.

7. Klaus Nagel, *On Subset Selection Rules with Certain Optimality Properties*, 1970.

8. Wen-Tao Huang, *Some Contributions to Sequential Selection and Ranking Procedures*, 1972.

9. Thomas J. Santner, *A Restricted Subset Selection Approach to Ranking and Selection Problems*, 1973.

10. Deng-Yuan Huang, *Some Contributions to Fixed Sample and Sequential Multiple Decision (Selection and Ranking) Theory*, 1974.

11. Raymond J. Carroll, *Some Contributions to the Theory of Parametric and Nonparametric Sequential Ranking and Selection*, 1974.

12. Wing-Yue Wong, *Some Contributions to Subset Selection Procedures*, 1976.

13. Ming-Wei Lu, *On Multiple Decision Problems*, 1976.

14. Yoon-Kwai Leong, *Some Results on Subset Selection Problems*, 1976.

15. Roger L. Berger, *Minimax, Admissible and Gamma-Minimax Multiple Decision Rules*, 1977.

16. Ashok K. Singh, *On Slippage Tests and Multiple Decision (Selection and Ranking) Procedures*, 1977.

17. Jason C. Hsu, *On Some Decision-Theoretic Contributions to the Subset Selection Problem*, 1977.

18. Woo-Chul Kim, *On Bayes and Γ-Minimax Subset Selection*, 1979.

19. Ping Hsiao, *Some Contributions to Gamma-Minimax and Empirical Bayes Selection Procedures*, 1979.

20. Hwa-Ming Yang, *Some Results in the Theory of Subset Selection Procedures*, 1980.

21. Lii-Yuh Leu, *Contributions to Multiple Decision Theory*, 1983.

22. TaChen Liang, *Some Contributions to Empirical Bayes, Sequential and Locally Optimal Subset Selection Rules*, 1984.

23. Joong K. Sohn, *Some Multiple Decision Procedures for Tukey's Generalized Lambda*, 1985.

24. Sang Hyun Han, *Contributions to Selection and Ranking Procedures with Special Reference to Logistic Populations*, 1987.

25. Sayaji N. Hande, *Contributions to Nonparametric Selection and Ranking Procedures*, 1992.

26. Yuning Liao, *On Statistical Selection Procedures Using Prior Information*, 1992.

27. Re-Bin Rau, *An Empirical Bayes Approach to Multiple Decision Procedures*, 1993.

28. Marcey Emery Abate, *The Use of Historical Data in Statistical Selection and Robust Product Design*, 1995.

Contributors

Balakrishnan, N. Department of Mathematics and Statistics, McMaster University, Hamilton, Ontario, Canada L8S 4K1
e-mail: *bala@mcmail.cis.mcmaster.ca*

Berger, James O. Institute of Statistics and Decision Sciences, Duke University, Durham, NC 27708-0251
e-mail: *jberger@stat.duke.edu*

Berger, Roger L. Department of Statistics, North Carolina State University, Box 8203, Raleigh, NC 27695-8203
e-mail: *berger@stat.ncsu.edu*

Boukai, Benzion Department of Mathematics, Indiana University–Purdue University, 402 N. Blackford Street, Indianapolis, IN 46202-3272
e-mail: *boukai@math.iupui.edu*

Carroll, Raymond J. Department of Statistics, Blocker Building, Room 447, Texas A&M University, College Station, TX 77843-3143
e-mail: *carroll@stat.tamu.edu*

Chaganty, N. Rao Department of Mathematics and Statistics, Old Dominion University, Norfolk, VA 23529-0077
e-mail: *nrc@math.odu.edu*

Deely, John Department of Statistics, Purdue University, West Lafayette, IN 47907-1399
e-mail: *jdeely@stat.purdue.edu*

Frankowski, Krzysztof Department of Computer Science, University of Minnesota, Minneapolis, MN 55455
e-mail: *kfrankow@cs.umn.edu*

Fritsch, Kathleen S. Department of Mathematics and Computer Science, The University of Tennessee at Martin, Martin, TN 38238

Gastaldi, Tommaso Dipartimento di Statistica, Probabilità e Statistiche Applicate, Università degli Studi di Roma "La Sapienze", Rome, Italy
e-mail: *gastaldi@pow2.sta.uniromal.it*

Ghosal, Subhashis Division of Theoretical Statistics and Mathematics, Indian Statistical Institute, 203 B. T. Road, Calcutta, West Bengal 700 035, India
e-mail: *tapas@isical.ernet.in*

Ghosh, Jayanta K. Division of Theoretical Statistics and Mathematics, Indian Statistical Institute, 203 B. T. Road, Calcutta, West Bengal 700 035, India
e-mail: *jkg@isical.ernet.in*

Goldsman, David M. School of Industrial and Systems Engineering, Georgia Institute of Technology, Atlanta, GA 30332-0205
e-mail: *sman@isye.gatech.edu*

Hayter, Anthony J. School of Industrial and Systems Engineering, Georgia Institute of Technology, Atlanta, GA 30332-0205
e-mail: *tony.hayter@isye.gatech.edu*

Hsu, Jason C. Department of Statistics, The Ohio State University, Cockins Hall, 1958 Neil Avenue, Columbus, OH 43210-1247
e-mail: *jch@stat.mps.ohio-state.edu*

Huang, Deng-Yuan Institute of Applied Statistics, Fu Jen Catholic University, 510 Chung Cheng Road, Hsinchuang, Taipei Hsien, Taiwan, Republic of China
e-mail: *stat1009@fujens.fju.tw*

Huang, Wen-Tao Institute of Statistical Science, Academia Sinica, Nankang, Taipei, Taiwan 11529, Republic of China
e-mail: *stwthuang@ccvax.sinica.edu.tw*

Hwang, J. T. Gene Department of Mathematics, Cornell University, White Hall, Ithaca, NY 14853-7901
e-mail: *hwang@math.cornell.edu*

Jeyaratnam, S. Department of Mathematics, Southern Illinois University, Carbondale, IL 62901-4408
e-mail: *sjeyarat@math.siu.edu*

Johnson, Wesley O. Division of Statistics, University of California, Davis, CA 95616
e-mail: *wojohnson@ucdavis.edu*

Kang, K. H. Department of Statistics, Seoul National University, San 56-1 Shinrim-dong, Kwanak Ku, Seoul 151-742, South Korea
e-mail: *khkang@plaza.snu.ac.kr*

Kastner, Thomas M. School of Industrial and Systems Engineering, Georgia Institute of Technology, Atlanta, GA 30332-0205

Kim, W. C. Department of Statistics, Seoul National University, San 56-1 Shinrim-dong, Kwanak Ku, Seoul 151-742, South Korea
e-mail: *wckim@alliant.snu.ac.kr*

Liang, TaChen Department of Mathematics, Wayne State University, Detroit, MI 48202
e-mail: *liang@math.wayne.edu*

Lin, Ching-Ching Department of Mathematics, Taiwan Normal University, Taipei, Taiwan, Republic of China

Ma, Yimin Department of Mathematics and Statistics, McMaster University, Hamilton, Ontario, Canada L8S 4K1

Malthouse, Edward C. Department of Marketing, Kellog Graduate School of Management, Northwestern University, Evanston, IL 60208
e-mail: *ecm@nwu.edu*

McDonald, Gary C. Department Head, Operations Research Department, GM Research & Development Center, MC 480-106-359, 30500 Mound Road, Box 9055, Warren, MI 48090-9055
e-mail: *gary_mcdonald@notes.gmr.com*

Miescke, Klaus J. Department of Mathematics, Statistics and Computer Science, University of Illinois at Chicago, 851 S. Morgan, Chicago, IL 60607-7045
e-mail: *klaus@uic.edu*

Moore, David S. Department of Statistics, Purdue University, West Lafayette, IN 47907-1399
e-mail: *dsm@stat.purdue.edu*

Olkin, Ingram Department of Statistics, Stanford University, Stanford, CA 94305-4065
e-mail: *iolkin@playfair.stanford.edu*

Palmer, W. Craig Staff Research Scientist, Operations Research Department, GM Research & Development Center, MC 480-106-359, 30500 Mound Road, Warren, MI 48090-9055
e-mail: *w._craig_palmer@notes.gmr.com*

Pan, Guohua Department of Mathematical Sciences, Oakland University, O'Dowd Hall, Rochester, MI 48309-4401
e-mail: *pan@oakland.edu*

Panchapakesan, S. Department of Mathematics, Southern Illinois University, Carbondale, IL 62901-4408
e-mail: *kesan@math.siu.edu*

Park, B. U. Department of Statistics, Seoul National University, San 56-1 Shinrimdong, Kwanak Ku, Seoul 151-742, South Korea

Park, Henry Department of Mathematics, Statistics and Computer Science, University of Illinois–Chicago, 851 S. Morgan, Chicago, IL 60607-7045
e-mail: *HPark1@uic.edu*

Prakasa Rao, B. L. S. Stat-Math. Unit, Indian Statistical Institute, 7 SJS Sansanwal Marg, New Delhi 110 016, India
e-mail: *blsp@isid.ernet.in*

Ramamoorthi, R. V. Department of Statistics and Probability, Michigan State University, Wells Hall, East Lansing, MI 48824
e-mail: *rama@assist.stt.msu.edu*

Rao, C. R. Department of Statistics, Penn State University, 326 Classroom Building, University Park, PA 16802-2111
e-mail: *crr1@psuvm.psu.edu*

Rukhin, Andrew L. Department of Mathematics, University of Maryland-Baltimore County, Baltimore, MD 21228-5329
e-mail: *rukhin@math.umbc.edu*

Santner, Thomas J. Department of Statistics, The Ohio State University, 141 Cockins Hall, 1958 Neil Avenue, Columbus, OH, 43210-1247
e-mail: *tjs@osustat.mps.ohio-state.edu*

Sen, Pranab K. Department of Biostatistics, University of North Carolina at Chapel Hill, Chapel Hill, NC 27599-7400
e-mail: *pksen@biostat.sph.unc.edu*

Sethuraman, J. Department of Statistics, Florida State University, Tallahassee, FL 32308-3033
e-mail: *sethu@stat.fsu.edu*

Sobel, Milton Department of Statistics & Applied Probability, University of California-Santa Barbara, Santa Barbara, CA 93106-3110
e-mail: *sobel@pstat.ucsb.edu*

Stefanski, L. A. Department of Statistics, North Carolina State University-Raleigh, Raleigh, NC 27695-8203
e-mail: *stefansk@stat.ncsu.edu*

Tamhane, Ajit S. Department of Industrial Engineering & Management Sciences, Northwestern University, 2006 Sheridan Road, Evanston, IL 60208-4070
e-mail: *ajit@iems.nwu.edu*

Tong, Y. L. Department of Mathematics, Georgia Institute of Technology, Atlanta, GA 30332-0160
e-mail: *tong@math.gatech.edu*

Viana, Marlos A. G. Department of Visual Sciences, College of Medicine, University of Illinois, 1855 West Taylor Street M/C 648, Chicago, IL 60612-7242
e-mail: *marlos.ag.viana@uic.edu*

Wang, Y. Department of Mathematics, Indiana University–Purdue University, 402 N. Blackford Street, Indianapolis, IN 46202-3272

Zhou, Zhenwei Department of Statistics, University of North Carolina at Chapel Hill, Chapel Hill, NC 27599-3260

Tables

Figures

PART I

BAYESIAN INFERENCE

1

Bayes for Beginners? Some Pedagogical Questions

David S. Moore

Purdue University, West Lafayette, IN

Abstract: Ought we to base beginning instruction in statistics for general students on the Bayesian approach to inference? In the long run, this question will be settled by progress (or lack of progress) in persuading users of statistical methods to choose Bayesian methods. This paper is primarily concerned with the *pedagogical* challenges posed by Bayesian reasoning. It argues, based on research in psychology and education and a comparison of Bayesian and standard reasoning, that Bayesian inference is harder to convey to beginners than the already hard reasoning of standard statistical inference.

Keywords and phrases: Bayesian inference, statistical education

1.1 Introduction

It is a pleasure to dedicate this paper to Shanti Gupta, who offered me my first position and contributed in many ways to my statistical education.

Bayesian methods are among the more active areas of statistical research. Moreover, a glance at recent journals shows that researchers have made considerable progress in applying Bayesian ideas and methods to specific problems arising in statistical practice. It is therefore not surprising that some Bayesians have turned their attention to the nature of introductory courses in statistics for general students. These students come to us from many fields, with the goal of learning to read and perhaps carry out statistical studies in their own disciplines. Several recent Bayesian textbooks [e.g., Albert (1996), Antleman (1996) and Berry (1996)] are aimed at such students, and their authors have argued in favor of a Bayesian approach in teaching beginners [Albert (1995, 1997) and Berry (1997)]. The arguments, put crudely to make the issues clear, are: (1) Bayesian methods are the only right methods, so we should teach them; (2) Bayesian inference is easier to understand than standard inference.

3

Although I do not accept argument (1), I have no wish to participate in the continuing debate about the "right" philosophy of inference. Inference from uncertain empirical data is a notoriously subtle issue; it is not surprising that thoughtful scholars disagree. I doubt that any of us will live to see a consensus about the reasoning of inference. Indeed, the eclectic approach favored by many practicing statisticians, who use Bayesian methods where appropriate but are unconvinced by universal claims, may well be a permanently justifiable response to the variety and complexity of statistical problems.

I do wish to dispute the second argument. I will give reasons why Bayesian reasoning is considerably more difficult to assimilate than the reasoning of standard inference (though, of course, neither is straightforward). I even doubt the common Bayesian claim that at least the *results* of Bayesian inference (expressed in terms of probabilities that refer to our conclusions in this one problem) are more comprehensible than standard results (expressed in terms of probabilities that refer to the methods we used). My viewpoint is that of a teacher concerned about issues of content and pedagogy in statistical education. I will point to research in education and psychology rather than in statistics.

I preface the paper's main argument by briefly stating in Section 1.2 what I consider to be decisive empirical and pragmatic reasons for not basing introductory courses for general students on Bayesian ideas at the present time. Those who think that the conditions I describe in Section 1.2 will change in the future can consider the rest of the paper as posing questions of pedagogy that we will have to face in the coming Bayesian era.

1.2   Unfinished Business: The Position of Bayesian Methods

Let us attempt to be empirical. Here are two questions that bear on our decision about teaching Bayes to beginners:

- Are Bayesian methods widely used in statistical practice?

- Are there standard Bayesian methods for standard problem settings?

Several recent surveys suggest that Bayesian methods are little used in current statistical practice. Rustagi and Wright (1995) report the responses of all 103 statisticians employed in Department of Energy National Laboratories. Asked to choose from a long list "the three statistical techniques that have been most important to your work/research," only four mentioned "Bayesian methods." The top responses are not surprising: "Regression analysis" (63), "Basic statistical methods" (37), "Analysis of variance" (26), and "Design of experiments" (26). Bayesian methods tied for 17th/18th among 20 methods mentioned by more than one respondent. Turning to medical research, Emerson and Colditz (1992) catalog the statistical methods used in articles in the 1989 volume of the *New England Journal of Medicine*. Of 115 "Original Articles," 45 use t-tests, 41 present contingency tables, 37 employ survival methods, and so on. Emerson and Colditz do not mention any use of Bayesian techniques.

The DOE respondents are professional statisticians working in a variety of applied areas. Medical research projects often engage trained statistical collaborators. These practitioners might be expected to employ more up-to-date methodology than those in fields that less often engage professional statisticians. Yet even here, Bayesian approaches have made very few inroads. The absence of Bayesian procedures in commercial statistical software systems is further evidence of lack of use. Although the statistical literature abounds in research papers developing and applying Bayesian ideas, these appear to be in the nature of demonstration pieces. I can find no empirical evidence of widespread use in actual practice.

Because Bayesian methods are relatively rarely used in practice, teaching them has an opportunity cost, depriving students of instruction about the standard methods that are in common use. It might be argued that we should teach the "right" methods, regardless of current practice. That is simply to invite students to go elsewhere, to turn to another of the many places on campus where basic statistics is taught. Aside from that pragmatic consideration, I believe that we have an obligation to meet the needs of our customers. Their need is to read and understand applications of statistics in their own disciplines. Those applications are not yet Bayesian.

Here is a second argument, also at least somewhat empirical. *Reading of current Bayesian literature strongly suggests that there are not yet standard Bayesian methods for standard problem settings.* There remains considerable disagreement among Bayesians on how to approach the inference settings usually considered in first courses.

Purists might insist on informative, subjective prior distributions—a position that says in effect that there can be no standard methods because there are no standard problems. Lindley (1971), for example, is a classical proponent of the subjective Bayesian position, but it is worth noting that arguments for the superiority of Bayesian reasoning usually start from this position. It is nonetheless now more common to concede that there is a need for standard methods to apply in standard settings. The usual approach is to use noninformative reference priors that are generated from the sampling distribution rather than from actual prior information. Not all Bayesians are comfortable with this triumph of pragmatic over normative thinking.

Let us accept the need for standard techniques, at least for beginners. Very well: which noninformative prior should we use? Berger, who favors this approach, admits [Berger (1985, p. 89)] that "Perhaps the most embarrassing feature of noninformative priors, however, is simply that there are often so many of them." He offers *four* choices when θ is the probability of success in the binomial setting, and says, "All four possibilities are reasonable." Robert (1994, p. 119) presents an example due to Berger and Bernardo showing that simply reordering the parameters in the one-way ANOVA setting leads to four different reference priors.

Bayesian hypothesis testing in particular appears to be work in progress. One can find current research [e.g., Kass and Wasserman (1995)] starting from the premise that practically useful Bayesian tests remain an important open problem. The excellent survey of Kass and Raftery (1995) convinces me that Bayesian testing is not yet ready for prime time.

I believe that these negative answers to the two questions posed at the beginning of this section are relatively empirical and "objective." They are spelled out in much greater detail in Moore (1997). It seems to me that the weakest conclusion possible is that it is *premature* to make Bayesian methods the focus of basic methodological courses. I am not unalterably philosophically opposed to Bayesian statistics. I would support teaching Bayesian methods in introductory courses if they became standard methods widely accepted and widely used in practice. As Keynes once said, "When the facts change, I change my mind. What do you do, sir?" The rest of this paper therefore concerns issues that provide both secondary reasons to avoid Bayes-for-beginners at present and pedagogical challenges in a Bayesian future.

1.3   The Standard Choice

If we are to compare the accessibility of Bayesian reasoning with that of standard inference, it is wise to first state what the standard is. Our topic is first courses on statistical methodology for beginners who come from other disciplines and are not on the road to becoming professional statisticians. I will therefore give a partial outline of a "standard" elementary statistics course for this audience.

> **A. Data analysis**. We begin with tools and tactics for exploring data. For a single measured variable, the central idea is a *distribution* of values. We meet several tools for graphic display of distributions, and we learn to look for the shape, center, and spread of a distribution. We learn to describe center and spread numerically and to choose among competing descriptions.

> **B. Data production**. We now distinguish a sample from the underlying population, and statistics from parameters. We meet sta-

tistical designs for producing sample data for inference about the underlying population or process. Randomized comparative experiments and random sampling have in common the deliberate use of chance mechanisms in data production. We motivate this as avoiding bias, then study the consequences by asking "What would happen if we did this many times?" The answer is that the statistic would vary. The pattern of variation is given by a distribution, the *sampling distribution*. We can produce sampling distributions by simulation and examine their shape, center, and spread using the tools and ideas of data analysis.

C. Formal inference. We want to draw a conclusion about a parameter, a fixed number that describes the population. To do this, we use a statistic, calculated from a sample and subject to variation in repeated sampling from the same population. Standard inference acts as if the sample comes from a randomized data production design. We consistently ask the question *"What would happen if we did this many times?"* and look at sampling distributions for answers. One common type of conclusion is, "If we drew many samples, the interval calculated by this method would catch the true population mean μ in 95% of all samples." Another is, "If we drew many samples from a population for which $\mu = 60$ is true, only 1.2% of all such samples would produce an $\bar{x}$ as far from 60 as this one did. That unlikely occurrence suggests that μ is not 60."

Inductive inference is a formidable task. There is no simple path. But there may be relatively simpler and more complex paths. Consider these characteristics of the "standard" outline above:

- A parameter is a fixed unknown number that describes the population. It is different in nature from a statistic, which describes a sample and varies when we take repeated samples.

- Inference is integrated with data analysis through the idea of a distribution. The central idea of a sampling distribution can be presented via simulation and studied using the tools of data analysis.

- Probability ideas are motivated by the design of data production, which uses balanced chance mechanisms to avoid bias. The issue of sampling variability arises naturally, and leads naturally to the key question, "What would happen if we took many samples?"

- Probability has a single meaning that is concrete and empirical: "What would happen if we did this many times?" We can demonstrate how probability behaves by actually doing many trials of chance phenomena, starting with physical trials and moving to computer simulation.

- Inference consistently asks "What would happen if we did this many times?" Although we use probability language to answer this question, we require almost no formal probability theory. Answers are based on sampling distributions, a concrete representation of the results of repeated sampling.

- For more able students, study of simulation and bootstrapping is a natural extension of the "do it many times" reasoning of standard inference.

This simple outline of standard statistics can legitimately be criticized as lacking generality—standard inference is limited by acting *as if* we did proper randomized data production, for example. For beginners, however, it is clarity rather than generality that we seek.

1.4   Is Bayesian Reasoning Accessible?

I find that the reasoning of Bayesian inference, though purportedly more general, is considerably more opaque than the reasoning of standard inference.

- A **parameter** does describe the population, but it is a random quantity that has a distribution. In fact, it has two distributions, prior and posterior. So, for example, μ and $\bar{x}$ are both random. Yet μ is not "random" in the same sense that $\bar{x}$ is random, because a distribution for μ reflects our uncertainty, while the sampling distribution of $\bar{x}$ reflects the possibility of actually taking several samples.

- **Probability** no longer has the single empirical meaning, "What would happen if we did this many times?" Subjective probabilities are conceptually simple, but they are not empirical and don't lend themselves to simulation. Because we hesitate to describe the sampling model for the data given the parameter entirely in terms of subjective probability, we must explain several interpretations of "probability," and we commonly mix them in the same problem.

- The core reasoning of Bayesian inference depends not merely on probability but on **conditional probability**, a notoriously difficult idea. Beginners must move from the prior distribution of the parameter and the conditional distribution of the data given the parameter to the conditional distribution of the parameter given the data. Keeping track of what we are conditioning on at each step is the key to unlocking Bayesian reasoning. Thus when we consider the sampling distribution of $\bar{x}$, we think of μ as fixed because we are conditioning on μ. This allows us to deal with the practical observation that if we took another sample from the same population, we would no doubt obtain a different $\bar{x}$. Once we have sample data, we regard $\bar{x}$ as fixed and condition on its observed value in order to update the distribution of μ. This makes sense, of course, but in my experience even mathematics majors have difficulty keeping the logic straight.

Let me be clear that I am not questioning the coherence or persuasiveness of Bayesian reasoning, only its ease of access. In a future Bayesian era, we shall all have to face the task of helping students clear these hurdles. At present, they hinder an attempt to include a small dose of Bayes in a course that (for the reasons noted in Section 1.2) must concentrate on standard approaches when dealing with inference. In either case, we ought to recognize that they pose genuine barriers to students.

I find, to be blunt, that many Bayesian plans for teaching beginners deal with these basic conceptual difficulties by either over-sophistication or denial. Let me give an example of each.

Many Bayesians insist that they have only one kind of probability, namely, subjective probability. That strikes me as over-sophisticated. Physical and personal probabilities are conceptually quite different. There is a clear practical distinction between a prior distribution that expresses our uncertainty about the value of μ and the sampling distribution that expresses the fact that if we took another sample we would observe a different value of $\bar{x}$. Saying that "do it many times" is just one route to subjective probability is intellectually convincing to sophisticates but doesn't deal with the beginner's difficulty. What is more, the priors used in our examples are often "noninformative" priors that attempt to represent a state of ignorance. So some prior distributions represent prior knowledge, while others, mathematically similar in kind and appearing to give equally detailed descriptions of the possible values of the parameter, represent ignorance. There are sophisticated ways to explain that "ignorance" means ignorance relative to the information supplied by the data—but we are discussing the accessibility of core ideas to unsophisticated students.

Here is a passage in which Albert (1997) seems to me to deal with the conceptual difficulties of conditional probabilities by denial:

> *Although one can be sympathetic with the difficulty of learning probability, it is unclear what is communicated about classical statistical inference if the student has only a modest knowledge of probability. If the student does not understand conditional probability, then how can she understand the computation of a p-value, which is a tail probability conditional on a particular hypothesis being true? What is the meaning of a sampling distribution if the student does not understand what model is being conditioned on?*

The concluding rhetorical question in this passage seems based on the view that all probabilities are conditional probabilities. That is over-sophisticated, but it also denies that there is a difference in difficulty between these questions:

A. Toss a balanced coin 10 times. What is the probability that exactly 4 heads come up?

B. Toss a balanced coin 10 times. Someone tells you that there were 2 or more heads. Given this information, what is the probability that exactly 4 heads come up?

That there is a "model being conditioned on" in Question A is language that a teacher not bound by precise Bayes-speak would avoid when addressing beginners. Question B, which involves *conditioning on an observed event*, is conceptually more complex than Question A, which does not. Speaking only of the formalities of probability, as opposed to its interpretation, students can grasp standard inference via probability at the level of Question A. Bayesian inference requires conditional probability of the kind needed for Question B. Why attempt to deny that the second path is harder?

Albert's point that P-values require conditioning displays a similar denial of the distinction between Questions A and B. To explain a P-value, we begin by saying, "Suppose for the sake of argument that the null hypothesis is true." To the Bayesian sophisticate, this is conditioning. To a teacher of beginners, however, it is like saying "Suppose that a coin is balanced." The supposition is the start of the reasoning we want our students to grasp: a result this extreme would be very unlikely to occur if H_0 were true; such a result is therefore evidence that H_0 is not true. This reasoning isn't easy (there is no easy road to inference), but it does not involve conditioning on an observed event. A grasp of P-values does not require, and is not much aided by, a systematic presentation of conditional probability. On the other hand, conditioning is so central to Bayesian reasoning that we must discuss it explicitly and very carefully. The distinction between Question A and Question B captures the distinction between what we must teach as background to standard (A) and Bayesian (B) inference.

There are numerous other complexities that the teacher of Bayesian methods must face (or choose to ignore). The use of default or reference priors opens a gap between Bayesian principle and Bayesian practice that is not easy to explain to beginners. The use of improper priors may sow puzzlement. The need to abandon what seemed satisfactory priors when we move from estimation to testing is annoying. And so on. I want, however, to concentrate on what appears to be the primary barrier to beginners' understanding of the core reasoning of Bayesian inference: the greater dependence on probability, especially conditional probability.

1.5   Probability and Its Discontents

Albert, in the passage I just cited, speaks for many teachers, Bayesian and non-Bayesian alike, when he says, "it is unclear what is communicated about classical statistical inference if the student has only a modest knowledge of probability." It is, unfortunately, a well-documented fact that the great majority of our beginning students *will*, despite our best efforts, have only a modest knowledge of probability.

Psychologists have been interested in our perception of randomness ever since the famous studies of Tversky and Kahneman (1983). Bar-Hillel and Wagenaar (1993) offer a recent survey. Much recent work has criticized Tversky and Kahneman's attempt to discover "heuristics" that help explain why our intuition of randomness is so poor and our reasoning about chance events is so faulty. That our intuition about random behavior *is* gravely defective, however, is a demonstrated fact. To give only a single example, most people accept an incorrect "law of small numbers" [so named by Tversky and Kahneman (1971)] that asserts that even short runs of random outcomes should show the kind of regularity that the laws of probability describe. When short runs are markedly irregular, we tend to seek some causal explanation rather than accepting the results of chance variation. As Tversky and Kahneman (1983, p. 313) say in summary, "intuitive judgments of all relevant marginal, conjunctive, and conditional probabilities are not likely to be coherent, that is, to satisfy the constraints of probability theory."

That people's intuitions about chance behavior are systematically faulty has implications for statistical education. Our students do not come to us as empty vessels into which we pour knowledge. They combine what we tell them with their existing knowledge and conceptions to construct a new state of knowledge, a process that Bayesians should find natural. The psychologists inform us that our students' existing conceptions of chance behavior are systematically defective: they do not conform to the laws of probability or to the actually observed behavior of chance phenomena. At this point, researchers on teaching and learning become interested; the teaching and learning of statistics and probability has been a hot field in mathematics education research for more than a decade. Psychologists attempt to describe how people think. Education research looks at the effects of our intervention (teaching) on students' thinking. Results of this research are summarized in Garfield (1995), Garfield and Ahlgren (1988), Kapadia and Borovcnik (1991) and Shaughnessy (1992).

The consensus of education research is, if anything, more discouraging than the findings of the psychologists. Even detailed study of formal probability, so that students can solve many formally posed problems, does little to correct students' misconceptions and so does little to equip them to use probabilistic reasoning flexibly in settings that are new to them. Garfield and Ahlgren (1988) conclude that "teaching a conceptual grasp of probability still appears to be a very difficult task, fraught with ambiguity and illusion." Some researchers [see Shaughnessy (1992, pp. 481–483)] have been able to change the misconceptions of some (by no means all) students by activities in which students must write down their predictions about outcomes of a random apparatus, then actually carry out many repetitions and explicitly compare the experimental results with their predictions. Some of these same researchers go so far as to claim that "not only is traditional instruction insufficient, it may even have negative effects on students' understanding of stochastics." Shaughnessy (1992, p. 484) also cites "strong evidence for the superiority of simulation methods over analytic methods in a course on probability." He stresses that changing ingrained misconceptions cannot be done quickly, but requires sustained efforts.

Research in education and psychology appears to confirm that conditional probabilities are particularly susceptible to misunderstanding. Garfield and Ahlgren (1988, p. 55) note that students find conditional probability confusing because "an important factor in misjudgment is misperception of the question being asked." Students find it very difficult to distinguish among $P(A|B)$, $P(A \text{ and } B)$, and $P(B|A)$ in plain-language settings. Shaughnessy (1992, pp. 473–476) discusses the difficulties associated with conditional probabilities at greater length. He agrees that "difficulties in selecting the event to be the conditioning event can lead to misconceptions of conditional probabilities." He also points to empirical studies suggesting that students may confuse conditioning with causality, are very reluctant to accept a "later" event as conditioning an "earlier" event, and are easily confused by apparently minor variations in the wording of conditional probability problems. The "Monty Hall problem" (that goat behind a door—a job for Bayes' theorem) and its kin remind us that it is conditional probability problems that so often give probability its air of infuriating unintuitive cleverness.

1.6   Barriers to Bayesian Understanding

It appears that we must accept these facts as describing the environment in which we must teach: beginning students find probability difficult; they find conditional probability particularly difficult; there is as yet no known way to relieve their difficulties that does not involve extensive hands-on activity and/or simulation over an extended time period. I believe that our experience as teachers generally conforms to these findings of systematic study.

The unusual difficulty of probability ideas, and the inability of study of formal probability theory to clarify these ideas for students, argue against a mathematically-based approach in teaching beginners. *Mathematical under-*

standing is not the only kind of understanding. Mathematical language helps
us to formulate, relate, and apply statistical ideas, but it does not help our
students nearly as much as we imagine. Recognition of the futility of a formal approach has been one factor in moving beginning statistics courses away
from the traditional probability-and-inference style toward a data analysis-data
production-concepts of inference-methods of inference style that pays more attention to data. I believe that the findings I have cited also point to very
substantial difficulties that stand in the way of effective Bayes-for-beginners
instruction:

- If our intuition of chance is systematically incoherent, is it wise to rely on
  subjective probability as a central idea in a first statistics course?

- If the only known ways of changing misconceptions about chance behavior involve confronting misconceptions via physical chance devices and
  simulation—that is, by asking "What would happen if we did this many
  times?"—ought we not to make the answer to that question our primary
  interpretation of probability?

- If we want to help students see that the laws of probability describe chance
  outcomes *only* in the long run—the law of large numbers is true, but the
  law of small numbers is false—how can we avoid confusion if our central
  notion of probability applies to even one-time events?

- If teaching correct probability is so difficult and requires such intensive
  work, ought we not to follow Garfield and Ahlgren (1988) in asking "how
  useful ideas of statistical inference can be taught independently of technically correct probability?" Ought we not at least seek to minimize the
  number of probability ideas required, in order to leave time for data-
  oriented statistics?

- If conditional probability is known to be particularly difficult, should we
  not hesitate to make it a central facet of introductory statistics?

Statistical inference is not conceptually simple. Standard and Bayesian inference each require a hard idea—sampling distributions for standard inference,
and conditional probability and updating via Bayes' rule for Bayesian inference.
I claim—not only as a personal opinion, but as a reasonable conclusion from the
research cited above—that the Bayesian idea is markedly more difficult for beginners to comprehend. Sampling distributions fit the "activity and simulation"
mode recommended by most experts on learning, and the absence of conditioning makes them relatively accessible both conceptually and to simulation. Once
students have grasped the "What would happen if we did this many times?"
method, we can hope that they will also grasp the main ideas of standard inference. Bayes-for-beginners, on the other hand, must either shortchange the

reasoning of inference or use two-way tables to very carefully introduce conditional probability and Bayes' theorem. Albert (1995) and Rossman and Short (1995) illustrate the care that is required.

Although I advocate a quite informal style in presenting statistical ideas to beginners, I recognize that different teachers prefer different levels of formality. I claim that at any level of formality,

> "State a prior distribution for the parameter and the conditional distribution of the data given the parameter. Update to obtain the conditional distribution of the parameter given the data."

is less accessible core reasoning than

> "What would happen if we did this many times?"

1.7   Are Bayesian Conclusions Clear?

We have seen that the need to understand probability, especially conditional probability, at a relatively profound level, is a barrier to understanding the reasoning of Bayesian inference. I believe that the same barrier stands in the way of understanding the results of inference. That is, Bayesian conclusions are perhaps not as clear to beginners as Bayesians claim.

It is certainly true, as Bayesians always point out, that users do not speak precisely in stating their understanding of the conclusions of standard inference. They often confuse probability statements about the *method* (standard inference) and probability statements about the *conclusion* (Bayesian inference). "I got this answer by a method that gives a correct answer 95% of the time" easily slides into "The probability that this answer is correct is 95%." If we regard this semantic confusion as important, we ought to ask whether the user of Bayes methods can explain without similar confusion what she means by "probability." The Bayesian conclusion is easy to state, but hard to explain. What is this "probability 95%"? Physical and personal probabilities are conceptually quite different, and the user of Bayes methods must be aware that probabilities are conditioned on different events at each stage. That a user gives a semantically correct Bayesian conclusion is not evidence that she understands that conclusion.

1.8 What Spirit for Elementary Statistics Instruction?

The continuing revolution in computing has changed the professional practice of statistics and our judgment of what constitutes interesting research in statistics. These changes are in turn changing the teaching of elementary statistics. We are moving from an over-emphasis on probability and formal inference toward a balanced presentation of data analysis, data production, and inference. See the report of the ASA/MAA joint curriculum committee [Cobb (1992)] for a clear statement of these trends. In particular, there is a consensus that introductions to statistics ought to involve constant interaction with real data in real problem settings. Real problem settings often have vaguely defined goals and require the exercise of judgment. The spirit of contemporary introductions to statistical practice is very different from the spirit of traditional "probability and statistics" courses.

Bayesian thinking fits uneasily with these trends. Exploratory data analysis allows the data to speak, and diagnostic procedures allow the data to criticize proposed models; Bayesians tend to say "no adhockery" and to start from models and structured outcomes rather than from data. Good designs for data production avoid disasters (voluntary response, confounding) and validate textbook models; many Bayesians question at least the role of randomization and sometimes the role of proper sample/experimental design in general. And the high opportunity cost of teaching conditional probability and Bayes' theorem in more-than-rote fashion forces cuts elsewhere.

Concentrating on Bayesian thinking is not in principle incompatible with data-oriented instruction. In practice, however, it is likely to turn elementary statistics courses back toward the older mode of concentrating on the parts of our discipline that can be reduced to mathematics. Avoiding this unfortunate retrogression is perhaps the most serious pedagogical challenge facing Bayes-for-beginners.

Acknowledgements. I am grateful to James Berger for discussions about Bayesian viewpoints, and to George Cobb and Paul Velleman for helpful comments on a first version of this paper.

References

1. Albert, J. A. (1995). Teaching inference about proportions using Bayes and discrete models, *Journal of Statistics Education* (electronic journal), **3**, No. 3.

2. Albert, J. A. (1996). *Bayesian Computation Using Minitab*, Wadsworth, MA: Duxbury Press.

3. Albert, J. A. (1997). Teaching Bayes' rule: a data oriented approach, *The American Statistician* (to appear).

4. Antleman, G. (1996). *Elementary Bayesian Statistics*, Cheltenham, England: Edward Elgar Publishing.

5. Bar-Hillel, M. and Wagenaar, W. A. (1993). The perception of randomness, In *A Handbook for Data Analysis in the Behavioral Sciences: Methodological Issues* (Eds., G. Keren and C. Lewis), pp. 369–393, Hillsdale, NJ: Erlbaum Associates.

6. Berger, J. O. (1985). *Statistical Decision Theory and Bayesian Analysis*, New York: Springer-Verlag.

7. Berry, D. A. (1996). *Statistics: A Bayesian Perspective*, Belmont, CA: Duxbury Press.

8. Berry, D. A. (1997). Teaching introductory Bayesian statistics with real applications in science: doing what comes naturally, *The American Statistican* (to appear).

9. Box, G. E. P. and Tiao, G. C. (1973). *Bayesian Inference in Statistical Analysis*, Reading, MA: Addison-Wesley.

10. Cobb, G. (1992). Teaching statistics, In *Heeding the Call for Change: Suggestions for Curricular Action* (Ed., L. A. Steen), pp. 3–43, Washington, D.C.: Mathematical Association of America.

11. Dickey, J. M. and Eaton, M. L. (1996). Review of *Bayesian Theory* by J. M. Bernardo and A. F. M. Smith, *Journal of the American Statistical Association*, **91**, 906–907.

12. Emerson, J. D. and Colditz, G. A. (1992). Use of statistical analysis in the *New England Journal of Medicine*, In *Medical Uses of Statistics* (Eds., J. C. Bailar and F. Mosteller), Second edition, pp. 45–57, Boston, MA: NEJM Press.

13. Garfield, J. (1995). How students learn statistics, *International Statistical Review*, **63**, 25–34.

14. Garfield, J. and Ahlgren, A. (1988). Difficulties in learning basic concepts in probability and statistics: implications for research, *Journal for Research in Mathematics Education*, **19**, 44–63.

15. Kapadia, R. and Borovcnik, M. (Eds.) (1991). *Chance Encounters: Probability in Education*, Dordrecht, The Netherlands: Kluwer.

16. Kass, R. E. and Raftery, A. E. (1995). Bayes factors, *Journal of the American Statistical Association*, **90**, 773–795.

17. Kass, R. E. and Wasserman, L. (1995). A reference Bayesian test for nested hypotheses and its relationship to the Schwarz criterion, *Journal of the American Statistical Association*, **90**, 928–934.

18. Lindley, D. V. (1971). *Bayesian Statistics: A Review*, Philadelphia, PA: SIAM.

19. Moore, D. S. (1997). Bayes for beginners? Some reasons to hesitate, *The American Statistican* (to appear).

20. Robert, C. (1994). *The Bayesian Choice: A Decision-Theoretic Motivation*, New York: Springer-Verlag.

21. Rossman, A. J. and Short, T. H. (1995). Conditional probability and education reform: Are they compatible? *Journal of Statistics Education* (electronic journal), **3**, No. 2.

22. Rustagi, J. S. and Wright, T. (1995). Employers' contributions to the training of professional statisticians, *Bulletin of the International Statistical Institute, Proceedings of the 50th Session*, LVI, Book 1, pp. 141–160.

23. Shaughnessy, J. M. (1992). Research in probability and statistics: reflections and directions, In *Handbook on Research on Mathematics Teaching and Learning* (Ed., D. A. Grouws), pp. 465–494, New York: Macmillan.

24. Tversky, A. and Kahneman, D. (1971). Belief in the law of small numbers, *Psychological Bulletin*, **76**, 105–110.

25. Tversky, A. and Kahneman, D. (1983). Extensional versus intuitive reasoning: the conjunction fallacy in probability judgment, *Psychological Review*, **90**, 293–315.

2

Normal Means Revisited

John Deely and Wes Johnson

Purdue University, West Lafayette, IN
University of California, Davis, CA

Abstract: The problem of ranking and selecting normal means as originally studied by Shanti Gupta is approached herein from a robust Bayesian perspective. This model uses the usual hierarchical Bayesian setup but does not require complete specifications of the hyperpriors. Instead, elicited prior information about the population of means as a group is used to specify a quantile class for the hyperpriors. Two criteria are suggested for ranking and selecting which provide insight not only to which population is best, but in addition give quantitative methods for deciding how much better one population is than another. Using these criteria, minimum and maximum values are calculated for the derived quantile class. Relative sizes of these evaluations and the distance between the max and min give insight as to the quality of the data and the sufficiency of the sample size. These concepts are illustrated with a numerical example.

Keywords and phrases: Normal means, hierarchical Bayesian, ranking and selection, robust Bayesian, quantile class, predictive distribution

2.1 Introduction

Shanti Gupta began a career in Ranking and Selection Methods when he wrote his thesis *On a Decision Rule for a Problem in Ranking Means* in 1956. His new approach to the selection problem was to derive procedures which selected a subset of "random" size in such a way that the probability of obtaining the "best" population was greater than a specified level. This formulation is to be contrasted to that initiated by Bechhofer (1954) in which a fixed size subset (generally size one) was selected. The basic problem with this new random approach involved deriving selection procedures which had certain desirable properties; namely, the selected subset should include the "best" population with reasonably high probability and the size of the selected subset should be as small as possible while still assuring the probability requirement. Thus began the quest to find an optimal selection procedure "t" which minimized $E[S|\boldsymbol{\vartheta}, t]$ subject to satisfying $P(CS|\boldsymbol{\vartheta}, t) \geq P^*$, where $\boldsymbol{\vartheta}$ is the k vector of unknown means, "t" denotes the particular selection procedure used and P^* is a pre-specified number close to unity.

It turns out that there is no one optimal solution for this problem. In fact, it has no solution in that context because for one configuration of the parameter space one procedure does better than another and vice versa for another configuration. See, for example, Paulson (1952) and Seal (1955). This impossibility thus led to an enormous amount of work which dealt with various formulations and models and their correspondent procedures. The fact that this area of research has flourished so much is due in no small measure to the influence and encouragement Shanti provided and a partial list of his contributions can be found in Gupta and Panchapakesan (1979) and Gupta and Huang (1981). He was always at the forefront encouraging development in a wide variety of areas. In that regard, it should be noted that both of the computations for $P(CS|\boldsymbol{\vartheta}, t)$ and $E[S|\boldsymbol{\vartheta}, t]$ were made initially in the frequentist sense (as though the problem facing the practitioner were going to be encountered over and over again under exactly the same experimental conditions) since modern Bayesian theory was just beginning to surface at that time. A classic paper by Dunnett (1960) was the first paper to deal with selection problems using a Bayesian flavor. But Shanti's perspective was so broad that all approaches were supported. In fact at a very early stage, in 1965, he even encouraged his first Ph.D. student (the first author) to write a thesis on *Multiple Decision Procedures from an Empirical Bayes Approach*. Since that time, a number of Bayesian and empirical Bayesian papers mainly from a decision theoretic point of view have appeared [see, for example, Goel and Rubin (1977)]. A recent paper by Berger and Deely (1988) does develop a more practical approach to this problem.

But in spite of all this work (both frequentist and Bayesian) on the normal means problem, there has not been a full scale adoption and application of them in the practical world. Practitioners are still using the old fashioned but more importantly inadequate AOV type analysis of data much of which really requires a procedure to rank the means. Of course, the computer facilities for these methods are well developed compared to the frequentist or Bayesian ranking procedures. In addition, frequentist ranking procedures require statements about the parameter space that may not always be practically realizable and whereas the Bayesian formulation is more practical, there is still the problem in that approach with the prior or lack of it for die hard frequentists.

The notion of exchangeability amongst the means in the hierarchical Bayesian (HB) formulation adds another dimension to the problem. Since the equipments, suppliers or processes generating the means to be compared are assumed to be somewhat similar, we treat the population means as exchangeable, an assumption which is conveniently modeled through a HB setup. Thus, the main feature of this approach is that it facilitates in a much more practical way the use of the type of prior information that practitioners are likely to have available; namely, information about the *group* of means as opposed to information about specific *individual* members of the group.

The first phase of this approach is contained in a paper by Berger and Deely (1988). With the advent of the Markov Chain Monte Carlo (MCMC) methods with emphasis on the Gibbs sampler a more general approach than taken in that paper is now computationally feasible. Even so, the HB approach uses either a non-informative or an elicited subjective hyperprior both of which may not be consistent with the type of prior information readily available about the group of means being studied. Specifically, it may be that the available prior information does not allow complete specification of the hyperpriors but on the other hand should not be ignored as in the non-informative case. Our approach to the hyperpriors assumes that they are determined *only* up to a *family* of distributions which depend on the available prior information. With this type of model and consequent analysis, we believe we are providing the practitioner with tools which effectively use the type of prior information that might be available in many situations; namely, information about the populations as a group as opposed to information about individual populations.

Thus, specification of a prior on the unknown population means using the HB model can easily incorporate this kind of information. The prior is a mixture of conditional distributions with the mixing distribution determined by what is known about the group of populations. Hence for the normal means problem, we can think of the individual means coming from a normal distribution with mean β and variance τ^2 which are distributed according to some hyperprior $h(\beta, \tau^2)$ where h is determined by the prior information. The robust Bayesian approach we take in this paper is to relax the requirement that h has to be completely specified to an assumption that the type of prior information available allows specification of h to belong to the Quantile class of distributions.

Specifically, we assume this family is the quantile class [cf. Lavine (1991)], where the elicited prior information specifies particular quantiles but nothing more about the hyperpriors. Using a technique from Lavine (1991), we then compute maximum and minimum values for the above criteria where these extrema are taken over the family of hyperpriors specified by the prior information. The utility of the prior information is assessed by computing the *difference* between the maximum and the minimum of the probabilities thus obtained; a small difference indicating a useful inference whereas large differences would be meaningless and thus not very useful. The effect of the sample size and the observed test data on inferences is assessed by consideration of the *magnitudes* of the criterion probabilities; for example, values of these probabilities near unity would indicate that the sample size and test data were very effective in determining a "best" member of the entire group, whereas small values would indicate ineffective test data. It is to be emphasized that a major implication of our treatment of the hyperpriors is that the prior distribution on the means implied through the HB model need not be completely specified. It is this latter feature which utilizes whatever partial prior information is available and in this sense guarantees the robustness of the suggested ranking procedures over all priors satisfying such information.

Finally, it should be noted that another feature of the Bayesian approach is the fact that the ranking criteria not only gives the ranks of the means but in addition can be used to determine *how much better* one population is than another. Further amplification of this feature is made in Section 2.2 where we define and discuss two specific criteria to be used in the ranking process. In Section 2.3, the details of the robust procedures are developed while numerical examples illustrating our methods are given in Section 2.4.

2.2 Selection Criteria

The suggested ranking criteria is based on Bayesian concepts arising from the posterior distribution. We focus on two concepts here.

Criterion 1: The posterior probability that any one of the means is larger than all of the others by an amount "*b*", i.e., compute for each $i = 1, 2, \ldots, k$, the quantity

$$P_i = P(\vartheta_i \geq \vartheta_j + b \text{ for ALL } j \neq i | \text{ data})$$

where b is a non-negative specified constant;

Criterion 2: The predictive distribution that any one of these populations will have a larger observation than all the others by an amount "*c*", i.e., compute for each $i = 1, 2, \ldots, k$, the quantity

$$PR_i = P(Y_i \geq Y_j + c \text{ for ALL } j \neq i | \text{ data})$$

where Y_i is a new observation from the i-th population and c is a non-negative specified constant.

Closer examination of the proposed ranking criteria reveals how they can be used to make quantitative rankings among the k individuals in the particular group being studied. Firstly, by computing each P_i and PR_i for $i = 1, 2, \ldots, k$, we can compare each member to all of the individuals remaining in the *total* group; from these computations, *subgroups* for further comparisons may be suggested. That is, suppose P_i and PR_i are very close to unity for a particular 'i'; this would indicate we have conclusively found the best amongst all k members. However, when this is not the case, it may be that a subgroup of just a few members may have their sum of P_i or PR_i very large in which case we would then be interested in comparisons amongst that subgroup only. It is easily seen that such iterations might eventually lead to simply a comparison of just two members of the original group of k. In general, we allow for the possibility of finding subgroups and making comparisons within those subgroups which includes ultimately all possible pairwise comparisons. There are clearly many possibilities each of which can be computed as required without effecting the validity of any other calculation. In addition, a salient feature of these computations is that the *degree of how much better* one member is than any other in the particular group being compared, be it one other or many, can be assessed by varying the quantities "*b*" and "*c*" in the formulas. This process lends itself to a type of "OC" analysis by plotting P_i against $\boldsymbol{b}$ and PR_i against $\boldsymbol{c}$. Note that Criterion 1 reduces to the Bayesian Probability of Correct Selection (PCS) when $b = 0$ but the fact that we allow b to take on positive values is an important improvement over the usual PCS criterion. Specific examples of these concepts will be given in Section 2.4.

The proposed criteria above are not new. The first criterion has been discussed extensively in Berger and Deely (1988) for the problem of ranking normal means and in a general context by Gupta and Yang (1985). The second criterion has been treated in a general context by Geisser (1971) and more specifically in Geisser and Johnson (1996).

2.3 Model and Computations

Let X_j, the sample mean for the j-th population based on n_j observations, be normally distributed with unknown mean θ_j and known variance σ_j^2/n_j. The prior on $\boldsymbol{\theta} = (\theta_1, \ldots, \theta_k)$ will be described by a two stage process as in the usual hierarchical Bayesian model. For the first stage, let $\theta_1, \ldots, \theta_k$ be conditionally i.i.d. with a normal distribution with mean β and variance τ^2. For the second stage, we assume only that β and τ^2 are independent with distributions h_1 and h_2 which are known to belong to a specific quantile class. This class is determined by the prior information available about the unknown means $\theta_1, \ldots, \theta_k$. Further discussion on this point will be made in the next section.

We can now proceed with the computational forms for Criteria 1 and 2. Firstly, it will be helpful to adopt the following notation. Let

$$A_i = \{\vartheta_i \geq \vartheta_j + b \text{ for all } j \neq i\} \text{ and } B_i = \{Y_i \geq Y_j + c \text{ for all } j \neq i\}.$$

From our model, it follows that conditional upon β and τ^2, the posterior density of θ_j is a normal distribution denoted by $\pi(\vartheta_j | x_j, \beta, \tau^2)$ with mean $\text{mpos}_j =$

$\alpha_j x_j + (1 - \alpha_j)\beta$ and variance $\mathrm{vpos}_j = \alpha_j \sigma_j^2/n_j$ where $\alpha_j = \tau^2\{\tau^2 + \sigma_j^2/n_j\}^{-1}$. We can then write the conditional joint posterior density of $\boldsymbol{\theta}$ as

$$\pi(\boldsymbol{\vartheta}|\boldsymbol{x}, \beta, \tau^2) = \prod_{j=1}^{k} \pi(\vartheta_j|x_j, \beta, \tau^2).$$

It will be convenient to denote the univariate normal cdf and pdf with mean μ and variance v by $G(\cdot|\mu, v)$ and $g(\cdot|\mu, v)$, respectively. In addition, we require the following form for the posterior distribution of β and τ^2 denoted by $h(\beta, \tau^2|\boldsymbol{x})$:

$$h(\beta, \tau^2|\boldsymbol{x}) = \frac{f(\boldsymbol{x}|\beta, \tau^2)\, h(\beta, \tau^2)}{\int f(\boldsymbol{x}|\beta, \tau^2)\, h(\beta, \tau^2)d\beta d\tau^2} \qquad (2.1)$$

where $h(\beta, \tau^2)$ denotes the hyperprior on β and τ^2. We will not assume a specific form for $h(\beta, \tau^2)$ but we will be able to use prior information about the group of populations to locate $h(\beta, \tau^2)$ in the quantile class. Let $\boldsymbol{H}$ denote this elicited class of hyperpriors. For the purposes of this paper, we will assume that the form of the prior information can be interpreted as follows: $h(\beta, \tau^2) = h_1(\beta) h_2(\tau^2)$ and that for β and τ^2, respectively, we are able to ascertain three regions R_1, R_2, R_3 and S_1, S_2, S_3 with respective prior probabilities p_1, p_2, p_3 and q_1, q_2, q_3. Thus,

$$\boldsymbol{H} = \{h\colon \int_{R_j} h_1(\beta)d\beta = p_j \text{ and } \int_{S_j} h_2(\tau^2)d\tau^2 = q_j \text{ for } j = 1, 2, 3\}.$$

Criterion 1: Using the above notation, we can then write

$$P_i = P(A_i|\boldsymbol{x}) = \int P(A_i|\boldsymbol{x}, \beta, \tau^2)h(\beta, \tau^2|\boldsymbol{x})d\beta d\tau^2, \qquad (2.2)$$

where

$$\begin{aligned} P(A_i|\boldsymbol{x}, \beta, \tau^2) &= \int P(A_i|\boldsymbol{x}, \beta, \tau^2, \vartheta_i)g(\vartheta_i|\mathrm{mpos}_i, \mathrm{vpos}_i)d\vartheta_i \\ &= \int_0^\infty \left\{\prod_{j \neq i} G(\vartheta_i - b|\mathrm{mpos}_j, \mathrm{vpos}_j)\right\} \\ &\quad \times g(\vartheta_i|\mathrm{mpos}_i, \mathrm{vpos}_i)d\vartheta_i. \end{aligned} \qquad (2.3)$$

Criterion 2: For this criterion, we firstly note that the predictive distribution of Y_j given β and τ^2 is normal with mean mpos_j and variance $u_j = \sigma_j^2 + \mathrm{vpos}_j$. Thus, we can write

$$PR_i = P(B_i|\boldsymbol{x}) = \int P(B_i|\boldsymbol{x}, \beta, \tau^2)h(\beta, \tau^2|\boldsymbol{x})d\beta d\tau^2 \qquad (2.4)$$

where

$$P(B_i|\boldsymbol{x}, \beta, \tau^2) = \int P(B_i|\boldsymbol{x}, \beta, \tau^2, y_i) g(y_i|\text{mpos}_i, u_i) dy_i$$

$$= \int_0^\infty \left\{ \prod_{j \neq i} G(y_i - b|\text{mpos}_j, u_j) \right\} g(y_i|\text{mpos}_i, u_i) dy_i. \tag{2.5}$$

Letting RC denote "Ranking Criterion" and using (2.1), we can then write both (2.2) and (2.4) as a function of the hyperprior h as

$$RC(h) = \frac{\int P(\beta, \tau^2) \, L(\beta, \tau^2) \, h(\beta, \tau^2) d\beta d\tau^2}{\int L(\beta, \tau^2) \, h(\beta, \tau^2) d\beta d\tau} = \frac{N(h)}{D(h)}, \tag{2.6}$$

where $P(\beta, \tau^2)$ is given by (2.3) and (2.5), respectively, for Criteria 1 and 2, and $L(\beta, \tau^2)$ is the second stage likelihood function given by

$$L(\beta, \tau^2) = f(\boldsymbol{x}|\beta, \tau^2) = \int \prod_{j=1}^k g(x_j|\vartheta_j, \sigma_j^2/n_j) \, g(\vartheta_j|\beta, \tau^2) d\vartheta_j$$

$$= \prod_{j=1}^k g(x_j|\beta, (\sigma_j^2/n_j) + \tau^2). \tag{2.7}$$

Our goal is to find maximum and minimum values for $RC(h)$ over the family $\boldsymbol{H}$. To accomplish this, we use the form (2.6) and invoke the Linearization Principle made popular recently by Lavine (1991) which allows us to write:

$$\max_{h \in H} \frac{N(h)}{D(h)} = \bar{a} \text{ iff } \bar{a} = \min\{a: \max_h [N(h) - a \, D(h)] \leq 0\} \tag{2.8}$$

$$\min_{h \in H} \frac{N(h)}{D(h)} = \underline{a} \text{ iff } \underline{a} = \min\{a: \min_h [N(h) - a \, D(h)] \leq 0\}. \tag{2.9}$$

The value of this principle can be appreciated by observing the fact that for a given "a" the quantities $\max\{N(h) - aD(h)]: h \in \boldsymbol{H}\}$ and $\min\{[N(h) - aD(h)]: h \in \boldsymbol{H}\}$ are easily computed. Specifically,

$$\max_h [N(h) - a \, D(h)] = \max_h \int \{[P(\beta, \tau^2) - a]L(\beta, \tau^2)\} h(\beta, \tau^2) d\beta d\tau^2$$

$$= \sum_{i=1}^3 \sum_{j=1}^3 \overline{PL}_{ij}(a) p_i q_j,$$

where

$$\overline{PL}_{ij}(a) = \max\{[P(\beta, \tau^2) - a]L(\beta, \tau^2): (\beta, \tau^2) \in R_i \cap S_j\}.$$

A similar calculation is used to obtain

$$\underline{PL}_{ij}(a) = \min\{[P(\beta, \tau^2) - a]L(\beta, \tau^2): \ (\beta, \tau^2) \in R_i \cap S_j\}.$$

By noting the forms of the functions P [in either (2.3) or (2.5)] and L in (2.7), it can be seen that for any given "a", values of $\overline{PL}_{ij}(a)$ and $\underline{PL}_{ij}(a)$ are easily obtained numerically. This, in turn, leads to the computation of the desired values $\bar{a}$ and $\underline{a}$. The numerical example in Section 2.4 illustrates these computations.

2.4   Numerical Example

Consider the example given in Moore and McCabe (1993, p. 756, Ex. 10.18) concerning a study of the effects of exercise on physiological and psychological variables. For illustrative purposes, we focus here on the psychological data which is summarized in Table 2.1. The Treatment consisted of a planned exercise program, the Control group were average type people while the Sedentary group were chosen by their inactivity. High scores indicate more depressed than low scores and here we are interested in computing which group is most (least) depressed and by how much. Thus if we think of θ_j as the unknown mean depression of the j-th group, then we can use either Criteria described in Section 2.1 to understand the effect of exercise (or lack thereof) on depression.

Table 2.1: Summary statistics for the psychological data

Group	n	sample mean	sample std. dev.
Treatment (T)	10	51.9	6.42
Control (C)	5	57.4	10.46
Joggers (J)	11	49.7	6.27
Sedentary (S)	10	58.2	9.49

Firstly, we want to consider the value of prior information about the group as opposed to information about each population. Of course, the type of prior information available in any particular situation can be very different. Here, we are simply indicating just one of many different elicitation scenarios. In any case, we do assume that the available prior information will be specific enough to indicate a particular quantile class which will then be used for the hyperpriors. For purposes of illustration, consider eliciting answers to the following questions:

(1) What is the smallest interval which contains all of the unknown means θ_i's? *Ans.* (35,70)

(2) What is the smallest interval containing the average of the θ_i's? *Ans.* (45,60)

(3) How confident are you that the distance between the maximum θ_i and the minimum θ_i is at least 10% of the range given in (1), i.e., the difference between the maximum and the minimum depression scores will be at least 3.5 units on the scale of measurements to be used. *Ans.* 90%

For Questions (1) and (2), we assume that we can place a probability of at least 0.95 on the answers. Using this elicited information, we can obtain regions in the (β, τ^2) space with their respective probabilities which then specifies the particular quantile class to be used for the hyperpriors on β and τ^2. Thus from (2), we have that $P\{45 < \beta < 60\} = .95$ since β is the prior mean of the θ_i's. Hence, we obtain three intervals R_1, R_2, R_3 on β as (0,45), (45,60) and (60,∞) with their respective probabilities of 0.025, 0.95, 0.025.

From (1), we can write $P\{\theta_{[k]} | -\theta_{[1]} < 35\} = .95$; from (3), we have $P\{\theta_{[k]} - \theta_{[1]} \geq 3.5\} = .9$. These two expressions can be used to obtain probability intervals for τ^2 by noting that

$$P\{\theta_{[k]} - \theta_{[1]} < c\} = P\{T - S < c/\tau\}$$

where T and S are the maximum and minimum, respectively, of k standard normal random variables. Using $k = 4$ with (1) and (3) gives the following results:

$$0.95 = P\{35/\tau \geq 3.65\} = P\{\tau^2 < 100\}$$

and

$$0.9 = P\{3.5/\tau < 1.1\} = P\{\tau^2 \geq 10\}.$$

This in turn gives three intervals S_1, S_2, S_3 on τ^2 as (0,10), (10,100) and (100,∞) with their respective probabilities of 0.1, 0.85, 0.05.

Putting the β and τ^2 intervals together gives nine regions with their correspondent probabilities. Thus we have determined which quantile class describes the hyperpriors for the elicited information. Table 2.2 indicates symbolically the regions and their respective probabilities.

Table 2.2: Nine regions in the (β, τ^2) space with their respective probabilities

	$R_3 = (60, \infty)$	0.0025	0.0213	0.0013
β	$R_2 = (45, 60)$	0.0950	0.8075	0.0475
	$R_1 = (0.45)$	0.0025	0.0213	0.0013
		$S_1 = (0, 10)$	$S_2 = (10, 100)$	$S_3 = (100, \infty)$
			τ^2	

For this configuration, the solutions to (2.8) and (2.9) when using Criterion 1 yield the (min, max) interval as (0.4575, 0.6919). An indication of the sensitivity

28   *John Deely and Wes Johnson*

to these values can be seen by changing the probabilities to

$$
\begin{array}{ccc}
.0025 & .0025 & .0025 \\
.0025 & .98 & .0025 \\
.0025 & .0025 & .0025
\end{array}
$$

We obtain (0.489, 0.702). If we keep the original probabilities, but tighten up the regions to make the middle interval on τ^2 to be (30, 40), then the computation gives (0.490, 0.697). Using this configuration and the increased probabilities, we obtain (0.5054, 0.6968). Changing the middle interval for τ^2 to (100, 500) and using the new probabilities as above, produces (0.5249, 0.6294). Thus, it can be seen that the type and size of the prior information has an effect but there is a general agreement amongst the resulting computations. In particular, when comparing all four populations, it is not overwhelmingly true that the Sedentary group suffers the most depression. This is not surprising when looking at the values for the Control group. However, when comparing the two populations, Sedentary and Joggers, there is overwhelming evidence that the Sedentary group are much more depressed than the Joggers. This data is reported in Table 2.3.

Table 2.3: Comparison of the two populations, Sedentary and Joggers

		Criterion 1			Criterion 2	
	b	min	max	c	min	max
	0	0.827	0.993	0	0.744	0.953
sample	1	0.640	0.984	1	0.574	0.929
sizes	2	0.406	0.966	2	0.487	0.898
10,11	5	0.101	0.829	5	0.244	0.746
	0	1.000	1.000	0	1.000	1.000
sample	5	0.978	0.999	5	0.937	0.985
sizes	8	0.304	0.656	8	0.362	0.612
100,110	10	0.016	0.083	10	0.060	0.164

In addition, it can be seen in Table 2.3 just how much more depressed the Sedentary group is. For example, when using Criterion 1 we can say that there is at least one unit difference with posterior probability between 0.640 and 0.984. Another aspect of these computations is illustrated by noting the effect of the sample size. The lower part of Table 2.3 has been calculated assuming the data for these two populations had been based on sample sizes of 100 and 110. If this had been the case, then the data would have indicated that there was a five unit difference with probability between 0.978 and 0.999.

Many other comparisons and calculations are suggested by the above. The computations are easily performed and a wide variety of inferences are possible.

Our purpose here, through these brief illustrations, has been to indicate how this methodology can be easily applied and how the results obtained give valuable new insight into the normal means ranking and selection problem.

References

1. Bechhofer, R. E. (1954). A single-sample multiple decision procedure for ranking means of normal populations with known variances, *Annals of Mathematical Statistics*, **25**, 16–39.

2. Berger, J. O. and Deely, J. J. (1988). A Bayesian approach to ranking and selection of related means with alternatives to AOV methodology, *Journal of the American Statistical Association*, **83**, 364–373.

3. Dunnett, C. W. (1960). On selecting the largest of k normal population means (with discussion), *Journal of the Royal Statistical Society, Series B*, **22**, 1–40.

4. Geisser, S. (1971). The inferential use of predictive distributions, In *Foundations of Statistical Inference* (Eds., V. P. Godambe and D. A. Sprott), Toronto: Holt, Rinehart and Winston.

5. Geisser, S. and Johnson, W. O. (1996). Sample size considerations in multivariate normal classification, In *Bayesian Analysis of Statistics and Econometrics* (Eds., D. A. Berry, K. M. Chaloner and J. W. Geweke), pp. 289–298, New York: John Wiley & Sons.

6. Goel, P. K. and Rubin, H. (1977). On selecting a subset containing the best population—A Bayesian approach, *Annals of Statistics*, **5**, 969–983.

7. Gupta, S. S. and Panchapakesan, S. (1979). *Multiple Decision Procedures: Theory and Methodology of Selecting and Ranking Populations*, New York: John Wiley & Sons.

8. Gupta, S. S. and Huang, D. Y. (1981). *Multiple Statistical Decision Theory: Recent Developments*, Lecture Notes in Statistics, **6**, New York: Springer-Verlag.

9. Gupta, S. S. and Yang, H. M. (1985). Bayes-P* subset selection procedures, *Journal of Statistical Planning*, **12**, 213–233.

10. Lavine, M. (1991). An approach to robust Bayesian analysis for multidimensional parameter spaces, *Journal of the American Statistical Association*, **86**, 400–403.

11. Moore, D. S. and McCabe, G. P. (1993). *Introduction to the Practice of Statistics*, Second edition, New York: W. H. Freeman and Company.

12. Paulson, E. (1952). On the comparison of several experimental categories with a control, *Annals of Mathematical Statistics*, **23**, 239–246.

13. Seal, K. C. (1955). On a class of decision procedures for ranking means of normal populations, *Annals of Mathematical Statistics*, **26**, 387–398.

3

Bayes m-Truncated Sampling Allocations for Selecting the Best Bernoulli Population

Klaus J. Miescke and Henry Park

University of Illinois at Chicago, IL

Abstract: Let $\mathcal{P}_1, \ldots, \mathcal{P}_k$ be $k \geq 2$ independent Bernoulli populations with success probabilities $\theta_1, \ldots, \theta_k$, respectively. Suppose we want to find the population with the largest success probability, using a Bayes selection procedure based on a prior density $\pi(\boldsymbol{\theta})$, which is the product of k known Beta densities, and a linear loss $L(\boldsymbol{\theta}, i)$, $\boldsymbol{\theta} = (\theta_1, \ldots, \theta_k)$, $i = 1, \ldots, k$. Assume that k independent samples of sizes $n_1, \ldots, n_k$, respectively, have been observed already at a first stage, and that m more observations are planned to be taken at a future second stage. The problem considered is how to allocate these m observations in a suitable manner among the k populations, given the information gathered so far. Several allocation schemes, which have been considered previously, are compared analytically as well as numerically with the optimum allocation rule which is based on backward optimization. The numerical results indicate that there exists a good approximation to the optimum allocation rule which is easier to implement.

Keywords and phrases: Bayes selection, best Bernoulli population, m-truncated sequential allocation, backward optimization

3.1 Introduction

Let $\mathcal{P}_1, \ldots, \mathcal{P}_k$ be $k \geq 2$ independent Bernoulli populations with success probabilities $\theta_1, \ldots, \theta_k \in [0, 1]$, respectively. Suppose we want to find that population which has the largest θ-value, using a Bayes selection rule which is based on a known prior density $\pi(\boldsymbol{\theta})$, $\boldsymbol{\theta} = (\theta_1, \ldots, \theta_k) \in [0, 1]^k$, and a given loss function $L(\boldsymbol{\theta}, i)$ for selecting population $\mathcal{P}_i$, $i \in \{1, 2, \ldots, k\}$ at $\boldsymbol{\theta} \in [0, 1]^k$. Assume that k independent samples of sizes $n_1, \ldots, n_k$, respectively, have been observed already at a first stage, and that m additional observations are planned to be taken at a second stage. Fixed and sequential sampling designs for allocating these m observations to the k populations have been considered by Gupta and Miescke (1996b). In this paper, the optimum sequential allocation, which is based on backward optimization, will be studied and compared with the former procedures.

The multi-stage approach has been utilized to the problem of selecting the best population in various settings and numerous papers. Overviews can be found in Bechhofer, Kiefer and Sobel (1968), Gupta and Panchapakesan (1979, 1991), and Miescke (1984). Shorter somewhat critical overviews with emphasis on types of *elimination* and *adaptive sampling* are presented in Gupta and Miescke (1984,1996b).

The basic idea of Bayesian sequential sampling allocation is to minimize the expected posterior Bayes risk on a collection of suitable sample allocation schemes. For simplicity of presentation, estimation and cost of sampling will not be considered here. The former would affect selection through the posterior expected loss due to estimation, whereas the latter would involve stopping rules. Details in this direction can be found in Gupta and Miescke (1993).

The simplest allocation rules are those which have fixed predetermined sample sizes $m_1, \ldots, m_k$ with $m_1 + \ldots + m_k = m$. Looking ahead m allocated but not yet drawn observations, using the expected posterior Bayes risk, given the prior and all the observations collected so far, and then minimizing it across all possible allocations of m observations, leads to the best of such allocation rules. To find the minimum expected posterior risk, one has first to know the single-stage Bayes selection rule for each of the allocations considered. This problem has been treated within the present setting, i.e. for binomial populations with independent Beta priors, in Abughalous and Miescke (1989). But once the optimum sampling allocation of m observations has been found, one may wonder why one should allocate all m observations at once, rather than allocate just a few observations (or one), learn more from them (it), and then continue with more appropriate allocations of the rest.

The most involved allocation rule, on the other hand, is the best: the Bayes m-truncated sequential allocation rule. After finding the Bayes terminal selections for every possible allocation of m observations and their outcomes, backward optimization is used to optimize successively every single allocation before. The latter is limited in practice by computing power, as will be explained in Section 3.4. Other approximately optimum procedures have been proposed and studied in Gupta and Miescke (1996b), which will also be briefly reconsidered in this paper. One of them allocates in an optimum way one observation at a time, pretending that it is the last one to be drawn before the final selection, and then iterates this process until m observations have been drawn.

Bayes m-truncated sequential rules are discussed in Berger (1985). Bayes *look ahead* techniques have been used already by Dunnett (1960) for means of normal populations, and by Govindarajulu and Katehakis (1991) for survey sampling. They are also discussed in various settings in Berger (1985), including relevant work by Amster (1963). Gupta and Miescke (1994,96a) have used this approach to select means of normal populations, and similar work has been done in Gupta and Miescke (1996b) for binomial populations. One difference worth mentioning is that for $k = 2$ normal, but not binomial, populations, the outcome of the observations from the first stage are irrelevant for any further optimum allocations.

The most commonly known selecting procedure is the so-called *natural selection rule*, which selects in terms of the largest sample mean. Although it is optimum in various ways whenever the number of observations is the same for each population, it loses much of its quality in other situations. This has been shown in Risko (1985) for $k = 2$, and in Abughalous and Miescke (1989) for $k \geq 2$. Similar findings are reported in Gupta and Miescke (1988) and in Bratcher and Bland (1975). Despite this fact, the natural selection rule still continues to enjoy rather unquestioned popularity.

Other types of adaptive sampling from k Bernoulli populations have been considered in the frequentist approach, with different objectives, by many authors. One particular approach, which was introduced by Bechhofer and Kulkarni (1982), has been followed up in several papers. It saves observations in the natural single stage procedure for equal sample sizes, without losing any power of performance. A thorough overview of this approach, and many related references, can be found in Bechhofer, Santner and Goldsman (1995).

In Section 3.2, Bayes look ahead fixed and m truncated sequential sampling allocation rules are described and discussed in a comparative manner. Applications of these general rules to the case of k Bernoulli populations with independent Beta priors and a linear loss is worked out in details in Section 3.3. Finally, in Section 3.4, numerical results from computer calculations and simulations, at the same parameter settings as in Gupta and Miescke (1996b), are presented, which provide the basis for a critical comparison of all sampling allocation schemes considered. It is found that three allocation rules are suitable approximations to the optimum rule based on backward optimization: the Bayes *look-ahead-one-observation-at-a-time* allocation rule, the optimum fixed sample sizes allocation rule, and a newly proposed rule which is easy to use and appears to be the best approximation.

3.2   General Outline of Bayes Sequential Sampling Allocations

All sampling allocations considered in this paper are based on the following general two-stage model which, after a standard reduction of the data by sufficiency, can be summarized as follows. At $\boldsymbol{\theta} = (\theta_1, \ldots, \theta_k) \in \Omega^k$, the parameter space, let X_i and Y_i be real valued sufficient statistics of the samples from population $\mathcal{P}_i$ at Stage 1 and Stage 2, resp., $i = 1, \ldots, k$, which altogether are assumed to be independent. A priori, the parameters are considered as realizations of a random variable $\boldsymbol{\Theta} = (\Theta_1, \ldots, \Theta_k)$ which follows a given prior distribution. Let the loss for selecting $\mathcal{P}_i$ at $\boldsymbol{\theta} \in \Omega^k$ be $L(\boldsymbol{\theta}, i)$, $i = 1, \ldots, k$. Cost of sampling, which would require the incorporation of a stopping rule, is not included in the loss to simplify the presentation of basic ideas. Modifications of the allocations discussed below to this more general setting are straightforward.

From now on it is assumed that sampling at Stage 1 has been completed already, and that $X_i = x_i$, based on sample size n_i, $i = 1, \ldots, k$, have been observed. For brevity, let $\boldsymbol{X} = (X_1, \ldots, X_k)$ and $\boldsymbol{x} = (x_1, \ldots, x_k)$. Likewise, let $Y_i = y_i$, based on sample size m_i, $i = 1, \ldots, k$, with $m_1 + \ldots + m_k = m$, denote the observations which have been observed at the end of Stage 2. For brevity let $\boldsymbol{Y} = (Y_1 \ldots, Y_k)$ and $\boldsymbol{y} = (y_1, \ldots, y_k)$. Later on, depending on the particular situation considered, we may also assume that only part of the data at Stage 2, say, $\tilde{Y}_i = \tilde{y}_i$, based on sample size $\tilde{m}_i \leq m_i$, $i = 1, \ldots, k$, have been observed. Hereby it is understood that the sampling information of $\tilde{Y}_i$ is contained in the sampling information of Y_i, $i = 1, \ldots, k$, i.e. the $\tilde{m}_i$ observations associated with $\tilde{Y}_i$ are part of the m_i observations associated with Y_i, $i = 1, \ldots, k$. Let $\tilde{\boldsymbol{Y}} = (\tilde{Y}_1 \ldots, \tilde{Y}_k)$ and $\tilde{\boldsymbol{y}} = (\tilde{y}_1, \ldots, \tilde{y}_k)$.

First we consider the situation at the end of Stage 2, where all data, $\boldsymbol{X} = \boldsymbol{x}$ and $\boldsymbol{Y} = \boldsymbol{y}$, have been observed. Every (it may not be unique) nonrandomized Bayes selection rule $d^*(\boldsymbol{x}, \boldsymbol{y})$ is determined by

$$E\{L(\boldsymbol{\Theta}, d^*(\boldsymbol{x}, \boldsymbol{y}))|\boldsymbol{X} = \boldsymbol{x}, \boldsymbol{Y} = \boldsymbol{y}\} = \min_{i=1,\ldots,k} E\{L(\boldsymbol{\Theta}, i)|\boldsymbol{X} = \boldsymbol{x}, \boldsymbol{Y} = \boldsymbol{y}\}. \quad (3.1)$$

The class of Bayes selections rules consists of all possible random choices among the respective nonrandomized Bayes selections. In the following we will use that particular Bayes rule which makes such choices with equal probabilities and call it *the* Bayes selection rule. It should be pointed out that (3.1) includes one-stage Bayes selections, by combining $\boldsymbol{X}$ and $\boldsymbol{Y}$ in a natural way into one set of data from the k populations.

An overview of the earlier literature of Bayes selection procedures is provided by Gupta and Panchapakesan (1979). Almost all of this earlier work is for equal sample sizes only. Nonsymmetric models have been considered later: the binomial case in Abughalous and Miescke (1989), the normal case in Gupta and Miescke (1988), and more involved models in Berger and Deely (1988) and in Fong and Berger (1993). In the sequel, at every particular sampling situation, it is assumed that the respective Bayes selection rule is known already, and that the focus is solely on sampling allocation.

Next we consider the situation at the end of Stage 1, and determine for it the best allocation, in the Bayes sense, of m observations at Stage 2. This can be achieved by finding $m_1, \ldots, m_k$ which yield

$$\min_{\substack{m_1,\ldots,m_k \\ m_1+\ldots+m_k=m}} E\{\min_{i=1,\ldots,k} E\{L(\boldsymbol{\Theta},i)|\boldsymbol{X}=\boldsymbol{x},\boldsymbol{Y}\}|\boldsymbol{X}=\boldsymbol{x}\}, \tag{3.2}$$

where the outer expectation is with respect to the conditional distribution of $\boldsymbol{Y}$, given $\boldsymbol{X}=\boldsymbol{x}$. In case of tied best allocations, it is assumed that one of them is chosen at random with equal probabilities, which is called from now on *the* Bayes allocation. A convenient way of handling (3.2) is to incorporate the information of $\boldsymbol{X}=\boldsymbol{x}$ into an updated prior [*cf.* Berger (1985, p. 445)], which simplifies (3.2) to

$$\min_{\substack{m_1,\ldots,m_k \\ m_1+\ldots+m_k=m}} E_{\boldsymbol{x}}(\min_{i=1,\ldots,k} E_{\boldsymbol{x}}\{L(\boldsymbol{\Theta},i)|\boldsymbol{Y}\}), \tag{3.3}$$

where the subscript $\boldsymbol{x}$ at expectations, and later on also at probabilities, indicate that the updated prior, based on the observations $\boldsymbol{X}=\boldsymbol{x}$, is used.

Now we consider the situation at an intermediate step of Stage 2 sampling, where only $\tilde{\boldsymbol{Y}} = \tilde{\boldsymbol{y}}$ has been observed with $1 < \tilde{m} < m$, where $\tilde{m} = \tilde{m}_1 + \ldots + \tilde{m}_k$. The best allocation of the remaining $m - \tilde{m}$ observations with $m_i \geq \tilde{m}_i$, $i = 1, \ldots, k$ achieves

$$\min_{\substack{m_1,\ldots,m_k \\ m_1+\ldots+m_k=m}} E_{\boldsymbol{x}}\left\{\min_{i=1,\ldots,k} E_{\boldsymbol{x}}\{L(\boldsymbol{\Theta},i)|\tilde{\boldsymbol{Y}}=\tilde{\boldsymbol{y}},\boldsymbol{Y}\}\Big|\tilde{\boldsymbol{Y}}=\tilde{\boldsymbol{y}}\right\}, \tag{3.4}$$

where the outer expectation is w.r.t. the conditional distribution of $\boldsymbol{Y}$, given $\tilde{\boldsymbol{Y}} = \tilde{\boldsymbol{y}}$.

Back at the end of Stage 1, the best allocation for drawing first $\tilde{m}$ and then $m - \tilde{m}$ observations at Stage 2, is found through backward optimization. First one has to determine for every possible sample size $\tilde{m}_1, \ldots, \tilde{m}_k$, and hereby for every possible outcome of $\tilde{\boldsymbol{Y}} = \tilde{\boldsymbol{y}}$, one allocation $m_i(\tilde{\boldsymbol{y}}, \tilde{m}_1, \ldots, \tilde{m}_k)$, $i = 1, \ldots, k$, which achieves (3.4), and then one has to find an allocation $\tilde{m}_i$, $i = 1, \ldots, k$, which achieves

$$\min_{\substack{\tilde{m}_1,\ldots,\tilde{m}_k \\ \tilde{m}_1+\ldots+\tilde{m}_k=\tilde{m}}} E_{\boldsymbol{x}}\left(\min_{\substack{m_1,\ldots,m_k \\ m_1+\ldots+m_k=m}} E_{\boldsymbol{x}}\left\{\min_{i=1,\ldots,k} E_{\boldsymbol{x}}\{L(\boldsymbol{\Theta},i)\mid\boldsymbol{Y}\}\Big|\tilde{\boldsymbol{Y}}\right\}\right). \tag{3.5}$$

One can see now that if all m observations are allocated at once, then chances may have been missed to utilize information at some intermediate sampling point for better sampling allocations from then on. This idea can be summarized as follows.

Theorem 3.2.1 *The best allocation scheme for drawing first $\tilde{m}$ and then $m - \tilde{m}$ observations at Stage 2 is at least as good as the best allocation of all m observations in one step, in the sense that the posterior Bayes risk (3.5) of the former is not larger than that one of the latter, given by (3.3). This process of stepwise optimum allocation can be iterated for further improvements. The best allocation rule is to draw, in m steps, one observation at a time, which are determined through m-step backward optimization.*

PROOF. Suppose we exchange in (3.5) the first expectation with the minimum adjacent to its right, where this minimum is now with respect to $m_i \geq \tilde{m}_i$, $i = 1, \ldots, k$ and $m_1 + \ldots + m_k = m$. This exchange results in the same value or a larger value. The two coupled minima can be combined into one minimum, and thus we get (3.3).

In (3.5) each of the first and the second minimum can be broken down in the same fashion into coupled minima with a consecutive exchange of minimum and conditional expectation. Again, using the basic fact that every relevant (conditional) expectation of a minimum is less than or equal to the minimum of the respective (conditional) expectation, we see that each such step leads to a value which is less than or equal to the one before. ∎

Breaking down the allocation of m observations into more than one step can be done in many ways, which are discussed in details in Gupta and Miescke (1996a,b). There, two allocations rules have been proposed which are easier to apply, yet reasonable approximations to the best allocation rule **BCK**, say, which is based on *m-step backward optimization*. The first is, with notation taken from there, **OPT**, which allocates all m observations in one step using (3.3). The second is **LAH**, which allocates one observation at a time, using (3.3) with properly updated prior, pretending each time that this is the last observation to be drawn. Thus for every single observation, it allocates it to that population which appears to be the most promising to improve the pretended final selection decision after it has been drawn. Populations tied for such single allocation are broken purely at random. This rule is called the *look-ahead-one-observation-at-a-time* allocation rule. It should be pointed out that **LAH** is different from **SOA**, the *state-of-the-art* allocation rule, which at each of the m steps finds that population which appears at that moment to be the best, e.g. which under the linear loss used in Section 3.3 is associated with the largest posterior mean of $\Theta_1, \ldots, \Theta_k$, and then allocate the next observation to it, with ties broken at random.

In the next section, these allocation rules will be studied in more detail for the case of k independent Bernoulli populations, under a linear loss and independent Beta priors. In the numerical comparisons of Section 3.4, two additional rules will be included: **EQL**, which assigns m/k (in case it is an integer) observations to each of the k populations, and **RAN**, which assigns in m steps one observation at a time, each purely at random. Finally, a new allocation rule **APP** will be proposed, which appears to be the closest approximation to **BCK** and is easy to use.

3.3 Allocations for Bernoulli Populations

The Bayes m-truncated and look ahead sampling allocation approach, which has been discussed in general terms in the previous section, will now be applied to independent Bernoulli sequences with independent Beta priors under the linear loss $L(\boldsymbol{\theta}, i) = \theta_{[k]} - \theta_i$, $i = 1, \ldots, k$, $\boldsymbol{\theta} \in [0,1]^k$, where $\theta_{[k]} = max\{\theta_1, \ldots, \theta_k\}$. Some reasons for adopting this type of loss, and not the 0-1 loss, in the present setting are given in Abughalous and Miescke (1989).

Given $\boldsymbol{\Theta} = \boldsymbol{\theta}$, using sufficiency, the observations at Stage 1 can be combined into $\boldsymbol{X} = (X_1, \ldots, X_k)$ with $X_i \sim \mathcal{B}(n_i, \theta_i)$, $i = 1, \ldots, k$, and the observations at Stage 2 can be combined into $\boldsymbol{Y} = (Y_1, \ldots, Y_k)$ with $Y_i \sim \mathcal{B}(m_i, \theta_i)$, $i = 1, \ldots, k$, where all $2k$ binomial random variables are independent. A priori, the parameters $\Theta_1, \ldots, \Theta_k$ are assumed to be independent Beta random variables with $\Theta_i \sim \mathcal{BE}(\alpha_i, \beta_i)$, where $\alpha_i, \beta_i > 0, i = 1, \ldots, k$, are known.

We assume that Stage 1 has been completed already, i.e. that $\boldsymbol{X} = \boldsymbol{x}$ has been observed, and focus on Stage 2. To simplify the presentation let the prior be updated with the information from $\boldsymbol{X} = \boldsymbol{x}$. This leads to $\Theta_i \sim \mathcal{BE}(\alpha_i + x_i, \beta_i + n_i - x_i)$, $i = 1, \ldots, k$, which in turn are independent.

The optimum allocation of all m observations at Stage 2, i.e. **OPT** can be found by using (3.3). In the present setting this leads [*cf.* Gupta and Miescke (1996b)] to sample sizes $m_1, \ldots, m_k$, subject to $m_1 + \ldots + m_k = m$, which yield the maximum posterior expectation of $max_{i=1,\ldots,k} E_{\boldsymbol{x}}(\Theta_i \mid \boldsymbol{Y})$, i.e.

$$\max_{\substack{m_1, \ldots, m_k \\ m_1 + \ldots + m_k = m}} E_{\boldsymbol{x}} \left(\max_{i=1,\ldots,k} \frac{a_i + Y_i}{a_i + b_i + m_i} \right), \qquad (3.6)$$

where for brevity, $a_i = \alpha_i + x_i$ and $b_i = \beta_i + n_i - x_i$, $i = 1, \ldots, k$, are used throughout this section. The expectation in (3.6) is with respect to the marginal distribution of $\boldsymbol{Y}$ at the end of Stage 1, which in Gupta and Miescke (1993) has been shown to be

$$P_{\boldsymbol{x}}(\boldsymbol{Y} = \boldsymbol{y}) = \prod_{i=1}^{k} \binom{m_i}{y_i} \frac{\Gamma(a_i + b_i)\Gamma(a_i + y_i)\Gamma(b_i + m_i - y_i)}{\Gamma(a_i)\Gamma(b_i)\Gamma(a_i + b_i + m_i)},$$
$$y_i = 0, \ldots, m_i, \ i = 1, \ldots, k, \qquad (3.7)$$

i.e. it is the product of k Pólya-Eggenberger distributions, [*cf.* Johnson and Kotz (1969, p. 230).] To summarize, **OPT** chooses, with equal probabilities, one of the allocations $(m_1, \ldots, m_k)$ which, subject to $m_1 + \ldots + m_k = m$, maximize

$$\sum_{\boldsymbol{y}} \max_{i=1,\ldots,k} \{ \frac{a_i + y_i}{a_i + b_i + m_i} \} P_{\boldsymbol{x}}(\boldsymbol{Y} = \boldsymbol{y}), \qquad (3.8)$$

where the distribution of $\boldsymbol{Y}$ is given by (3.7).

As for **OPT**, with any allocation rule for m observations the expected posterior loss under the linear loss is equal to the difference of $E_{\boldsymbol{x}}(\Theta_{[k]})$, the expectation of the maximum of k independent Beta random variables $\Theta_i \sim \mathcal{BE}(a_i, b_i)$, $i = 1, \ldots, k$, and, say, the expected posterior gain of that particular rule, which for **OPT** is given by (3.6). Therefore, it suffices to deal only with the latter in the following, where besides **OPT**, the five other allocation rules, which are described at the end of the previous section, will be considered: **BCK**, **LAH**, **SOA**, **EQL**, and **RAN**.

In Section 3.4, numerical comparisons will be made between these six allocation rules for the case of $k = 3$. Hereby, the look ahead expected posterior gains of **BCK**, **OPT** and **EQL** are computed directly, whereas those of the other three allocation rules are determined through computer simulations. As it will be seen there, **OPT** and **LAH** perform similarly well, close to **BCK**, but better that the remaining three rules.

In Gupta and Miescke (1996b), **LAH** has been the favorite rule because of the easiness of its usage. This allocation rule has been shown to have some interesting properties. Each of its m steps involves allocation of one observation, using (3.6) with $m = 1$ and a respective updated prior. Without loss of generality, let us consider how it allocates the first of the m observations. This is done by choosing with equal probabilities one of those populations $\mathcal{P}_j$, $j \in \{1, \ldots, k\}$, for which the look ahead one observation expected posterior gain g_j, say, is maximized:

$$
\begin{aligned}
g_j &= E_{\boldsymbol{x}}\left(\max\left\{\frac{a_j + Y_j}{a_j + b_j + 1}, \max_{i \neq j}\{\mu_i\}\right\}\right) \\
&= \max\left\{\frac{a_j + 1}{a_j + b_j + 1}, \max_{i \neq j}\{\mu_i\}\right\}\mu_j \\
&\quad + \max\left\{\frac{a_j + 0}{a_j + b_j + 1}, \max_{i \neq j}\{\mu_i\}\right\}(1 - \mu_j), \quad\quad (3.9)
\end{aligned}
$$

where $\mu_t = a_t/(a_t + b_t)$ is the posterior mean of Θ_t, $t = 1, \ldots, k$, at the end of Stage 1.

Let $\mu_{[1]} \leq \mu_{[2]} \leq \cdots \leq \mu_{[k]}$ denote the ordered values of these posterior means $\mu_1, \ldots, \mu_k$. Let $\mathcal{P}_{(t)}$ be any population associated with $\mu_{[t]}$, $t = 1, \ldots, k$, but in some specific way that all k populations are included. Moreover, let $a_{(t)}$, $b_{(t)}$, $m_{(t)}$, $g_{(t)}$, and $\epsilon_{(t)}$ in turn be associated with $\mathcal{P}_{(t)}$, where $\epsilon_{(t)} = 1/(a_{(t)} + b_{(t)})$ for brevity, $t = 1, \ldots, k$. Then **LAH** allocates the first observation, with equal probabilities, to one of those populations which are tied for a maximum expected posterior gain, i.e. for the maximum of the k quantities

$$
g_{(k)} = \mu_{(k)} + \max\left\{0, (1 - \mu_{(k)})\mu_{(k-1)} - \frac{\mu_{(k)}(1 - \mu_{(k)})}{1 + \epsilon_{(k)}}\right\}
$$

$$g_{(j)} = \mu_{(k)} + \max\left\{ 0, (1 - \mu_{(k)}) \mu_{(j)} - \frac{\mu_{(j)}(1 - \mu_{(j)})}{1 + \epsilon_{(j)}} \right\},$$

$$j = 1, \ldots, k - 1, \qquad (3.10)$$

which can be derived from (3.9) as shown in Gupta and Miescke (1996b). All of the gains in (3.10) are at least as large as $\mu_{(k)}$, since making one more observation can never decrease the Bayes posterior risk achieved already, which in the present situation is equal to $\mu_{(k)}$.

To summarize the use of **LAH**, the first allocation is made, with equal probabilities, to one of those populations $\mathcal{P}_{(t)}$ with $g_{(t)} = \max\{g_{(1)}, \ldots, g_{(k)}\}$, $t = 1, \ldots, k$, where $g_{(1)}, \ldots g_{(k)}$ are given by (3.10). All consecutive allocations are made analogously, with $\mu_{(t)}$ and $\epsilon_{(t)}$, $t = 1, \ldots, k$, properly updated by all previous allocations and observations.

Finding all populations which are tied for the maximum value of the expected posterior gains given by (3.10) can be done through paired comparisons. One of these is different from all others: the comparison of $g_{(k-1)}$ and $g_{(k)}$ is made only through the respective fractions in (3.10). A similar phenomenon occurs in the normal case, as shown in Gupta and Miescke (1996a), where the comparison of the *state-of-the-art* population and its *runner-up* is different from all other comparisons. Further results on **LAH**, and another useful representation of (3.10) can also be found there.

3.4 Numerical Results and Comparisons

Numerical comparisons, in terms of their respective expected posterior gains, of five of the six allocation rules considered in the previous sections have been presented in Gupta and Miescke (1996b) for $k = 3$ populations. In this section, numerical results on **BCK**, the best allocation rule, are added. These results could not be produced until recently, after Microsoft Visual Basic Version 4.0 had been released. The reason is that a six-dimensional array with a common subscript range of $1, 2, \ldots, 15$, i.e. more than 10^7 variables, which are needed for the program to evaluate **BCK**, could not be handled before properly.

The numerical results reported in the tables have been calculated on an IBM-type Pentium 66MHz computer with 16 MB RAM, using Microsoft Quick-BASIC Version 4.5 for the five allocation rules mentioned above, and using Microsoft Visual Basic V. 4.0 for **BCK**. Calculations for allocation rules **EQL** and **OPT** have been performed directly, using routines BICO, FACTLN, and GAMMLN from Sprott (1991). Every single result in the tables for allocation rules **RAN**, **SOA**, and **LAH** is the average over 100,000 computer simulation runs, using the random number generator RND of QuickBASIC, where the seed has been reset with RANDOMIZE(TIMER) at the beginning of each of these

100,000 runs. Calculations for **BCK** have been performed directly, using backward recursion. As to the precision, each expected posterior gain reported in the tables is, due to rounding, accurate only up to $\pm$ .0001. All programs used for this purpose are available in ASCII code from the authors upon request. The rules considered are:

RAN   Assign one observation at a time, each purely at random.
EQL   Assign $m/3$ observations to each population $\mathcal{P}_1$, $\mathcal{P}_2$, $\mathcal{P}_3$.
SOA   Assign one observation at a time, following *state-of-the-art*.
LAH   Assign one observation at a time using (3.6) with $m = 1$.
OPT   Assign m_t observations to $\mathcal{P}_t$, $t = 1, 2, 3$, using (3.6).
BCK   Assign one observation at a time, using *backward optimization*.

The values considered for m are $1, 3, 9$, and 15. For $m = 1$ **EQL** has been set to take its observation from $\mathcal{P}_1$, rather than leaving the respective spaces empty in the tables. As to **SOA**, *state-of-the-art* means the largest posterior expectation of $\Theta_1, \ldots, \Theta_k$ at any moment. **RAN**, **SOA**, and **LAH** are breaking ties at random with equal probabilities, whenever they occur. The numerical comparisons are made by means of three examples, each with specifically chosen values for a_i and b_i, $i = 1, 2, 3$, which are reported in the tables. In case of ties for **OPT**, the lexicographically first choice is displayed there.

The values of $a_i = \alpha_i + x_i$ and $b_i = \beta_i + n_i - x_i$, $i = 1, 2, 3$, can be interpreted conveniently as follows. Let us assume from now on that $\alpha_i = \beta_i = 1$, $i = 1, 2, 3$, i.e. that the prior is noninformative, and that at Stage 1 $n_i = a_i + b_i - 2$ observations from $\mathcal{P}_i$, with $x_i = a_i - 1$ successes and $n_i - x_i$ failures, and thus with success rate $\bar{x}_i = x_i/n_i = (a_i - 1)/(a_i + b_i - 2)$ have been drawn, $i = 1, 2, 3$.

Each of the three examples starts, for $m = 1, 3, 9, 15$, with a specific choice of $a_1, a_2, a_3, b_1, b_2, b_3$, where $n_1 < n_2 < n_3$, and continues with seven related configurations, which are the result of exchanging the values of a_i and b_i, i.e. the numbers of successes and failures. Altogether, this allows to cover various interesting settings which are discussed in Gupta and Miescke (1996b).

Example 3.4.1 Let $\alpha_i = \beta_i = 1$, $i = 1, 2, 3$, and assume first that at Stage 1 samples of the following types have been observed: $n_1 = 10$ and $x_1 = 4$ from $\mathcal{P}_1$, $n_2 = 15$ and $x_2 = 6$ from $\mathcal{P}_2$, and $n_3 = 20$ and $x_3 = 8$ from $\mathcal{P}_3$, with respective success rates $\bar{x}_1 = \bar{x}_2 = \bar{x}_3 = .4$. The associated configuration $(a_1, a_2, a_3, b_1, b_2, b_3) = (5, 7, 9, 7, 10, 13)$ is shown in Table 3.1 as the first one for each $m = 1, 3, 9, 15$. The seven other parameter configurations are obtained by exchanging the number of successes and failures, i.e. the values of a_i and b_i, within one or more populations.

Example 3.4.2 Let $\alpha_i = \beta_i = 1$, $i = 1, 2, 3$, and assume first that at Stage 1 samples of the following types have been observed: $n_1 = 5$ and $x_1 = 3$ from $\mathcal{P}_1$, $n_2 = 14$ and $x_2 = 9$ from $\mathcal{P}_2$, and $n_3 = 18$ and $x_3 = 11$ from $\mathcal{P}_3$, with respective success rates $\bar{x}_1 = .6$, $\bar{x}_2 = .6429$, and $\bar{x}_3 = .6111$. The associated configuration

$(a_1, a_2, a_3, b_1, b_2, b_3) = (4, 10, 12, 3, 6, 8)$ is shown in Table 3.2 as the first one for each $m = 1, 3, 9, 15$. The seven other parameter configurations are obtained, as in the previous example, by exchanging the values of a_i and b_i within one or more populations. This example includes configurations where $\bar{x}_1$, $\bar{x}_2$, and $\bar{x}_3$ are close to each other.

Example 3.4.3 Let $\alpha_i = \beta_i = 1$, $i = 1, 2, 3$, and assume first that at Stage 1 samples of the following types have been observed: $n_1 = 5$ and $x_1 = 3$ from $\mathcal{P}_1$, $n_2 = 12$ and $x_2 = 7$ from $\mathcal{P}_2$, and $n_3 = 17$ and $x_3 = 9$ from $\mathcal{P}_3$, with respective success rates $\bar{x}_1 = .6$, $\bar{x}_2 = .5833$, and $\bar{x}_1 = .5294$. The associated configuration $(a_1, a_2, a_3, b_1, b_2, b_3) = (4, 8, 10, 3, 6, 9)$ is shown in Table 3 as the first one for each $m = 1, 3, 9, 15$. The seven other parameter configurations are obtained in the same manner as before.

The results of the three examples have been discussed, except for **BCK**, in Gupta and Miescke (1996b). To summarize these findings, **OPT** and **LAH** compare favorably with the rules **RAN**, **EQL**, and **SOA**. The overall performance of **LAH** is close to the one of **OPT**. That **LAH** is not always as good as **OPT** shows that it cannot be a version of the rule $(\mathcal{R}_{m,1}, \mathcal{R}_{m-1,1}, \ldots, \mathcal{R}_{2,1}, \mathcal{R}_1)$ considered there, since the latter is at least as good as **OPT**. The allocation rule $(\mathcal{R}_{m,1}, \mathcal{R}_{m-1,1}, \ldots, \mathcal{R}_{2,1}, \mathcal{R}_1)$ proceeds as follows. First it takes one observation from one of those population included by **OPT** based on m, then after updating the prior, it takes one observation from one of those populations included by **OPT** based on $m - 1$, and so on. Since it was not clear which one of such populations one should choose, this rule had not been studied further.

The new results added in the tables are the two columns associated with **BCK**. They show that indeed **LAH** and **OPT** are good approximations to **BCK**. The last column indicates with a 1(0) from which population the first observation should (should not) be drawn using **BCK**. Comparing this with the preceding column for **OPT** one can see the following striking fact. In all but one situation (Table 3.3, $m = 3$, fifth configuration), the population to which **OPT** allocates the largest sample size is one of those with which **BCK** would start. This suggests to use the following version **APP**, say, of $(\mathcal{R}_{m,1}, \mathcal{R}_{m-1,1}, \ldots, \mathcal{R}_{2,1}, \mathcal{R}_1)$: Allocate one observation at a time. If q observations remain to be drawn, compute the sample sizes of **OPT** based on q, and then take the observation from that population which is associated with the largest sample size. This allocation rule **APP** is as good as **OPT** and seems to be very close to **BCK** in its performance. Since **OPT** is much easier to compute than **BCK** for larger values of m, this new allocation rule **APP** is recommended for practical use.

Table 3.1: Expected posterior gains* in Example 3.4.1

$a_1a_2a_3b_1b_2b_3$	RAN	EQL	SOA	LAH	OPT	BCK	Allocations OPT	BCK
$m = 1$								
5 7 9 7 10 13	.4282	.4325	.4325	.4325	.4325	.4325	1 0 0	100
5 7 13 7 10 9	.5909	.5909	.5908	.5909	.5909	.5909	0 1 0	111
5 10 9 7 7 13	.5882	.5882	.5882	.5882	.5882	.5882	1 0 0	111
5 10 13 7 7 9	.5981	.5909	.6004	.6028	.6028	.6028	0 1 0	010
7 7 9 5 10 13	.5833	.5833	.5833	.5833	.5833	.5833	1 0 0	111
7 7 13 5 10 9	.5982	.6052	.5982	.6052	.6052	.6052	1 0 0	100
7 10 9 5 7 13	.5972	.6041	.5996	.6041	.6041	.6041	1 0 0	100
7 10 13 5 7 9	.6028	.6052	.6004	.6052	.6052	.6052	1 0 0	100
$m = 3$								
5 7 9 7 10 13	.4384	.4394	.4410	.4432	.4435	.4436	2 1 0	100
5 7 13 7 10 9	.5909	.5909	.5909	.5908	.5909	.5909	0 3 0	111
5 10 9 7 7 13	.5882	.5882	.5882	.5882	.5882	.5882	2 1 0	111
5 10 13 7 7 9	.6034	.6028	.6068	.6077	.6081	.6081	0 2 1	011
7 7 9 5 10 13	.5833	.5833	.5831	.5833	.5833	.5833	1 2 0	111
7 7 13 5 10 9	.6047	.6052	.6063	.6112	.6119	.6119	3 0 0	100
7 10 9 5 7 13	.6045	.6041	.6084	.6105	.6109	.6109	3 0 0	100
7 10 13 5 7 9	.6119	.6101	.6111	.6184	.6172	.6183	2 1 0	110
$m = 9$								
5 7 9 7 10 13	.4540	.4543	.4554	.4582	.4575	.4598	5 3 1	100
5 7 13 7 10 9	.5910	.5910	.5910	.5912	.5918	.5918	9 0 0	100
5 10 9 7 7 13	.5885	.5885	.5882	.5886	.5892	.5892	6 3 0	110
5 10 13 7 7 9	.6125	.6122	.6159	.6159	.6176	.6177	0 9 0	010
7 7 9 5 10 13	.5836	.5837	.5846	.5839	.5845	.5846	8 1 0	100
7 7 13 5 10 9	.6153	.6157	.6173	.6212	.6241	.6241	9 0 0	100
7 10 9 5 7 13	.6161	.6161	.6196	.6207	.6226	.6227	9 0 0	100
7 10 13 5 7 9	.6274	.6272	.6273	.6322	.6305	.6340	6 3 0	100
$m = 15$								
5 7 9 7 10 13	.4631	.4638	.4628	.4663	.4660	.4684	9 5 1	100
5 7 13 7 10 9	.5915	.5917	.5912	.5919	.5930	.5931	12 0 3	101
5 10 9 7 7 13	.5892	.5896	.5895	.5904	.5909	.5910	9 6 0	110
5 10 13 7 7 9	.6178	.6181	.6210	.6208	.6231	.6233	0 12 3	010
7 7 9 5 10 13	.5847	.5851	.5864	.5859	.5864	.5867	11 4 0	100
7 7 13 5 10 9	.6216	.6222	.6232	.6262	.6292	.6292	13 0 2	101
7 10 9 5 7 13	.6224	.6231	.6260	.6258	.6291	.6292	11 4 0	110
7 10 13 5 7 9	.6364	.6369	.6354	.6399	.6390	.6423	9 6 0	110

* accurate up to $\pm.0001$

Table 3.2: Expected posterior gains* in Example 3.4.2

							Allocations	
$a_1a_2a_3b_1b_2b_3$	RAN	EQL	SOA	LAH	OPT	BCK	OPT	BCK
$m = 1$								
4 10 12 3 6 8	.6270	.6250	.6294	.6294	.6294	.6294	0 1 0	010
4 10 8 3 6 12	.6250	.6250	.6248	.6250	.6250	.6250	0 0 1	111
4 6 12 3 10 8	.6048	.6143	.6000	.6143	.6143	.6143	1 0 0	100
4 6 8 3 10 12	.5714	.5714	.5713	.5714	.5714	.5714	0 0 1	111
3 10 12 4 6 8	.6271	.6250	.6294	.6294	.6294	.6294	0 1 0	010
3 10 8 4 6 12	.6250	.6250	.6250	.6250	.6250	.6250	0 0 1	111
3 6 12 4 10 8	.6000	.6000	.6000	.6000	.6000	.6000	0 0 1	111
3 6 8 4 10 12	.4333	.4429	.4428	.4429	.4429	.4429	1 0 0	100
$m = 3$								
4 10 12 3 6 8	.6359	.6359	.6345	.6418	.6429	.6468	3 0 0	100
4 10 8 3 6 12	.6322	.6329	.6311	.6377	.6429	.6429	3 0 0	100
4 6 12 3 10 8	.6135	.6143	.6115	.6238	.6238	.6238	2 0 1	111
4 6 8 3 10 12	.5714	.5714	.5713	.5714	.5714	.5714	0 1 2	111
3 10 12 4 6 8	.6307	.6319	.6345	.6338	.6355	.6355	0 3 0	010
3 10 8 4 6 12	.6250	.6250	.6249	.6250	.6250	.6250	0 1 2	111
3 6 12 4 10 8	.6000	.6000	.6000	.6001	.6000	.6000	1 0 2	111
3 6 8 4 10 12	.4437	.4468	.4453	.4539	.4540	.4540	2 1 0	110
$m = 9$								
4 10 12 3 6 8	.6531	.6551	.6478	.6586	.6591	.6627	6 3 0	110
4 10 8 3 6 12	.6453	.6466	.6419	.6512	.6554	.6554	8 1 0	110
4 6 12 3 10 8	.6281	.6295	.6240	.6369	.6410	.6410	9 0 0	100
4 6 8 3 10 12	.5744	.5742	.5772	.5748	.5780	.5784	9 0 0	100
3 10 12 4 6 8	.6391	.6400	.6434	.6423	.6440	.6443	0 7 2	010
3 10 8 4 6 12	.6260	.6263	.6251	.6270	.6290	.6290	8 1 0	110
3 6 12 4 10 8	.6019	.6019	.6000	.6034	.6066	.6066	9 0 0	100
3 6 8 4 10 12	.4624	.4642	.4624	.4691	.4701	.4729	7 2 0	100
$m = 15$								
4 10 12 3 6 8	.6636	.6646	.6559	.6680	.6676	.6715	8 6 1	110
4 10 8 3 6 12	.6527	.6534	.6481	.6577	.6612	.6613	13 2 0	110
4 6 12 3 10 8	.6356	.6371	.6296	.6421	.6458	.6459	15 0 0	100
4 6 8 3 10 12	.5779	.5781	.5801	.5788	.5812	.5819	13 1 1	100
3 10 12 4 6 8	.6458	.6462	.6483	.6480	.6494	.6507	0 10 5	010
3 10 8 4 6 12	.6281	.6284	.6252	.6295	.6316	.6316	11 4 0	110
3 6 12 4 10 8	.6046	.6049	.6004	.6065	.6091	.6092	12 0 3	100
3 6 8 4 10 12	.4730	.4741	.4703	.4775	.4780	.4807	10 4 1	100

* accurate up to ±.0001

Table 3.3: Expected posterior gains* in Example 3.4.3

							Allocations	
$a_1a_2a_3b_1b_2b_3$	RAN	EQL	SOA	LAH	OPT	BCK	OPT	BCK
m = 1								
4 8 10 3 6 9	.5870	.6020	.5949	.6020	.6020	.6020	1 0 0	100
4 8 9 3 6 10	.5871	.6020	.5949	.6020	.6020	.6020	1 0 0	100
4 6 10 3 8 9	.5750	.5827	.5827	.5827	.5827	.5827	1 0 0	100
4 6 9 3 8 10	.5714	.5714	.5714	.5714	.5714	.5714	0 0 1	111
3 8 10 4 6 9	.5713	.5714	.5714	.5714	.5714	.5714	0 0 1	111
3 8 9 4 6 10	.5714	.5714	.5714	.5714	.5714	.5714	0 0 1	111
3 6 10 4 8 9	.5263	.5263	.5263	.5263	.5263	.5263	0 0 1	111
3 6 9 4 8 10	.4775	.4850	.4737	.4850	.4850	.4850	1 0 0	100
m = 3								
4 8 10 3 6 9	.6000	.6037	.6071	.6113	.6122	.6122	3 0 0	100
4 8 9 3 6 10	.5991	.6020	.6048	.6107	.6122	.6122	3 0 0	100
4 6 10 3 8 9	.5814	.5827	.5892	.5924	.5940	.5940	3 0 0	100
4 6 9 3 8 10	.5738	.5714	.5776	.5759	.5802	.5802	3 0 0	100
3 8 10 4 6 9	.5745	.5752	.5767	.5764	.5770	.5778	0 3 0	100
3 8 9 4 6 10	.5717	.5714	.5717	.5726	.5748	.5748	3 0 0	100
3 6 10 4 8 9	.5286	.5263	.5263	.5310	.5351	.5351	3 0 0	100
3 6 9 4 8 10	.4861	.4871	.4793	.4964	.4962	.4967	3 0 0	100
m = 9								
4 8 10 3 6 9	.6185	.6203	.6222	.6260	.6271	.6288	7 2 0	100
4 8 9 3 6 10	.6158	.6173	.6195	.6233	.6254	.6258	9 0 0	100
4 6 10 3 8 9	.5951	.5963	.6001	.6030	.6068	.6069	9 0 0	100
4 6 9 3 8 10	.5829	.5838	.5871	.5875	.5908	.5915	9 0 0	100
3 8 10 4 6 9	.5852	.5856	.5861	.5878	.5876	.5911	5 4 0	100
3 8 9 4 6 10	.5782	.5787	.5763	.5806	.5815	.5833	5 4 0	100
3 6 10 4 8 9	.5379	.5386	.5302	.5425	.5457	.5466	9 0 0	100
3 6 9 4 8 10	.5037	.5050	.4946	.5110	.5096	.5134	7 2 0	100
m = 15								
4 8 10 3 6 9	.6296	.6303	.6300	.6348	.6358	.6375	9 6 0	110
4 8 9 3 6 10	.6251	.6261	.6265	.6305	.6327	.6334	9 6 0	110
4 6 10 3 8 9	.6033	.6043	.6054	.6092	.6119	.6125	14 0 1	100
4 6 9 3 8 10	.5904	.5911	.5936	.5941	.5955	.5972	12 0 3	100
3 8 10 4 6 9	.5932	.5935	.5914	.5955	.5949	.5987	7 8 0	110
3 8 9 4 6 10	.5841	.5849	.5799	.5868	.5875	.5895	9 6 0	110
3 6 10 4 8 9	.5453	.5462	.5349	.5492	.5500	.5528	13 0 2	100
3 6 9 4 8 10	.5146	.5153	.5035	.5191	.5179	.5220	9 4 2	100

* accurate up to ±.0001

Acknowledgements. The first author wishes to thank Professor Shanti S. Gupta for his cooperation, support, and encouragement over the past twenty years.

References

1. Abughalous, M. M. and Miescke, K. J. (1989). On selecting the largest success probability under unequal sample sizes, *Journal of Statistical Planning and Inference*, **21**, 53–68.

2. Amster, S. J. (1963). A modified Bayesian stopping rule, *Annals of Mathematical Statistics*, **34**, 1404–1413.

3. Bechhofer, R. E., Kiefer, J., and Sobel, M. (1968). *Sequential Identification and Ranking Procedures*, Chicago, IL: The University of Chicago Press.

4. Bechhofer, R. E. and Kulkarni, R. V. (1982). Closed adaptive sequential procedures for selecting the best of $k \geq 2$ Bernoulli populations, In *Statistical Decision Theory and Related Topics - III* (Eds., S. S. Gupta and J. O. Berger), Vol. 1, pp. 61–108, New York, NY: Academic Press.

5. Bechhofer, R. E., Santner, T. J., and Goldsman, D. M. (1995). *Design and Analysis of Experiments for Statistical Selection, Screening, and Multiple Comparisons,* New York, NY: John Wiley & Sons.

6. Berger, J. O. (1985). *Statistical Decision Theory and Bayesian Analysis,* Second edition, New York, NY: Springer Verlag.

7. Berger, J. O. and Deely, J. (1988). A Bayesian approach to ranking and selection of related means with alternatives to AOV methodology, *Journal of the American Statistical Association*, **83**, 364–373.

8. Bratcher, T. L. and Bland, R. P. (1975). On comparing binomial probabilities from a Bayesian viewpoint, *Communications in Statistics*, **4**, 975–985.

9. Dunnett, C. W. (1960). On selecting the largest of k normal population means (with discussion), *Journal of the Royal Statistical Society*, **22**, 1–40.

10. Fong, D. K. H. and Berger, J. O. (1993). Ranking, estimation, and hypothesis testing in unbalanced two-way additive models - A Bayesian approach, *Statistics and Decisions*, **11**, 1–24.

11. Govindarajulu, Z. and Katehakis, M. N. (1991). Dynamic allocation in survey sampling, *American Journal of Mathematical and Management Sciences*, **11**, 199–221.

12. Gupta, S. S. and Miescke, K. J. (1984). Sequential selection procedures - A decision theoretic approach, *Annals of Statistics*, **12**, 336–350.

13. Gupta, S. S. and Miescke, K. J. (1988). On the problem of finding the largest normal mean under heteroscedasticity, In *Statistical Decision Theory and Related Topics - IV* (Eds. S. S. Gupta and J. O. Berger), Vol. 2, pp. 37–49, New York, NY: Springer Verlag.

14. Gupta, S. S. and Miescke, K. J. (1993). On combining selection and estimation in the search of the largest binomial parameter, *Journal of Statistical Planning and Inference*, **36**, 129–140.

15. Gupta, S. S. and Miescke, K. J. (1994). Bayesian look ahead one-stage sampling allocations for selecting the largest normal mean, *Statistical Papers*, **35**, 169–177.

16. Gupta, S. S. and Miescke, K. J. (1996a). Bayesian look ahead one-stage sampling allocations for selection of the best population, *Journal of Statistical Planning and Inference*, **54**, 229–244.

17. Gupta, S. S. and Miescke, K. J. (1996b). Bayesian look ahead one-stage sampling allocations for selection of the best Bernoulli population, In *Festschrift in Honor of Madan L. Puri* (Eds. E. Brunner and Denker), Vilnius, Lithuania: TEV Typesetting Services (to appear).

18. Gupta, S. S. and Panchapakesan, S. (1979). *Multiple Decision Procedures: Theory and Methodology of Selecting and Ranking Populations*, New York, NY: John Wiley & Sons.

19. Gupta, S. S. and Panchapakesan, S. (1991). Sequential ranking and selection procedures, In *Handbook of Sequential Analysis,* (Eds. B. K. Ghosh and P. K. Sen), pp. 363–380, New York, NY: Marcel Dekker.

20. Johnson, N. L. and Kotz, S. (1969). *Discrete Distributions*, Boston, MA: Houghton and Mifflin.

21. Miescke, K. J. (1984). Recent results on multi-stage selection procedures, In *Proceedings of the Seventh Conference on Probability Theory,* (Ed. M. Iosifescu), pp. 259–268, Bucharest, România: Editura Academiei Republicii Socialiste România.

22. Risko, K. R. (1985). Selecting the better binomial population with unequal sample sizes, *Communications in Statistics—Theory and Methods*, **14**, 123–158.

23. Sprott, J. C. (1991). *Numerical Recipes: Routines and Examples in BA-SIC*, Cambridge, MA: Cambridge University Press.

4

On Hierarchical Bayesian Estimation and Selection for Multivariate Hypergeometric Distributions

TaChen Liang

Wayne State University, Detroit, MI

Abstract: In this paper, we deal with the problem of simultaneous estimation for n independent multivariate hypergeometric distributions $\pi(M_i, m_i, s_i)$, $i = 1, \ldots, n$, and simultaneous selection of the most probable event for each of the n $\pi(M_i, m_i, s_i)$. In order to model the uncertainty of the unknown parameters s_i, $i = 1, \ldots, n$, and to incorporate information from the n $\pi(M_i, m_i, s_i)$, a two-stage prior distribution on the parameters s_i, $i = 1, \ldots, n$, is introduced. With this hierarchical structure of prior distributions, we derive simultaneous hierarchical Bayesian estimators for s_i, $i = 1, \ldots, n$, and simultaneous selection rule for selecting the most probable event for each of the n $\pi(M_i, m_i, s_i)$. We compare the performance of the hierarchical Bayesian procedures to those of the "pure" Bayesian procedures. The relative regret Bayes risks of the hierarchical Bayesian procedures are used as measures of the optimality of the procedures. It is shown that, for the estimation problem, the relative regret Bayes risk of the hierarchical Bayesian estimator converges to zero at a rate of order $O(n^{-1})$; and for the selection problem, the relative regret Bayes risk of the hierarchical Bayesian selection rule converges to zero at a rate of order $O(e^{-n\beta})$ for some positive constant β.

Keywords and phrases: Hierarchical prior, most probable event, multivariate hypergeometric distribution, simultaneous estimation, simultaneous selection

4.1 Introduction

The multivariate hypergeometric distribution arises in survey sampling without replacement from a finite population. Consider a population of M individuals, of which s_i are of type i, $i = 1, 2, \ldots, k$, with $\sum_{i=1}^{k} s_i = M$. Suppose, a sample of size $m < M$ is chosen, without replacement, from among these M individuals. Let $X_1, \ldots, X_k$ represent the number of individuals of types $1, 2, \ldots, k$, respectively, in the sample. Then, $\boldsymbol{X} = (X_1, \ldots, X_k)$ has the joint probability function $f(\boldsymbol{x}|\boldsymbol{s}, M, m)$ given by

$$f(\boldsymbol{x}|\boldsymbol{s}, M, m) = \left\{ \prod_{j=1}^{k} \binom{s_j}{x_j} \right\} \Big/ \binom{M}{m}, \tag{4.1}$$

where $\boldsymbol{s} = (s_1, \ldots, s_k)$, $\boldsymbol{x} = (x_1, \ldots, x_k)$ is such that $\sum_{i=1}^{k} x_i = m$, and $\max(0, s_i + m - M) \leq x_i \leq \min(s_i, m)$, $i = 1, \ldots, k$. Such a finite population with the probability model (4.1) is called a multivariate hypergeometric distribution with parameters $(M, m, \boldsymbol{s})$, and is denoted by $\pi(M, m, \boldsymbol{s})$; see, for example, Johnson, Kotz and Balakrishnan (1997, Chapter 39).

In a multivariate hypergeometric distribution $\pi(M, m, \boldsymbol{s})$, let $s_{[1]} \leq \cdots \leq s_{[k]}$ denote the ordered values of the parameters $s_1, \ldots, s_k$. Suppose the exact pairing between the ordered and the unordered parameters is unknown. A category with $s_i = s_{[k]}$ is considered as the most probable event and a category with $s_j = s_{[1]}$ is considered as the least probable event of the population $\pi(M, m, \boldsymbol{s})$. In many situations, one may be interested in the selection of the most probable event. For example, consider k candidates $c_1, \ldots, c_k$ running for one post in an election, each supported by one of k different political parties. Suppose each voter can select only one candidate. Let M denote the number of voters and let s_i, $i = 1, \ldots, k$, denote the number of voters who prefer candidate c_i. The candidate c_i with $s_i = s_{[k]}$ may be considered as the most probable candidate. There are usually many election polls before an election. Based on the result of polls, one may be interested in estimating the values of s_i and finding the most probable candidate.

When $k = 2$, the probability model described in (4.1) turns out to be a hypergeometric distribution. Cressie and Scheult (1985) and Walter and Hamedani (1988) have studied the empirical Bayes estimation for such a hypergeometric distribution. In the area of ranking and selection, Bartlett and Govindarajulu (1970) have studied the problem of selecting the best population from among ℓ (≥ 2) independent hypergeometric distributions via the subset selection approach. Balakrishnan and Ma (1997) have proposed empirical Bayes rules for selecting the best hypergeometric population. Gupta and Liang (1994) have studied the problem of acceptance sampling for hypergeometric distributions using the empirical Bayes and hierarchical empirical Bayes approaches. Interested readers may refer to Gupta and Panchapakesan (1979) and Bechhofer, Santner and Goldsman (1995) for detailed general discussions pertaining to the area of ranking and selection.

In this paper, we deal with n independent multivariate hypergeometric distributions say $\pi(M_i, m_i, \boldsymbol{s}_i)$, $i = 1, \ldots, n$. We study the problems of simultaneous estimation of the parameters $\boldsymbol{s}_i$, $i = 1, \ldots, n$, with squared error loss and simultaneous selection of the most probable event for each of the n multivariate hypergeometric distributions. It should be noted that Balakrishnan and Ma (1996) have studied the problem of selecting the most and the least probable events for one multivariate hypergeometric distribution using the empirical Bayes approach.

Here, we assume that for each $i = 1, \ldots, n$, $\boldsymbol{s}_i$ is a realization of a random vector $\boldsymbol{S}_i = (S_{i1}, \ldots, S_{ik})$ which has a multinomial$(M_i, \boldsymbol{p})$ prior distribution with common but unknown probability vector $\boldsymbol{p} = (p_1, \ldots, p_k)$, where $0 \le p_i \le 1$, $i = 1, \ldots, k$ and $\sum_{i=1}^{k} p_i = 1$. In order to model the uncertainty of the unknown hyperparameter $\boldsymbol{p}$ and to incorporate information from the n multivariate hypergeometric distributions, we introduce a Dirichlet distribution on the hyperparameter $\boldsymbol{p}$ as the second-stage prior distribution. With this hierarchical structure of the prior distributions, we derive the simultaneous hierarchical Bayesian estimation for $\boldsymbol{s}_i$, $i = 1, \ldots, n$, and simultaneous hierarchical Bayesian selection rule for selecting the most probable event for each of the n multivariate hypergeometric distributions. We then compare the performance of the hierarchical Bayesian procedures with that of the pure Bayesian procedures that will be used if the value of the hyperparameter $\boldsymbol{p}$ was known. The relative regret Bayes risks of the hierarchical Bayesian procedures are used as measures of optimality. We then study the asymptotic optimality of the hierarchical Bayesian procedures. It is shown that for the estimation problem, the relative regret Bayes risk of the hierarchical Bayesian estimator converges to zero at a rate of order $O(n^{-1})$; and for the selection problem, the relative regret Bayes risk of the hierarchical Bayesian selection rule converges to zero at a rate of order $O(e^{-\beta n})$ for some positive constant β.

4.2   Formulation of the Problems

Consider n independent multivariate hypergeometric distributions with k-categories $\pi(M, m_i, \boldsymbol{s}_i)$, where $\boldsymbol{s}_i = (s_{i1}, \ldots, s_{ik})$, $i = 1, \ldots, n$. For each i, let $\boldsymbol{X}_i = (X_{i1}, \ldots, X_{ik})$ denote the observations that arise from a multivariate hypergeometric distribution $\pi(M_i, m_i, \boldsymbol{s}_i)$. Then, given $\boldsymbol{s}_i, \boldsymbol{X}_i$ has the probability function

$$f_i(\boldsymbol{x}_i | \boldsymbol{s}_i, M_i, m_i) = \left\{ \prod_{j=1}^{k} \binom{s_{ij}}{x_{ij}} \right\} \Big/ \binom{M_i}{m_i}, \tag{4.2}$$

where $\boldsymbol{x}_i = (x_{i1}, \ldots, x_{ik})$, $\max(0, s_{ij} + m_i - M_i) \le x_{ij} \le \min(s_{ij}, m_{ij})$, $j = 1, \ldots, k$, and $\sum_{j=1}^{k} x_{ij} = m_i$.

It is assumed that for each $i = 1, \ldots, n$, the parameter $\boldsymbol{s}_i$ is a realization of an integer-valued random vector $\boldsymbol{S}_i = (S_{i1}, \ldots, S_{ik})$ and $\boldsymbol{S}_1, \ldots, \boldsymbol{S}_n$ are mutually independent, and for each i, $\boldsymbol{S}_i$ has a Multinomial$(M_i, \boldsymbol{p})$ prior distribution with common but unknown probability vector $\boldsymbol{p} = (p_1, \ldots, p_k)$, where $0 \leq p_i \leq 1$, $i = 1, \ldots, k$ and $\sum_{i=1}^{k} p_i = 1$. That is, $\boldsymbol{S}_i$ has the prior probability function

$$g_i(\boldsymbol{s}_i | M_i, \boldsymbol{p}) = \frac{M_i!}{\prod_{j=1}^{k} s_{ij}!} \prod_{j=1}^{k} p_j^{s_{ij}}, \tag{4.3}$$

for $\boldsymbol{s}_i = (s_{i1}, \ldots, s_{ik})$ such that $0 \leq s_{ij} \leq M_i$, $j = 1, \ldots, k$, and $\sum_{j=1}^{k} s_{ij} = M_i$.

The statistical model described through (4.2) and (4.3) is called a *multivariate hypergeometric–multinomial model*; see Balakrishnan and Ma (1996). It follows that given $\boldsymbol{p}$, $\boldsymbol{X}_i$ has a marginal probability function

$$
\begin{aligned}
f_i(\boldsymbol{x}_i | \boldsymbol{p}, m_i) &= \sum_{\boldsymbol{s}_i} f_i(\boldsymbol{x}_i, M_i, m_i) g_i(\boldsymbol{s}_i | M_i, \boldsymbol{p}) \\
&= \frac{m_i!}{\prod_{j=1}^{k} x_{ij}!} \prod_{j=1}^{k} p_j^{x_{ij}},
\end{aligned}
\tag{4.4}
$$

where the summation in (4.4) is taken over the parameter space $\Omega_i = \{\boldsymbol{s}_i | 0 \leq s_{ij} \leq M_i, \ j = 1, \ldots, k \text{ and } \sum_{j=1}^{k} s_{ij} = M_i\}$. That is, given $\boldsymbol{p}$, $\boldsymbol{X}_i$ follows a marginal Multinomial$(m_i, \boldsymbol{p})$ distribution. Also, given $\boldsymbol{X}_i = \boldsymbol{x}_i$ and $\boldsymbol{p}$, the posterior probability function of $\boldsymbol{S}_i$ is:

$$
\begin{aligned}
g_i(\boldsymbol{s}_i | \boldsymbol{x}_i, \boldsymbol{p}, M_i, m_i) &= \frac{f_i(\boldsymbol{x}_i | \boldsymbol{s}_i, M_i, m_i) g_i(\boldsymbol{s}_i | M_i, \boldsymbol{p})}{f_i(\boldsymbol{x}_i | \boldsymbol{p}, m_i)} \\
&= \frac{(M_i - m_i)!}{\prod_{j=1}^{k} (s_{ij} - x_{ij})!} \prod_{j=1}^{k} p_j^{s_{ij} - x_{ij}}.
\end{aligned}
\tag{4.5}
$$

Thus, given $\boldsymbol{X}_i = \boldsymbol{x}_i$ and $\boldsymbol{p}$, $\boldsymbol{S}_i - \boldsymbol{X}_i$ follows a Multinomial$(M_i - m_i, \boldsymbol{p})$ distribution.

It should also be noted that under the statistical model described above, $\boldsymbol{X}_1, \ldots, \boldsymbol{X}_n$ are mutually independent.

In this paper, we consider the following two decision problems.

A. Simultaneous estimation of s_i, $i = 1, \ldots, n$

Let $\boldsymbol{\varphi}_i = (\varphi_{i1}, \ldots, \varphi_{ik})$ denote an estimator of $\boldsymbol{s}_i = (s_{i1}, \ldots, s_{ik})$, $i = 1, \ldots, n$. We consider the squared error loss defined by

$$L(\boldsymbol{\varphi}, \boldsymbol{s}) = \sum_{i=1}^{n} \sum_{j=1}^{k} (\varphi_{ij} - s_{ij})^2, \tag{4.6}$$

for $\boldsymbol{s} = (\boldsymbol{s}_1, \ldots, \boldsymbol{s}_n)$ and $\boldsymbol{\varphi} = (\boldsymbol{\varphi}_1, \ldots, \boldsymbol{\varphi}_n)$.

B. Simultaneous selection of the most probable event

For each i, let $\boldsymbol{a}_i = (a_{i1}, \ldots, a_{ik})$ denote an action for the selection problem involved in the multivariate hypergeometric distribution $\pi(M_i, m_i, \boldsymbol{s}_i)$, where $a_{ij} = 0$ or 1 for $j = 1, \ldots, k$, and $\sum_{j=1}^{k} a_{ij} = 1$. When $a_{ij} = 1$ for some j, it means that category j of the $\pi(M_i, m_i, \boldsymbol{s}_i)$ is selected as the most probable event and when $a_{ij} = 0$, it means that category j is excluded and is not considered as the most probable event of the $\pi(M_i, m_i, \boldsymbol{s}_i)$. When $\boldsymbol{s}_i$ is the true state of nature, the loss of taking action $\boldsymbol{a}_i$ is defined by

$$L_i(\boldsymbol{s}_i, \boldsymbol{a}_i) = s_{i[k]} - \sum_{j=1}^{k} a_{ij} s_{ij}, \tag{4.7}$$

where $s_{i[k]} = \max_{1 \leq j \leq k} s_{ij}$. Let $\boldsymbol{s} = (\boldsymbol{s}_1, \ldots, \boldsymbol{s}_n)$ and $\boldsymbol{a} = (\boldsymbol{a}_1, \ldots, \boldsymbol{a}_n)$. The loss for the simultaneous selection problem is then defined as

$$L(\boldsymbol{s}, \boldsymbol{a}) = \sum_{i=1}^{n} L_i(\boldsymbol{s}_i, \boldsymbol{a}_i). \tag{4.8}$$

4.2.1 Bayesian estimation of $s_i, i = 1, \ldots, n$

It is known that under the squared error loss, the Bayes estimator of the parameter $\boldsymbol{s}_i = (s_{i1}, \ldots, s_{ik})$ is the posterior mean $\boldsymbol{\varphi}_i^B(\boldsymbol{x}_i) = (\varphi_{i1}^B(\boldsymbol{x}_i), \ldots, \varphi_{ik}^B(\boldsymbol{x}_i))$ where, by (4.5),

$$\begin{aligned} \varphi_{ij}^B(\boldsymbol{x}_i) &= E[S_{ij}|\boldsymbol{x}_i, \boldsymbol{p}] \\ &= x_{ij} + (M_i - m_i)p_j. \end{aligned} \tag{4.9}$$

4.2.2 Bayesian selection rule

Let $\mathcal{X}$ be the sample space generated by $\boldsymbol{X} = (\boldsymbol{X}_1, \ldots, \boldsymbol{X}_n)$. For each i, let $\boldsymbol{\delta}_i = (\delta_{i1}, \ldots, \delta_{ik})$ be a selection rule for the i-th component problem involving the $\pi(M_i, m_i, \boldsymbol{s}_i)$. That is, $\boldsymbol{\delta}_i$ is a mapping from the sample space into $[0, 1]^k$ such that for each $\boldsymbol{x} \in \mathcal{X}, 0 \leq \delta_{ij}(\boldsymbol{x}) \leq 1$, $j = 1, \ldots, k$, and $\sum_{j=1}^{k} \delta_{ij}(\boldsymbol{x}) = 1$, and $\delta_{ij}(\boldsymbol{x})$ is the probability of selecting category j of the $\pi(M_i, m_i, \boldsymbol{s}_i)$ as the most probable event given that $\boldsymbol{X} = \boldsymbol{x}$ is observed. Let $\boldsymbol{\delta} = (\boldsymbol{\delta}_1, \ldots, \boldsymbol{\delta}_n)$ and let $\mathcal{D}$ be the set of all selection rules $\boldsymbol{\delta}$ defined as above. Let $G_{\boldsymbol{p}}$ denote the joint prior probability function of $(\boldsymbol{S}_1, \ldots, \boldsymbol{S}_n)$. Note that $G_{\boldsymbol{p}}$ can be obtained from (4.3). For each $\boldsymbol{\delta} \in \mathcal{D}$, let $R(G_{\boldsymbol{p}}, \boldsymbol{\delta})$ denote the Bayes risk associated with the selection rule $\boldsymbol{\delta}$. Then, $R(G_{\boldsymbol{p}}) = \inf_{\boldsymbol{\delta} \in \mathcal{D}} R(G_{\boldsymbol{p}}, \boldsymbol{\delta})$ is the minimum Bayes risk. By the loss function in (4.8), the Bayes risk $R(G_{\boldsymbol{p}}, \boldsymbol{\delta})$ can be represented as

$$R(G_{\boldsymbol{p}}, \boldsymbol{\delta}) = \sum_{i=1}^{n} R_i(G_{\boldsymbol{p}}, \boldsymbol{\delta}_i), \tag{4.10}$$

where

$$
\begin{aligned}
R_i(G\boldsymbol{p}, \boldsymbol{\delta}_i) &= \sum_{\boldsymbol{s}} \sum_{\boldsymbol{x}} [s_{i[k]} - \sum_{j=1}^{k} \delta_{ij}(\boldsymbol{x}) s_{ij}] \prod_{\ell=1}^{n} [f_\ell(\boldsymbol{x}_\ell | \boldsymbol{s}_\ell, M_\ell, m_\ell) g_\ell(\boldsymbol{s}_\ell | M_\ell, \boldsymbol{p})] \\
&= C_i - \sum_{\boldsymbol{x}} [\sum_{j=1}^{k} \delta_{ij}(\boldsymbol{x}) \varphi_{ij}^{G}(\boldsymbol{x})] \prod_{\ell=1}^{n} f_\ell(\boldsymbol{x}_\ell | \boldsymbol{p}, m_\ell), \qquad (4.11)
\end{aligned}
$$

and

$$
\begin{aligned}
\varphi_{ij}^{G}(\boldsymbol{x}) &= E[S_{ij} | \boldsymbol{x}, \underline{p}] \\
&= x_{ij} + (M_i - m_i) p_j \equiv \varphi_{ij}^{B}(\boldsymbol{x}_i), \qquad (4.12) \\
C_i &= \sum_{\boldsymbol{s}_i} s_{i[k]} g_i(\boldsymbol{s}_i | M_i, \boldsymbol{p}).
\end{aligned}
$$

Note that $\varphi_{ij}^{G}(\boldsymbol{x})$ depends on $\boldsymbol{x}$ only through the observation $\boldsymbol{x}_i$ associated with the $\pi(M_i, m_i, \boldsymbol{s}_i)$ since $\boldsymbol{X}_1, \ldots, \boldsymbol{X}_n$ are mutually independent and $\varphi_{ij}^{G}(\boldsymbol{x}) = \varphi_{ij}^{B}(\boldsymbol{x}_i)$.

For each i, and each observation $\boldsymbol{x}_i$, let

$$
A_i^{B}(\boldsymbol{x}_i) = \left\{ j | \varphi_{ij}^{B}(\boldsymbol{x}_i) = \max_{1 \le \ell \le k} \varphi_{i\ell}^{B}(\boldsymbol{x}_i) \right\}, \qquad (4.13)
$$

and

$$
j_i^{B} \equiv j_i^{B}(\boldsymbol{x}_i) = \min\{ j | j \in A_i^{B}(\boldsymbol{x}_i) \}. \qquad (4.14)
$$

Then, a Bayesian selection rule $\boldsymbol{\delta}^{B} = (\boldsymbol{\delta}_1^{B}, \ldots, \boldsymbol{\delta}_n^{B})$, where $\boldsymbol{\delta}_i^{B} = (\delta_{i1}^{B}, \ldots, \delta_{ik}^{B})$, $i = 1, \ldots, n$, is as follows: For each $\boldsymbol{x} \in \mathcal{X}$, and each $i = 1, \ldots, n$,

$$
\delta_{ij}^{B}(\boldsymbol{x}_i) = \delta_{ij}^{B}(\boldsymbol{x}) = \begin{cases} 1 & \text{if } j = j_i^{B}, \\ 0 & \text{otherwise} \end{cases} \qquad (4.15)
$$

Note that for the i-th component selection problem, the Bayesian selection rule $\boldsymbol{\delta}_i^{B} = (\delta_{i1}^{B}, \ldots, \delta_{ik}^{B})$ depends on $\boldsymbol{x}$ only through $\boldsymbol{x}_i$.

4.3   A Hierarchical Bayesian Approach

From (4.9)–(4.15), it can be seen that both the Bayesian estimator $\boldsymbol{\varphi}_i^{B}$ and the Bayesian selection rule $\boldsymbol{\delta}^{B}$ depend on the unknown hyperparameter $\boldsymbol{p}$. Thus, it is not possible to apply these Bayesian procedures for the decision problems at hand. In the following, in order to model the uncertainty of the unknown hyperparameter $\boldsymbol{p}$ and to incorporate information from the n $\pi(M_i, m_i, \boldsymbol{s}_i)$ distributions, we introduce a Dirichlet distribution with probability density function $h(\boldsymbol{p}|\boldsymbol{\alpha})$ on the hyperparameter $\boldsymbol{p}$ as the second stage prior distribution, where

$$h(\boldsymbol{p}|\boldsymbol{\alpha}) = \frac{\Gamma(\alpha_0)}{\prod_{j=1}^{k} \Gamma(\alpha_j)} \prod_{j=1}^{k} p_j^{\alpha_j - 1}, \qquad (4.16)$$

where $\boldsymbol{\alpha} = (\alpha_1, \ldots, \alpha_k)$, $0 < \alpha_j < \infty$, $j = 1, \ldots, k$, and $\alpha_0 = \sum_{j=1}^{k} \alpha_j$. Such a Dirichlet distribution with parameter $\boldsymbol{\alpha}$ is denoted by Dirichlet($\boldsymbol{\alpha}$).

Let $f(\boldsymbol{x}_1, \ldots, \boldsymbol{x}_n; \boldsymbol{s}_1, \ldots, \boldsymbol{s}_n; \boldsymbol{p}) = h(\boldsymbol{p}|\boldsymbol{\alpha}) \prod_{i=1}^{n} [f_i(\boldsymbol{x}_i|\boldsymbol{s}_i, M_i, m_i) g_i(\boldsymbol{s}_i|M_i, \boldsymbol{p})]$.
Then,

$$
\begin{aligned}
f(\boldsymbol{x}_1, \ldots, \boldsymbol{x}_n; \boldsymbol{s}_1, \ldots, \boldsymbol{s}_n) &= \int f(\boldsymbol{x}_1, \ldots, \boldsymbol{x}_n; \boldsymbol{s}_1, \ldots, \boldsymbol{s}_n; \boldsymbol{p}) d\boldsymbol{p} \\
&= \left\{ \prod_{i=1}^{n} \frac{m_i!(M_i - m_i)!}{\prod_{j=1}^{k} [x_{ij}!(s_{ij} - x_{ij})!]} \right\} \\
&\quad \times \frac{\Gamma(\alpha_0)}{\prod_{j=1}^{k} \Gamma(\alpha_j)} \frac{\prod_{j=1}^{k} \Gamma(\sum_{i=1}^{n} s_{ij} + \alpha_j)}{\Gamma(M_1 + \cdots + M_n + \alpha_0)} .
\end{aligned}
$$

Also,

$$
\begin{aligned}
f(\boldsymbol{x}_1, \ldots, \boldsymbol{x}_n) &= \int h(\boldsymbol{p}|\boldsymbol{\alpha}) \prod_{\ell=1}^{n} f_\ell(\boldsymbol{x}_\ell|\boldsymbol{p}, m_i) d\boldsymbol{p} \\
&= \left\{ \prod_{i=1}^{n} \frac{m_i!}{\prod_{j=1}^{k} x_{ij}!} \right\} \frac{\Gamma(\alpha_0)}{\prod_{j=1}^{k} \Gamma(\alpha_j)} \frac{\prod_{j=1}^{k} \Gamma(\sum_{i=1}^{n} x_{ij} + \alpha_j)}{\Gamma(m_1 + \cdots + m_n + \alpha_0)} .
\end{aligned}
$$

Therefore, the posterior probability function of $(\boldsymbol{S}_1, \ldots, \boldsymbol{S}_n)$, given $(\boldsymbol{X}_1, \ldots, \boldsymbol{X}_n)$ $= (\boldsymbol{x}_1, \ldots, \boldsymbol{x}_n)$, is

$$
\begin{aligned}
g(\boldsymbol{s}_1, \ldots, \boldsymbol{s}_n | \boldsymbol{x}_1, \ldots, \boldsymbol{x}_n) &= \frac{f(\boldsymbol{x}_1, \ldots, \boldsymbol{x}_n; \boldsymbol{s}_1, \ldots, \boldsymbol{s}_n)}{f(\boldsymbol{x}_1, \ldots, \boldsymbol{x}_n)} \\
&= \left[\prod_{i=1}^{n} \frac{(M_i - m_i)!}{\prod_{j=1}^{k} (s_{ij} - x_{ij})!} \right] \prod_{j=1}^{k} \left[\frac{\Gamma(\sum_{i=1}^{n} s_{ij} + \alpha_j)}{\Gamma(\sum_{i=1}^{n} x_{ij} + \alpha_j)} \right] \\
&\quad \times \frac{\Gamma(m_1 + \cdots + m_n + \alpha_0)}{\Gamma(M_1 + \cdots + M_n + \alpha_0)} .
\end{aligned}
$$

Then, under the squared error loss, the hierarchical Bayesian estimation of s_{ij} is $\varphi_{ij}^{H}(\boldsymbol{x})$, the posterior mean of S_{ij} given $\boldsymbol{X} = \boldsymbol{x}$, computed under the hierarchical structure of prior distributions described through (4.3) and (4.16). That is,

$$
\begin{aligned}
\varphi_{ij}^{H}(\boldsymbol{x}) &= E^{H}[S_{ij}|\boldsymbol{X} = \boldsymbol{x}] \\
&= \sum_{\boldsymbol{s}} s_{ij} g(\boldsymbol{s}_1, \ldots, \boldsymbol{s}_n | \boldsymbol{x}_1, \ldots, \boldsymbol{x}_n) \\
&= x_{ij} + (M_i - m_i) \frac{\sum_{\ell=1}^{n} x_{\ell j} + \alpha_j}{m_1 + \cdots + m_n + \alpha_0} . \qquad (4.17)
\end{aligned}
$$

For each selection rule $\boldsymbol{\delta}$, let $R(H, \boldsymbol{\delta})$ denote its associated Bayes risk computed with respect to the hierarchical prior distributions of (4.3) and (4.16). Then,

$$R(H, \boldsymbol{\delta}) = \sum_{i=1}^{n} R_i(H, \boldsymbol{\delta}_i), \tag{4.18}$$

where

$$R_i(H, \boldsymbol{\delta}_i) = C^* - \sum_{\boldsymbol{x}} \left[\sum_{j=1}^{k} \delta_{ij}(\boldsymbol{x}) \varphi_{ij}^H(\boldsymbol{x}) \right] f(\boldsymbol{x}_1, \ldots, \boldsymbol{x}_n), \tag{4.19}$$

and

$$C^* = \sum_{\boldsymbol{s}_i} s_{i[k]} \int g_i(\boldsymbol{s}_i | M_i, \boldsymbol{p}) h(\boldsymbol{p}|\boldsymbol{\alpha}) d\boldsymbol{p}.$$

For each $i = 1, \ldots, n$, and each observation $\boldsymbol{x}$, define

$$A_i^H(\boldsymbol{x}) = \{j | \varphi_{ij}^H(\boldsymbol{x}) = \max_{1 \le \ell \le k} \varphi_{i\ell}^H(\boldsymbol{x})\}, \tag{4.20}$$

and

$$j_i^H \equiv j_i^H(\boldsymbol{x}) = \min\{j | j \in A_i^H(\boldsymbol{x})\}. \tag{4.21}$$

Then, the hierarchical Bayesian selection rule is $\boldsymbol{\delta}^H = (\boldsymbol{\delta}_1^H, \ldots, \boldsymbol{\delta}_n^H)$ where for each i, $\boldsymbol{\delta}_i^H = (\delta_{i1}^H, \ldots, \delta_{ik}^H)$ and

$$\delta_{ij}^H(\boldsymbol{x}) = \begin{cases} 1 & \text{if } j = j_i^H, \\ 0 & \text{otherwise.} \end{cases} \tag{4.22}$$

Remark 4.3.1 Balakrishnan and Ma (1997) have constructed an empirical Bayes estimator for $\boldsymbol{s}_n$ using the parametric empirical Bayes approach. For the hierarchical Bayesian estimator $\boldsymbol{\varphi}^H$, if we let $\alpha_0 \to 0$, then $\lim_{\alpha_0 \to 0} \varphi_{ij}^H(\boldsymbol{x}) = x_{ij} + (M_i - m_i) \frac{x_{1j} + \cdots + x_{nj}}{m_1 + \cdots + m_n}$, which has a form similar to the empirical Bayes estimator proposed by Balakrishnan and Ma (1997).

Remark 4.3.2 Let $\boldsymbol{\varphi} = (\boldsymbol{\varphi}_1, \ldots, \boldsymbol{\varphi}_n)$ be an estimator of the parameter $\boldsymbol{s} = (\boldsymbol{s}_1, \ldots, \boldsymbol{s}_n)$, where $\boldsymbol{\varphi}_i = (\varphi_{i1}, \ldots, \varphi_{ik})$ and $\boldsymbol{s}_i = (s_{i1}, \ldots, s_{ik})$, $i = 1, \ldots, n$. Under the squared error loss, the Bayes risk of the estimator $\boldsymbol{\varphi}$ is

$$r(G_{\boldsymbol{p}}, \boldsymbol{\varphi}) = \sum_{i=1}^{n} \sum_{j=1}^{k} E[\varphi_{ij}(\boldsymbol{X}) - S_{ij}]^2, \tag{4.23}$$

where the expectation in (4.23) is taken with respect to the probability models in (4.2) and (4.3).

An estimator φ is said to be inadmissible if there exists an estimator, say φ^*, such that $r(G\boldsymbol{p}, \varphi^*) \leq r(G\boldsymbol{p}, \varphi)$ for all $\boldsymbol{p} \in \mathcal{E} = \{(p_1, \ldots, p_k) \mid 0 \leq p_i \leq 1,$ $i = 1, \ldots, k$ and $\sum_{i=1}^{k} p_i = 1\}$ and $r(G\boldsymbol{p}, \varphi^*) < r(G\boldsymbol{p}, \varphi)$ for all $\boldsymbol{p} \in \mathcal{E}_1 \sqsubset \mathcal{E}$ for which $\int_{\mathcal{E}_1} d\boldsymbol{p} > 0$. φ is said to be admissible if it is not inadmissible.

Since $\varphi^H = (\varphi_1^H, \ldots, \varphi_n^H)$ is the Bayes estimator of $\boldsymbol{s} = (\boldsymbol{s}_1, \ldots, \boldsymbol{s}_n)$ with respect to the hierarchical prior distributions of (4.2) and (4.16), one can see that φ^H is an admissible estimator.

Remark 4.3.3 A selection rule $\boldsymbol{\delta}$ is said to be inadmissible if there exists a selection rule, say $\boldsymbol{\delta}^*$ such that $R(G\boldsymbol{p}, \boldsymbol{\delta}^*) \leq R(G\boldsymbol{p}, \boldsymbol{\delta})$ for all $\boldsymbol{p} \in \mathcal{E}$ and $R(G\boldsymbol{p}, \boldsymbol{\delta}^*) < R(G\boldsymbol{p}, \boldsymbol{\delta})$ for all $\boldsymbol{p} \in \mathcal{E}_2 \sqsubset \mathcal{E}$ for which $\int_{\mathcal{E}_2} d\boldsymbol{p} > 0$. Otherwise, $\boldsymbol{\delta}$ is said to be admissible.

Since $\boldsymbol{\delta}^H = (\boldsymbol{\delta}_1^H, \ldots, \boldsymbol{\delta}_n^H)$ is the Bayes selection rule with respect to the hierarchical prior distributions of (4.3) and (4.16), $\boldsymbol{\delta}^H$ is admissible.

4.4 Optimality of the Hierarchical Bayesian Procedures

In the following, it is assumed that the true statistical model is the one described through (4.2) and (4.3). We want to study the performance of the hierarchical Bayesian procedure compared with that of the corresponding Bayesian procedures that will be used if the values of the hyperparameter $\boldsymbol{p}$ was known.

4.4.1 Optimality of φ^H

For an estimator φ, $E[\varphi_{ij}(\boldsymbol{X}) - S_{ij}]^2 \geq E[\varphi_{ij}^B(\boldsymbol{X}) - S_{ij}]^2$ for all $j = 1, \ldots, k$, and $i = 1, \ldots, n$, since $\varphi^B(\boldsymbol{x})$ is the Bayes estimator of $\boldsymbol{S}$ under the squared error loss. Therefore, $r(G\boldsymbol{p}, \varphi) \geq r(G\boldsymbol{p}, \varphi^B)$. The nonnegative difference $d(G\boldsymbol{p}, \varphi) = r(G\boldsymbol{p}, \varphi) - r(G\boldsymbol{p}, \varphi^B)$ is called the regret Bayes risk of the estimator φ. The ratio $\rho(G\boldsymbol{p}, \varphi) = d(G\boldsymbol{p}, \varphi)/r(G\boldsymbol{p}, \varphi^B)$ is called the relative regret Bayes risk of the estimator φ. We will use the relative regret Bayes risk as a measure of performance of the estimators.

Note that under the statistical models in (4.2) and (4.3) and the assumptions mentioned in Section 4.2, $X_{1j}, \ldots, X_{nj}$ are mutually independent, and marginally, $X_{\ell j} \sim \text{Binomial}(m_\ell, p_j)$, $\ell = 1, \ldots, n$. Hence, $X_{1j} + \cdots X_{nj} \sim \text{Binomial}(m_1 + \cdots + m_n, p_j)$.

By the above fact, (4.9) and (4.17), a straightforward computation yields

$$
\begin{aligned}
0 &\leq E[\varphi_{ij}^H(\boldsymbol{X}) - S_{ij}]^2 - E[\varphi_{ij}^B(\boldsymbol{X}) - S_{ij}]^2 \\
&= E[\varphi_{ij}^H(\boldsymbol{X}) - \varphi_{ij}^B(\boldsymbol{X})]^2
\end{aligned}
$$

$$
\begin{aligned}
&= (M_i - m_i)^2 E\left[\frac{X_{1j} + \cdots + X_{nj} + \alpha_j}{m_1 + \cdots + m_n + \alpha_0} - p_j\right]^2 \\
&= (M_i - m_i)^2 E\left[\frac{X_{1j} + \cdots + X_{nj} - (m_1 + \cdots + m_n)p_j}{m_1 + \cdots + m_n + \alpha_0}\right. \\
&\quad\left. + \frac{\alpha_j - \alpha_0 p_j}{m_1 + \cdots + m_n + \alpha_0}\right]^2 \\
&= (M_i - m_i)^2 \left\{ E\left[\frac{X_{1j} + \cdots + X_{nj} - (m_1 + \cdots + m_n)p_j}{m_1 + \cdots + m_n + \alpha_0}\right]^2 \right. \\
&\quad\left. + \frac{(\alpha_j - \alpha_0 p_j)^2}{(m_1 + \cdots + m_n + \alpha_0)^2} \right\} \\
&= (M_i - m_i)^2 \left\{ \frac{(m_1 + \cdots + m_n)p_j(1 - p_j)}{(m_1 + \cdots + m_n + \alpha_0)^2} + \frac{(\alpha_j - \alpha_0 p_j)^2}{(m_1 + \cdots + m_n + \alpha_0)^2} \right\}.
\end{aligned}
$$

(4.24)

Therefore,

$$
\begin{aligned}
&r(G\boldsymbol{p}, \boldsymbol{\varphi}^H) - r(G\boldsymbol{p}, \boldsymbol{\varphi}^B) \\
&= \sum_{j=1}^{n}\sum_{j=1}^{k} E[\varphi_{ij}^H(\boldsymbol{X}) - \varphi_{ij}^B(\boldsymbol{X})]^2 \\
&= \sum_{i=1}^{n}\sum_{j=1}^{k} (M_i - m_i)^2 \frac{[(m_1 + \cdots + m_n)p_j(1 - p_j) + (\alpha_j - \alpha_0 p_j)^2]}{[m_1 + \cdots + m_n + \alpha_0]^2} \\
&= \frac{1}{(m_1 + \cdots + m_n + \alpha_0)^2}\left\{ (m_1 + \cdots + m_n)\sum_{j=1}^{k} p_j(1 - p_j) \right. \\
&\quad\left. + \sum_{j=1}^{k}(\alpha_j - \alpha_0 p_j)^2 \right\} \sum_{i=1}^{n}(M_i - m_i)^2.
\end{aligned}
$$

(4.25)

Also, since given $\boldsymbol{X} = \boldsymbol{x}, S_{ij} - x_{ij} \sim \text{Binomial}(M_i - m_i, p_j)$, we obtain

$$
\begin{aligned}
E[\varphi_{ij}^B(\boldsymbol{X}) - S_{ij}]^2 &= E[X_{ij} + (M_i - m_i)p_j - S_{ij}]^2 \\
&= E[E[(S_{ij} - X_{ij} - (M_i - m_i)p_j)^2|\boldsymbol{X}]] \\
&= E[(M_i - m_i)p_j(1 - p_j)] \\
&= (M_i - m_i)p_j(1 - p_j).
\end{aligned}
$$

Therefore,

$$r(G\boldsymbol{p}, \boldsymbol{\varphi}^B) = \sum_{i=1}^{n} \sum_{j=1}^{k} E[\varphi_{ij}^B(\boldsymbol{X}) - S_{ij}]^2$$

$$= [(M_1 + \cdots + M_n) - (m_1 + \cdots + m_n)] \left[\sum_{j=1}^{k} p_j(1 - p_j) \right]$$

$$(4.26)$$

Hence, we have

$$\rho(G\boldsymbol{p}, \boldsymbol{\varphi}^H)$$
$$= \frac{[(m_1 + \cdots + m_n) \sum_{j=1}^{k} p_j(1-p_j) + \sum_{j=1}^{k} (\alpha_j - \alpha_0 p_j)^2][\sum_{i=1}^{n}(M_i - m_i)^2]}{(m_1 + \cdots + m_n + \alpha_0)^2[(M_1 + \cdots + M_n) - (m_1 + \cdots + m_n)][\sum_{j=1}^{k} p_j(1-p_j)]}. \quad (4.27)$$

We assume the following condition.

Condition C: $m_* \leq m_i < M_i \leq M^*$ for all $i = 1, 2, \ldots$, where m_* and M^* are positive constants, independent of n.

We summarize the above result in the form of a theorem as follows.

Theorem 4.4.1 *Under the model described in Section 4.2 and Condition C, we have*

$$\rho(G\boldsymbol{p}, \boldsymbol{\varphi}^H) = O\left(\frac{1}{m_1 + \cdots + m_n} \right) = O\left(\frac{1}{n} \right).$$

PROOF. From (4.27) and under Condition C, we have

$$\rho(G\boldsymbol{p}, \boldsymbol{\varphi}^H)$$
$$\leq \frac{(M^* - m_*)^2}{(m_1 + \cdots + m_n + \alpha_0)} + \frac{(M^* - m_*)^2[\sum_{j=1}^{k}(\alpha_j - \alpha_0 p_j)^2]}{(m_1 + \cdots + m_n + \alpha_0)^2[\sum_{j=1}^{k} p_j(1 - p_j)]}$$

$$= O\left(\frac{1}{m_1 + \cdots + m_n} \right)$$

$$= O\left(\frac{1}{n} \right).$$

Therefore, the proof is complete. ∎

4.4.2   Optimality of δ^H

For a selection rule $\boldsymbol{\delta} = (\delta_1, \ldots, \delta_n), R_i(G\boldsymbol{p}, \delta_i) \geq R_i(G\boldsymbol{p}, \delta_i^B), i = 1, \ldots, n$, since $\boldsymbol{\delta}^B = (\delta_1^B, \ldots, \delta_n^B)$ is a Bayesian selection rule. Hence, $D(G\boldsymbol{p}, \boldsymbol{\delta}) = R(G\boldsymbol{p}, \boldsymbol{\delta}) - R(G\boldsymbol{p}, \boldsymbol{\delta}^B) \geq 0$. The relative regret Bayes risk of the selection rule

δ is defined as $\rho(G\boldsymbol{p}, \delta) = D(G\boldsymbol{p}, \delta)/R(G\boldsymbol{p}, \delta^B)$. In the following, $\rho(G\boldsymbol{p}, \delta^H)$ will be used as a measure of performance of the selection rule δ^H.

For each $i = 1, \ldots, n$, let $\boldsymbol{X}(i) = (\boldsymbol{X}_1, \ldots, \boldsymbol{X}_{i-1}, \boldsymbol{X}_{i+1}, \ldots, \boldsymbol{X}_n)$ and let E_i denote the expectation taken with respect to the probability measure generated by $\boldsymbol{X}(i)$.

By (4.11), (4.12) and (4.15), for each $i = 1, \ldots, n$

$$
\begin{aligned}
R_i & (G\boldsymbol{p}, \delta_i^H) - R_i(G\boldsymbol{p}, \delta_i^B) \\
= & \sum_{\boldsymbol{x}} \left[\sum_{j=1}^k \delta_{ij}^B(\boldsymbol{x}) \varphi_{ij}^B(\boldsymbol{x}) \right] \prod_{\ell=1}^n f_\ell(\boldsymbol{x}_\ell | \boldsymbol{p}, m_\ell) \\
& - \sum_{\boldsymbol{x}} \left[\sum_{j=1}^k \delta_{ij}^H(\boldsymbol{x}) \varphi_{ij}^B(\boldsymbol{x}) \right] \prod_{\ell=1}^n f_\ell(\boldsymbol{x}_\ell | \boldsymbol{p}, m_\ell) \\
= & \sum_{\boldsymbol{x}_i} \left[\sum_{j=1}^k \delta_{ij}^B(\boldsymbol{x}_i) \varphi_{ij}^B(\boldsymbol{x}_i) \right] f_i(\boldsymbol{x}_i | \boldsymbol{p}, m_i) \\
& - \sum_{\boldsymbol{x}} \left[\sum_{j=1}^k \delta_{ij}^H(\boldsymbol{x}) \varphi_{ij}^B(\boldsymbol{x}_i) \right] \prod_{\ell=1}^n f_\ell(\boldsymbol{x}_\ell | \boldsymbol{p}, m_\ell) \\
= & \sum_{\boldsymbol{x}_i} \left[\sum_{j=1}^k \delta_{ij}^B(\boldsymbol{x}_i) \varphi_{ij}^B(\boldsymbol{x}_i) \right] f_i(\boldsymbol{x}_i | \boldsymbol{p}, m_i) \\
& - \sum_{\boldsymbol{x}_i} \left[\sum_{j=1}^k E_i[\delta_{ij}^H(\boldsymbol{x}_i, \boldsymbol{X}(i))] \varphi_{ij}^B(\boldsymbol{x}_i) \right] f_i(\boldsymbol{x}_i | \boldsymbol{p}, m_i) \\
= & \sum_{\boldsymbol{x}_i} \left[\sum_{j=1}^k (\delta_{ij}^B(\boldsymbol{x}_i) - E_i[\delta_{ij}^H(\boldsymbol{x}_i, \boldsymbol{X}(i))]) \varphi_{ij}^B(\boldsymbol{x}_i) \right] f_i(\boldsymbol{x}_i | \boldsymbol{p}, m_i). \quad (4.28)
\end{aligned}
$$

For each $\boldsymbol{x}_i$, let $C_i(\boldsymbol{x}_i) = \{j = 1, \ldots, k | j \notin A_i^B(\boldsymbol{x}_i)\}$, where $A_i^B(\boldsymbol{x}_i)$ is the set defined in (4.13). By the definitions of $A_i^B(\boldsymbol{x}_i)$ and $C_i(\boldsymbol{x}_i)$, $\varphi_{i\ell}^B(\boldsymbol{x}_i) - \varphi_{ij}^B(\boldsymbol{x}_i) > 0$ for $\ell \in A_i^B(\boldsymbol{x}_i)$ and $j \in C_i(\boldsymbol{x}_i)$. Let $\mathcal{X}_i$ be the space of $\boldsymbol{X}_i$. Since $\mathcal{X}_i$ is a finite space, $\min_{\boldsymbol{x}_i \in \mathcal{X}_i} \{ \varphi_{i\ell}^B(\boldsymbol{x}_i) - \varphi_{ij}^B(\boldsymbol{x}_i) | \ell \in A_i^B(\boldsymbol{x}_i), j \in C_i(\boldsymbol{x}_i) \} = 2c_i > 0$. Also, $0 \le \varphi_{ij}^B(\boldsymbol{x}_i) \le M_i$.

Let $T_{i\ell} = \sum_{a=1, a \neq i}^n X_{a\ell}$. By (4.12) and (4.17), a straightforward computation yields that for each $\boldsymbol{x}_i \in \mathcal{X}_i$,

$$
\begin{aligned}
\sum_{j=1}^k & [\delta_{ij}^B(\boldsymbol{x}_i) - E_i[\delta_{ij}^H(\boldsymbol{x}_i, \boldsymbol{X}(i))]] \varphi_{ij}^B(\boldsymbol{x}_i) \\
\le & \sum_{\ell \in A_i^B(\boldsymbol{x}_i)} \sum_{j \in C_i(\boldsymbol{x}_i)} P\{ \varphi_{i\ell}^H(\boldsymbol{x}_i, \boldsymbol{X}(i)) \le \varphi_{ij}^H(\boldsymbol{x}_i, \boldsymbol{X}(i)) \} [\varphi_{i\ell}^B(\boldsymbol{x}_i) - \varphi_{ij}^B(\boldsymbol{x}_i)] \\
\le & M_i \sum_{\ell \in A_i^B(\boldsymbol{x}_i)} \sum_{j \in C_i(\boldsymbol{x}_i)} P\{ [\varphi_{i\ell}^H(\boldsymbol{x}_i, \boldsymbol{X}(i)) - \varphi_{i\ell}^B(\boldsymbol{x}_i)]
\end{aligned}
$$

$$- [\varphi_{ij}^H(\boldsymbol{x}_i, \boldsymbol{X}(i)) - \varphi_{ij}^B(\boldsymbol{x}_i)] < -2c_i\}$$

$$\leq M_i \sum_{\ell \in A_i^B(\boldsymbol{x}_i)} \sum_{j \in C_i(\boldsymbol{x}_i)} (P\{\varphi_{i\ell}^H(\boldsymbol{x}_i, \boldsymbol{X}(i)) - \varphi_{i\ell}^B(\boldsymbol{x}_i) < -c_i\}$$

$$+ P\{\varphi_{ij}^H(\boldsymbol{x}_i, \boldsymbol{X}(i)) - \varphi_{ij}^B(\boldsymbol{x}_i) > c_i\})$$

$$\leq M_i \sum_{\ell=1}^k P\{|\varphi_{i\ell}^H(\boldsymbol{x}_i, \boldsymbol{X}(i)) - \varphi_{i\ell}^B(\boldsymbol{x}_i)| > c_i\}$$

$$= M_i \sum_{\ell=1}^k P\left\{\left|\frac{T_{i\ell} + x_{i\ell} + \alpha_\ell}{m_1 + \cdots + m_n + \alpha_0} - p_\ell\right| > \frac{c_i}{M_i - m_i}\right\}. \tag{4.29}$$

Using the inequality in (4.29) into (4.28), we obtain

$$R_i(G\boldsymbol{p}, \boldsymbol{\delta}_i^H) - R_i(G\boldsymbol{p}, \boldsymbol{\delta}_i^B)$$

$$\leq M_i \sum_{\boldsymbol{x}_i} \left[\sum_{\ell=1}^k P\left\{\left|\frac{T_{i\ell} + x_{i\ell} + \alpha_\ell}{m_1 + \cdots + m_n + \alpha_0} - p_\ell\right| > \frac{c_i}{M_i - m_i}\right\}\right] f_i(\boldsymbol{x}_i|\boldsymbol{p}, m_i)$$

$$= M_i \sum_{\ell=1}^k P\left\{\left|\frac{T_{i\ell} + X_{i\ell} + \alpha_\ell}{m_1 + \cdots + m_n + \alpha_0} - p_\ell\right| > \frac{c_i}{M_i - m_i}\right\}. \tag{4.30}$$

By Markov's inequality, (4.24) and by noting that $T_{i\ell} = \sum_{j=1, j\neq i}^n X_{j\ell}$, we get

$$P\left\{\left|\frac{T_{i\ell} + X_{i\ell} + \alpha_\ell}{m_1 + \cdots + m_n + \alpha_0} - p_\ell\right| > \frac{c_i}{M_i - m_i}\right\}$$

$$\leq \frac{(M_i - m_i)^2}{c_i^2} E\left[\left|\frac{T_{i\ell} + X_{i\ell} + \alpha_\ell}{m_1 + \cdots + m_n + \alpha_0} - p_\ell\right|^2\right]$$

$$= \frac{(M_i - m_i)^2}{c_i^2} \frac{[(m_1 + \cdots + m_n)p_\ell(1 - p_\ell) + (\alpha_\ell - \alpha_0 p_\ell)^2]}{(m_1 + \cdots + m_n + \alpha_0)^2}. \tag{4.31}$$

Combining (4.30) and (4.31), we obtain

$$R_i(G\boldsymbol{p}, \boldsymbol{\delta}_i^H) - R_i(G\boldsymbol{p}, \boldsymbol{\delta}_i^B)$$

$$\leq \frac{M_i(M_i - m_i)^2}{c_i^2} \sum_{\ell=1}^k \frac{[(m_1 + \cdots + m_n)p_\ell(1 - p_\ell) + (\alpha_\ell - \alpha_0 p_\ell)^2]}{(m_1 + \cdots m_n + \alpha_0)^2},$$

and

$$R(G\boldsymbol{p}, \boldsymbol{\delta}^H) - R(G\boldsymbol{p}, \boldsymbol{\delta}^B)$$

$$= \sum_{i=1}^n [R_i(G\boldsymbol{p}, \boldsymbol{\delta}_i^H) - R_i(G\boldsymbol{p}, \boldsymbol{\delta}_i^B)]$$

$$\leq \frac{\sum_{\ell=1}^k [(m_1 + \cdots + m_n)p_\ell(1 - p_\ell) + (\alpha_\ell - \alpha_0 p_\ell)^2]}{(m_1 + \cdots + m_n + \alpha_0)^2}$$

$$\times \left[\sum_{i=1}^n \frac{M_i(M_i - m_i)^2}{c_i^2}\right]. \tag{4.32}$$

It should be noted that under the statistical model described in Section 4.2, if $M_i = M_j$ and $m_i = m_j$, then $R_i(Gp, \delta_i^B) = R_j(Gp, \delta_j^B)$. Under Condition C, M^* is a finite number independent of n. Therefore, under Condition C,

$$\min_{1 \leq i \leq n} c_i = c_0 > 0, \quad \min_{1 \leq i \leq n} R_i(Gp, \delta_i^B) = \tau > 0, \quad R(Gp, \delta^B) \geq n\tau,$$

and c_0 and τ are independent of n. Hence, under Condition C, we obtain

$$\begin{aligned}
\rho(Gp, \delta^H) &= D(Gp, \delta^H)/R(Gp, \delta^B), \\
&\leq \frac{1}{n\tau} \frac{[(m_1 + \cdots + m_n)\sum_{\ell=1}^{k} p_\ell(1 - p_\ell) + \sum_{\ell=1}^{k}(\alpha_\ell - \alpha_0 p_\ell)^2]}{(m_1 + \cdots + m_n)^2} \\
&\quad \times \left[\sum_{i=1}^{n} \frac{(M_i - m_i)^2}{c_i}\right] \\
&\leq \frac{C_1}{(m_1 + m_2 + \cdots + m_n + \alpha_0)} + \frac{C_2}{(m_1 + m_2 + \cdots + m_n + \alpha_0)^2} \\
&= O\left(\frac{1}{m_1 + \cdots + m_n}\right) \\
&= O\left(\frac{1}{n}\right) \quad (4.33)
\end{aligned}$$

where C_1 and C_2 are some positive numbers independent of n.

Furthermore, when n is sufficiently large such that both $d_{i\ell n} > 0$ and $e_{i\ell n} > 0$, where $d_{i\ell n} = (m_1 + \cdots + m_n + \alpha_0)(\frac{c_i}{M_i - m_i}) - \alpha_\ell + \alpha_0 p_\ell$, and $e_{i\ell n} = (m_1 + \cdots + m_n + \alpha_0)(\frac{c_i}{M_i - m_i}) + \alpha_\ell - \alpha_0 p_\ell$, then

$$\begin{aligned}
&P\left\{\left|\frac{T_{i\ell} + X_{i\ell} + \alpha_\ell}{m_1 + \cdots + m_n + \alpha_0} - p_\ell\right| > \frac{c_i}{M_i - m_i}\right\} \\
&= P\left\{\frac{X_{1\ell} + \cdots + X_{n\ell}}{m_1 + \cdots + m_n + \alpha_0} - p_\ell < -\frac{c_i}{M_i - m_i}\right\} \\
&\quad + P\left\{\frac{X_{1\ell} + \cdots + X_{n\ell}}{m_1 + \cdots + m_n + \alpha_0} - p_\ell > \frac{c_i}{M_i - m_i}\right\} \\
&= P\{X_{1\ell} + \cdots + X_{n\ell} - (m_1 + \cdots + m_n)p_\ell < -e_{i\ell n}\} \\
&\quad + P\{X_{1\ell} + \cdots + X_{n\ell} - (m_1 + \cdots + m_n)p_\ell > d_{i\ell n}\} \\
&\leq \exp\left\{-2(m_1 + \cdots + m_n)\left(\frac{e_{i\ell n}}{m_1 + \cdots + m_n}\right)^2\right\} \\
&\quad + \exp\left\{-2(m_1 + \cdots + m_n)\left(\frac{d_{i\ell n}}{m_1 + \cdots + m_n}\right)^2\right\} \\
&= \exp\left\{-2(m_1 + \cdots + m_n)\left[\frac{c_2}{M_i - m_i} + \frac{\alpha_\ell - \alpha_0 p_\ell}{m_1 + \cdots + m_n}\right]^2\right\} \\
&\quad + \exp\left\{-2(m_1 + \cdots + m_n)\left[\frac{c_i}{M_i - m_i} + \frac{\alpha_0 p_\ell - \alpha_\ell}{m_1 + \cdots + m_n}\right]^2\right\}.
\end{aligned}$$

$$(4.34)$$

In (4.34), the last inequality is obtained using Hoeffding's inequality.
Combining (4.30) and (4.34), we obtain

$$
R_i(G\boldsymbol{p}, \boldsymbol{\delta}_i^H) - R_i(G\boldsymbol{p}, \boldsymbol{\delta}_i^B)
$$
$$
\leq M_i \sum_{\ell=1}^{k} \left\{ \exp\left(-2(m_1 + \cdots + m_n)\left[\frac{c_i}{M_i - m_i} + \frac{\alpha_\ell - \alpha_0 p_\ell}{m_1 + \cdots + m_n}\right]^2\right) \right.
$$
$$
\left. + \exp\left(-2(m_1 + \cdots + m_n)\left[\frac{c_i}{M_i - m_i} + \frac{\alpha_0 p_\ell - \alpha_\ell}{m_1 + \cdots + m_n}\right]^2\right) \right\},
$$

and

$$
R(G\boldsymbol{p}, \boldsymbol{\delta}^H) - R(G\boldsymbol{p}, \boldsymbol{\delta}^B)
$$
$$
\leq \sum_{i=1}^{n} M_i \sum_{\ell=1}^{k} \left\{ \exp\left(-2(m_1 + \cdots + m_n)\left[\frac{c_i}{M_i - m_i} + \frac{\alpha_\ell - \alpha_0 p_\ell}{m_1 + \cdots + m_n}\right]^2\right) \right.
$$
$$
\left. + \exp\left(-2(m_1 + \cdots + m_n)\left[\frac{c_i}{M_i - m_i} + \frac{\alpha_0 p_\ell - \alpha_\ell}{m_1 + \cdots + m_n}\right]^2\right) \right\}. \quad (4.35)
$$

Let $\beta = \min_{1 \leq i \leq n}\{\frac{c_i}{M_i - m_i}\}$. Under Condition C, $\beta > 0$ and is independent of n. Then, from (4.35) and the fact that $R(G\boldsymbol{p}, \boldsymbol{\delta}^B) \geq n\tau$, we obtain

$$
\rho(G\boldsymbol{p}, \boldsymbol{\delta}^H) = D(G\boldsymbol{p}, \boldsymbol{\delta}^H)/R(G\boldsymbol{p}, \boldsymbol{\delta}^B)
$$
$$
\leq \frac{M^*}{\tau} \sum_{\ell=1}^{k} \left\{ \exp\left(-2(m_1 + \cdots + m_n)(\beta + \frac{\alpha_\ell - \alpha_0 p_\ell}{m_1 + \cdots + m_n})^2\right) \right.
$$
$$
\left. + \exp\left(-2(m_1 + \cdots + m_n)(\beta + \frac{\alpha_0 p_\ell - \alpha_\ell}{m_1 + \cdots + m_n})^2\right) \right\}
$$
$$
= O\left(e^{-2(m_1 + \cdots + m_n)\beta}\right). \quad (4.36)
$$

We summarize the above results in the form of a theorem as follows.

Theorem 4.4.2 *Under the model described in Section 4.2 and Condition C, we have*

$$
\rho(G\boldsymbol{p}, \boldsymbol{\delta}^H) \leq \frac{C_1}{m_1 + \cdots + m_n + \alpha_0} + \frac{C_2}{(m_1 + \cdots + m_n + \alpha_0)^2}
$$

for some positive constants C_1 and C_2 independent of n.
Furthermore, for n sufficiently large, we have

$$
\rho(G\boldsymbol{p}, \boldsymbol{\delta}^H) = O\left(e^{-2(m_1 + \cdots + m_n)\beta}\right)
$$

for some positive value β independent of n.

Remark: We have presented methods of constructing hierarchical Bayesian estimators and selection procedures for multivariate hypergeometric distributions. For each of the two decision problems, the associated procedure is shown to be asymptotically optimal, at a rate of convergence of order $O(n^{-1})$ and $O(e^{-n\beta})$ for some positive value β, respectively, where n is the number of underlying populations involved in the decision problem. When n is small, these convergence rates are slow even though the two procedures φ^H and δ^H still possess the admissibility. Further study is needed for the case when n is small.

References

1. Balakrishnan, N. and Ma, Y. (1996). Empirical Bayes rules for selecting the most and least probable multivariate hypergeometric event, *Statistics & Probability Letters*, **27**, 181–188.

2. Balakrishnan, N. and Ma, Y. (1997). On empirical Bayes rules for selecting the best hypergeometric population, *Sequential Analysis* (to appear).

3. Bartlett, N. S. and Govindarajulu, Z. (1970). Selecting a subset containing the best hypergeometric population, *Sankhyā, Series B*, **32**, 341–352.

4. Bechhofer, R. E., Santner, T. J. and Goldsman, D. M. (1995). *Design and Analysis of Experiments For Statistical Selection, Screening, and Multiple Comparisons*, New York: John Wiley & Sons.

5. Cressie, N. and Scheult, A. (1985). Empirical Bayes estimation in sampling inspection, *Biometrika*, **72**, 451–458.

6. Gupta, S. S. and Liang, T. (1994). On empirical Bayes selection rules for sampling inspection, *Journal of Statistical Planning and Inference*, **38**, 43–64.

7. Gupta, S. S. and Panchapakesan, S. (1979). *Multiple Decision Procedures: Theory and Methodology of Selecting and Ranking Populations*. New York: John Wiley & Sons.

8. Johnson, N. L., Kotz, S. and Balakrishnan, N. (1997). *Discrete Multivariate Distributions*, New York: John Wiley & Sons.

9. Walter, G. G. and Hamedani, G. G. (1988). Empiric Bayes estimation of hypergeometric probability, *Metrika*, **35**, 127–143.

5

Convergence Rates of Empirical Bayes Estimation and Selection for Exponential Populations With Location Parameters

N. Balakrishnan and Yimin Ma

McMaster University, Hamilton, Ontario, Canada

Abstract: In this paper, we discuss the problems of empirical Bayes estimation for the exponential distributions with location parameters and the empirical Bayes rule for selecting the best of the exponential populations. We first derive the Bayes estimators and the Bayes selection rule and then construct empirical Bayes estimators for the location parameters and the empirical Bayes selection rule. Finally, we examine the asymptotic optimality properties of the proposed empirical Bayes estimators and the empirical Bayes selection rule.

Keywords and phrases: Bayes rule, best population, convergence rates, empirical Bayes rule, exponential population, location parameter

5.1 Introduction

The empirical Bayes approach, formulated by Robbins (1955), is appropriate when one is confronted repeatedly and independently with the same decision problem. This approach has been used extensively for various statistical problems by many authors including Robbins (1963, 1964), Johns and Van Ryzin (1971, 1972), Singh (1979), Gupta and Liang (1986, 1988), and Gupta, Liang and Rau (1994).

The exponential distribution with location (or threshold) parameter arises in many areas of applications including reliability and life-testing, survival analysis, and engineering problems; for example, see Balakrishnan and Basu (1995). In these literatures, the location or the threshold parameter is interpreted to be the guaranteed life-time. For this distributional model, Singh and Prasad

(1989) and Prasad and Singh (1990) have discussed the empirical Bayes estimation under squared error loss. They have also discussed some asymptotic properties of their empirical Bayes estimators under the assumption of the class of all prior distributions having support in a compact interval of the real line. Lin and Leu (1994) have studied the problem of selecting the best population from k independent exponential populations with different location parameters through the empirical Bayes approach; however, they have not examined the convergence rates of their empirical Bayes selection rule.

In this paper, we consider the problems of the empirical Bayes estimation for the location parameters without the assumption of the compact support for the prior distributions and the empirical Bayes selection of the best of k exponential populations with different location parameters. In Section 5.2, we formulate the selection problem and derive the Bayes estimators for the location parameters and the Bayes selection rule. In Section 5.3, we construct the empirical Bayes estimators for the location parameters and the empirical Bayes selection rule. Finally, we discuss the asymptotic optimality properties of these empirical Bayes estimators and the empirical Bayes selection rule in Sections 5.4 and 5.5, respectively, and give an example to illustrate the obtained results in Section 5.6.

5.2   Bayes Estimators and Bayes Selection Rule

Consider k independent populations $\pi_1, \ldots, \pi_k$, where an observation x_i from π_i has an exponential distribution with location parameter θ_i and scale parameter σ as follows:

$$f_i(x_i|\theta_i) = \frac{1}{\sigma} \exp\left\{-\frac{x_i - \theta_i}{\sigma}\right\}, \qquad x_i > \theta_i,\ \theta_i, \sigma > 0 . \tag{5.1}$$

The density function (5.1) provides a model for life length data when we assume a minimum guaranteed life-time θ_i, which is here a location (or threshold) parameter. It is assumed that all the k populations have a common known scale parameter σ. The θ_i's are unknown and let $\theta_{[1]} \leq \cdots \leq \theta_{[k]}$ be the ordered parameters of $\theta_1, \ldots, \theta_k$. A population π_i with $\theta_i = \theta_{[k]}$ is considered as the best population. Our interest is to select the population π_i associated with the largest guaranteed life-time $\theta_i = \theta_{[k]}$. Note that the best population also has the largest mean life-time $\theta_{[k]} + \sigma$.

Let $\Omega = \{\boldsymbol{\theta} \mid \boldsymbol{\theta} = (\theta_1, \ldots, \theta_k),\ \theta_i > 0,\ i = 1, \ldots, k\}$ be the parameter space. It is assumed that the parameter $\boldsymbol{\theta}$ has a prior distribution $G(\boldsymbol{\theta})$ with a joint distribution function $G(\boldsymbol{\theta}) = \prod_{i=1}^{k} G_i(\theta_i)$. Note that θ_i's are assumed to be independently distributed.

Let $A = \{i \mid i = 1, \ldots, k\}$ be the action space; when action i is taken, it means that population π_i is selected as the best population. For the parameter

$\boldsymbol{\theta}$ and action i, the loss function is defined by

$$L(\boldsymbol{\theta}, i) = \theta_{[k]} - \theta_i, \tag{5.2}$$

the difference between the best and the selected population. This loss function is very common in Bayes and empirical Bayes selection problems.

Let X be the sample space generated by $\boldsymbol{x} = (x_1, \ldots, x_k)$. A selection rule $\boldsymbol{d} = (d_1, \ldots, d_k)$ is a mapping from the sample space X to $[0,1]^k$ such that for each $\boldsymbol{x} \in X$, the function $\boldsymbol{d}(\boldsymbol{x}) = (d_1(\boldsymbol{x}), \ldots, d_k(\boldsymbol{x}))$ satisfies $0 \leq d_i(\boldsymbol{x}) \leq 1$, $i = 1, \ldots, k$, and $\sum_{i=1}^k d_i(\boldsymbol{x}) = 1$. Note that $d_i(\boldsymbol{x})$, $i = 1, \ldots, k$, is the probability of selecting the population π_i as the best population when $\boldsymbol{x}$ is observed.

Let D be the set of all selection rules. For each $\boldsymbol{d} \in D$, let $r(G, \boldsymbol{d})$ denote the associated Bayes risk. Then from the loss function (5.2), the Bayes risk associated with $\boldsymbol{d}$ can be written as

$$\begin{aligned}
r(G, \boldsymbol{d}) &= \int \int \sum_{i=1}^k L(\boldsymbol{\theta}, i) d_i(\boldsymbol{x}) f(\boldsymbol{x} \mid \boldsymbol{\theta}) dG(\boldsymbol{\theta}) d\boldsymbol{x} \\
&= C - \int \left\{ \sum_{i=1}^k d_i(\boldsymbol{x}) \varphi_i(x_i) \right\} f(\boldsymbol{x}) d\boldsymbol{x},
\end{aligned} \tag{5.3}$$

where

$$f(\boldsymbol{x}) = \prod_{i=1}^k f_i(x_i), \quad f_i(x_i) = \int_0^{x_i} f(x_i \mid \theta_i) dG_i(\theta_i), \tag{5.4}$$

$$C = \int \int \theta_{[k]} f(\boldsymbol{x} \mid \boldsymbol{\theta}) dG(\boldsymbol{\theta}) d\boldsymbol{x} = \int \theta_{[k]} dG(\boldsymbol{\theta}), \tag{5.5}$$

and

$$\varphi_i(x_i) = E(\theta_i \mid x_i) = \frac{\int_0^{x_i} \theta_i f_i(x_i \mid \theta_i) dG_i(\theta_i)}{f_i(x_i)}. \tag{5.6}$$

Note that $\varphi_i(x_i)$, $i = 1, \ldots, k$, is the Bayes estimator of location parameter θ_i under the squared error loss.

The minimum Bayes risk among the class D is defined by $r(G) \equiv \inf_{\boldsymbol{d} \in D} r(G, \boldsymbol{d})$; any selection rule $\boldsymbol{d}$ such that $r(G, \boldsymbol{d}) = r(G)$ is called a Bayes selection rule. For each $\boldsymbol{x} \in X$, let

$$A(\boldsymbol{x}) = \{i \mid \varphi_i(x_i) = \max_{1 \leq j \leq k} \varphi_j(x_j)\} \tag{5.7}$$

and

$$i^* \equiv i^*(\boldsymbol{x}) = \min\{i \mid i \in A(\boldsymbol{x})\}; \tag{5.8}$$

then from (5.3), a Bayes selection rule $\boldsymbol{d}_G = (d_{1G}, \ldots, d_{kG})$ is given by

$$d_{iG}(\boldsymbol{x}) = \begin{cases} 1 & \text{if } i = i^* \\ 0 & \text{otherwise} . \end{cases} \tag{5.9}$$

5.3   Empirical Bayes Estimators and Empirical Bayes Selection Rule

Since the Bayes estimators $\varphi_i(x_i)$, $i = 1, \ldots, k$, and Bayes selection rule $\boldsymbol{d}_G$ are both dependent on the prior distributions $G_i(\theta_i)$, $i = 1, \ldots, k$, which may not be known, it is impossible to apply the Bayes estimators and Bayes selection rule in practice. Hence, we adopt now the empirical Bayes approach.

Let x_{ij} denote the observations from population π_i at stage j, $j = 1, \ldots, n$. It is assumed that conditional on θ_{ij}, x_{ij} follows an exponential distribution with location parameter θ_{ij}, as follows

$$x_{ij} \mid \theta_{ij} \sim \frac{1}{\sigma} \exp\left\{-\frac{x_{ij} - \theta_{ij}}{\sigma}\right\} \qquad x_{ij} > \theta_{ij}, \ \theta_{ij}, \sigma > 0 . \tag{5.10}$$

Denote $\boldsymbol{\theta}_j = (\theta_{1j}, \ldots, \theta_{kj})$ and assume that $\boldsymbol{\theta}_j$, $j = 1, \ldots, n$, are i.i.d. with the prior distribution $G(\boldsymbol{\theta}) = \prod_{i=1}^{k} G_i(\theta_i)$. Let $\boldsymbol{x}_j = (x_{1j}, \ldots, x_{kj})$ denote the observations at the jth stage, $j = 1, \ldots, n$. We also let $\boldsymbol{x}_{n+1} = \boldsymbol{x} = (x_1, \ldots, x_k)$ denote the observations at the present stage.

Under this statistical framework, for each $i = 1, \ldots, k$, by a method similar to the one used in Lemma 4.1 of Fox (1978) we can derive

$$\varphi_i(x_i) = E(\theta_i | x_i) = x_i - \psi_i(x_i), \tag{5.11}$$

where

$$\psi_i(x_i) = \frac{\int_0^{x_i} e^{-\frac{1}{\sigma}(x_i - t_i)} dF_i(t_i)}{f_i(x_i)} \overset{d}{=} \frac{w_i(x_i)}{f_i(x_i)} \tag{5.12}$$

with $f_i(x_i)$ being the density function of x_i and $F_i(x_i)$ the corresponding cumulative distribution function. Note that under the statistical model (5.1), $0 \le \varphi_i(x_i) \le x_i$, then we have $0 \le \psi_i(x_i) \le x_i$.

For each $i = 1, \ldots, k$, based on the past data $x_{i1}, \ldots, x_{in}$, we define

$$w_{in}(x_i) = \frac{1}{n} \sum_{j=1}^{n} e^{-\frac{1}{\sigma}(x_i - x_{ij})} I_{(0,x_i)}(x_{ij}) \tag{5.13}$$

as the estimator of $w_i(x_i)$. In order to estimate $f_i(x_i)$, we employ kernel functions used by Johns and Van Ryzin (1972) and Singh (1977, 1979). Let k_r be the class of all Borel-measurable real-valued bounded functions k vanishing off $(0, 1)$ such that

$$\int k(y)dy = 1, \quad \int y^\ell k(y)dy = 0 \text{ for } \ell = 1, \ldots, r - 1, \tag{5.14}$$

where r is an arbitrary but fixed positive integer. Define

$$f_{in}(x_i) = \frac{1}{nh_n} \sum_{j=1}^{n} k\left(\frac{x_{ij} - x_i}{h_n}\right), \tag{5.15}$$

where h_n is a positive function of n such that $h_n \to 0$ and $nh_n \to \infty$ as $n \to \infty$.

Utilizing $w_{in}(x_i)$ and $f_{in}(x_i)$, we propose the empirical Bayes estimators $\varphi_{in}(x_i)$ for location parameters θ_i, $i = 1, 2, \ldots, k$, as

$$\begin{aligned} \varphi_{in}(x_i) &= x_i - 0 \vee \left(\frac{w_{in}(x_i)}{f_{in}(x_i)}\right) \wedge x_i \\ &\stackrel{d}{=} x_i - \psi_{in}(x_i), \end{aligned} \tag{5.16}$$

where $a \vee b = \max(a, b)$ and $a \wedge b = \min(a, b)$.

Next, we propose an empirical Bayes selection rule $\boldsymbol{d}_n^* = (d_{1n}^*, \ldots, d_{kn}^*)$ for the selection problem under study as follows:

For each $\boldsymbol{x} \in X$, let

$$A_n^*(\boldsymbol{x}) = \{i \mid \varphi_{in}(x_i) = \max_{1 \le j \le k} \varphi_{jn}(x_j)\}, \tag{5.17}$$

where $\varphi_{in}(x_i)$ is given by (5.16), and

$$i_n^* \equiv i_n^*(\boldsymbol{x}) = \min\{i \mid i \in A_n^*(\boldsymbol{x})\}; \tag{5.18}$$

then define the empirical Bayes rule as

$$d_{in}^*(\boldsymbol{x}) = \begin{cases} 1 & \text{if } i = i_n^* \\ 0 & \text{otherwise.} \end{cases} \tag{5.19}$$

5.4 Asymptotic Optimality of the Empirical Bayes Estimators

Under the squared error loss function, for each $i = 1, \ldots, k$, the Bayes risk of the proposed empirical Bayes estimators $\varphi_{in}(x_i)$ and the Bayes estimators $\varphi_i(x_i)$ are respectively

$$R_i(G_i, \varphi_{in}(x_i)) = E(\theta_i - \varphi_{in}(x_i))^2, \tag{5.20}$$

and

$$R_i(G_i) = R_i(G_i, \varphi(x_i)) = E(\theta_i - \varphi_i(x_i))^2. \tag{5.21}$$

Since $\varphi_i(x_i)$ is the Bayes estimator of θ_i, obviously

$$E_{in}\{R_i(G_i, \varphi_{in}(x_i))\} - R_i(G_i) \geq 0 , \qquad (5.22)$$

where E_{in} denotes the expectation with respect to $(x_{i1}, \ldots, x_{in})$. It can be shown that

$$
\begin{aligned}
E_{in}\{R_i(G_i, \varphi_{in}(x_i))\} - R_i(G_i) &= \int f_i(x_i) E_n(\varphi_{in}(x_i) - \varphi_i(x_i))^2 dx_i \\
&= \tilde{E}_{in}(\varphi_{in}(x_i) - \varphi_i(x_i))^2 \qquad (5.23)
\end{aligned}
$$

where $\tilde{E}_{in}$ denotes the expectation with respect to $(x_i, x_{i1}, \ldots, x_{in})$.

Definition 5.4.1 A sequence of empirical Bayes estimators $\{\varphi_{in}\}$ is said to be asymptotically optimal at least of order α_n relative to the prior G_i if

$$E_{in}\{R_i(G_i, \varphi_{in}(x_i))\} - R_i(G_i) \leq O(\alpha_n) \qquad \text{as } n \to \infty ,$$

where $\{\alpha_n\}$ is a sequence of positive numbers satisfying $\lim_{n\to\infty} \alpha_n = 0$.

In order to investigate the asymptotic optimality of the proposed empirical Bayes estimators $\varphi_{in}(x_i)$, we present some useful lemmas.

Lemma 5.4.1 Let y, $z \neq 0$ and $L \geq 0$ be real numbers, and Y, Z be two real valued random variables; then for any $0 < \tau \leq 2$,

$$E\left(\left|\frac{y}{z} - \frac{Y}{z}\right| \wedge L\right)^\tau \leq 2|z|^{-\tau}\left\{E|y - Y|^\tau + \left(\left|\frac{y}{z}\right| + L\right)^\tau E|z - Z|^\tau\right\} . \quad (5.24)$$

PROOF. This is Lemma 3.1 in Singh and Wei (1992). ∎

Lemma 5.4.2 For $i = 1, 2, \ldots, k$, let $f_{in}(x_i)$ be defined by (5.15); if for $r \geq 1$, the r-th derivatives of $f_i(x_i)$ exists, then for any $\varepsilon > 0$

(a) $E_{in}(|f_{in}(x_i) - f_i(x_i)|^2) \leq O(h_n^{2r})[f_{i\varepsilon}^{(r)}(x_i)]^2 + O(n^{-1}h_n^{-1})f_{i\varepsilon}(x_i)$ (5.25)

where $f_{i\varepsilon}(x_i) = \sup_{0 \leq u_i \leq \varepsilon} f(x_i + u_i)$, $f_{i\varepsilon}^{(r)}(x_i) = \sup_{0 \leq u_i \leq \varepsilon} |f^{(r)}(x_i + u_i)|$;
(b) for any $0 < \delta \leq 2$, when $h_n = n^{-1/(2r+1)}$,

$$E_{in}(|f_{in}(x_i) - f_i(x_i)|^\delta) \leq O\left(n^{-\frac{\delta r}{2r+1}}\right)\{[f_{i\varepsilon}(x_i)]^{\delta/2} + [f_{i\varepsilon}^{(r)}(x_i)]^\delta\}. \quad (5.26)$$

PROOF. (a) is easily proved by Theorem 3.3 in Singh (1977). If $h_n = n^{-1/(2r+1)}$, from (a), we have

$$E_{in}(|f_{in}(x_i) - f_i(x_i)|^2) \leq O\left(n^{-\frac{2r}{2r+1}}\right)\{f_{i\varepsilon}(x_i) + [f_{i\varepsilon}^{(r)}(x_i)]^2\};$$

then for any $0 < \delta \leq 2$, by Hölder's inequality

$$
\begin{aligned}
E_{in}(|f_{in}(x_i) - f_i(x_i)|^\delta) &\leq \left[E_{in}(|f_{in}(x_i) - f_i(x_i)|^2)\right]^{\delta/2} \\
&\leq cn^{-\frac{\delta r}{2r+1}}\left\{[f_{i\varepsilon}(x_i)]^{\delta/2} + [f_{i\varepsilon}^{(r)}(x_i)]^\delta\right\}
\end{aligned}
$$

and thus (b) is true.   ■

The following theorem is one of the main results in this paper concerning convergence rate of the empirical Bayes estimators. In the rest of this paper, c_1, c_2, c_3 and c always stand for some positive constants, and they may be different even with the same notations.

Theorem 5.4.1 *For $i = 1, \ldots, k$, let $\{\varphi_{in}(x_i)\}$ be the sequence of empirical Bayes estimators defined by (5.16); if for $r \geq 1$, the r-th derivatives of $f_i(x_i)$ exist and for $\varepsilon > 0$, $0 < \delta < 2$,*

(a)
$$
\int_0^\infty x_i^{2-\delta}[f_i(x_i)]^{1-\delta}\left[\int_0^{x_i} e^{-\frac{2}{\sigma}(x_i-t_i)}dF_i(t_i)\right]^{\delta/2} dx_i < \infty
$$

(b)
$$
\int_0^\infty x_i^2[f_i(x_i)]^{1-\delta}\left\{[f_{i\varepsilon}(x_i)]^{\delta/2} + [f_{i\varepsilon}^{(r)}(x_i)]^\delta\right\} dx_i < \infty,
$$

then with the choice of $h_n = n^{-1/(2r+1)}$, we have

$$
E_{in}\{R_i(G_i, \varphi_{in}(X_i))\} - R_i(G_i) \leq O\left(n^{-\frac{\delta r}{2r+1}}\right). \tag{5.27}
$$

PROOF. For each $i = 1, \ldots, k$, from the definition of $w_{in}(x_i)$, we know that $w_{in}(x_i)$ is an unbiased estimator of $w_i(x_i)$; then we have

$$
\begin{aligned}
E_{in}(|w_{in}(x_i) - w_i(x_i)|^2) &= \text{Var}(x_{in}) = \frac{1}{n}\,\text{Var}\left\{e^{-\frac{1}{\sigma}(x_i-t_i)}I_{(0,x_i)}(t_i)\right\} \\
&\leq \frac{1}{n}\,E\left(e^{-\frac{1}{\sigma}(x_i-t_i)}I_{(0,x_i)}(t_i)\right)^2 \\
&= \frac{1}{n}\int_0^{x_i} e^{-\frac{2}{\sigma}(x_i-t_i)}dF_i(t_i)
\end{aligned}
$$

and by Hölder's inequality

$$
\begin{aligned}
E_{in}(|w_{in}(x_i) - w_i(x_i)|^\delta) &\leq \left[E_{in}(|w_{in}(x_i) - w_i(x_i)|^2)\right]^{\delta/2} \\
&= (n^{-\delta/2})\left[\int_0^{x_i} e^{-\frac{2}{\sigma}(x_i-t_i)}dF_i(t_i)\right]^{\delta/2}.
\end{aligned}
$$

Now by Lemma 5.4.1 and Lemma 5.4.2,

$$
E_{in}\{R_i(G_i, \varphi_{in}(x_i))\} - R_i(G_i) = \tilde{E}_{in}(\varphi_{in}(x_i) - \varphi_i(x_i))^2
$$

$$
\begin{aligned}
&= \quad \tilde{E}_{in}(\psi_{in}(x_i) - \psi_i(x_i))^2 \\
&\leq \quad \tilde{E}_{in}\left(\left|\frac{w_{in}(x_i)}{f_{in}(x_i)} - \frac{w_i(x_i)}{f_i(x_i)}\right| \wedge x_i\right)^2 \\
&\leq \quad \tilde{E}_{in}\left[x_i^{2-\delta}\left(\left|\frac{w_{in}(x_i)}{f_{in}(x_i)} - \frac{w_i(x_i)}{f_i(x_i)}\right| \wedge x_i\right)^\delta\right] \\
&\leq \quad E\left\{2x_i^{2-\delta}|f_i(x_i)|^{-\delta}[E_{in}(|w_{in} - w_i|^\delta) + cx_i^\delta E_{in}(|f_{in} - f_i|^\delta)]\right\} \\
&\leq \quad c_1 E\left[x_i^{2-\delta}f_i^{-\delta}\left(\int_0^{x_i} e^{-\frac{2}{\sigma}(x_i-t_i)}dF_i(t_i)\right)^{\delta/2}\right]\left(n^{-\frac{\delta}{2}}\right) \\
&\quad + c_2 E\left\{x_i^2 f_i^{-\delta}[(f_{i\varepsilon}(x_i))^{\delta/2} + (f_{i\varepsilon}^{(r)}(x_i))^\delta]\right\}\left(n^{-\frac{\delta r}{2r+1}}\right) \\
&\leq \quad O\left(n^{-\frac{\delta r}{2r+1}}\right)
\end{aligned}
$$

by assumptions (a) and (b). This proves the theorem.   ∎

5.5 Asymptotic Optimality of the Empirical Bayes Selection Rule

Now we investigate the convergence rate of the proposed empirical Bayes selection rule $\{d_n^*\}$. By the formula (5.3), the Bayes risk associated with the selection rule $\{d_n^*\}$ and the Bayes selection rule d_G are given by

$$
r(G, d_n^*) = c - \int\left\{\sum_{i=1}^k d_{in}^*(\boldsymbol{x})\varphi_i(x_i)\right\}f(\boldsymbol{x})d\boldsymbol{x} \tag{5.28}
$$

and

$$
r(G) = r(G, d_G) = c - \int\left\{\sum_{i=1}^k d_{iG}(\boldsymbol{x})\varphi_i(x_i)\right\}f(\boldsymbol{x})d\boldsymbol{x} , \tag{5.29}
$$

respectively. It is obvious that

$$
E_n\{r(G, d_n^*)\} - r(G) \geq 0 \tag{5.30}
$$

because the Bayes rule d_G achieves the minimum Bayes risk $r(G)$ and the expectation E_n is taken with respect to $(\boldsymbol{x}_1, \ldots, \boldsymbol{x}_n)$.

From the following straightforward computation, we have

$$
\begin{aligned}
0 &\leq \quad E_n\{r(G, d_n^*)\} - r(G) \\
&= \quad \int\left\{\sum_{i=1}^k (d_{iG}(\boldsymbol{x}) - E_n(d_{in}^*(\boldsymbol{x}))\varphi_i(x_i)\right\}f(\boldsymbol{x})d\boldsymbol{x}
\end{aligned}
$$

$$= \int E_n \{ d_{i^*G}(\boldsymbol{x}) \varphi_{i^*}(x_{i^*}) - d_{i_n^* n}(\boldsymbol{x}) \varphi_{i_n^* n}(x_{i_n^*}) \} f(\boldsymbol{x}) d\boldsymbol{x}$$

$$= \int E_n \left\{ \sum_{i=1}^{k} \sum_{j=1}^{k} I_{(i^*=i,\, i_n^*=j)} (\varphi_i(x_i) - \varphi_j(x_j)) \right\} f(\boldsymbol{x}) d\boldsymbol{x}$$

$$= \int \left\{ \sum_{i=1}^{k} \sum_{j=1}^{k} P\{i^*=i,\, i_n^*=j\} (\varphi_i(x_i) - \varphi_j(x_j)) \right\} f(\boldsymbol{x}) d\boldsymbol{x}$$

$$\leq \sum_{i \neq j} \int \int \left\{ \Pr\left(|\varphi_{in}(x_i) - \varphi_i(x_i)| > \frac{1}{2} |\varphi_i(x_i) - \varphi_j(x_j)| \right) \right.$$

$$+ \left. \Pr\left(|\varphi_{jn}(x_j) - \varphi_j(x_j)| > \frac{1}{2} |\varphi_i(x_i) - \varphi_j(x_j)| \right) \right\}$$

$$\times |\varphi_i(x_i) - \varphi_j(x_j)| f(x_i) f(x_j) \, dx_i dx_j \, . \tag{5.31}$$

Definition 5.5.1 A sequence of empirical Bayes selection rules $\{d_n\}$ is said to be asymptotically optimal at least of order β_n relative to the unknown prior distribution G if

$$E_n\{r(G, \boldsymbol{d}_n)\} - r(G) \leq O(\beta_n) \text{ as } n \to \infty$$

where β_n is a sequence of positive numbers such that $\lim_{n \to \infty} \beta_n = 0$.

The following theorem is another main result in this paper concerning the convergence rate of the empirical Bayes selection rule $\{\boldsymbol{d}_n^*\}$.

Theorem 5.5.1 *Let $\{\boldsymbol{d}_n^*\}$ be the sequence of empirical Bayes selection rules defined by (5.19); if for $r \geq 1$, the r-th derivatives of $f_i(x_i)$ exist and for $\varepsilon > 0$, $0 < \delta < 2$,*

(a)
$$\int_0^\infty [f_i(x_i)]^{1-\delta} x_i^{2-\delta} \left[\int_0^{x_i} e^{-\frac{2}{\sigma}(x_i - t_i)} dF_i(t_i) \right]^{\delta/2} dx_i < \infty$$

(b)
$$\int_0^\infty [f_i(x_i)]^{1-\delta} x_i^2 \left\{ [f_{i\varepsilon}(x_i)]^{\delta/2} + [f_{i\varepsilon}^{(r)}(x_i)]^\delta \right\} dx_i < \infty,$$

then with the choice of $h_n = n^{-1/(2r+1)}$, we have

$$E_n\{r(G, \boldsymbol{d}_n^*)\} - r(G) \leq O\left(n^{-\frac{\delta r}{2(2r+1)}} \right). \tag{5.32}$$

PROOF. Let
$$X_{ijn} = \left\{ (x_i, x_j) \big| |\varphi_i(x_i) - \varphi_j(x_j)| \leq \varepsilon_n \right\}$$

where $\varepsilon_n > 0$ and $\varepsilon_n \to 0$ as $n \to \infty$. Then by Markov's inequality, and Theorem 5.4.1, we have

$$E_n\{r(G, \boldsymbol{d}_n^*)\} - r(G)$$

$$\leq \sum_{i \neq j} 2\varepsilon_n \int \int_{X_{ijn}} f_i(x_i) f_j(x_j) dx_i dx_j$$

$$+ \sum_{i \neq j} \int \int_{X_{ijn}^c} \left\{ \Pr(|\varphi_{in}(x_i) - \varphi_i(x_i)| > \frac{1}{2} |\varphi_i(x_i) - \varphi_j(x_j)|) \right.$$

$$\left. + \Pr(|\varphi_{jn}(x_j) - \varphi_j(x_j)| > \frac{1}{2} |\varphi_i(x_i) - \varphi_j(x_j)|) \right\}$$

$$\times |\varphi_i(x_i) - \varphi_j(x_j)| f_i(x_i) f_j(x_j) dx_i \, dx_j$$

$$\leq c_1 \varepsilon_n + c_2 \sum_{i \neq j} \int \int_{X_{ijn}^c} \left(\frac{E_{in} |\varphi_{in}(x_i) - \varphi_i(x_i)|^2}{|\varphi_i(x_i) - \varphi_j(x_j)|} \right.$$

$$\left. + \frac{E_{jn} |\varphi_{jn}(x_j) - \varphi_j(x_j)|^2}{|\varphi_i(x_i) - \varphi_j(x_j)|} \right) f_i(x_i) f_j(x_j) dx_i \, dx_j$$

$$\leq c_1 \varepsilon_n + c_2 \sum_{i \neq j} \int \int_{X_{ijn}^c} \frac{1}{\varepsilon_n} E_{in} |\varphi_{in}(x_i) - \varphi_i(x_i)|^2 f(x_i) f(x_j) dx_i dx_j$$

$$\leq c_1 \varepsilon_n + c_2 \frac{1}{\varepsilon_n} \sum_{i=1}^k \tilde{E}_{in}(|\varphi_{in}(x_i) - \varphi_i(x_i)|^2)$$

$$\leq c_1 \varepsilon_n + c_2 \varepsilon_n^{-1} \left(n^{-\frac{\delta r}{2r+1}} \right).$$

Thus, letting $\varepsilon_n = n^{-\frac{\delta r}{2(2r+1)}}$, we obtain

$$E_n\{r(G, \boldsymbol{d}_n^*)\} - r(G) \leq O\left(n^{-\frac{\delta r}{2(2r+1)}} \right).$$

■

5.6   An Example

Finally, we discuss an example to illustrate the results obtained in Theorems 5.4.1 and 5.5.1. Without loss of generality, we assume that common known scale parameter $\sigma = 1$ in (5.1), i.e., for $i = 1, \ldots, k$,

$$f_i(x_i|\theta_i) = \exp\{-(x_i - \theta_i)\}, \qquad x_i > \theta_i, \ \theta_i > 0. \tag{5.33}$$

Let $G_i(\theta_i)$ be a prior distribution with density function as

$$\frac{dG_i(\theta_i)}{d\theta_i} = \left(1 - \frac{1}{\tau}\right) \frac{\theta_i}{\tau^2} e^{-\frac{\theta_i}{\tau}} + \left(\frac{1}{\tau}\right) \frac{1}{\tau} e^{-\frac{\theta_i}{\tau}}, \qquad \theta_i > 0, \ \tau > 1. \tag{5.34}$$

With such prior distribution on θ_i, we have

$$
\begin{aligned}
f_i(x_i) &= \int_0^{x_i} f_i(x_i|\theta_i)dG_i(\theta_i) \\
&= e^{-x_i}\left[\int_0^{x_i}\left(1-\frac{1}{\tau}\right)\frac{\theta_i}{\tau^2}\,e^{-\frac{\theta_i}{\tau}}e^{\theta_i}d\theta_i + \int_0^{x_i}\frac{1}{\tau^2}e^{-\frac{\theta_i}{\tau}}e^{\theta_i}d\theta_i\right] \\
&= \frac{x_i}{\tau^2}\,e^{-\frac{x_i}{\tau}}, \qquad x_i > 0,\ \tau > 1;
\end{aligned}
\tag{5.35}
$$

then

$$
\begin{aligned}
\int_0^{x_i} e^{-2(x_i-t_i)}f_i(t_i)dt_i &= \int_0^{x_i} e^{-2x_i}\frac{t_i}{\tau^2}\,e^{(2-\frac{1}{\tau})t_i}dt_i \\
&\le c\,\frac{x_i}{\tau^2}\,e^{-\frac{x_i}{\tau}} = cf_i(x_i)
\end{aligned}
$$

and

$$
f_i^{(r)}(x_i) = c_1\left(\frac{1}{\tau}\,e^{-\frac{x_i}{\tau}}\right) + c_2\left(\frac{x_i}{\tau^2}\,e^{-\frac{x_i}{\tau}}\right)
$$

where c_1, c_2 and c are dependent on r and τ. Since $xe^{-\frac{x}{\tau}}$ is increasing when $0 < x < \tau$ and decreasing when $x \ge \tau$, then, for any $\varepsilon > 0$,

$$
f_{i\varepsilon}(x_i) = \sup_{0\le u_i\le\varepsilon} f_i(x_i+u_i) \le \begin{cases} f_i(x_i) & \text{if } x_i \ge \tau \\ f_i(\tau) & \text{if } x_i < \tau \end{cases}
$$

and

$$
f_{i\varepsilon}^{(r)}(x_i) = \sup_{0\le u_i\le\varepsilon}|f_i^{(r)}(x_i+u_i)| \le \begin{cases} c_1e^{-\frac{x_i}{\tau}}+c_2x_ie^{-\frac{x_i}{\tau}} & \text{if } x_i \ge \tau \\ c_1+c_2f_i(\tau) & \text{if } x_i < \tau. \end{cases}
$$

Thus, for any $0 < \delta < 2$,

$$
\begin{aligned}
&\int_0^\infty [f_i(x_i)]^{1-\delta}x_i^{2-\delta}\left[\int_0^{x_i} e^{-2(x_i-t_i)}dF_i(t_i)\right]^{\delta/2}dx_i \\
&\quad\le c\int_0^\infty x_i^{2-\delta}[f_i(x_i)]^{1-\delta/2}dx_i \\
&\quad= c\int_0^\infty x_i^{3(2-\delta)/2}\,e^{-\frac{(2-\delta)x_i}{2\tau}}dx_i < \infty
\end{aligned}
\tag{5.36}
$$

and

$$
\begin{aligned}
&\int_0^\infty x_i^2[f_i(x_i)]^{1-\delta}\{[f_{i\varepsilon}(x_i)]^{\delta/2}+[f_{i\varepsilon}^{(r)}(x_i)]^\delta\}dx_i \\
&\quad\le c_1\int_0^\tau x_i^2\left(x_i^{1-\delta}e^{-\frac{(1-\delta)x_i}{\tau}}\right)dx_i \\
&\qquad + c_2\int_\tau^\infty x_i^2[f_i(x_i)]^{1-\delta}\{[f_i(x_i)]^{\delta/2}+c_3[f_i(x_i)]^\delta\}dx_i \\
&\quad= c_1\int_0^\tau x_i^{3-\delta}e^{-\frac{(1-\delta)x_i}{\tau}}dx_i + c_2\int_\tau^\infty x_i^2[f_i(x_i)]^{1-\delta/2}dx_i \\
&\qquad + c_3\int_\tau^\infty x_i^2 f_i(x_i)dx_i < \infty.
\end{aligned}
\tag{5.37}
$$

Therefore, conditions (a) and (b) in Theorems 5.4.1 and 5.5.1 are satisfied for the prior distribution (5.34) for any δ arbitrarily close to 2, so that the convergence rates can be arbitrarily close to $O(n^{-1})$ and $O(n^{-1/2})$, respectively, if integer r is sufficiently large.

Acknowledgments. The first author thanks the Natural Sciences and Engineering Research Council of Canada for funding this research while the second author thanks the Canadian International Development Agency for providing financial support for graduate studies at McMaster University.

References

1. Balakrishnan, N. and Basu, A. P. (Eds.) (1995). *The Exponential Distribution: Theory, Methods and Applications,* Langhorne, Pennsylvania: Gordon and Breach Publishers.

2. Fox, R. (1978). Solutions to empirical Bayes squared error loss estimation problems, *Annals of Statistics* **6**, 846–853.

3. Gupta, S. S. and Liang, T. (1986). Empirical Bayes rules for selecting good binomial population, In *Adaptive Statistical Procedures and Related Topics* (Ed., J. Van Ryzin), pp. 110–128, IMS Lecture Notes—Monograph Series, Vol. **8.**

4. Gupta, S. S. and Liang, T. (1988). Empirical Bayes rules for selecting the best binomial population, In *Statistical Decision Theory and Related Topics–IV* (Eds., S. S. Gupta and J. O. Berger), pp. 213–224, New York: Springer-Verlag

5. Gupta, S. S., Liang, T., and Rau, R. (1994). Empirical Bayes rules for selecting the best normal population compared with a control, *Statistics & Decisions,* **12,** 125–147.

6. Johns, M. V. and Van Ryzin, J. (1971). Convergence rates for empirical Bayes two-action problem I. Discrete case, *Annals of Mathematical Statistics,* **42,** 1521–1539.

7. Johns, M. V. and Van Ryzin, J. (1972). Convergence rates for empirical Bayes two-action problem II. Continuous case, *Annals of Mathematical Statistics,* **43,** 934–947.

8. Lin, I. C. and Leu, L. Y. (1994). Empirical Bayes rules for selecting the best exponential population with location parameter, *Communications in Statistics—Theory and Methods,* **23,** 1797–1809.

9. Prasad, B. and Singh, R. S. (1990). Estimation of prior distribution and empirical Bayes estimation in a non-exponential family, *Journal of Statistical Planning and Inference, 24,* 81–86.

10. Robbins, H. (1955). An empirical Bayes approach to statistics, *Proceedings of the Third Berkeley Symposium on Mathematical Statistics and Probability,* Volume **1,** pp. 157–163, Berkeley, CA: University of California Press.

11. Robbins, H. (1963). The empirical Bayes approach to testing statistical hypothesis, *Reviews of the International Statistical Institute, 31,* 195–208.

12. Robbins, H. (1964). The empirical Bayes approach to statistical decision problems, *Annals of Mathematical Statistics, 35,* 1–10.

13. Singh, R. S. (1977). Improvement on some known nonparametric uniformly consistent estimators of derivatives of a density, *Annals of Statistics,* **5,** 394–399.

14. Singh, R. S. (1979). Empirical Bayes estimation in Lebesgue exponential families with rates near the best possible rate, *Annals of Statistics,* **7,** 890–902.

15. Singh, R. S. and Prasad, B. (1989). Uniformly strongly consistent prior-distribution and empirical Bayes estimators with asymptotic optimality and rates in a non-exponential family, *Sankhyā, Series A,* **51,** 334–342.

16. Singh, R. S. and Wei, L. (1992). Empirical Bayes with rates and best rates of convergence in $u(x)c(\theta)\exp(-x/\theta)$-family: estimation case, *Annals of the Institute of Statistical Mathematics,* **44,** 435–449.

6

Empirical Bayes Rules for Selecting the Best Uniform Populations

Wen-Tao Huang and TaChen Liang

Academia Sinica, Taipei, Taiwan, R.O.C.
Wayne State University, Detroit, MI

Abstract: The problem of selecting the best among k (≥ 2) uniform distributions is studied via the empirical Bayes approach. A Bayes selection procedure d^* with respect to an estimated prior distribution is proposed for empirical Bayes selection procedure. We show the consistency of the estimated prior distribution and establish the asymptotic optimality of the procedure d^*. It is found that under certain conditions, the convergence rate is of order $O((\ln n/n)^{1/4})$ where n is the number of accumulated past observations at hand.

Keywords and phrases: Antitonic regression, asymptotic optimality Bayes risk, empirical Bayes, prior distribution, convergence rate

6.1 Introduction

Consider k independent populations $\pi_1, \ldots, \pi_k$, where for each i, population π_i is characterized by the value of a parameter of interest, say θ_i. Let $\theta_{[1]} \leq \ldots \leq \theta_{[k]}$ denote the ordered values of the parameters $\theta_1, \ldots, \theta_k$. It is assumed that the exact pairing between the ordered and the unordered parameters is unknown. A population π_i with $\theta_i = \theta_{[k]}$ is called the best population. In many practical situations, the experimenter is interested in the selection of the best population. The problem of selecting the best population was studied in papers pioneered by Bechhofer (1954) using the indifference zone approach and by Gupta (1956) employing the subset selection approach. A discussion of their differences and various modifications that have taken place since then can be found in Gupta and Panchapakesan (1979).

Consider a situation in which an experimenter is repeatedly dealing with the same selection problem independently. In such instances, the experimenter may

employ the empirical Bayes idea of Robbins (1956) by formulating the component problems in the sequence as Bayes selection problems with respect to an
unknown prior distribution, and then using the accumulated past observations
to improve the decision at each stage. Empirical Bayes selection procedures
have been derived for subset selection goals by Deely (1965) and Huang (1975).
Recently, Gupta and Hsiao (1983), Gupta and Leu (1983) and Gupta and Liang
(1986, 1988) have studied some selection problems using the empirical Bayes
approach. Many such empirical Bayes selection procedures have been shown
to be asymptotically optimal in the sense that the risk for the n-th selection
problem converges to the optimal Bayes risk which would have been obtained
if the prior distribution was fully known and the Bayes procedure with respect
to this prior distribution was used.

This paper is organized as follows. In Section 6.2, we describe the statistical
framework of the concerned empirical Bayes selection problem. An estimator for
the unknown prior distribution is proposed in Section 6.3; the corresponding
consistency property is also established. We then use the related Bayes selection procedure of the estimated prior distribution as the proposed empirical
Bayes selection procedure. The associated asymptotic optimality is investigated
in Section 6.4. Finally, in Section 6.5, we consider a particular loss function
$L(\boldsymbol{\theta}, i) = \theta_{[k]} - \theta_i$, where i denotes the index of the selected population, and
study the rate of convergence of the proposed empirical Bayes selection pro-
cedure. It is shown that under certain conditions the associated convergence
rate is of order $O((\ln n/n)^{1/4})$, where n is the number of accumulated past
experience at hand.

6.2   Empirical Bayes Framework of the Selection Problem

Suppose that there are k treatments for a certain disease and the upper bound of
effect for the i-th treatment follows an unknown distribution G_i $(i = 1, 2, \ldots, k)$.
The effectiveness of the i-th treatment has been tested on n subjects. For different treatments, different groups of subjects are used and it can be so arranged
that the effects of different treatments are independent. Let θ_i denote the unknown parameter that the effect of the i-th treatment can reach at most. Let Y_{ij}
denote an observable result of the i-th treatment on the subject j. It is natural
to assume that for fixed i, Y_{ij}, $j = 1, 2, \ldots, n$, are i.i.d. from a uniform distribution $U(0, \theta_i)$. Based on those random observations Y_{ij}, we are interested in
determining which treatment has the largest effect θ_i. This becomes a ranking
and selection problem for k uniform populations. In a non-Bayesian approach,
Barr and Rizvi (1966) and McDonald (1976,1979) have considered some particular statistics for the selection procedures. A comprehensive bibliography on

the subject can be found in Dudewicz and Koo (1982).

Consider k independent populations $\pi_1, \ldots, \pi_k$. For each $i = 1, \ldots, k$, let X_i denote a random observation arising from population π_i. Assume that X_i has a uniform distribution with pdf $f_i(x|\theta_i) = \theta_i^{-1} I_{(0,\theta_i)}(x)$, where the value of the parameter θ_i is positive but unknown. Let $\theta_{[1]} \leq \ldots \leq \theta_{[k]}$ denote the ordered values of the parameters $\theta_1, \ldots, \theta_k$. It is assumed that the exact pairing between the ordered and the unordered parameters is unknown. A population π_i with $\theta_i = \theta_{[k]}$ is considered as the best population. Our goal is to select the best population. We formulate this selection problem in the empirical Bayes framework as follows.

Let $\mathcal{A} = \{i|i = 1, \ldots, k\}$ be the action space. When action i is taken, it means that population π_i is selected as the best population. Let $\Omega = \{\boldsymbol{\theta} = (\theta_1, \ldots, \theta_k)|\theta_i > 0, i = 1, \ldots, k\}$ be the parameter space. For the parameter $\boldsymbol{\theta}$ and the action i, the loss function $L(\boldsymbol{\theta}, i)$ is defined as:

$$L(\boldsymbol{\theta}, i) = \ell(\theta_{[k]} - \theta_i), \tag{6.1}$$

where $\ell(x)$ is a nondecreasing function of the variable x for $x \geq 0$ with $\ell(0) = 0$.

For each $i = 1, \ldots, k, \theta_i$ is viewed as a realization of a random variable Θ_i having an unknown continuous prior distribution G_i on $(0, \infty)$. It is assumed that the k random variables $\Theta_1, \ldots, \Theta_k$ are mutually independent. Therefore, $(\Theta_1, \ldots, \Theta_k)$ has a joint prior distribution $\boldsymbol{G}(\boldsymbol{\theta}) = \prod_{i=1}^{k} G_i(\theta_i)$ over the parameter space Ω. It is also assumed that $\int \ell(\theta_{[k]}) d\boldsymbol{G}(\boldsymbol{\theta}) < \infty$ to guarantee the finiteness of the Bayes risks.

Let $\mathcal{X}$ be the sample space generated by $\boldsymbol{X} = (X_1, \ldots, X_k)$. A selection procedure $\boldsymbol{d} = (d_1, \ldots, d_k)$ is defined to be a mapping from the sample space $\mathcal{X}$ to the product space $[0, 1]^k$ such that for each observation $\boldsymbol{x} = (x_1, \ldots, x_k) \in \mathcal{X}$, the function $\boldsymbol{d}(\boldsymbol{x}) = (d_1(\boldsymbol{x}), \ldots, d_k(\boldsymbol{x}))$ satisfies that $0 \leq d_i(\boldsymbol{x}) \leq 1, i = 1, \ldots, k$, and $\sum_{i=1}^{k} d_i(\boldsymbol{x}) = 1$. Note that $d_i(\boldsymbol{x})$ is the probability of selecting population π_i as the best population when $\boldsymbol{x}$ is observed.

Let D be the class of all the selection procedures defined previously. For each $\boldsymbol{d} \in D$, let $r(\boldsymbol{G}, \boldsymbol{d})$ denote the associated Bayes risk. Then $r(\boldsymbol{G}) \equiv \inf_{\boldsymbol{d} \in D} r(\boldsymbol{G}, \boldsymbol{d})$ is the minimum Bayes risk among the class D. Any selection procedure $\boldsymbol{d}$ such that $r(\boldsymbol{G}, \boldsymbol{d}) = r(\boldsymbol{G})$ is called a Bayes selection procedure.

In this statistical model, for selection procedure $\boldsymbol{d}$, the associated Bayes risk $r(\boldsymbol{G}, \boldsymbol{d})$ can be written as:

$$
\begin{aligned}
r(\boldsymbol{G}, \boldsymbol{d}) &= \int_{\mathcal{X}} \sum_{i=1}^{k} d_i(\boldsymbol{x}) [\int_{\Omega} \ell(\theta_{[k]} - \theta_i) \boldsymbol{f}(\boldsymbol{x}|\boldsymbol{\theta}) d\boldsymbol{G}(\boldsymbol{\theta})] d\boldsymbol{x} \\
&= \int_{\mathcal{X}} \sum_{i=1}^{k} d_i(\boldsymbol{x}) [\int_{\Omega} \ell(\theta_{[k]} - \theta_i) d\boldsymbol{G}(\boldsymbol{\theta}|\boldsymbol{x})] \boldsymbol{f}(\boldsymbol{x}) d\boldsymbol{x}, \tag{6.2}
\end{aligned}
$$

where $\boldsymbol{f}(\boldsymbol{x}|\boldsymbol{\theta}) = \prod_{i=1}^{k} f_i(x_i|\theta_i)$, $\boldsymbol{f}(\boldsymbol{x}) = \prod_{i=1}^{k} f_i(x_i)$, $f_i(x_i) = \int f_i(x_i|\theta) dG_i(\theta)$

$= \int_{x_i}^{\infty} \theta^{-1} dG_i(\theta)$ is the marginal pdf of the random variable X_i, and $\mathbf{G}(\boldsymbol{\theta}|\boldsymbol{x})$ is the posterior joint distribution of $\boldsymbol{\Theta} = (\Theta_1, \ldots, \Theta_k)$ given $\boldsymbol{X} = \boldsymbol{x}$.

From (6.2), a randomized Bayes selection procedure $\boldsymbol{d_G} = (d_{1\boldsymbol{G}}, \ldots, d_{k\boldsymbol{G}})$ can be obtained as follows:

For each $\boldsymbol{x} \in \mathcal{X}$, let

$$\triangle_i(\boldsymbol{x}) = \int_{\Omega} \ell(\theta_{[k]} - \theta_i) \boldsymbol{f}(\boldsymbol{x}|\boldsymbol{\theta}) d\boldsymbol{G}(\boldsymbol{\theta}), i = 1, \ldots, k. \tag{6.3}$$

Also, let

$$A(\boldsymbol{x}) = \{i | \triangle_i(\boldsymbol{x}) = \min_{1 \leq j \leq k} \triangle_j(\boldsymbol{x})\}. \tag{6.4}$$

We then define $d_{i\boldsymbol{G}}, i = 1, \ldots, k$, as follows: For $\boldsymbol{x} \in \mathcal{X}$,

$$d_{i\boldsymbol{G}}(\boldsymbol{x}) = \begin{cases} |A(\boldsymbol{x})|^{-1} & \text{if } i \in A(\boldsymbol{x}), \\ 0 & \text{otherwise,} \end{cases} \tag{6.5}$$

where $|A(\boldsymbol{x})|$ denotes the cardinality of the set $A(\boldsymbol{x})$.

It is to be noted that, in fact, there would not be any ties occurring in (6.4) since the prior is continuous. So, here the set $A(\boldsymbol{x})$ is a singleton and $|A(\boldsymbol{x})|^{-1} = 1$.

In the empirical Bayes framework, it is assumed that certain past observations are available. For each $j = 1, 2, \ldots, n$, we let $\boldsymbol{X}_j = (X_{1j}, \ldots, X_{kj})$ denote the j-th previous observations obtained from the k populations, where $\boldsymbol{X}_1, \ldots, \boldsymbol{X}_n$ are i.i.d. with marginal pdf $\boldsymbol{f}(\boldsymbol{x})$. Hence, for each $i = 1, \ldots, k, X_{i1},$ $\ldots, X_{in}$ are i.i.d. with marginal pdf $f_i(x)$

Let $\mathcal{X}(n)$ denote the sample space generated by $\boldsymbol{X}(n) = (\boldsymbol{X}_1, \ldots, \boldsymbol{X}_n)$, and let $\boldsymbol{x}(n)$ denote an observed value of $\boldsymbol{X}(n)$. An empirical Bayes selection procedure $\boldsymbol{d}_n = (d_{1n}, \ldots, d_{kn})$ is defined to be a mapping from $\mathcal{X} \times \mathcal{X}(n)$ into the product space $[0, 1]^k$ such that for each fixed $(\boldsymbol{x}, \boldsymbol{x}(n)) \in \mathcal{X} \times \mathcal{X}(n), 0 \leq d_{in}(\boldsymbol{x}, \boldsymbol{x}(n)) \leq 1$ for all $i = 1, \ldots, k$, and $\sum_{i=1}^{k} d_{in}(\boldsymbol{x}, \boldsymbol{x}(n)) = 1$. That is, $d_{in}(\boldsymbol{x}, \boldsymbol{x}(n))$ is the probability of selecting population π_i as the best population when $(\boldsymbol{x}, \boldsymbol{x}(n))$ is observed. For $\boldsymbol{x}(n)$ to be fixed, the empirical Bayes selection procedure $\boldsymbol{d}_n$ can be viewed as a selection procedure as defined previously. For given $\boldsymbol{X}(n) = \boldsymbol{x}(n)$, we denote the conditional Bayes risk of the empirical Bayes selection procedure $\boldsymbol{d}_n$ by $r(\boldsymbol{G}, \boldsymbol{d}_n|\boldsymbol{x}(n))$, and the overall Bayes risk of $\boldsymbol{d}_n$ by $r(\boldsymbol{G}, \boldsymbol{d}_n)$. That is, from (6.2),

$$r(\boldsymbol{G}, \boldsymbol{d}_n|\boldsymbol{x}(n)) = \int_{\mathcal{X}} \sum_{i=1}^{k} d_{in}(\boldsymbol{x}, \boldsymbol{x}(n)) [\int_{\Omega} \ell(\theta_{[k]} - \theta_i) \boldsymbol{f}(\boldsymbol{x}|\boldsymbol{\theta}) d\boldsymbol{G}(\boldsymbol{\theta})] d\boldsymbol{x}$$

$$= \int_{\mathcal{X}} \sum_{i=1}^{k} d_{in}(\boldsymbol{x}, \boldsymbol{x}(n)) [\int_{\Omega} \ell(\theta_{[k]} - \theta_i) d\boldsymbol{G}(\boldsymbol{\theta}|\boldsymbol{x})] \boldsymbol{f}(\boldsymbol{x}) d\boldsymbol{x}$$

and

$$r(\boldsymbol{G}, \boldsymbol{d}_n) = E[r(\boldsymbol{G}, \boldsymbol{d}_n | \boldsymbol{X}(n))]$$

where the expectation E is taken with respect to $\boldsymbol{X}(n)$.

Note that $r(\boldsymbol{G}, \boldsymbol{d}_n | \boldsymbol{x}(n)) - r(\boldsymbol{G}) \geq 0$, hence we use the nonnegative difference $r(\boldsymbol{G}, \boldsymbol{d}_n) - r(\boldsymbol{G})$ as a measure to evaluate the performance of the empirical Bayes selection procedure $\boldsymbol{d}_n$.

Definition 6.2.1 (a)   A sequence of empirical Bayes selection procedures $\{\boldsymbol{d}_n\}_{n=1}^{\infty}$ is said to be asymptotically optimal relative to the prior distribution $\boldsymbol{G}$ if $\lim_{n \to \infty} r(\boldsymbol{G}, \boldsymbol{d}_n) = r(\boldsymbol{G})$.

(b)   A sequence of empirical Bayes selection procedures $\{\boldsymbol{d}_n\}_{n=1}^{\infty}$ is said to be asymptotically optimal of order β_n relative to the prior distribution $\boldsymbol{G}$ if $r(\boldsymbol{G}, \boldsymbol{d}_n) - r(\boldsymbol{G}) = O(\beta_n)$ where $\{\beta_n\}$ is a sequence of positive numbers such that $\lim_{n \to \infty} \beta_n = 0$.

Gupta and Hsiao (1983) and Gupta and Leu (1983) investigated empirical Bayes selection procedures for uniform populations. They both considered the Bayes selection procedures in terms of the unknown prior distribution, and then use the accumulated past observations to estimate the behavior of the Bayes selection procedures directly. However, our approach here is: We first construct an estimator for the unknown prior distribution on the basis of the accumulated data, and then adopt the Bayes selection procedure of the estimated prior distribution as our empirical Bayes selection procedure. The advantage of this approach is that the proposed empirical Bayes selection procedure will always be Bayes with respect to a (the estimated) prior distribution, and it may share certain properties possessed by the true Bayes selection procedure, which might not be available by estimating the behavior of the Bayes selection procedure directly.

6.3   Estimation of the Prior Distribution

For each $i = 1, \ldots, k$, let $F_i(x)$ be the marginal cumulative distribution function associated with the marginal pdf $f_i(x)$. Fox (1978) obtained a relation between the prior distribution G_i and the marginal functions F_i and f_i as follows:

$$G_i(x) = F_i(x) - x f_i(x). \tag{6.6}$$

Susarla and O'Bryan (1979) used the relation (6.6) to construct a consistent estimator of the prior distribution G_i; the proposed estimator itself, as mentioned in that paper, is not a distribution function since it may take negative values.

Note that $f_i(x) = 0$ for $x < 0$ and $f_i(x) = \int_x^\infty \theta^{-1} dG_i(\theta)$ for $x \geq 0$, which is a nonincreasing function of x for $x \geq 0$. Therefore, $F_i(0) = 0$ and $F_i(x)$ is a concave function on $[0, \infty)$. This property and the relation (6.6) will be used for the construction of an estimator of the prior distribution G_i.

Let $F_{in}(x)$ be the empirical distribution based on the past observations $X_{i1}, \ldots, X_{in}$. Though $F_{in}(x)$ is a consistent estimator of the marginal distribution function $F_i(x)$, it is not concave in general, although $F_i(x)$ is. We consider a smoothed version of $F_{in}(x)$.

6.3.1   Construction of the prior estimator

Let $\{\alpha_n\}$ be a sequence of decreasing positive numbers so that $\lim_{n \to \infty} \alpha_n = 0$. For each n, let $C_{n,j} = j\alpha_n, j = 0, 1, 2, \ldots$. Define function $f_{in}^*(C_{n,j}), j = 0, 1, 2, \ldots$, recursively as follows:

$$f_{in}^*(C_{n,0}) = \max_{j \geq 1}\{\frac{F_{in}(C_{n,j})}{j\alpha_n}\};$$

$$f_{in}^*(C_{n,j}) = \max_{m \geq j}\{\frac{F_{in}(C_{n,m}) - \alpha_n \sum_{\ell=0}^{j-1} f_{in}^*(C_{n,\ell})}{(m - j + 1)\alpha_n}\}, \quad j = 1, 2 \ldots \quad (6.7)$$

We extend the definition of the function f_{in}^* on the real line as:

$$f_{in}^*(t) = \begin{cases} 0 & \text{if } t < 0; \\ f_{in}^*(C_{n,j}) & \text{if } C_{n,j} \leq t < C_{n,j+1}, \quad j = 0, 1, \ldots. \end{cases} \quad (6.8)$$

We then define a function $F_{in}^*(x)$ on the real line as:

$$F_{in}^*(x) = \int_{-\infty}^x f_{in}^*(t) dt. \quad (6.9)$$

It should be noted that $\{f_{in}^*(C_{n,j})\}_{j=0}^\infty$ is the antitonic regression of $\{[F_{in}(C_{n,j+1}) - F_{in}(C_{n,j})]/\alpha_n\}_{j=0}^\infty$ with equal weight α_n, see Puri and Singh (1990). Therefore, $f_{in}^*(C_{n,j})$ is nonnegative and nonincreasing in j for $j = 0, 1, 2, \ldots$, and by (6.8), $f_{in}^*(t)$ is nonnegative and nonincreasing in t for $t \geq 0$. Also, $f_{in}^*(t) = 0$ for $t \geq C_{n,j^*}$, where $j^* = \min\{j | \max_{1 \leq m \leq n} X_{im} < C_{n,j}\}$. From (6.8) and (6.9), the function $F_{in}^*(x)$ is nondecreasing in x with $F_{in}^*(0) = 0$. Also, $F_{in}^*(x)$ is concave on $[0, \infty)$. From Barlow *et al.* (1972),

$$F_{in}^*(C_{n,j}) \geq F_{in}(C_{n,j}) \text{ for all } j = 0, 1, 2, \ldots$$
$$F_{in}^*(C_{n,j^*}) = F_{in}(C_{n,j^*}) = 1$$
$$\sup_{j \geq 0} |F_{in}^*(C_{n,j}) - F_i(C_{n,j})| \leq \sup_{j \geq 0} |F_{in}(C_{n,j}) - F_i(C_{n,j})|. \quad (6.10)$$

Hence, $F_{in}^*(x)$ can be viewed as a distribution function and $f_{in}^*(x)$ as the associated probability density function.

Motivated by the relation (6.6), we define

$$G_{in}^*(x) = F_{in}^*(x) - x f_{in}^*(x) \tag{6.11}$$

and propose it as an estimator for the prior distribution G_i. Note that $G_{in}^*(0) = 0$, $G_{in}^*(C_{n,j^*}) = 1$ and $G_{in}^*(x)$ is nondecreasing in x which can be seen by the properties associated with the function $f_{in}^*(x)$ just discussed previously. Hence, $G_{in}^*(x)$ is a distribution function.

6.3.2 Consistency of the estimator

In this subsection, we investigate the consistency of the estimator $G_{in}^*(x)$. We claim the following result.

Theorem 6.3.1 *(a) Suppose there exists a positive M such that $\int_x^y \frac{1}{\theta} dG_i(\theta) \leq M(y-x)$ for all $0 \leq x < y < \infty$. Let the sequence $\{\alpha_n\}$ be chosen such that $\alpha_n \geq n^{-\alpha}$ where $0 < \alpha < \frac{1}{2}$ and $\lim_{n\to\infty} \alpha_n = 0$. Then,*

$$G_{in}^*(x) \longrightarrow G_i(x) \quad a.e.$$

(b) Under assumption (a) and furthermore, if $G_i(Q) = 1$ for some finite positive number Q, then

$$\sup_{x>0} |G_{in}^*(x) - G_i(x)| \longrightarrow 0 \quad a.e.$$

Remark: Let $h_i(x) = \int_x^\infty \frac{1}{\theta} dG_i(x)$. Then the condition on $G_i(x)$ in Theorem 6.3.1 becomes $(h_i(x) - h_i(y))/(y-x) \leq M$ for all $0 \leq x < y < \infty$, i.e. $h_i(x)$ satisfies a Lipschitz condition of order 1. If $G_i(x)$ has a density $g_i(x)$, it becomes $g_i(x)/x \leq M$ for all $x > 0$. The class of densities satisfying this condition includes $\Gamma(\alpha, \beta)$ with $\alpha \geq 2$, Inverse Gamma(α, β) and Beta(α, β) with $\alpha \geq 1$, and others.

Before we prove the theorem, we first present some preliminary results.

Lemma 6.3.1 (Schuster (1969)) *For each fixed positive number ε,*

$$P\{\sup_{x>0} |F_{in}(x) - F_i(x)| \geq \varepsilon\} \leq c \exp(-2n\varepsilon^2),$$

where the constant c is independent of the distribution F_i.

Lemma 6.3.2 *Under the assumptions of Theorem 6.3.1 part(a), the following holds.*

(a) For each $j = 1, 2, \ldots$,
$$\left| \frac{F_i(C_{n,j}) - F_i(C_{n,j-1})}{\alpha_n} - f_i(C_{n,j-1}) \right| \leq M\alpha_n;$$
$$\left| \frac{F_i(C_{n,j}) - F_i(C_{n,j-1})}{\alpha_n} - f_i(C_{n,j}) \right| \leq M\alpha_n.$$

(b) For each $x \in [C_{n,j-1}, C_{n,j}), j = 1, 2, \ldots$, and fixed $\varepsilon > 0$, let
$$S_{in}(x, \varepsilon, j) = \alpha_n[G_i(x) + \varepsilon] - C_{n,j}F_i(C_{n,j-1}) + C_{n,j-1}F_i(C_{n,j}),$$
$$T_{in}(x, \varepsilon, j) = \alpha_n[G_i(x) - \varepsilon] - C_{n,j}F_i(C_{n,j-1}) + C_{n,j-1}F_i(C_{n,j}).$$

Then, there exists an $N(x, \varepsilon)$ such that for $n > N(x, \varepsilon)$, we have: $S_{in}(x, \varepsilon, j) > \alpha_n \varepsilon / 2$ and $T_{in}(x, \varepsilon, j) < -\alpha_n \varepsilon$.

PROOF. (a)   By the nonincreasing property of the marginal pdf $f_i(x)$,

$$
|\frac{F_i(C_{n,j}) - F_i(C_{n,j-1})}{\alpha_n} - f_i(C_{n,j-1})| \leq f_i(C_{n,j-1}) - f_i(C_{n,j})
$$

$$
= \int_{C_{n,j-1}}^{C_{n,j}} \frac{1}{\theta} dG_i(\theta)
$$

$$
\leq M \alpha_n,
$$

where the second inequality is obtained due to the assumption. The other part of (a) can be proved similarly.

(b) Note that

$$
S_{in}(x, \varepsilon, j) = \alpha_n [G_i(x) - G_i(C_{n,j-1})]
$$

$$
+ \alpha_n \{ \varepsilon + C_{n,j-1} [\frac{F_i(C_{n,j}) - F_i(C_{n,j-1})}{\alpha_n} - f_i(C_{n,j-1})] \},
$$

where $G_i(x) - G_i(C_{n,j-1}) \geq 0$, since G_i is a distribution function and $x \geq C_{n,j-1}$. From part (a) there exists some $N(x, \varepsilon)$. So that

$$
|C_{n,j-1} [\frac{F_i(C_{n,j}) - F_i(C_{n,j-1})}{\alpha_n} - f_i(C_{n,j-1})]| \leq xM\alpha_n < \frac{\varepsilon}{2} \text{ as } n > N(x, \varepsilon).
$$

Therefore, as $n > N(x, \varepsilon)$, $S_{in}(x, \varepsilon, j) > \frac{\alpha_n \varepsilon}{2}$.

Also,

$$
T_{in}(x, \varepsilon, j) = \alpha_n [G_i(x) - G_i(C_{n,j})] - \alpha_n \{ \varepsilon + [\frac{F_i(C_{n,j}) - F_i(C_{n,j-1})}{\alpha_n}
$$

$$
- f_i(C_{n,j})] \}
$$

$$
< -\alpha_n \varepsilon,
$$

since $x < C_{n,j}$ and so $G_i(x) - G_i(C_{n,j}) \leq 0$, and $\frac{F_i(C_{n,j}) - F_i(C_{n,j-1})}{\alpha_n} - f_i(C_{n,j}) \geq 0$.

PROOF OF THEOREM 6.3.1(A).

For each $x > 0$, let $[C_{n,j-1}, C_{n,j})$ be the interval containing the point x. For the fixed $\varepsilon > 0$,

$$
\{ |G_{in}^*(x) - G_i(x)| > \varepsilon \}
$$

$$
= \{ G_{in}^*(x) - G_i(x) < -\varepsilon \} \cup \{ G_{in}^*(x) - G_i(x) > \varepsilon \}. \tag{6.12}
$$

From (6.8) and (6.9), one can obtain the following:

$$F_{in}^*(x) = F_{in}^*(C_{n,j-1}) + (x - C_{n,j-1})f_{in}^*(x);$$
$$f_{in}^*(x) = [F_{in}^*(C_{n,j}) - F_{in}^*(C_{n,j-1})]/\alpha_n. \qquad (6.13)$$

Define $R_{in}(y) = F_{in}^*(y) - F_i(y)$. Suppose that $C_{n,j-1} > 0$. From (6.10), (6.11) and (6.13), for $n > N(x,\varepsilon)$, straightforward computation leads to

$$
\begin{aligned}
&\{G_{in}^*(x) - G_i(x) > \varepsilon\} \\
=\ &\{C_{n,j}F_{in}^*(C_{n,j-1}) - C_{n,j-1}F_{in}^*(C_{n,j}) > \alpha_n[G_i(x) + \varepsilon]\} \\
=\ &\{C_{n,j}R_{in}(C_{n,j-1}) - C_{n,j-1}R_{in}(C_{n,j}) > S_{in}(x,\varepsilon,j)\} \\
\subset\ &\{C_{n,j}R_{in}(C_{n,j-1}) - C_{n,j-1}R_{in}(C_{n,j}) > \alpha_n\varepsilon/2\} \\
\subset\ &\{R_{in}(C_{n,j-1}) > \alpha_n\varepsilon/(4C_{n,j})\ \text{ or }\ R_{in}(C_{n,j}) < -\alpha_n\varepsilon/(4C_{n,j-1})\} \\
\subset\ &\{\sup_{m\geq 1} |R_{in}(C_{n,m})| > \alpha_n\varepsilon/(4(x+\alpha_n))\} \\
\subset\ &\{\sup_{m\geq 1} |F_{in}(C_{n,m}) - F_i(C_{n,m})| > \alpha_n\varepsilon/(4(x+\alpha_n))\} \\
\subset\ &\{\sup_{y>0} |F_{in}(y) - F_i(y)| > \alpha_n\varepsilon/(4(x+\alpha_n))\}. \qquad (6.14)
\end{aligned}
$$

Similarly, for $n > N(x,\varepsilon)$, we can obtain the following:

$$
\begin{aligned}
&\{G_{in}^*(x) - G_i(x) < -\varepsilon\} \\
=\ &\{C_{n,j}R_{in}(C_{n,j-1}) - C_{n,j-1}R_{in}(C_{n,j}) < T_{in}(x,\varepsilon,j)\} \\
\subset\ &\{C_{n,j}R_{in}(C_{n,j-1}) - C_{n,j-1}R_{in}(C_{n,j}) < -\alpha_n\varepsilon\} \\
\subset\ &\{R_{in}(C_{n,j-1}) < -\alpha_n\varepsilon/(2C_{n,j})\ \text{ or }\ R_{in}(C_{n,j}) > \alpha_n\varepsilon/(2C_{n,j-1})\} \\
\subset\ &\{\sup_{y>0} |F_{in}(y) - F_i(y)| > \alpha_n\varepsilon/(2(x+\alpha_n))\}. \qquad (6.15)
\end{aligned}
$$

By (6.14) and (6.15), we then have

$$\{|G_{in}^*(x) - G_i(x)| > \varepsilon\} \subset \{\sup_{y>0} |F_{in}(y) - F_i(y)| > \alpha_n\varepsilon/(4(x+\alpha_n))\}. \qquad (6.16)$$

Note that (6.16) still holds when $C_{n,j-1} = 0$. Therefore by Lemmas 6.3.1 there is some $N(x,\varepsilon)$ so that for $n > N(x,\varepsilon)$,

$$
\begin{aligned}
&P\{|G_{in}^*(x) - G_i(x)| > \varepsilon\} \\
\leq\ &P\{\sup_{y>0} |F_{in}(y) - F_i(y)| > \alpha_n\varepsilon/(4(x+\alpha_n))\} \\
\leq\ &c\exp[-2n\alpha_n^2\varepsilon^2/(4(x+\alpha_n))^2] \\
\leq\ &c\exp[-2\ln n] \\
=\ &\frac{c}{n^2} \qquad (6.17)
\end{aligned}
$$

which implies that

$$\sum_{n=1}^{\infty} P\{|G_{in}^*(x) - G_i(x)| > \varepsilon\} < \infty. \qquad (6.18)$$

By the Borel-Cantelli Lemma, we conclude that

$$P\{|G_{in}^*(x) - G_i(x)| > \varepsilon \ \ i.o.\} = 0,$$

where $i.o.$ denotes infinitely often.

(b)   Suppose that $G_i(Q) = 1$ for some positive number Q. By the definition of $F_{in}^*(x)$ and $f_{in}^*(x)$, $f_{in}^*(x) = 0$ and $F_{in}^*(x) = 1$ if $x \geq Q$. Therefore $G_{in}^*(x) = 1$ if $x \geq Q$. Hence,

$$P\{|G_{in}^*(x) - G_i(x)| > \varepsilon\} = 0 \ \ \text{if } x \geq Q.$$

Examining the proof of part (a), we can see that for any $x \in (0, Q)$, and for sufficiently large n, $n > N(\varepsilon)$, say, the following holds:

$$\{|G_{in}^*(x) - G_i(x)| > \varepsilon\} \subset \left\{\sup_{y>0}|F_{in}(y) - F_i(y)| > \frac{\alpha_n\varepsilon}{4(Q + \alpha_n)}\right\}.$$

Therefore, for $n > N(\varepsilon)$

$$\{\sup_{x>0}|G_{in}^*(x) - G_i(x)| > \varepsilon\}$$

$$= \bigcup_{0<x<Q} \{|G_{in}^*(x) - G_i(x)| > \varepsilon\}$$

$$\subset \left\{\sup_{y>0}|F_{in}(y) - F_i(y)| > \frac{\alpha_n\varepsilon}{4(Q + \alpha_n)}\right\}.$$

We also have,

$$P\{\sup_{x>0}|G_{in}^*(x) - G_i(x)| > \varepsilon\}$$

$$\leq P\left\{\sup_{y>0}|F_{in}(y) - F_i(y)| > \frac{\alpha_n\varepsilon}{4(Q + \alpha_n)}\right\}$$

$$\leq c \exp\left\{-\frac{n\alpha_n^2\varepsilon^2}{8(Q + \alpha_n)^2}\right\}$$

$$\leq \frac{c}{n^2}.$$

Hence,

$$\sum_{n=1}^{\infty} P\{\sup_{x>0}|G_{in}^*(x) - G_i(x)| > \varepsilon\} < \infty$$

and thus

$$P\{\sup_{x>0}|G_{in}^*(x) - G_i(x)| > \varepsilon \ \ i.o.\} = 0.$$

6.4   The Proposed Empirical Bayes Selection Procedure

Define $\boldsymbol{G}_n^*(\boldsymbol{\theta}) = \prod_{i=1}^{k} G_{in}^*(\theta_i)$, and for each $\boldsymbol{x} \in \mathcal{X}$ and each $n = 1, 2, \ldots$, let

$$\triangle_{in}(\boldsymbol{x}) = \int_{\Omega} \ell(\theta_{[k]} - \theta_i) \boldsymbol{f}(\boldsymbol{x}|\boldsymbol{\theta}) d\boldsymbol{G}_n^*(\boldsymbol{\theta}), i = 1, \ldots, k \tag{6.19}$$

and let

$$A_n(\boldsymbol{x}) = \{i | \triangle_{in}(\boldsymbol{x}) = \min_{1 \le j \le k} \triangle_{jn}(\boldsymbol{x})\}. \tag{6.20}$$

Since the empirical estimate of the prior is discrete, so ties may occur and thus randomization has to be taken into account.

We propose a randomized empirical Bayes selection procedure $\boldsymbol{d}_n^* = (d_{1n}^*, \ldots, d_{kn}^*)$ defined as follows: For each $i = 1, \ldots, k$,

$$d_{in}^*(\boldsymbol{x}, \boldsymbol{X}(n)) = \begin{cases} |A_n(\boldsymbol{x})|^{-1} & \text{if } i \in A_n(\boldsymbol{x}); \\ 0 & \text{otherwise.} \end{cases} \tag{6.21}$$

Note that the definition of the empirical Bayes selection procedure $\boldsymbol{d}_n^*$ is a mimic of the Bayes selection procedure $\boldsymbol{d_G}$. In the remainder of this section, we study the asymptotic optimality of the empirical Bayes selection procedure $\boldsymbol{d}_n^*$. First we have the following.

Lemma 6.4.1 *Let $\triangle_i(\boldsymbol{x})$ and $\triangle_{in}(\boldsymbol{x})$ be defined by (6.3) and (6.19), respectively. Suppose that*

(a) $\ell(\theta_{[k]}) \boldsymbol{f}(\boldsymbol{x}|\boldsymbol{\theta})$ *is a bounded function of the variable $\boldsymbol{\theta}$, and*
(b)   *the assumption of Theorem 6.3.1 (a) holds for each $i = 1, \ldots, k$.*

Then, $\triangle_{in}(\boldsymbol{x}) \xrightarrow{P} \triangle_i(\boldsymbol{x})$ (w.r.t. $\boldsymbol{X}(n)$) for each $i = 1, \ldots, k$.

PROOF. This is a direct application of Theorem 6.3.1 (a) and Proposition 18 of Royden (1963, cf. p. 232) by noting that $\ell(x)$ is a nonnegative, nondecreasing function of $x, x \ge 0$.

Next, we have the following inequality:

$$\begin{aligned} 0 \quad &\le \quad r(\boldsymbol{G}, \boldsymbol{d}_n^*) - r(\boldsymbol{G}) \\ &\le \quad \int_{\mathcal{X}} \sum_{i=1}^{k} \sum_{j \in E_i(\boldsymbol{x})} \delta_{ij}(\boldsymbol{x}) P\{\triangle_{in}(\boldsymbol{x}) \le \triangle_{jn}(\boldsymbol{x})\} d\boldsymbol{x} \\ &\quad + \int_{\mathcal{X}} \sum_{i=1}^{k} \sum_{j \in H_i(\boldsymbol{x})} [-\delta_{ij}(\boldsymbol{x})] P\{\triangle_{jn}(\boldsymbol{x}) \le \triangle_{in}(\boldsymbol{x})\} d\boldsymbol{x}, \quad (6.22) \end{aligned}$$

where $\delta_{ij}(\boldsymbol{x}) = \triangle_i(\boldsymbol{x}) - \triangle_j(\boldsymbol{x})$, $E_i(\boldsymbol{x}) = \{j|\delta_{ij}(\boldsymbol{x}) > 0\}$ and $H_i(\boldsymbol{x}) = \{j|\delta_{ij}(\boldsymbol{x}) < 0\}$.

Suppose that $\int \ell(\theta_{[k]})dG(\boldsymbol{\theta}) < \infty$. To show $r(\boldsymbol{G}, \boldsymbol{d}_n^*) - r(\boldsymbol{G}) \to 0$ as $n \to \infty$, by the Lebesgue Convergence Theorem, it suffices to prove that: For each $\boldsymbol{x} \in \mathcal{X}$,

$$P\{\triangle_{in}(\boldsymbol{x}) \le \triangle_{jn}(\boldsymbol{x})\} \to 0 \quad \text{for each } j \in E_i(\boldsymbol{x}), \text{ and}$$
$$P\{\triangle_{jn}(\boldsymbol{x}) \le \triangle_{in}(\boldsymbol{x})\} \to 0 \quad \text{for each } j \in H_i(\boldsymbol{x}).$$

For $j \in E_i(\boldsymbol{x}), \delta_{ij}(\boldsymbol{x}) > 0$. Then,

$$P\{\triangle_{in}(\boldsymbol{x}) \le \triangle_{jn}(\boldsymbol{x})\}$$
$$\le P\{\triangle_{in}(\boldsymbol{x}) - \triangle_i(\boldsymbol{x}) < -\delta_{ij}(\boldsymbol{x})/2\} + P\{\triangle_{jn}(\boldsymbol{x}) - \triangle_j(\boldsymbol{x}) > \delta_{ij}(\boldsymbol{x})/2\}.$$

If both $\triangle_{in}(\boldsymbol{x}) \xrightarrow{P} \triangle_i(\boldsymbol{x})$ and $\triangle_{jn}(\boldsymbol{x}) \xrightarrow{P} \triangle_j(\boldsymbol{x})$, then $P\{\triangle_{in}(\boldsymbol{x}) \le \triangle_{jn}(\boldsymbol{x})\} \to 0$ as $n \to \infty$.

Similarly, for $j \in H_i(\boldsymbol{x}), P\{\triangle_{jn}(\boldsymbol{x}) \le \triangle_{in}(\boldsymbol{x})\} \to 0$ as $n \to \infty$ if both $\triangle_{in}(\boldsymbol{x}) \xrightarrow{P} \triangle_i(\boldsymbol{x})$ and $\triangle_{jn}(\boldsymbol{x}) \xrightarrow{P} \triangle_j(\boldsymbol{x})$.

Finally, it should be noted that Lemma 6.4.1 provides a sufficient condition under which $\triangle_{in}(\boldsymbol{x}) \xrightarrow{P} \triangle_i(\boldsymbol{x})$ for each $i = 1, \ldots, k$.

We summarize the above discussions as follows.

Theorem 6.4.1 *Suppose that*

(a) $\int_\Omega \ell(\theta_{[k]})dG(\boldsymbol{\theta}) < \infty$;
(b) for each $\boldsymbol{x} \in \mathcal{X}, \ell(\theta_{[k]})\boldsymbol{f}(\boldsymbol{x}|\boldsymbol{\theta})$ is a bounded function of the variable $\boldsymbol{\theta}$;
(c) there exists a positive constant M such that
     $\int_x^y \frac{1}{\theta}dG_i(\theta) \le M(y - x)$ *for all $0 \le x < y < \infty$, for each $i = 1, \ldots, k$;*
(d) the sequence $\{\alpha_n\}$ is chosen so that $\alpha_n \ge n^{-\alpha}$, where
     $0 < \alpha < \frac{1}{2}$, *and $\lim_{n\to\infty} \alpha_n = 0$.*

Then, $r(\boldsymbol{G}, \boldsymbol{d}_n^*) - r(\boldsymbol{G}) \to 0$ as $n \to \infty$.

6.5   Rates of Convergence

Consider the loss function $\ell(\theta_{[k]} - \theta_i) = \theta_{[k]} - \theta_i$, the difference between the parameters of the best and the selected population. We are interested in deriving the rates of convergence of the empirical Bayes selection procedure $\boldsymbol{d}_n^*$. The evaluation is made under the assumptions (a) and (b) of Theorem 6.3.1.

Note that

$$\triangle_i(\boldsymbol{x}) = \int_\Omega \theta_{[k]} \boldsymbol{f}(\boldsymbol{x}|\boldsymbol{\theta}) dG(\boldsymbol{\theta}) - \int_\Omega \theta_i \boldsymbol{f}(\boldsymbol{x}|\boldsymbol{\theta}) dG(\boldsymbol{\theta})$$

$$= \int_\Omega \theta_{[k]} \boldsymbol{f}(\boldsymbol{x}|\boldsymbol{\theta}) dG(\boldsymbol{\theta}) - [1 - G_i(x_i)] \prod_{j=1,j\neq i}^k f_j(x_j),$$

and

$$\triangle_{in}(\boldsymbol{x}) = \int_\Omega \theta_{[k]} \boldsymbol{f}(\boldsymbol{x}|\boldsymbol{\theta}) dG_n^*(\boldsymbol{\theta}) - [1 - G_{in}^*(x_i)] \prod_{j=1,j\neq i}^k f_{jn}^*(x_j).$$

For $1 \leq i < j \leq k$, let

$$\tau_{i,j}(x_i, x_j) = [1 - G_i(x_i)]f_j(x_j) - [1 - G_j(x_j)]f_i(x_i),$$

$$\tau_{i,j,n}(x_i, x_j) = [1 - G_{in}^*(x_i)]f_{jn}^*(x_j) - [1 - G_{jn}^*(x_j)]f_{in}^*(x_i),$$

$$\mathcal{X}_{i,j}(1) = \{(x_i, x_j)|x_i \geq 0, x_j \geq 0, \tau_{i,j}(x_i, x_j) > 0\}$$

$$\mathcal{X}_{i,j}(2) = \{(x_i, x_j)|x_i \geq 0, x_j \geq 0, \tau_{i,j}(x_i, x_j) < 0\}.$$

Let ε_n be a positive number. Define

$$B_{ij}(\varepsilon_n) = \{(x_i, x_j)|x_i \geq 0, x_j \geq 0, |\tau_{i,j}(x_i, x_j)| > \varepsilon_n\},$$

$$B_{ij}^c(\varepsilon_n) = \{(x_i, x_j)|x_i \geq 0, x_j \geq 0, |\tau_{i,j}(x_i, x_j)| \leq \varepsilon_n\}.$$

Let $\varphi_i(x_i)$ denote the posterior mean of the random variable Θ_i given that $X_i = x_i$ is observed.

From (6.22), using a suitable partition of sets, tedious but straightforward computations lead to

$$0 \leq r(\boldsymbol{G}, \boldsymbol{d}_n^*) - r(\boldsymbol{G})$$

$$\leq \int_{\mathcal{X}} \sum_{1 \leq i < j \leq k} |\varphi_i(x_i) - \varphi_j(x_j)| P\{\prod_{\ell=1}^k f_{\ell n}^*(x_\ell) = 0\} \boldsymbol{f}(\boldsymbol{x}) d\boldsymbol{x}$$

$$+ \sum_{1 \leq i < j \leq k} \int_{R_{ij}^c(\boldsymbol{\varepsilon}_n)} |\tau_{i,j}(x_i, x_j)| dx_i dx_j$$

$$+ \sum_{1 \leq i < j \leq k} \int_{\mathcal{X}_{ij}(1) \cap B_{ij}(\varepsilon_n)} \tau_{i,j}(x_i, x_j) P\{\tau_{i,j,n}(x_i, x_j) \leq 0\} dx_i dx_j$$

$$+ \sum_{1 \leq i < j \leq k} \int_{\mathcal{X}_{ij}(2) \cap B_{ij}(\varepsilon_n)} [-\tau_{i,j}(x_i, x_j)] P\{\tau_{i,j,n}(x_i, x_j) \geq 0\} dx_i dx_j.$$

Lemma 6.5.1 *Suppose there exists a finite positive number Q such that $G_i(Q) = 1$ for each $i = 1, \ldots, k$. Then,*

(a) $\sum\sum_{1 \leq i < j \leq k} \int_{B_{ij}^c(\varepsilon_n)} |\tau_{i,j}(x_i, x_j)| dx_i dx_j = O(\varepsilon_n)$, *and*

(b) $\int_{\mathcal{X}} \sum\sum_{1 \leq i < j \leq k} |\varphi_i(x_i) - \varphi_j(x_j)| P\{\prod_{\ell=1}^k f_{\ell n}^*(x_\ell) = 0\} \boldsymbol{f}(\boldsymbol{x}) d\boldsymbol{x} = O(n^{-1})$.

PROOF. (a) is trivial.

Note that by the definition of $f_{jn}^*(x_j)$, $f_{jn}^*(x_j) = 0$ implies $\max_{1 \leq m \leq n} X_{jm} \leq x_j$. Hence,

$$
\begin{aligned}
P\{\prod_{\ell=1}^{k} f_{\ell n}^*(x_\ell) = 0\} &= P\{f_{\ell n}^*(x_\ell) = 0 \text{ for some } \ell = 1, \ldots, k\} \\
&\leq P\{\max_{1 \leq m \leq n} X_{\ell m} \leq x_\ell \text{ for some } \ell = 1, \ldots, k\} \\
&\leq \sum_{\ell=1}^{k} [F_\ell(x_\ell)]^n.
\end{aligned}
$$

Under the assumption, $0 \leq \varphi_\ell(x_\ell) \leq Q$ for all $0 \leq x_\ell \leq Q$. Therefore,

$$
\begin{aligned}
&\int_{\mathcal{X}} \sum \sum_{1 \leq i < j \leq k} |\varphi_i(x_i) - \varphi_j(x_j)| P\{\prod_{\ell=1}^{k} f_{\ell n}^*(x_\ell) = 0\} \boldsymbol{f}(\boldsymbol{x}) d\boldsymbol{x} \\
&= O(\int_{\mathcal{X}} P\{\prod_{\ell=1}^{k} f_{\ell n}^*(x_\ell) = 0\} \boldsymbol{f}(\boldsymbol{x}) d\boldsymbol{x}) \\
&= O(\sum_{\ell=1}^{k} \int_0^Q [F_\ell(x_\ell)]^n f_\ell(x_\ell) dx_\ell) \\
&= O(n^{-1}).
\end{aligned}
$$

Lemma 6.5.2 *Suppose that the assumptions (a) and (b) of Theorem 6.3.1 hold for each $i = 1, \ldots, k$. Let $\varepsilon_n = \alpha_n \max(4(M+2), 16MQ(Q + \alpha_n + 1))$. Then, for n sufficiently large, we have the following*

(a) For each $(x_i, x_j) \in \mathcal{X}_{ij}(1) \cap B_{ij}(\varepsilon_n)$,
   $P\{\tau_{i,j,n}(x_i, x_j) \leq 0\} \leq 2c \exp\{-2n\alpha_n^4\}$.
   For each $(x_i, x_j) \in \mathcal{X}_{ij}(2) \cap B_{ij}(\varepsilon_n)$,
   $P\{\tau_{i,j,n}(x_i, x_j) \geq 0\} \leq 2c \exp\{-2n\alpha_n^4\}$.

where c is some universal constant.
(b) $\int_{\mathcal{X}_{ij}(1) \cap B_{ij}(\varepsilon_n)} \tau_{ij}(x_i, x_j) P\{\tau_{i,j,n}(x_i, x_j) \leq 0\} dx_i dx_j = O(\exp[-2n\alpha_n^4])$,
and

   $\int_{\mathcal{X}_{ij}(2) \cap B_{ij}(\varepsilon_n)} [-\tau_{i,j}(x_i, x_j)] P\{\tau_{i,j,n}(x_i, x_j) \geq 0\} dx_i dx_j = O(\exp[-2n\alpha_n^4])$.

PROOF.

(a) Under the assumptions, for each $i = 1, \ldots, k$, and $0 \leq x \leq Q$,

$$
f_i(x) = \int_x^Q \frac{1}{\theta} dG_i(\theta) \leq \int_0^Q \frac{1}{\theta} dG_i(\theta) \leq MQ.
$$

For $(x_i, x_j) \in \mathcal{X}_{ij}(1) \cap B_{ij}(\varepsilon_n), \tau_{i,j}(x_i, x_j) > \varepsilon_n$. Therefore,

$$
\begin{aligned}
&\{\tau_{i,j,n}(x_i, x_j) \leq 0\} \\
\subset\ &\{\tau_{i,j,n}(x_i, x_j) - \tau_{i,j}(x_i, x_j) < -\varepsilon_n\} \\
\subset\ &\{[1 - G_{in}^*(x_i)]f_{jn}^*(x_j) - [1 - G_i(x_i)]f_j(x_j) < -\varepsilon_n/2\} \\
&\cup\{[1 - G_{jn}^*(x_j)]f_{in}^*(x_i) - [1 - G_j(x_j)]f_i(x_i) > \varepsilon_n/2\} \\
\subset\ &\{[1 - G_{in}^*(x_i)][f_{jn}^*(x_j) - f_j(x_j)] < -\varepsilon_n/4\} \\
&\cup\{f_j(x_j)[G_{in}^*(x_i) - G_i(x_i)] > \varepsilon_n/4\} \\
&\cup\{[1 - G_{jn}^*(x_j)][f_{in}^*(x_i) - f_i(x_i)] > \varepsilon_n/4\} \\
&\cup\{f_i(x_i)[G_{jn}^*(x_j) - G_j(x_j)] < -\varepsilon_n/4\} \\
\subset\ &\{f_{jn}^*(x_j) - f_j(x_j) < -\varepsilon_n/4\} \cup \{G_{in}^*(x_i) - G_i(x_i) > \varepsilon_n/(4MQ)\} \\
&\cup\{f_{in}^*(x_i) - f_i(x_i) > \varepsilon_n/4\} \\
&\cup\{G_{jn}^*(x_j) - G_j(x_j) < -\varepsilon_n/(4MQ)\}. \qquad (6.23)
\end{aligned}
$$

From (6.14) and (6.15), and the definition of ε_n, for $0 \leq x_i \leq Q$, we have

$$
\begin{aligned}
&\{G_{in}^*(x_i) - G_i(x_i) > \varepsilon_n/(4MQ)\} \\
\subset\ &\{\sup_{y>0} |F_{in}(y) - F_i(y)| > \frac{\alpha_n \varepsilon_n}{16MQ(Q + \alpha_n)}\} \\
\subset\ &\{\sup_{y>0} |F_{in}(y) - F_i(y)| > \alpha_n^2\}, \qquad (6.24)
\end{aligned}
$$

and

$$
\begin{aligned}
&\{G_{jn}^*(x_j) - G_j(x_j) < -\varepsilon_n/(4MQ)\} \\
\subset\ &\{\sup_{y>0} |F_{jn}(y) - F_j(y)| > \alpha_n^2\}. \qquad (6.25)
\end{aligned}
$$

For $0 \leq x_i \leq Q$, let $[C_{n,\ell-1}, C_{n,\ell})$ be the interval containing the point x_i. From (6.13), for n sufficiently large, we have

$$
\begin{aligned}
&\{f_{in}^*(x_i) - f_i(x_i) > \varepsilon_n/4\} \\
=\ &\{F_{in}^*(C_{n,\ell}) - F_{in}^*(C_{n,\ell-1}) > [\varepsilon_n/4 + f_i(x_i)]\alpha_n\} \\
=\ &\{R_{in}(C_{n,\ell}) - R_{in}(C_{n,\ell-1}) > \alpha_n[\varepsilon_n/4 + f_i(x_i) - \frac{F_i(C_{n,\ell}) - F_i(C_{n,\ell-1})}{\alpha_n}]\} \\
\subset\ &\{R_{in}(C_{n,\ell}) - R_{in}(C_{n,\ell-1}) > 2\alpha_n^2\} \\
\subset\ &\{\sup_{\ell \geq 1} |R_{in}(C_{n,\ell})| > \alpha_n^2\} \\
\subset\ &\{\sup_{y>0} |F_{in}(y) - F_i(y)| > \alpha_n^2\}, \qquad (6.26)
\end{aligned}
$$

where the first inclusion relation in (6.26) is obtained by the definition of ε_n and the fact that

$$
|f_i(x_i) - \frac{F_i(C_{n,\ell}) - F_i(C_{n,\ell-1})}{\alpha_n}| \leq f_i(C_{n,j-1}) - f_i(C_{n,j}) \leq M\alpha_n,
$$

which is guaranteed under the assumptions.

Similarly, for $0 \leq x_j \leq Q$ and n sufficiently large,

$$\{f_{jn}^*(x_j) - f_j(x_j) < -\varepsilon_n/4\} \subset \{\sup_{y>0} |F_{jn}(y) - F_j(y)| > \alpha_n^2\}. \tag{6.27}$$

Through (6.23)-(6.27) and by Lemma 6.3.1, we conclude: For sufficiently large n,

$$P\{\tau_{i,j,n}(x_i, x_j) \leq 0\} \leq 2c \exp[-2n\alpha_n^4]. \tag{6.28}$$

Similarly, the following can be proved: For each $(x_i, x_j) \in \mathcal{X}_{ij}(2) \cap B_{ij}(\varepsilon_n)$, and for sufficiently large n,

$$P\{\tau_{i,j,n}(x_i, x_j) \geq 0\} \leq 2c \exp[-2n\alpha_n^4]. \tag{6.29}$$

Note that the upper bounds in (6.28) and (6.29) are independent of (x_i, x_j).
(b) Part (b) is a direct result of part (a) and the fact that

$$\begin{aligned}
|\tau_{i,j}(x_i, x_j)| &= |[1 - G_i(x_i)]f_j(x_j) - [1 - G_j(x_j)]f_i(x_i)| \\
&\leq \max([1 - G_i(x_i)]f_j(x_j), [1 - G_j(x_j)]f_i(x_i)) \\
&\leq MQ < \infty.
\end{aligned}$$

We summarize the above discussions in the following.

Theorem 6.5.1 *Suppose that*

(a) $G_i(Q) = 1$ for some finite positive number Q for each $i = 1, \ldots, k$ and
(b) $\int_x^y \frac{1}{\theta} dG_i(\theta) \leq M(y - x)$ for all $0 \leq x < y \leq Q$ for some finite positive number M for each $i = 1, \ldots, k$.

Take $\alpha_n = (\ln n/n)^{1/4}$. Then,

$$r(\boldsymbol{G}, \boldsymbol{d}_n^*) - r(\boldsymbol{G}) = O(\alpha_n).$$

PROOF. From (6.23) and Lemmas 6.5.1 and 6.5.2,

$$\begin{aligned}
0 &\leq r(\boldsymbol{G}, \boldsymbol{d}_n^*) - r(\boldsymbol{G}) \\
&= O(\varepsilon_n) + O(n^{-1}) + O(\exp[-2n\alpha_n^4]) \\
&= O(\alpha_n) + O(n^{-1}) + O(n^{-2}) \\
&= O(\alpha_n),
\end{aligned}$$

where $\varepsilon_n = \alpha_n \max(4(M + 2), 16MQ(Q + \alpha_n + 1))$. Hence, the proof of the theorem is completed.

Acknowledgements. The authors are thankful to the referee for a careful reading and corrections. This research is partially supported by National Science Council of Taiwan through the grant NSC80-0208-M001-24.

References

1. Barlow, R. E., Bartholomew, D. J., Bremner, J. M. and Brunk, H. D. (1972). *Statistical Inference under Order Restrictions*, New York: John Wiley & Sons.

2. Barr, D. R. and Rizvi, M. M. (1966). Ranking and selection problems of uniform distributions, *Trabajos Estadist*, **17**, 15–31.

3. Bechhofer, R. E. (1954). A single-sample multiple-decision procedure for ranking means of normal populations with known variances, *Annals of Mathematical Statistics*, **25**, 16–39.

4. Deely, J. J. (1965). Multiple decision procedures from an empirical Bayes approach. Ph.D. Thesis (Mimeo. Ser. No. 45), Department of Statistics, Purdue University, West Lafayette, Indiana.

5. Dudewicz, E. J. and Koo, J. O.(1982). *The Complete Categorized Guide to Statistical Selection and Ranking Procedures*, Columbus, Ohio: American Sciences Press.

6. Fox, R. J. (1978). Solutions to empirical Bayes squared error loss estimation problems, *Annals of Statistics*, **6**, 846–854.

7. Gupta, S. S. (1956). On a decision rule for a problem in ranking means. Ph.D. Thesis (Mimeograph Series No. 150), Institute of Statistics, University of North Carolina, Chapel Hill, North Carolina.

8. Gupta, S. S. and Hsiao, P. (1983). Empirical Bayes rules for selecting good populations, *Journal of Statistical Planning and Inference*, **8**, 87–101.

9. Gupta, S. S. and Leu, L. Y. (1983). On Bayes and empirical Bayes rules for selecting good populations, *Technical Report # 83-37*, Department of Statistics, Purdue University, West Lafayette, Indiana.

10. Gupta, S. S. and Liang, T. (1986). Empirical Bayes rules for selecting good binomial populations, In *Adaptive Statistical Procedures and Related Topics* (Ed., J. Van Ryzin), IMS Lecture Notes-Monograph Series, Vol. 8, 110–128.

11. Gupta, S. S. and Liang, T. (1988). Empirical Bayes rules for selecting the best binomial population, In *Statistical Decision Theory and Related Topics-IV* (Eds., S. S. Gupta and J. O. Berger), Volume 1, pp. 213–224, New York: Springer-Verlag.

12. Gupta, S. S. and Panchapakesan, S. (1979). *Multiple Decision Procedures*, New York: John Wiley & Sons.

13. Huang, W. T. (1975). Bayes approach to a problem of partitioning k normal populations, *Bull. Inst. Math. Academia Sinica*, **3**, 87–97.

14. McDonald, G. C. (1976). The distribution of a variate based on independent ranges from a uniform population, *Technometrics*, **18**, 343–349.

15. McDonald, G. C. (1976). Subset selection rules based on quasi-ranges for uniform populations, *Sankhyā, Series B*, **40**, 163–191.

16. Puri, P. S. and Singh, H. (1990). On recursive formulas for isotonic regression useful for statistical inference under order restrictions, *Journal of Statistical Planning and Inference*, **24**, 1–11.

17. Robbins, H. (1956). An empirical Bayes approach to statistics, *Proceedings Third Berkeley Symp. Math. Statist. Probab.*, *1*, University of California Press, 157–163.

18. Royden H. I. (1963). *Real Analysis*, New York: Macmillan.

19. Schuster, E. F. (1969). Estimation of a probability density function and its derivatives, *Annals of Mathematical Statistics*, **40**, 1187–1195.

20. Susarla, V. and O'Bryan, T. (1979). Empirical Bayes interval estimates involving uniform distributions, *Communications in Statistics—Theory and Methods*, **8**, 385–397.

PART II
DECISION THEORY

7

Adaptive Multiple Decision Procedures for Exponential Families

Andrew L. Rukhin

University of Maryland Baltimore County, Baltimore, MD

Abstract: The asymptotic behavior of multiple decision procedures is studied when the underlying distributions belong to an exponential family. An adaptive procedure must be asymptotically optimal for each value of the unknown nuisance parameter, on which it does not depend. A necessary and sufficient condition for the existence of such a procedure is discussed. The regions of the parameter space, where the traditional overall maximum likelihood rule is adaptive, are described. The examples of a normal family and a family of gamma-distributions are considered in detail.

Keywords and phrases: Adaptive procedures, Bayes risk, error probabilities, exponential family, multiple decision problem, overall maximum likelihood rule

7.1 Introduction

Let $x = (x_1, x_2, \ldots, x_n)$ be a random sample drawn from one of different probability distributions $F_1, \ldots, F_g$ with densities $f_1, \ldots, f_g$. We will assume that these distributions are mutually absolutely continuous, i.e., that all densities f_i can be chosen to be positive on the same set. In some situations one can find prior probabilities $\omega_1, \ldots, \omega_g$ of the sample distributions $P_i = F_i \otimes \cdots \otimes F_i$, $i = 1, \ldots, g$. The performance of a multiple decision rule $\delta(x)$ taking values in the set $\{1, \ldots, g\}$ is measured by the error probabilities $P_i(\delta \neq i)$ with the Bayes risk $\sum_i \omega_i P_i(\delta \neq i)$ or the minimax risk $\max_i P_i(\delta \neq i)$.

For mutually absolutely continuous probability distributions P and Q let

$$H_s(P,Q) = \log E^Q \left[\frac{dP}{dQ}(X) \right]^s \tag{7.1}$$

be the logarithm of the Hellinger type integral.

Obviously H_s is a convex analytic function of s defined on an interval containing the closed interval $[0,1]$ with the derivative at $s = 0$ being $-K(Q, P)$ and at $s = 1$ equal to $K(P, Q)$. Here

$$K(P, Q) = E^P \log \left[\frac{d\,P}{d\,Q}(X) \right] \tag{7.2}$$

is the (Kullback–Leibler) information number. Hence the minimum of H_s is attained in the interval $(0, 1)$.

These quantities specify the exponential rate of the risk decay in the multiple decision problem (see Chernoff, 1956, Renyi, 1970, Krafft and Puri, 1974).

Theorem 7.1.1 *Assume that $\omega_i > 0$ for all i. Then for any multiple decision rule $\delta = \delta(x)$ based on the random sample $x = (x_1, x_2, \cdots, x_n)$ from the family $\mathcal{P} = \{P_i = F_i \otimes \cdots \otimes F_i, \ i = 1, \ldots, g\}$ one has*

$$
\begin{aligned}
\liminf \frac{1}{n} \log \max_i P_i(\delta \neq i) \ &= \ \liminf \frac{1}{n} \log \sum_i \omega_i P_i(\delta \neq i) \\
&\geq \ \max_{i \neq k} \inf_{0 < s < 1} H_s\left(F_i, F_k\right) = \rho(\mathcal{P}). \tag{7.3}
\end{aligned}
$$

For the Bayes rule

$$\{\delta_B(x) = i\} = \left\{ \prod_{j=1}^{n} \omega_i f_i(x_j) = \max_k \prod_{j=1}^{n} \omega_k f_k(x_j) \right\}$$

or the maximum likelihood rule

$$\{\hat{\delta}(x) = i\} = \left\{ \prod_{j=1}^{n} f_i(x_j) = \max_k \prod_{j=1}^{n} f_k(x_j) \right\},$$

(7.3) is the equality.

In a more realistic scenario probability distributions P_i of a random sample are not known exactly, but only up to a (nuisance) parameter α taking values in a set $\mathcal{A}$. In other words, we assume a collection of probability distribution families $\mathcal{P}_\alpha = \left(P_1^\alpha, \ldots, P_g^\alpha \right), \alpha \in \mathcal{A}$.

For example, a repeated message may be sent through the one of channels indexed by α, or by using one of different languages forming the set $\mathcal{A}$. The goal of the statistician is to recover the message, no matter which channel or which language has been used. One can think about $\mathcal{A}$ as the set of individuals with different handwritings. A sequence of written letters is to be recognized independent of the individual who wrote them.

Thus, independently of the true value of the nuisance parameter α, one would like to use an efficient multiple decision rule. This objective is formalized with help of Theorem 7.1 by the following definition.

A classification rule δ_a is called *adaptive* if for all α

$$\liminf \frac{1}{n} \log \max_i P_i^\alpha(\delta_a \neq i) = \max_{i \neq k} \inf_{s>0} H_s(F_i^\alpha, F_k^\alpha) = \rho(\mathcal{P}_\alpha) = \rho_\alpha$$

and δ_a does not depend on α. In the Bayes setting with fixed prior probabilities ω_i this definition could be modified by replacing the minimax risk by the Bayes risk.

In the presence of unknown nuisance parameters an adaptive rule must exhibit the same asymptotic optimal behavior as when these parameters were given. Of course, it should not depend on these unknown parameters.

A survey of the work in the area of semiparametric inference, which is built around the concept of adaptation for a continuous parameter introduced by Stein (1956), can be found in the monograph by Bickel *et al.* (1993).

The existence condition and the form of an adaptive procedure were derived in Rukhin (1982, 1984). An adaptive classification procedure exists if and only if

$$\max_{\alpha,\beta} \max_{i \neq k} \inf_{s>0} \left[H_s(F_i^\alpha, F_k^\beta) - s\rho_\alpha - (1-s)\rho_\beta \right]$$

$$= \max_\alpha \max_{i \neq k} \inf_{s>0} [H_s(F_i^\alpha, F_k^\alpha) - \rho_\alpha] = 0. \tag{7.4}$$

The heuristic interpretation of (7.4) is that the existence of an adaptive procedure means that any decision problem $\mathcal{P}_\alpha$ is at least as difficult as problems formed by P_i^α, P_k^β, $i \neq k, \alpha \neq \beta$.

Also, as is shown in Theorem 2 in Rukhin (1984), the rule:

$$\{\delta_a(x) = i\} = \left\{ \max_\alpha \prod_{j=1}^n e^{-\rho_\alpha} f_i^\alpha(x_j) = \max_k \max_\alpha \prod_{j=1}^n e^{-\rho_\alpha} f_k^\alpha(x_j) \right\} \tag{7.5}$$

is adaptive if there are adaptive rules.

The traditional way to eliminate a nuisance parameter is by using the uniform prior for this parameter. In our problem this method may lead to non-adaptive procedures. Indeed, the resulting "naive" overall maximum likelihood classification rule

$$\{\delta_0(x) = i\} = \left\{ \max_\alpha \prod_{j=1}^n f_i^\alpha(x_j) = \max_k \max_\alpha \prod_{j=1}^n f_k^\alpha(x_j) \right\}$$

may not be adaptive when (7.4) holds.

The proof of the following corollaries is also given in Rukhin (1984).

Corollary 7.1.1 *An adaptive procedure exists if and only if for all $\alpha \neq \beta$ and all $i \neq k$*

$$\inf_{0<s<1} \left[H_s(F_i^\alpha, F_k^\beta) - s(\rho_\alpha - \rho_\beta) \right] \leq \rho_\beta. \tag{7.6}$$

In other words, an adaptive procedure exists if and only if

$$\max_{\alpha:\rho_\alpha \geq \rho_\beta} \max_{i \neq k} \inf_{0<s<1} \left[H_s(F_i^\alpha, F_k^\beta) - s(\rho_\alpha - \rho_\beta) \right] \leq \rho_\beta.$$

Corollary 7.1.2 *If for some $\alpha \neq \beta$ and $i \neq k$, $F_i^\alpha = F_k^\beta$, then an adaptive procedure cannot exist.*

The Corollary 7.1.2 supports the heuristic interpretation of (7.4) according to which an adaptive procedure exists if and only if the distributions from any $\mathcal{P}_\alpha$ are "at least as close" as the distributions P_i^α and P_k^β, $i \neq k, \alpha \neq \beta$.

The goal of this paper is to specify these results when the distributions F_i^α form an exponential family. This is always the case when the set $\mathcal{A}$ is finite, say, $\mathcal{A} = \{1, \dots, A\}$. In this situation the following result holds.

Proposition 7.1.1 *Let, for some $b_1, \dots, b_A$*

$$\{\delta_b(x) = i\} = \left\{ \max_\alpha \prod_{j=1}^n e^{b_\alpha} f_i^\alpha(x_j) = \max_k \max_\alpha \prod_{j=1}^n e^{b_\alpha} f_k^\alpha(x_j) \right\}.$$

If

$$\max_{i \neq k} \max_{\beta \neq \alpha} \inf_{s_\gamma \geq 0, \gamma = 1, \dots, A} \left[\log E_i^\alpha \prod_\gamma \left[\frac{f_k^\beta}{f_i^\gamma}(X) \right]^{s_\gamma} + \sum_\gamma s_\gamma(b_\beta - b_\gamma) - \rho_\alpha \right] \leq 0,$$

then δ_b is adaptive.

7.2 Adaptation for Exponential Families

Let the distributions F_i^α be members of an exponential family, that is, let the densities f_i^α have the form

$$f_i^\alpha(u) = \exp\{\theta_i^\alpha u - \chi(\theta_i^\alpha)\}, \quad i = 1, \dots, g.$$

Because of Corollary 7.1.2 the distributions P_i^α will be supposed to be different so that the common support of these distributions contains at least two points, and the function χ is strictly convex over the natural parameter space $\Theta = \{\theta : \chi(\theta) < \infty\}$. Mainly for the notational simplicity, we assume here that θ is real so that Θ is an interval.

One has

$$\rho_\alpha = \max_{i \neq k} \inf_{s>0} \left[\chi \left(s\theta_i^\alpha + (1-s)\theta_k^\alpha \right) - s\chi \left(\theta_i^\alpha \right) - (1-s)\chi \left(\theta_k^\beta \right) \right]. \tag{7.7}$$

By differentiating the left-hand side of (7.7) one notices that for a fixed α and $i \neq k$, the unique minimum is attained at $s = s_{ik}^\alpha$ such that

$$\left(\theta_i^\alpha - \theta_k^\alpha \right) \chi' \left(s\theta_i^\alpha + (1-s)\theta_k^\alpha \right) = \chi \left(\theta_i^\alpha \right) - \chi \left(\theta_k^\alpha \right).$$

Let $\sigma_{ik}^\alpha = s_{ik}^\alpha \theta_i^\alpha + (1 - s_{ik}^\alpha)\theta_k^\alpha$, so that

$$\chi' (\sigma_{ik}^\alpha) = \frac{\chi \left(\theta_i^\alpha \right) - \chi \left(\theta_k^\alpha \right)}{\theta_i^\alpha - \theta_k^\alpha}.$$

Then

$$\begin{aligned}
\rho_\alpha &= \max_{i \neq k} \left[\chi \left(\sigma_{ik}^\alpha \right) - s_{ik}^\alpha \chi \left(\theta_i^\alpha \right) - (1 - s_{ik}^\alpha)\chi \left(\theta_k^\alpha \right) \right] \\
&= \max_{i \neq k} \left[\chi \left(\sigma_{ik}^\alpha \right) - \chi \left(\theta_k^\alpha \right) - \left(\sigma_{ik}^\alpha - \theta_k^\alpha \right) \chi' \left(\sigma_{ik}^\alpha \right) \right]. \tag{7.8}
\end{aligned}$$

According to Corollary 7.1.1 an adaptive rule exists if and only if for any β

$$\max_{\alpha:\rho_\alpha \geq \rho_\beta} \max_{i \neq k} \inf_{s>0} \left[\chi(s\theta_i^\alpha + (1-s)\theta_k^\beta) - s\chi(\theta_i^\alpha) - (1-s)\chi(\theta_k^\beta) \right.$$

$$\left. -s\rho_\alpha - (1-s)\rho_\beta \right] \leq 0. \tag{7.9}$$

Now assuming that $\theta_i^\alpha \neq \theta_k^\beta$, the minimum in (7.9) is attained at $s = s_{ik}^{\alpha\beta}$ such that

$$\left(\theta_i^\alpha - \theta_k^\beta \right) \chi' \left(s\theta_i^\alpha + (1-s)\theta_k^\beta \right) = \chi \left(\theta_i^\alpha \right) + \rho_\alpha - \chi \left(\theta_k^\beta \right) - \rho_\beta.$$

This value belongs to the interval $(0, 1)$ if and only if

$$\theta_i^\alpha \wedge \theta_k^\beta < \sigma_{ik}^{\alpha\beta} = s_{ik}^{\alpha\beta} \theta_i^\alpha + \left(1 - s_{ik}^{\alpha\beta} \right) \theta_k^\beta < \theta_i^\alpha \vee \theta_k^\beta, \tag{7.10}$$

and this condition is necessary and sufficient for δ_a to be a consistent procedure.
Since

$$\chi' \left(\sigma_{ik}^{\alpha\beta} \right) = \frac{\chi \left(\theta_i^\alpha \right) + \rho_\alpha - \chi \left(\theta_k^\beta \right) - \rho_\beta}{\theta_i^\alpha - \theta_k^\beta},$$

one has $\chi' \left(\sigma_{ik}^{\alpha\beta} \right) = \chi' \left(\sigma_{ki}^{\beta\alpha} \right)$, so that $\sigma_{ik}^{\alpha\beta} = \sigma_{ki}^{\beta\alpha}$.
Using these formulas one derives the following version of (7.9): for any β

$$\max_{\alpha:\rho_\alpha \geq \rho_\beta} \max_{i \neq k} \left[\chi \left(\sigma_{ik}^{\alpha\beta} \right) - \chi \left(\theta_k^\beta \right) - \left(\sigma_{ik}^{\alpha\beta} - \theta_k^\beta \right) \chi' \left(\sigma_{ik}^{\alpha\beta} \right) \right] \leq \rho_\beta$$

and (7.10) holds.

To make the adaptation condition more explicit we define for a fixed θ the function $W_\theta(t)$, $t \in \Theta$,

$$W_\theta(t) = \chi(t) - \chi(\theta) - (t - \theta)\chi'(t).$$

Then $W_\theta'(t) = (\theta - t)\chi''(t)$, so that as a function of t, W_θ increases when $t \leq \theta$, reaches its maximum 0 at $t = \theta$ and decreases for $t \geq \theta$. Also $W_{\theta_k^\beta}\left(\sigma_{ik}^\beta\right) = W_{\theta_i^\beta}\left(\sigma_{ik}^\beta\right)$. Observe that $W_\theta(t) = -K(F_t, F_\theta)$ with K defined by (7.2).

According to (7.8)

$$\rho_\alpha = \max_{i \neq k} W_{\theta_i^\alpha}\left(\sigma_{ik}^\alpha\right).$$

An examination of (7.10) shows that this condition means that for any $\alpha \neq \beta$

$$\max_{i \neq k} W_{\theta_k^\beta}\left(\theta_i^\alpha\right) < \rho_\beta - \rho_\alpha < -\max_{i \neq k} W_{\theta_k^\beta}\left(\theta_i^\alpha\right).$$

These inequalities are met if

$$\max_{\alpha,\beta:\rho_\alpha \geq \rho_\beta} \max_{i \neq k} \left[W_{\theta_k^\beta}\left(\theta_i^\alpha\right) - \rho_\beta + \rho_\alpha\right] < 0. \tag{7.11}$$

The adaptation condition can be rewritten in terms of the functions W_θ as the combination of (7.11) and the inequalities

$$\max_{\alpha:\rho_\alpha \geq \rho_\beta} \max_{i \neq k} W_{\theta_k^\beta}\left(\sigma_{ik}^{\alpha\beta}\right) \leq \rho_\beta = \max_{i \neq k} W_{\theta_k^\beta}\left(\sigma_{ik}^\beta\right). \tag{7.12}$$

Observe that

$$W_{\theta_k^\beta}\left(\sigma_{ik}^{\alpha\beta}\right) - W_{\theta_i^\alpha}\left(\sigma_{ik}^{\alpha\beta}\right) = \rho_\beta - \rho_\alpha, \tag{7.13}$$

which directly shows that if $\rho_\alpha \geq \rho_\beta$, then $W_{\theta_k^\beta}\left(\sigma_{ik}^{\alpha\beta}\right) \leq W_{\theta_i^\alpha}\left(\sigma_{ik}^{\alpha\beta}\right)$. Also $W_{\theta_k^\beta}\left(\theta_i^\alpha\right) < \rho_\beta - \rho_\alpha$ means that $\sigma_{ik}^{\alpha\beta} < \theta_i^\alpha$, when $\theta_i^\alpha > \theta_k^\beta$, or that $\sigma_{ik}^{\alpha\beta} < \theta_i^\alpha$, when $\theta_i^\alpha < \theta_k^\beta$.

We shall need the following quantities $\tau_{ik}^{\alpha\beta}$ defined by the formula

$$\chi'\left(\tau_{ik}^{\alpha\beta}\right) = \frac{\chi\left(\theta_i^\alpha\right) - \chi\left(\theta_k^\beta\right)}{\theta_i^\alpha - \theta_k^\beta}, \tag{7.14}$$

which shows that $\tau_{ik}^{\alpha\beta} = \tau_{ki}^{\beta\alpha}$ and $\tau_{ik}^{\alpha\alpha} = \sigma_{ik}^\alpha$. Clearly $\tau_{ik}^{\alpha\beta}$ automatically belongs to the interval with the end-points θ_i^α and θ_k^β, and $W_{\theta_k^\beta}\left(\tau_{ik}^{\alpha\beta}\right) = W_{\theta_i^\alpha}\left(\tau_{ik}^{\alpha\beta}\right)$. Notice that $\left(\theta_i^\alpha - \theta_k^\beta\right)\left(\tau_{ik}^{\alpha\beta} - \sigma_{ik}^{\alpha\beta}\right) = \rho_\alpha - \rho_\beta$, so that if, say, $\rho_\alpha > \rho_\beta$ and $\theta_i^\alpha > \theta_k^\beta$, then necessarily $\tau_{ik}^{\alpha\beta} > \sigma_{ik}^{\alpha\beta}$.

Observe that for a fixed θ, the function $[\chi(t) - \chi(\theta)]/(t - \theta)$ is strictly increasing in t. Indeed its derivative, $-W_\theta(t)/(t-\theta)^2$, is non-negative. Therefore, the inequality $\tau_{ik}^{\alpha\beta} > \sigma_{ik}^\alpha$ means that $\theta_i^\alpha < \theta_k^\beta$.

When $A = 2$ the conditions (7.12) and (7.11) can be written in a more specific form: if $\rho_1 \leq \rho_2$, then

$$W_{\theta_1^1}\left(\sigma_{12}^{12}\right) \vee W_{\theta_2^1}\left(\sigma_{21}^{12}\right) \leq \rho_1,$$

$$W_{\theta_1^1}\left(\theta_2^2\right) \vee W_{\theta_2^1}\left(\theta_1^2\right) < \rho_1 - \rho_2,$$

and if $\rho_2 \leq \rho_1$,

$$W_{\theta_1^2}\left(\sigma_{21}^{12}\right) \vee W_{\theta_2^2}\left(\sigma_{12}^{12}\right) \leq \rho_2,$$

$$W_{\theta_1^2}\left(\theta_2^1\right) \vee W_{\theta_2^2}\left(\theta_1^1\right) < \rho_2 - \rho_1.$$

We are interested in the adaptation region R^a formed by pairs (θ_1^2, θ_2^2), for which this adaptation condition holds.

Proposition 7.2.1 1. *Let*

$$R_0 = \{\rho_1 < \rho_2, W_{\theta_1^1}\left(\theta_2^2\right) \vee W_{\theta_2^1}\left(\theta_1^2\right) < \rho_1 - \rho_2\}$$
$$\bigcup\{\rho_2 \leq \rho_1, W_{\theta_1^2}\left(\theta_2^1\right) \vee W_{\theta_2^2}\left(\theta_1^1\right) < \rho_2 - \rho_1\}.$$

Then $(\theta_1^2, \theta_2^2) \in R_0$ is a necessary adaptation condition, i.e. there are no adaptive rules in the region R_0^c.

To be specific assume that $\theta_1^1 < \theta_2^1$. Then

2. In the region $R_1 = \{\theta_1^2 \leq \theta_2^2, \theta_1^1 \leq \theta_2^2, \theta_1^2 \leq \theta_2^1\}$ an adaptive procedure exists if and only if $\sigma_{12}^1 = \sigma_{12}^2$;

3. In the region $R_2 = \{\theta_2^1 > \theta_2^2, \theta_1^2 > \theta_1^1, \theta_1^2 \geq \theta_2^2\}$ no adaptive procedure exists.

PROOF. 1. According to (7.11) the rule δ_a is not even consistent in the region R_0^c, so it cannot be adaptive and no adaptive procedure exists.

2. In the region R_1, when an adaptive procedure exists, one must have $\theta_1^1 \leq \sigma_{12}^{12} \leq \theta_2^2$ and $\theta_1^1 \leq \sigma_{12}^1$. Therefore, the inequality $W_{\theta_1^1}\left(\sigma_{12}^{12}\right) \leq W_{\theta_1^1}\left(\sigma_{12}^1\right)$ means that $\sigma_{12}^1 \leq \sigma_{12}^{12} = \sigma_{21}^{12}$. Similarly for θ-values from this region the inequality $W_{\theta_2^1}\left(\sigma_{21}^{12}\right) \leq W_{\theta_2^1}\left(\sigma_{12}^1\right)$ is equivalent to $\sigma_{12}^1 \geq \sigma_{21}^{12}$, the inequality $W_{\theta_1^2}\left(\sigma_{21}^{12}\right) \leq W_{\theta_1^2}\left(\sigma_{12}^2\right)$ is tantamount to $\sigma_{12}^2 \leq \sigma_{21}^{12}$, and $W_{\theta_2^2}\left(\sigma_{21}^{12}\right) \leq W_{\theta_2^2}\left(\sigma_{12}^2\right)$ if and only if $\sigma_{12}^2 \geq \sigma_{21}^{12}$. Combining these facts, we see that the adaptation condition is equivalent to the following inequalities: $\sigma_{21}^{12} \leq \sigma_{12}^1 \leq \sigma_{21}^{12} \leq \sigma_{12}^2 \leq \sigma_{21}^{12}$, i.e. $\sigma_{21}^{12} = \sigma_{12}^1 = \sigma_{12}^2 = \sigma_{21}^{12}$.

Notice now that the identity $\sigma_{12}^1 = \sigma_{12}^2$ implies that $\sigma_{21}^{12} = \sigma_{21}^{21} = \sigma_{12}^1$. Indeed, for example, one has

$$\rho_1 = \chi(\sigma_{12}^1) - \chi(\theta_2^1) - (\sigma_{12}^1 - \theta_2^1)\chi'(\sigma_{12}^1)$$

and

$$\rho_2 = \chi(\sigma_{12}^2) - \chi(\theta_1^2) - (\sigma_{12}^2 - \theta_1^2)\chi'(\sigma_{12}^2).$$

Thus if $\sigma_{12}^1 = \sigma_{12}^2$, then

$$\rho_2 - \rho_1 = \chi(\theta_2^1) - \chi(\theta_1^2) - (\theta_2^1 - \theta_1^2)\chi'(\sigma_{12}^1)$$

and $\chi'(\sigma_{12}^1) = \chi'(\sigma_{12}^{21})$.

3. If an adaptive procedure exists in the region $\{\theta_1^2 \geq \theta_2^1, \theta_1^1 \geq \theta_2^2\}$, one must have $\theta_2^1 \leq \sigma_{21}^{12} \leq \theta_1^2$, so that the inequality $W_{\theta_1^2}(\sigma_{21}^{12}) \leq W_{\theta_1^2}(\sigma_{12}^{12}) = \rho_2$ holds if and only if $\sigma_{21}^{21} \leq \sigma_{12}^1$. Also since $\theta_2^2 \leq \sigma_{21}^{21} \leq \theta_1^1$, $W_{\theta_2^2}(\sigma_{21}^{21}) \leq W_{\theta_2^2}(\sigma_{12}^2) = \rho_2$ means that $\sigma_{12}^2 \leq \sigma_{21}^{21}$. Thus the adaptation assumption leads to the contradiction, $\theta_1^1 \geq \sigma_{21}^{21} \geq \sigma_{12}^2 \geq \sigma_{21}^{12} \geq \theta_2^1$.

In the part of the region R_2, where $\theta_1^1 \leq \theta_2^2 \leq \theta_1^2 \leq \theta_2^1$, one has $\rho_1 < \rho_2$. Indeed, $W_{\theta_1^1}(\sigma_{12}^1) \leq W_{\theta_2^2}(\sigma_{12}^2)$, since $\theta_1^1 \leq \theta_2^2$ and $\theta_1^1 \leq \sigma_{12}^1$, $\theta_2^2 \leq \sigma_{12}^2$. Thus the adaptation condition in this case takes the form $W_{\theta_1^1}(\sigma_{21}^{21}) \leq W_{\theta_1^1}(\sigma_{12}^1)$ and $W_{\theta_2^1}(\sigma_{21}^{12}) \leq W_{\theta_2^1}(\sigma_{12}^1)$. These inequalities mean that $\sigma_{21}^{21} \geq \sigma_{12}^1 \geq \sigma_{21}^{12}$, which leads to the contradiction, $\theta_1^2 \leq \theta_2^2$.

Assuming that $\rho_1 > \rho_2$, where $\theta_2^2 < \theta_1^1$ and $\theta_1^2 < \theta_2^1$, one obtains $\sigma_{21}^{12} > \theta_1^2$, so that the inequality $W_{\theta_2^1}(\sigma_{21}^{12}) \leq W_{\theta_2^1}(\sigma_{12}^1)$ implies $\sigma_{21}^{12} < \sigma_{12}^1$. Since $\rho_1 > \rho_2$, one has $\chi'(\tau_{21}^{12}) < \chi'(\sigma_{21}^{12})$. The last inequality shows that $\tau_{21}^{12} < \sigma_{21}^{12} < \sigma_{12}^1$. As was observed, the inequality $\tau_{21}^{12} < \sigma_{12}^1$ means that $\theta_1^1 > \theta_1^2$, contradicting to our assumption.

The case, when $\rho_1 < \rho_2$, is treated similarly, and the region $\{\theta_2^2 < \theta_1^2, \theta_1^1 < \theta_1^2\}$ does not present any new complications. ∎

We turn now to the adaptation issue of δ_0 and will demonstrate that for any α

$$\max_{i \neq k} \max_{\beta, \beta \neq \alpha} \inf_{s_\gamma \geq 0, \gamma = 1, \ldots, A} \left\{ \log E_i^\alpha \prod_\gamma \left[\frac{f_k^\beta}{f_i^\gamma}(X) \right]^{s_\gamma} \right\}$$
$$= \max_{i \neq k} \max_{\beta, \beta \neq \alpha} \min_\gamma \inf_{s \geq 0} \left\{ \chi\left(\theta_i^\alpha + s(\theta_k^\beta - \theta_i^\gamma)\right) - \chi(\theta_i^\alpha) - s\left[\chi(\theta_k^\beta) - \chi(\theta_i^\gamma)\right] \right\}.$$

Indeed assume that the gradient of the (convex) function of $\vec{s} = (s_1, \ldots, s_A)$ above vanishes at some $\vec{s}$ with at least two strictly positive coordinates, say, s_λ and s_π. Then

$$\chi'\left(\theta_i^\alpha + \sum_\gamma s_\gamma(\theta_k^\beta - \theta_i^\gamma)\right) = \frac{\chi(\theta_k^\beta) - \chi(\theta_i^\lambda)}{\theta_k^\beta - \theta_i^\lambda} = \frac{\chi(\theta_k^\beta) - \chi(\theta_i^\pi)}{\theta_k^\beta - \theta_i^\pi}.$$

It follows that $\theta_i^\lambda = \theta_i^\pi$, which because of Corollary 7.1.2 contradicts the assumed existence of an adaptive rule.

By Proposition 7.1.1, the adaptation condition for δ_0 can be rewritten in the following form: for any β

$$\max_{i\neq k}\ \max_{\alpha:\rho_\alpha\geq\rho_\beta}\ \min_{\gamma}\ \inf_{s>0}\left[\chi(\theta_k^\beta + s(\theta_i^\alpha - \theta_k^\gamma)) - \chi(\theta_k^\beta) - s\chi(\theta_i^\alpha) + s\chi(\theta_k^\gamma)\right] \leq \rho_\beta.$$

The infimum above is attained when $\theta_k^\beta + s(\theta_i^\alpha - \theta_k^\gamma) = \sigma_{ik}^{\alpha\gamma}$, which corresponds to positive s if and only if

$$\left(\tau_{ik}^{\alpha\gamma} - \theta_k^\beta\right)\left(\theta_i^\alpha - \theta_k^\gamma\right) > 0. \tag{7.15}$$

Denote by $\Gamma_{ik}^{\alpha\beta}$ the set of all γ for which (7.15) holds. Then always $\beta \in \Gamma_{ik}^{\alpha\beta}$.
   Thus δ_0 is adaptive if and only if

$$\max_{\alpha:\rho_\alpha\geq\rho_\beta}\ \max_{i\neq k}\ \min_{\gamma\in\Gamma_{ik}^{\alpha\beta}} W_{\theta_k^\beta}\left(\tau_{ik}^{\alpha\gamma}\right) \leq \rho_\beta = \max_{i\neq k} W_{\theta_k^\beta}\left(\sigma_{ik}^\beta\right). \tag{7.16}$$

When $A = 2$, for $\gamma = \alpha$, (7.15) takes the form

$$\left(\sigma_{ik}^\alpha - \theta_k^\beta\right)\left(\theta_i^\alpha - \theta_k^\alpha\right) > 0,$$

and for $\gamma = \beta$ it holds automatically.
   The next result gives the explicit form of adaptation subregions R^a and R^0 of δ_a and δ_0 outside the union of $R_0^c \bigcup R_1 \bigcup R_2$ treated in Proposition 7.2.1. We define the sets

$$R_3 = \{\theta_1^2 < \theta_2^2 < \theta_1^1, \rho_2 < \rho_1\},$$

$$R_4 = \{\theta_1^2 < \theta_2^2 < \theta_1^1, \rho_2 \geq \rho_1\},$$

$$R_5 = \{\theta_2^2 < \theta_1^2 < \theta_1^1, \rho_2 > \rho_1\},$$

$$R_6 = \{\theta_2^2 < \theta_1^2 < \theta_1^1, \rho_2 \leq \rho_1\},$$

for which, as is easy to see, $\sigma_{12}^2 < \sigma_{12}^1$, and the sets

$$R_7 = \{\theta_2^1 < \theta_1^2 < \theta_2^2, \rho_2 < \rho_1\},$$

$$R_8 = \{\theta_2^1 < \theta_1^2 < \theta_2^2, \rho_2 \geq \rho_1\},$$

$$R_9 = \{\theta_2^1 < \theta_2^2 < \theta_1^2, \rho_2 > \rho_1\},$$

$$R_{10} = \{\theta_2^1 < \theta_2^2 < \theta_1^2, \rho_2 \leq \rho_1\},$$

where $\sigma_{12}^2 > \sigma_{12}^1$.

Proposition 7.2.2 *Assume that $\theta_1^1 < \theta_2^1$. The adaptation subregions R_i^a of $R_i, i = 3, \ldots, 10$ and the δ_0–adaptation subregions R_i^0 have the following form*

Region R_i	Adaptation Region R_i^a	δ_0 − Adaptation Region R_i^0
R_3	$\{W_{\theta_2^2}(\sigma_{12}^{12}) \leq \rho_2, \sigma_{12}^{12} < \theta_1^1\}$	$\{W_{\theta_2^2}(\tau_{12}^{12}) \leq \rho_2\}$
R_4	$\{W_{\theta_1^1}(\sigma_{12}^{12}) \leq \rho_1, \sigma_{12}^{12} > \theta_2^2\}$	$\{W_{\theta_1^1}(\tau_{12}^{12}) \leq \rho_1\}$
R_5	$\{W_{\theta_1^1}(\sigma_{12}^2) \leq \rho_1\}$	$\{W_{\theta_1^1}(\sigma_{12}^2) \leq \rho_1\}$
R_6	$\{W_{\theta_1^2}(\sigma_{12}^1) \leq \rho_2\}$	$\{W_{\theta_1^2}(\sigma_{12}^1) \leq \rho_2\}$
R_7	$\{W_{\theta_1^2}(\sigma_{21}^{12}) \leq \rho_2, \sigma_{21}^{12} > \theta_2^2\}$	$\{W_{\theta_1^2}(\tau_{21}^{12}) \leq \rho_2\}$
R_8	$\{W_{\theta_2^1}(\sigma_{21}^{12}) \leq \rho_1, \sigma_{21}^{12} < \theta_1^2\}$	$\{W_{\theta_2^1}(\tau_{21}^{12}) \leq \rho_1\}$
R_9	$\{W_{\theta_2^1}(\sigma_2^{21}) \leq \rho_1\}$	$\{W_{\theta_2^1}(\sigma_{12}^2) \leq \rho_1\}$
R_{10}	$\{W_{\theta_2^2}(\sigma_{12}^1) \leq \rho_2\}$	$\{W_{\theta_2^2}(\sigma_{12}^1) \leq \rho_2\}$

PROOF. Returning to the discussion before Proposition 7.2.1 we see that δ_a is adaptive in the region R_3 if and only if $W_{\theta_1^2}(\theta_2^1) < \rho_2 - \rho_1$, $W_{\theta_2^2}(\theta_1^1) < \rho_2 - \rho_1$, $W_{\theta_2^2}(\sigma_{12}^{12}) \leq \rho_2$ and $W_{\theta_1^2}(\sigma_{21}^{12}) \leq \rho_2 = W_{\theta_1^2}(\sigma_{12}^2)$. Since $\sigma_{12}^2 > \theta_1^2$ and

$$\chi'\left(\sigma_{21}^{12}\right) = \frac{\chi(\theta_2^1) - \chi(\theta_1^2) + \rho_1 - \rho_2}{\theta_2^1 - \theta_1^2}$$

$$\geq \frac{\chi(\theta_2^1) - \chi(\theta_1^2)}{\theta_2^1 - \theta_1^2} = \chi'\left(\tau_{21}^{12}\right) \geq \chi'\left(\theta_1^2\right),$$

one can rewrite the last adaptation condition as $\sigma_{12}^2 < \sigma_{21}^{12}$ or as

$$\chi'\left(\sigma_{12}^2\right) < \chi'\left(\sigma_{21}^{12}\right) = \frac{\chi(\theta_2^1) - \chi(\theta_1^2) + \rho_1 - \rho_2}{\theta_2^1 - \theta_1^2}.$$

This inequality is implied by the inequality $\sigma_{12}^2 < \tau_{21}^{12}$, which merely means that $\theta_2^2 < \theta_2^1$. In the region R_3, $\theta_2^1 > \sigma_{12}^2$, so that $W_{\theta_1^2}(\theta_2^1) < W_{\theta_1^2}(\sigma_{12}^2) = \rho_2 < \rho_2 - \rho_1$,

Thus the adaptation condition for R_3 reduces to the conditions $W_{\theta_2^2}(\theta_1^1) < \rho_2 - \rho_1$, which is the same as $\sigma_{12}^{12} < \theta_1^1$, and $W_{\theta_2^2}(\sigma_{12}^{12}) \leq \rho_2$. The form of regions R_4^a, R_7^a and R_8^a is obtained in a similar fashion.

In the region R_5, $\sigma_{12}^{12} < \tau_{12}^{12} < \theta_1^1$, and the adaptation condition from (7.12) reduces to the inequality $W_{\theta_1^1}(\sigma_{12}^{12}) \leq \rho_1$. The consistency condition (7.11) means that $W_{\theta_1^1}(\theta_2^2) < \rho_1 - \rho_2$ or that $\sigma_{12}^{12} \geq \theta_2^2$. We prove that these inequalities are equivalent to the condition $W_{\theta_1^1}(\sigma_{12}^2) \leq \rho_2$.

The inequality $W_{\theta_1^1}(\sigma_{12}^2) \leq \rho_1$ implies that $W_{\theta_1^1}(\sigma_{12}^{12}) \leq \rho_1$. Also according to the same inequality, $W_{\theta_1^1}(\theta_2^2) \leq W_{\theta_1^1}(\sigma_{12}^2) \leq \rho_1 < \rho_1 - \rho_2$. Now we show that under adaptation and consistency conditions $W_{\theta_1^1}(\sigma_{12}^2) \leq \rho_1$. Because of (7.13) $W_{\theta_2^2}(\sigma_{12}^{12}) \leq \rho_2 = W_{\theta_2^2}(\sigma_{12}^2)$. Thus, since $\sigma_{12}^2 \geq \theta_2^2$, the latter inequality means that $\sigma_{12}^{12} \geq \sigma_{12}^2$, and this shows that $W_{\theta_1^1}(\sigma_{12}^2) \leq W_{\theta_1^1}(\sigma_{12}^2) \leq \rho_1$ establishing the claimed equivalence.

In R_6, the equivalence of the conditions $W_{\theta_1^2}\left(\sigma_{12}^1\right) \le \rho_2$ and $W_{\theta_1^2}\left(\sigma_{21}^{12}\right) \le \rho_2$, $W_{\theta_1^2}\left(\theta_2^1\right) \le \rho_2 - \rho_1$ is proved quite similarly. In the region R_9 $W_{\theta_2^1}\left(\sigma_{12}^2\right) \le \rho_1$ if and only if $W_{\theta_2^1}\left(\sigma_{21}^{12}\right) \le \rho_1$, $W_{\theta_2^1}\left(\theta_1^2\right) \le \rho_1 - \rho_2$, and in R_{10}, $W_{\theta_2^2}\left(\sigma_{12}^1\right) \le \rho_2$ is tantamount to inequalities $W_{\theta_2^2}\left(\sigma_{12}^{12}\right) \le \rho_2$, $W_{\theta_2^2}\left(\theta_1^1\right) \le \rho_2 - \rho_1$. In these regions the adaptation condition is common for both δ_a and δ_0, and we turn now to the adaptation issue of the latter procedure.

In the region R_3, $(\sigma_{12}^1 - \theta_2^2)\left(\theta_1^1 - \theta_2^1\right) > 0$, is an impossible inequality, so that $\Gamma_{12}^{12} = \{2\}$. For δ_0 to be adaptive, one must have $W_{\theta_2^2}\left(\tau_{12}^{12}\right) \le \rho_2$. It is easy to see that $\Gamma_{12}^{21} = \{1,2\}$ and $W_{\theta_1^2}\left(\sigma_{12}^1\right) \wedge W_{\theta_1^2}\left(\tau_{21}^{12}\right) = W_{\theta_1^2}\left(\sigma_{12}^1\right)$ because of the inequalities $\theta_1^2 \le \tau_{21}^{12} \le \sigma_{12}^1$. Also $W_{\theta_1^2}\left(\sigma_{12}^1\right) \le \rho_2 = W_{\theta_1^2}\left(\sigma_{12}^2\right)$ if and only if $\sigma_{12}^1 \ge \sigma_{12}^2$, which holds in the region R_3. Thus the adaptation condition (7.16) for $\alpha = 1, \beta = 2, i = 2, k = 1$, is satisfied automatically, and the one for $i = 1, k = 2$, means that $W_{\theta_2^2}\left(\tau_{12}^{12}\right) \le \rho_2$. Since $\rho_1 > \rho_2$, one has $\theta_2^2 \le \tau_{12}^{12} \le \sigma_{12}^{12}$, so that the adaptation condition for δ_a implies $W_{\theta_2^2}\left(\sigma_{12}^{12}\right) \le \rho_2$. Also because of the inequality $\tau_{12}^{12} < \theta_1^1$, this condition implies that $W_{\theta_2^2}\left(\theta_1^1\right) < W_{\theta_2^2}\left(\tau_{12}^{12}\right) \le \rho_2 < \rho_2 - \rho_1$. Thus, as was to be expected, δ_0 is adaptive only if the adaptation condition for δ_a is met.

In the region R_4, $\Gamma_{12}^{21} = \{1,2\}$ and $\Gamma_{21}^{21} = \{1\}$. The condition

$$W_{\theta_2^1}\left(\sigma_{12}^2\right) \vee W_{\theta_2^1}\left(\tau_{21}^{21}\right) = W_{\theta_2^1}\left(\sigma_{12}^2\right) \le \rho_1$$

holds in R_4 automatically, and the adaptation condition for δ_0 takes the form $W_{\theta_1^1}\left(\tau_{12}^{12}\right) \le \rho_1$.

For the region R_5, $\Gamma_{12}^{21} = \{2\}$ and $\Gamma_{21}^{21} = \{1,2\}$. The condition $W_{\theta_2^1}\left(\tau_2^{12}\right) \le \rho_1$, holds since $\tau_{12}^2 \le \sigma_1^{12} \le \theta_2^1$. In R_5 one has $\theta_1^1 > \tau_{12}^{12} > \tau_2^{12}$, with the resulting δ_0-adaptation condition, $W_{\theta_1^1}\left(\tau_{12}^2\right) \wedge W_{\theta_1^1}\left(\tau_{12}^{12}\right) = W_{\theta_1^1}\left(\sigma_{12}^1\right) \le \rho_1$.

Similarly in the region R_6, $\Gamma_{12}^{21} = \{1,2\}$, $\Gamma_{12}^{12} = \{2\}$, and the adaptation condition reduces to $W_{\theta_1^2}\left(\sigma_{12}^1\right) \le \rho_2$. The remaining regions are treated in the same way. ∎

To conclude this section we look at the form of the estimators δ_a and δ_0. One has

$$\{\delta_a(x_1,\ldots,x_n) = 1\} = \{\left(\theta_1^1\bar{x} - \chi\left(\theta_1^1\right) - \rho_1\right) \vee \left(\theta_1^2\bar{x} - \chi\left(\theta_1^2\right) - \rho_2\right)$$
$$> \left(\theta_2^1\bar{x} - \chi\left(\theta_2^1\right) - \rho_1\right) \vee \left(\theta_2^2\bar{x} - \chi\left(\theta_2^2\right) - \rho_2\right)\}$$

and

$$\{\delta_0(x_1,\ldots,x_n) = 1\} = \{\left(\theta_1^1\bar{x} - \chi\left(\theta_1^1\right)\right) \vee \left(\theta_1^2\bar{x} - \chi\left(\theta_1^2\right)\right)$$
$$> \left(\theta_2^1\bar{x} - \chi\left(\theta_2^1\right)\right) \vee \left(\theta_2^2\bar{x} - \chi\left(\theta_2^2\right)\right)\}.$$

The next Proposition gives the explicit form of these rules in the regions $R_i^a, i = 3,\ldots,10$.

Proposition 7.2.3 *The set $\{\delta_a(x_1, \ldots, x_n) = 1\}$ has the form*

Adaptation Region	$\{\delta_a(x_1, \ldots, x_n) = 1\}$
$R_3^a \bigcup R_4^a$	$\left\{\bar{x} < \chi'(\sigma_{12}^2)\right\} \bigcup \left\{\chi'(\sigma_{12}^{12}) \leq \bar{x} < \chi'(\sigma_{12}^1)\right\}$
$R_5^a \bigcup R_6^a$	$\left\{\chi'(\sigma_{12}^2) \leq \bar{x} < \chi'(\sigma_{12}^1)\right\}$
$R_7^a \bigcup R_8^a$	$\left\{\bar{x} < \chi'(\sigma_{12}^1)\right\} \bigcup \left\{\chi'(\sigma_{21}^{12}) \leq \bar{x} < \chi'(\sigma_{12}^2)\right\}$
$R_9^a \bigcup R_{10}^a$	$\left\{\bar{x} < \chi'(\sigma_{12}^1)\right\} \bigcup \left\{\chi'(\sigma_{12}^2) \leq \bar{x}\right\}$

and the set $\{\delta_0(x_1, \ldots, x_n) = 1\}$ has the form

Region	$\{\delta_0(x_1, \ldots, x_n) = 1\}$
$R_3 \bigcup R_4$	$\{\bar{x} < \chi'(\sigma_{12}^2)\} \bigcup \{\chi'(\tau_{12}^{12}) \leq \bar{x} < \chi'(\sigma_{12}^1)\}$
$R_5 \bigcup R_6$	$\{\chi'(\sigma_{12}^2) \leq \bar{x} < \chi'(\sigma_{12}^1)\}$
$R_7 \bigcup R_8$	$\{\bar{x} < \chi'(\sigma_{12}^1)\} \bigcup \{\chi'(\tau_{21}^{12}) \leq \bar{x} < \chi'(\sigma_{12}^2)\}$
$R_9 \bigcup R_{10}$	$\{\bar{x} < \chi'(\sigma_{12}^1)\} \bigcup \left\{\chi'(\sigma_{12}^2) \leq \bar{x}\right\}.$

We omit the straightforward proof of this proposition.

Proposition 7.2.3 shows that in the regions R_5, R_6, R_9 and R_{10} the rules δ_a and δ_0 coincide, so that they automatically are adaptive simultaneously as implied by Proposition 7.2.2.

7.3   Gamma Distributions Family

The distribution F_i^α is a gamma–distribution with a positive parameter η_i^α and the fixed number degrees of freedom p. Thus

$$f_i^\alpha(u) = (\eta_i^\alpha)^p \frac{e^{-\eta_i^\alpha u} u^{p-1}}{\Gamma(p)} , \qquad u > 0.$$

Then $\theta_i^\alpha = -\eta_i^\alpha < 0$ and $\chi(\theta_i^\alpha) = -p \log |\theta_i^\alpha|$. To simplify the following calculations we put $p = 1$, although the forms of the adaptation regions $R_3 - R_{10}$ are the same for any positive p.

It follows that for $\theta, t < 0$

$$W_\theta(t) = G\left(\frac{|\theta|}{|t|}\right),$$

where for positive u

$$G(u) = 1 + \log u - u.$$

According to (7.7)

$$\frac{1}{|\sigma_{ik}^\alpha|} = \frac{\log \eta_i^\alpha - \log \eta_k^\alpha}{\eta_i^\alpha - \eta_k^\alpha},$$

and with $g(u) = \log u/(u-1)$

$$\rho_\alpha = \max_{i \neq k} G\left(\frac{\log \eta_i^\alpha - \log \eta_k^\alpha}{\eta_i^\alpha/\eta_k^\alpha - 1}\right) = \max_{i \neq k} G\left(g\left(\frac{\eta_i^\alpha}{\eta_k^\alpha}\right)\right).$$

Notice that $G(u)$ is a unimodal function with mode 0 at $u = 1$ and

$$G\left(g(u)\right) = G\left(g(u^{-1})\right).$$

Also

$$\frac{1}{|\sigma_{ik}^{\alpha\beta}|} = \frac{\log \eta_k^\beta - \log \eta_i^\alpha + \rho_\alpha - \rho_\beta}{\eta_i^\alpha - \eta_k^\beta},$$

with a similar formula for $\tau_{ik}^{\alpha\beta}$.

Proposition 7.2.2 leads to the form of regions R_i^a and R_i^0, $i = 3, \ldots, 10$. To facilitate their description assume for the sake of concreteness that $\theta_1^1 = -2, \theta_2^1 = -1$, so that

$$\rho_1 = G(2\log 2) = G(\log 2) = 1 + \log \log 2 - \log 2.$$

Notice that $\rho_1 \leq \rho_2$ if and only if $1/2 \leq \eta_2^2/\eta_1^2 \leq 2$, which follows directly from the formula for ρ_α.

In the region R_3^a, $2 < \eta_2^2 < \eta_1^2/2$. The inequality $W_{\theta_2^2}(\sigma_{12}^{12}) \leq \rho_2$ means that

$$G\left(\frac{\eta_2^2}{|\sigma_{12}^{12}|}\right) \leq G\left(\frac{\eta_2^2}{|\sigma_{12}^2|}\right) = G\left(g\left(\frac{\eta_2^2}{\eta_1^2}\right)\right).$$

Since in R_3^a, $|\sigma_{12}^{12}| \leq \eta_2^2$ and $\eta_2^2 \leq \eta_1^2$, one has $g\left(\frac{\eta_2^2}{\eta_1^2}\right) > 1$. By unimodality of G, this inequality means that

$$\frac{\eta_2^2}{|\sigma_{12}^{12}|} \geq g\left(\frac{\eta_2^2}{\eta_1^2}\right)$$

or that

$$\frac{\eta_2^2[\log(\eta_2^2/2) + \rho_1 - \rho_2]}{\eta_2^2 - 2} \geq \frac{\eta_2^2/\eta_1^2)}{\eta_2^2/\eta_1^2 - 1}.$$

The consistency condition of δ_a has the form

$$\frac{\eta_2^2}{|\sigma_{12}^{12}|} < \frac{\eta_2^2}{2},$$

or

$$\frac{\log(\eta_2^2/2) + \rho_1 - \rho_2}{\eta_2^2 - 2} \leq \frac{1}{2},$$

and the adaptation condition for δ_0 is

$$\frac{\eta_2^2}{|\tau_{12}^{12}|} \geq g\left(\frac{\eta_2^2}{\eta_1^2}\right)$$

or

$$\frac{\eta_2^2 \log(\eta_2^2/2)}{\eta_2^2 - 2} \geq \frac{\eta_2^2/\eta_1^2)}{\eta_2^2/\eta_1^2 - 1} \; .$$

In the region R_4, the δ_0-adaptation condition $\log 2 \geq \frac{2}{|\tau_{12}^{12}|}$ means that $\frac{2 \log(\eta_2^2/2)}{\eta_2^2 - 2} \leq \log 2$ or that $\eta_2^2 \geq 4$. In the region R_8, a similar condition $\frac{1}{|\tau_{21}^{12}|} \geq 2 \log 2$ reduces to $\eta_1^2 \leq 1/2$.

The forms of other regions R_i^a and R_i^0 are obtained similarly. One has:

Region R_i	Adaptation Region R_i^a	δ_0 $-$ Adaptation Region R_i^0
R_3	$\frac{\eta_2^2}{2} > \frac{\eta_2^2}{|\sigma_{12}^{12}|} \geq g\left(\frac{\eta_2^2}{\eta_1^2}\right)$	$\frac{\eta_2^2}{|\tau_{12}^{12}|} \geq g\left(\frac{\eta_2^2}{\eta_1^2}\right)$
R_4	$\log 2 \geq \frac{2}{|\sigma_{12}^{12}|} > \frac{2}{\eta_2^2}$	$\eta_2^2 \geq 4$
R_5	$\log 2 \geq \frac{2}{|\sigma_{12}^{12}|}$	$\log 2 \geq \frac{2}{|\sigma_{12}^{12}|}$
R_6	$\eta_1^2 \log 2 \geq g\left(\frac{\eta_1^2}{\eta_2^2}\right)$	$\eta_1^2 \log 2 \geq g\left(\frac{\eta_1^2}{\eta_2^2}\right)$
R_7	$g\left(\frac{\eta_1^2}{\eta_2^2}\right) \geq \frac{\eta_1^2}{|\sigma_{21}^{12}|} > \eta_1^2$	$g\left(\frac{\eta_1^2}{\eta_2^2}\right) \geq \frac{\eta_1^2}{|\tau_{21}^{12}|}$
R_8	$\frac{1}{\eta_1^2} > \frac{1}{|\sigma_{21}^{12}|} \geq 2 \log 2$	$\eta_1^2 \leq 1/2$
R_9	$\frac{1}{|\sigma_{12}^{12}|} \geq 2 \log 2$	$\frac{1}{|\sigma_{12}^{12}|} \geq 2 \log 2$
R_{10}	$\eta_2^2 \log 2 \geq g\left(\frac{\eta_2^2}{\eta_1^2}\right)$	$\eta_2^2 \log 2 \geq g\left(\frac{\eta_2^2}{\eta_1^2}\right)$

According to Proposition 7.2.3, the rule δ_0 has the form

$$\{\delta_0(x_1, \ldots, x_n) = 1\}$$
$$= \left\{ \min\left[2\bar{x} - \log 2, \eta_1^2 \bar{x} - \log \eta_1^2\right] < \min\left[\bar{x}, \eta_2^2 \bar{x} - \log \eta_2^2\right] \right\}.$$

For example, $\eta_1^2 = 16, \eta_2^2 = 5$ belongs to the region R_3^a, but not to R_3^0, and the rule δ_0

$$\{\delta_0(x_1, \ldots, x_n) = 1\} = \{\bar{x} < 0.1057..\} \cup \{0.3054.. \leq \bar{x} \leq 0.6931..\},$$

is not adaptive. In our situation, $\rho_1 = -0.0597..$ and $\rho_2 = -0.1660...$

The rule δ_a has the form

$$\{\delta_a(x_1, \ldots, x_n) = 1\} = \left\{\bar{x} < \chi'(\sigma_{12}^2)\right\} \cup \left\{\chi'(\sigma_{12}^{12}) \leq \bar{x} < \chi'(\sigma_{12}^1)\right\}$$
$$= \{\bar{x} < 0.1057..\} \cup \{0.3409.. \leq \bar{x} \leq 0.6931..\},$$

and it is adaptive.

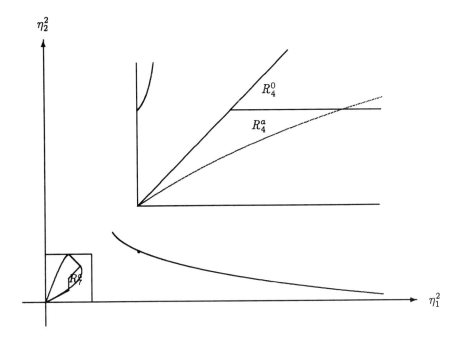

Figure 7.1: The adaptation regions for δ_0 and δ_a with exponential distributions when $\eta_1^1 = 2$, $\eta_2^1 = 1$

For equal prior probabilities, its minimum Bayes risk, for $\alpha = 1$,

$$b_n(\delta_a) = \frac{P_1^1(\delta_a = 2) + P_2^1(\delta_a = 2)}{2} \ ,$$

has the following asymptotic behavior,

$$\frac{\log b_n(\delta_a)}{n} \sim \frac{\log b_n(\delta_0)}{n} \to \rho_1 = -.0597...$$

However, when $\alpha = 2$,

$$\frac{\log b_n(\delta_a)}{n} \to G\left(5\frac{\log 2.5 + \rho_1 - \rho_2}{3}\right) = -.1660\ldots,$$

while the corresponding limit for δ_0 is much closer to zero, namely,

$$\frac{\log b_n(\delta_0)}{n} \to G\left(5\frac{\log 2.5}{3}\right) = -.1037\ldots.$$

7.4 Normal Family

The distribution F_i^α is the multivariate normal with the mean η_i^α and the nonsingular covariance matrix R, which does not depend on i or α. Then $\theta_i^\alpha = R^{-1}\eta_i^\alpha$ and with $<.,.>$ denoting the inner product in the coreesponfing Euclidean space

$$\chi(\theta_i^\alpha) = \frac{1}{2} < R^{-1}\eta_i^\alpha, \eta_i^\alpha > = \frac{1}{2} < R\theta_i^\alpha, \theta_i^\alpha > = \frac{1}{2}||\theta_i^\alpha||_R^2,$$

say. According to (7.7),

$$\rho_\alpha = \frac{1}{2} \max_{i\neq k} \inf_{s>0} \left[||s\theta_i^\alpha + (1-s)\theta_k^\alpha||_R^2 - 2s||\theta_i^\alpha||_R^2 - 2(1-s)||\theta_k^\alpha||_R^2 \right]$$

$$= -\frac{1}{8} \min_{i\neq k} ||\theta_i^\alpha - \theta_k^\alpha||_R^2. \tag{7.17}$$

Now we determine the explicit form of (7.9). For fixed $\alpha \neq \beta, i \neq k$ and $\theta_i^\alpha \neq \theta_k^\beta$ the minimizer in (7.9) has the form

$$s_{ik}^{\alpha\beta} = \frac{1}{2} + \frac{\rho_\alpha - \rho_\beta}{||\theta_i^\alpha - \theta_k^\beta||_R^2} \ .$$

The condition $0 < s_{ik}^{\alpha\beta} < 1$, which is necessary for adaptation because of (7.6), means that δ_a is consistent and corresponds to the region R_0 discussed in Section 2. It can be rewritten in the form $s_{ik}^{\alpha\beta}\left(1 - s_{ik}^{\alpha\beta}\right) > 0$, which takes place if and only if

$$2|\rho_\alpha - \rho_\beta| < ||\theta_i^\alpha - \theta_k^\beta||_R^2. \tag{7.18}$$

One has

$$\min_s \left[\frac{1}{2}||s\theta_i^\alpha + (1-s)\theta_k^\beta||_R^2 - \frac{s||\theta_i^\alpha||_R^2}{2} - \frac{(1-s)||\theta_k^\beta||_R^2}{2} - s\rho_\alpha - (1-s)\rho_\beta \right]$$

$$= \frac{1}{2} \left[-\frac{(\rho_\alpha - \rho_\beta)^2}{||\theta_i^\alpha - \theta_k^\beta||_R^2} - \rho_\alpha - \rho_\beta - \frac{||\theta_i^\alpha - \theta_k^\beta||_R^2}{4} \right] .$$

Thus the condition

$$\inf_{s>0} \left[\frac{1}{2}||s\theta_i^\alpha + (1-s)\theta_k^\beta||_R^2 - s||\theta_i^\alpha||_R^2 - (1-s)||\theta_k^\beta||_R^2 - s\rho_\alpha - (1-s)\rho_\beta \right] \leq 0$$

means that

$$\frac{||\theta_i^\alpha - \theta_k^\beta||_R}{2} + \frac{|\rho_\alpha - \rho_\beta|}{||\theta_i^\alpha - \theta_k^\beta||_R} \geq [|\rho_\alpha| + |\rho_\beta| + |\rho_\alpha - \rho_\beta|]^{1/2} = \sqrt{2|\rho_\alpha| \vee |\rho_\beta|}.$$

This inequality signifies that either

$$||\theta_i^\alpha - \theta_k^\beta||_R \geq \sqrt{2|\rho_\alpha| \vee |\rho_\beta|} + \sqrt{2|\rho_\alpha| \wedge |\rho_\beta|} = \sqrt{2|\rho_\alpha|} + \sqrt{2|\rho_\beta|} \qquad (7.19)$$

or

$$||\theta_i^\alpha - \theta_k^\beta||_R \leq \sqrt{2|\rho_\alpha| \vee |\rho_\beta|} - \sqrt{2|\rho_\alpha| \wedge |\rho_\beta|}.$$

The latter inequality contradicts (7.18), whereas (7.19) implies (7.18). Therefore (7.19) provides the necessary and sufficient condition for the existence of an adaptive rule in this example.

Under this condition the adaptive rule δ_a has the form

$$\begin{aligned}
\{\delta_a(x_1, \ldots, x_n) = i\} &= \{\min_\alpha \left[< R^{-1}(\bar{x} - \eta_i^\alpha), \bar{x} - \eta_i^\alpha > +2\rho_\alpha \right] \\
&= \min_{\alpha,k} \left[< R^{-1}(\bar{x} - \eta_k^\alpha), \bar{x} - \eta_k^\alpha > +2\rho_\alpha \right] \}
\end{aligned}$$

with ρ_α given in (7.17).

The overall maximum likelihood rule δ_0 has the form

$$\begin{aligned}
\{\delta_0(x_1, \ldots, x_n) = i\} &= \{\min_\alpha < R^{-1}(\bar{x} - \eta_i^\alpha), \bar{x} - \eta_i^\alpha > \\
&= \min_{\alpha,k} < R^{-1}(\bar{x} - \eta_k^\alpha), \bar{x} - \eta_k^\alpha > \}
\end{aligned}$$

The adaptation region R^0 for the rule δ_0 is described by the condition that for any $\alpha \neq \beta$ and $i \neq k$ there exists γ and $m \neq k$ such that

$$\min_{s>0} \left[||\theta_i^\alpha + s(\theta_k^\beta - \theta_m^\gamma)||_R^2 + s||\theta_m^\gamma||_R^2 - s||\theta_k^\beta||_R^2 \right] \leq \frac{||\theta_i^\alpha||_R^2}{2} .$$

The analogue of (7.15) has the form

$$||\theta_k^\beta||_R^2 - ||\theta_m^\gamma||_R^2 > 2 < R \left(\theta_k^\beta - \theta_i^\alpha \right), \theta_i^\alpha >, \qquad (7.20)$$

which is true automatically if $\gamma = \alpha$. The adaptation condition for δ_0 has the form

$$\min_{k \neq i} \min_{\beta : \rho_\beta > \rho_\alpha} \max_{\gamma, m \neq k} \left[\frac{||\theta_m^\gamma + \theta_k^\beta||_R^2}{4} - < R \left(\theta_k^\beta + \theta_m^\gamma - \theta_i^\alpha \right), \theta_i^\alpha > \right] \geq -2\rho_\alpha. \quad (7.21)$$

Let $g = 2$, $F_1^1 = N(-1, 1)$, $F_2^1 = N(1, 1)$, and $F_1^2 = N(\xi, 1)$, $F_2^2 = N(\psi, 1)$ with some ξ and ψ. According to (7.17) $\rho_1 = -1/2, \rho_2 = -(\xi - \psi)^2/8$, and (7.19) takes the form $|1 - \xi| \wedge |1 + \psi| \geq 1 + |\xi - \psi|/2$. The adaptation region R^a consists of two wedges in the (ξ, ψ)- plane, $\{\xi \leq 3\psi, \xi + \psi \geq 4, 3\xi - \psi \geq 4\} \cup \{\psi \geq 3\xi, \xi + \psi \leq -4, 3\psi - \xi \leq -4\}$, plus the line $\{\xi + \psi = 0, \xi < 0\}$ and minus the line $\{\xi = \psi\}$. See Figure 7.2.

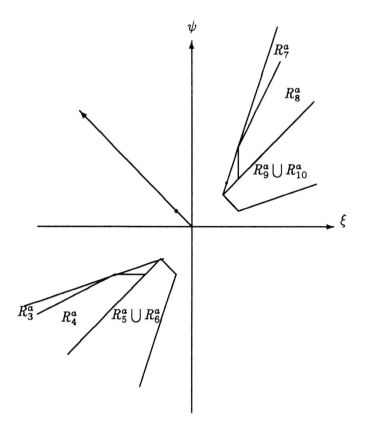

Figure 7.2: The adaptation regions R^a and R^0 for the rules δ_a and δ_0 in the normal example

The inequalities (7.21) and (7.20) lead to a smaller adaptation region R^0 for δ_0, namely, $\{3 \leq \xi < \psi, 2\xi - \psi \geq 1\} \cup \{2 \leq \xi \leq 3, \xi < \psi, 3\xi - \psi \geq 4\} \cup \{\xi > \psi, 3\psi \geq \xi, \xi + \psi \geq 4\}$ and a similar region obtained by replacing (ξ, ψ) by $(-\psi, -\xi)$.

Proposition 7.2.3 shows that when $\psi > \xi \geq (\psi - \xi)/2$ (which corresponds to the region $R_7^a \cup R_8^a$)

$$\{\delta_a(x_1, \ldots, x_n) = 1\} = \{\bar{x} < 0\} \cup \left\{ \frac{4\xi^2 - (\xi - \psi)^2}{8(\xi - 1)} \leq \bar{x} \leq \frac{\xi + \psi}{2} \right\},$$

and

$$\{\delta_0(x_1, \ldots, x_n) = 1\} = \{\bar{x} < 0\} \cup \left\{ \frac{\xi + 1}{2} \leq \bar{x} \leq \frac{\xi + \psi}{2} \right\}.$$

For $\xi > \psi > 1$ (i.e. in $R_9^a \cup R_{10}^a$), according to Proposition 7.2.3 these rules coincide.

It follows that for instance, when $\xi = 2\frac{1}{4}, \psi = 2\frac{3}{4}$, an adaptive rule exists. Namely the procedure δ_a, which in this situation takes the form

$$\{\delta_a(x_1, \ldots, x_n) = 1\} = \{\bar{x} < 0\} \cup \left\{2 \leq \bar{x} \leq \frac{5}{2}\right\},$$

is adaptive.

For equal prior probabilities its minimum Bayes risk $b_n(\delta_a)$ when $\alpha = 1$ has the following asymptotic behavior:

$$\frac{1}{2}\left[3\Phi\left(-\sqrt{n}\right) + \Phi\left(-\frac{7\sqrt{n}}{2}\right) - \Phi\left(-3\sqrt{n}\right) - \Phi\left(-\frac{3\sqrt{n}}{2}\right)\right]$$

$$\sim \frac{3e^{-n/2}}{2\sqrt{2\pi n}}$$

with Φ denoting the distribution function of the standard normal distribution.

For $\alpha = 2$, $b_n(\delta_a)$ has the form

$$\frac{1}{2}\left[3\Phi\left(-\frac{\sqrt{n}}{4}\right) + \Phi\left(-\frac{11\sqrt{n}}{4}\right) - \Phi\left(-\frac{9\sqrt{n}}{4}\right) - \Phi\left(-\frac{3\sqrt{n}}{4}\right)\right]$$

$$\sim \frac{e^{-n/32}}{2\sqrt{2\pi n}}.$$

Therefore, in both of these cases δ_a is asymptotically efficient, i.e. $\lim n^{-1} \log b(n)$ for $\alpha = 1$ is $\rho_1 = -1/2$ and for $\alpha = 2$ one has $\rho_2 = -1/32$. However δ_0, which has the from

$$\{\delta_0(x_1, \ldots, x_n) = 1\} = \{\bar{x} < 0\} \cup \left\{\frac{13}{8} \leq \bar{x} \leq \frac{5}{2}\right\},$$

is not adaptive.

When $\alpha = 1$

$$b_n(\delta_0) \sim \frac{e^{-25n/128}}{\sqrt{2\pi n}},$$

and for $\alpha = 2$ $b_n(\delta_0) \sim e^{-n/32}/\sqrt{2\pi n}$. Thus for $\alpha = 2$, this rule is asymptotically fully efficient, but for $\alpha = 1$ the asymptotic performance of δ_0 is distinctly suboptimal.

References

1. Bickel, P. A, Klaassen, C. A. J., Ritov, J. and Wellner, J. A. (1993). *Efficient and Adaptive Estimation for Semiparametric Models*, Baltimore: Johns Hopkins University Press.

2. Chernoff, H. (1956). Large-sample theory: Parametric case, *Annals of Mathematical Statistics*, **27**, 1–22.

3. Krafft, O. and Puri, M. L. (1974). The asymptotic behavior of the minimax risk for multiple decision problems, *Sankhya A*, **36**, 1–12.

4. Renyi, A. (1970). On some problems of statistics from the point of view of information theory, In *Proceedings of the Colloquium on Information Theory*–**2**, pp. 343–357, Budapest, Hungary: Bolyai Mathematical Society.

5. Rukhin, A. L. (1982). Adaptive procedures for a finite number of probability distributions families, In *Proceedings of Third Purdue Symposium on Decision Theory* –**1** (Eds. S. S. Gupta and J. O. Berger), pp. 269–286, New York: Academic Press.

6. Rukhin, A. L. (1984). Adaptive classification procedures, *Journal of the American Statistical Association*, **79**, 415–422.

7. Stein, C. (1957). Efficient nonparametric testing and estimation, In *Proceedings of Third Berkeley Symposium on Mathematical Statistics and Probability*–**1**, pp. 187–196, Berkeley: University of California Press.

8

Non-Informative Priors Via Sieves and Packing Numbers

S. Ghosal, J. K. Ghosh and R. V. Ramamoorthi

Indian Statistical Institute, Calcutta, India
Indian Statistical Institute, Calcutta, India
Michigan State University, East Lansing, MI

Abstract: In this paper, we propose methods for the construction of a non-informative prior through the uniform distributions on approximating sieves. In parametric families satisfying regularity conditions, it is shown that Jeffreys' prior is obtained. The case with nuisance parameters is also considered. In the infinite dimensional situation, we show that such a prior leads to consistent posterior.

Keywords and phrases: Non-informative prior, sieve, packing number, Jeffreys' prior, posterior consistency

8.1 Introduction

There has been some revival of interest in non-informative and automatic priors for quick, automatic Bayesian analysis as well as for providing a sort of reference point for subjective Bayesian analysis, vide Berger and Bernardo (1992) and Bernardo (1979). Some recent references are Berger and Bernardo (1989), Tibshirani (1989), Zellner (1990), Ghosh and Mukerjee (1992), Kass and Wasserman (1992), Datta and Ghosh (1995a,b), Ghosh (1994), Datta and Ghosh (1995, 1996) and Zellner (1990). These papers deal with a single parametric model. There has also been interest in such priors when one has several nested or non-nested parametric models; see, for example, Spiegelhalter and Smith (1982), Berger and Pericchi (1994, 1996) and O'Hagan (1995). Kass and Wasserman (1996) is a recent survey on non-informative and related priors.

Except for a non-informative choice of the base measure $\alpha(\cdot)$ for a Dirichlet process prior, very little is known about non-informative priors in non-parametric or infinite dimensional problems. In the following pages, we propose and study a few methods of generating such priors using sieves and packing numbers. We also apply our ideas to parametric models.

Suppose we have a model $\mathcal{P}$, equipped with a metric ρ. Our preferred metric will be the Hellinger metric. To keep the discussion simple, we will initially assume that $\mathcal{P}$ is compact. This assumption can then be relaxed in at least some σ-compact cases in a standard way as indicated in Section 8.2. Our starting point is a sequence ε_i diminishing to zero and sieves $\mathcal{F}_i$, where $\mathcal{F}_i$ is a finite set whose elements are separated from each other by at least ε_i and has cardinality $D(\varepsilon_i, \mathcal{P})$, the largest m for which there are $P_1, \ldots, P_m \in \mathcal{P}$ with $\rho(P_j, P_{j'}) > \varepsilon_i$, $j \neq j'$, $j, j' = 1, \ldots, m$. Clearly given any $P \in \mathcal{P}$, there exists $P' \in \mathcal{F}_i$ such that $\rho(P, P') \leq \varepsilon_i$. Thus $\mathcal{F}_i$ approximates $\mathcal{P}$ within ε_i and removal of a single point from $\mathcal{F}_i$ will destroy this property.

In the first method, we fix the sample size n, i.e., we assume we have n i.i.d. $X_i \sim P \in \mathcal{P}$. We choose $\varepsilon_{i(n)}$ satisfying (1.5) of Wong and Shen (1995) or in some other suitable way. It is then convenient to think of $\mathcal{F}_{i(n)}$ as a discrete or finite approximation to $\mathcal{P}$, which is as fine as is compatible with our resource measured by the sample size n. Of course, greater the value of n, the richer the approximating set we will treat as our proxy model. In the first method, our noninformative prior is simply the uniform distribution on $\mathcal{F}_{i(n)}$.

This seems to accord well with Basu's (1975) recommendation in the parametric case to approximate the parameter space Θ by a finite set and then put a uniform distribution. It is also intuitively plausible that the complexity or richness of a model $\mathcal{F}_{i(n)}$ may be allowed to depend on the sample size. Unfortunately, it follows from Heath and Sudderth (1978) that our inference based on a sample size dependent prior cannot be coherent in their sense, if one has to accept bets given any data $X_1, \ldots, X_m$, where m is arbitrary. We therefore consider two other approaches which are coherent at least in the compact case.

In the second approach, we consider the sequence of uniform distributions π_i on $\mathcal{F}_i$ and consider any weak limit point π^* of $\{\pi_i\}$ as a non-informative prior. If π^* is unique, it is simply the uniform distribution defined and studied by Dembski (1990). It is shown in Section 8.3 that this approach leads to Jeffreys' prior in parametric problems satisfying regularity conditions. In Ghosh and Ramamoorthi (1997), where limit points π^* were proposed as non-informative priors, the result in Section 8.3 was stated as a conjecture. In Section 8.3, we also consider the parametric case where $\vartheta = (\theta, \varphi)$, θ alone is the parameter of interest and the conditional prior distribution of φ given θ, namely $\pi(\varphi|\theta)$, is given. Using the natural metric in this problem, we derive a non-informative marginal prior for θ which is similar, but not identical with the reference prior of Bernardo (1979) or Berger and Bernardo (1989) and the probability matching prior satisfying the partial differential equation mentioned in Ghosh and Mukerjee (1992).

In the infinite dimensional case, evaluation of the limit points may prove to be impossible. However, the first approach may be used and $\pi_{i(n)}$ may be treated as an approximation to a limit point π^*.

We now come to the third approach. Here we consider a hierarchical prior which picks up the index or hyperparameter with probability λ_i and then uses π_i. In Section 8.4, we prove in a class of infinite dimensional problems that under a certain condition on the rate of decay of λ_i, we obtain posterior consistency, which is a very weak form of robustness generally considered important by Bayesians. Preliminary theoretical considerations suggest that the posterior in the third approach will be close to the posterior in the first approach for large n at least when the λ_i's decay suitably to make the posterior attain the optimal rate of convergence of Wong and Shen (1995). We expect to report on this later.

We offer these approaches as tentative ideas to be tried out in several problems. Computational and other considerations may require replacing $\mathcal{F}_i$ by other sieves which need not be finite, an index i which may take value in a continuum, and distributions on $\mathcal{F}_i$ which are not uniform. These relaxations will create a very large class of non-parametric priors, not necessarily non-informative, from which it may be convenient to elicit a prior. We would like to think of this as a fourth method. It includes such priors as a random histogram of width h introduced by Gasperini (1992) , and a Dirichlet convoluted with $N(0, h)$, introduced by Lo (1984) . These latter priors have received a lot of attention recently [see West (1992), West, Mueller and Escobar (1994)] . The parameter h here can be viewed as indexing a sieve. It is possible to obtain results on the consistency of the posterior for these and similar priors occurring in density estimation problems. These results will be presented elsewhere.

Our use of packing numbers was influenced by Wong and Shen (1995) and Kolmogorov and Tihomirov (1961). Sieves and bracketing entropy ideas have also been used in the context of posterior consistency by Barron, Schervish and Wasserman (1996).

8.2 Preliminaries

Let K be a compact metric space with a metric ρ. A finite subset S of K is called ε-dispersed if $\rho(x, y) \geq \varepsilon$ for all $x, y \in S$, $x \neq y$. A maximal ε-dispersed set is called an ε-net. An ε-net with maximum possible cardinality is said to be an ε-lattice and the cardinality of an ε-lattice is called the packing number (or ε-capacity) of K and is denoted by $D(\varepsilon, K) = D(\varepsilon, K; \rho)$. As K is totally bounded, $D(\varepsilon, K)$ is finite.

Define the ε-probability P_ε on K by

$$P_\varepsilon(X) = \frac{D(\varepsilon, X)}{D(\varepsilon, K)}, \quad X \subset K. \tag{8.1}$$

It follows that $0 \leq P_\varepsilon(\cdot) \leq 1$, $P_\varepsilon(\emptyset) = 0$, $P_\varepsilon(K) = 1$, $P_\varepsilon(\cdot)$ is subadditive, and for $X, Y \subset K$ which are separated at least by ε, $P_\varepsilon(X \cup Y) = P_\varepsilon(X) + P_\varepsilon(Y)$.

For an ε-lattice S_ε in K, the discrete uniform distribution μ_ε (say) on S_ε can be thought as an approximate uniform distribution on K. Because K is compact, μ_ε will have subsequential weak limits. If all the subsequential limits are the same, then K is called uniformizable and the common limit point is called the uniform probability on K.

The following result, due to Dembski (1990), will be used in Section 8.3.

Theorem 8.2.1 ((Dembski).) *Let (K, ρ) be a compact metric space. Then the following assertions hold:*

(a) *If K is uniformizable with uniform probability μ, then $\lim_{\varepsilon \to 0} P_\varepsilon(X) = \mu(X)$ for all $X \subset K$ with $\mu(\partial X) = 0$.*

(b) *If $\lim_{\varepsilon \to 0} P_\varepsilon(X)$ exists on some convergence determining class in K, then K is uniformizable.*

Remark 8.2.1 It is often not possible to evaluate the limit points, particularly in the case of an infinite dimensional family. Moreover, the limit point may sometimes be a degenerate measure [see, Example 2 of Dembski (1990)], which is undesirable as a non-informative prior. However, it is easy to see that if the growth of the packing number of every nonempty open set U is like that of K itself (i.e., $D(\varepsilon, U) \sim \mu(U) D(\varepsilon, K)$ as $\varepsilon \to 0$ for some $\mu(U) > 0$), then K cannot have any accumulation point and there is a uniform probability ν which is positive for each open set U and is equal to $\mu(U)$ for open U with $\nu(\partial U) = 0$. Moreover, if for all balls $B(x; \varepsilon)$, $x \in K$, $\mu(B(x; \varepsilon)) \to 0$ as $\varepsilon \to 0$, then ν is non-atomic.

To extend these ideas to non-compact σ-compact spaces, one can take a sequence of compact subsets $K_n \uparrow K$ having uniform probability μ_n; the above specification of μ is consistent in view of Proposition 2 of Dembski (1990). Any positive Borel measure μ satisfying

$$\mu(\cdot \cap K_n) = \frac{\mu_n(\cdot \cap K_n)}{\mu_n(K_1)} \tag{8.2}$$

may be thought as an (improper) uniform distribution on K. Such a measure, if exists, is also uniquely determined up to a positive multiple; see Lemma 2 of Dembski (1990).

8.3 Jeffreys' Prior

Let X_i's be i.i.d. with density $f(\cdot; \boldsymbol{\theta})$ (with respect to a σ-finite measure ν) where $\boldsymbol{\theta} \in \Theta$ and Θ is an open subset of $\mathbf{R}^d$. We assume that $\{f(\cdot; \boldsymbol{\theta}) : \boldsymbol{\theta} \in \Theta\}$ is a regular parametric family, i.e., there exist $\boldsymbol{\psi}(\cdot; \boldsymbol{\theta}) \in (L^2(\nu))^d$ (d-fold product of $L^2(\nu)$) such that for any compact $K \subset \Theta$,

$$\sup_{\boldsymbol{\theta} \in K} \int |f^{1/2}(x; \boldsymbol{\theta} + \mathbf{h}) - f^{1/2}(x; \boldsymbol{\theta}) - \mathbf{h}^T \boldsymbol{\psi}(x; \boldsymbol{\theta})|^2 \nu(dx) = o(\|\mathbf{h}\|^2) \qquad (8.3)$$

as $\|\mathbf{h}\| \to 0$. Define the Fisher information by the relation

$$\mathbf{I}(\boldsymbol{\theta}) = 4 \int \boldsymbol{\psi}(x; \boldsymbol{\theta})(\boldsymbol{\psi}(x; \boldsymbol{\theta}))^T \nu(dx) \qquad (8.4)$$

and assume that $\mathbf{I}(\boldsymbol{\theta})$ is positive definite and the map $\boldsymbol{\theta} \mapsto \mathbf{I}(\boldsymbol{\theta})$ is continuous. Further, assume the following stronger form of identifiability: On every compact subset $K \subset \Theta$,

$$\inf\left\{\int (f^{1/2}(x; \boldsymbol{\theta}_1) - f^{1/2}(x; \boldsymbol{\theta}_2))^2 \nu(dx) : \boldsymbol{\theta}_1, \boldsymbol{\theta}_2 \in K, \|\boldsymbol{\theta}_1 - \boldsymbol{\theta}_2\| \geq \varepsilon\right\} > 0, \quad \varepsilon > 0.$$

For i.i.d. observations, it is natural to equip Θ with the Hellinger distance defined by

$$H(\boldsymbol{\theta}_1, \boldsymbol{\theta}_2) = \left(\int |f^{1/2}(x; \boldsymbol{\theta}_1) - f^{1/2}(x; \boldsymbol{\theta}_2)|^2 \nu(dx)\right)^{1/2}. \qquad (8.5)$$

The following result is the main theorem of this section.

Theorem 8.3.1 *Fix a compact subset K of Θ. Then for all $Q \subset K$ with* $\mathrm{vol}(\partial Q) = 0$, *we have*

$$\lim_{\varepsilon \to 0} \frac{D(\varepsilon, Q)}{D(\varepsilon, K)} = \frac{\int_Q \sqrt{\det \mathbf{I}(\boldsymbol{\theta})} d\boldsymbol{\theta}}{\int_K \sqrt{\det \mathbf{I}(\boldsymbol{\theta})} d\boldsymbol{\theta}}. \qquad (8.6)$$

By using Theorem 8.2.1, we conclude from Theorem 8.3.1 that the Jeffreys measure μ on Θ defined by

$$\mu(Q) \propto \int_Q \sqrt{\det \mathbf{I}(\boldsymbol{\theta})} d\boldsymbol{\theta}, \quad Q \subset \Theta, \qquad (8.7)$$

is the (possibly improper) non-informative prior on Θ in the sense of the second approach described in the introduction.

PROOF OF THEOREM 8.3.1. Fix $0 < \eta < 1$. Cover K by J cubes of length η. In each cube, consider the interior cube with length $\eta - \eta^2$.

Since by continuity, the eigenvalues of $\mathbf{I}(\boldsymbol{\theta})$ are uniformly bounded away from zero and infinity on K, by standard arguments [see, e.g., Ibragimov and Has'minskii (1981, Theorem I.7.6)], it follows from (8.3) that there exist $M > m > 0$ such that

$$m\|\boldsymbol{\theta}_1 - \boldsymbol{\theta}_2\| \leq H(\boldsymbol{\theta}_1, \boldsymbol{\theta}_2) \leq M\|\boldsymbol{\theta}_1 - \boldsymbol{\theta}_2\|, \quad \boldsymbol{\theta}_1, \boldsymbol{\theta}_2 \in K. \tag{8.8}$$

Given $\eta > 0$, choose $\varepsilon > 0$ so that $\varepsilon/(2m) < \eta^2$. Any two interior cubes are thus separated at least by ε/m in terms of Euclidean distance, and so by ε in terms of the Hellinger distance.

For $Q \subset K$, let Q_j be the intersection of Q with the j-th cube and Q'_j be the intersection with the j-th interior cube, $j = 1, \ldots, J$. Thus

$$Q_1 \cup \cdots \cup Q_J = Q \supset Q'_1 \cup \cdots \cup Q'_J. \tag{8.9}$$

Hence

$$\sum_{j=1}^{J} D(\varepsilon, Q'_j; H) \leq D(\varepsilon, Q; H) \leq \sum_{j=1}^{J} D(\varepsilon, Q_j; H). \tag{8.10}$$

In particular, with $Q = K$, we obtain

$$\sum_{j=1}^{J} D(\varepsilon, K'_j; H) \leq D(\varepsilon, K; H) \leq \sum_{j=1}^{J} D(\varepsilon, K_j; H), \tag{8.11}$$

where K_j, K'_j are analogously defined.

From the j-th cube, choose $\boldsymbol{\theta}_j \in K$. By an argument similar to that used in the derivation of (8.8), we have for all $\boldsymbol{\theta}, \boldsymbol{\theta}'$ in the j-th cube,

$$\frac{\underline{\lambda}(\eta)}{2}\sqrt{(\boldsymbol{\theta} - \boldsymbol{\theta}')^T \mathbf{I}(\boldsymbol{\theta}_j)(\boldsymbol{\theta} - \boldsymbol{\theta}')} \leq H(\boldsymbol{\theta}, \boldsymbol{\theta}') \leq \frac{\overline{\lambda}(\eta)}{2}\sqrt{(\boldsymbol{\theta} - \boldsymbol{\theta}')^T \mathbf{I}(\boldsymbol{\theta}_j)(\boldsymbol{\theta} - \boldsymbol{\theta}')}, \tag{8.12}$$

where $\underline{\lambda}(\eta)$ and $\overline{\lambda}(\eta)$ tend to 1 as $\eta \to 0$.

Let

$$\underline{H}_j(\boldsymbol{\theta}, \boldsymbol{\theta}') = \frac{1}{2}\underline{\lambda}(\eta)\sqrt{(\boldsymbol{\theta} - \boldsymbol{\theta}')^T \mathbf{I}(\boldsymbol{\theta}_j)(\boldsymbol{\theta} - \boldsymbol{\theta}')}$$

and

$$\overline{H}_j(\boldsymbol{\theta}, \boldsymbol{\theta}') = \frac{1}{2}\overline{\lambda}(\eta)\sqrt{(\boldsymbol{\theta} - \boldsymbol{\theta}')^T \mathbf{I}(\boldsymbol{\theta}_j)(\boldsymbol{\theta} - \boldsymbol{\theta}')}.$$

Then from (8.12), we have

$$\begin{aligned} D(\varepsilon, Q_j; H) &\leq D(\varepsilon, Q_j; \underline{H}_j), \\ D(\varepsilon, Q'_j; H) &\geq D(\varepsilon, Q'_j; \overline{H}_j). \end{aligned} \tag{8.13}$$

By using the second part of Theorem IX of Kolmogorov and Tihomirov (1961), for some constants τ_j, τ_j' and an absolute constant A_d (depending only on the dimension d), we have

$$D(\varepsilon, Q_j; \underline{H}_j) \sim A_d \, \tau_j \, \mathrm{vol}(Q_j) \sqrt{\det \mathbf{I}(\boldsymbol{\theta}_j)} (\underline{\lambda}(\eta))^{-d} \varepsilon^{-d} \tag{8.14}$$

and

$$D(\varepsilon, Q_j'; \overline{H}_j) \sim A_d \, \tau_j \, \mathrm{vol}(Q_j') \sqrt{\det \mathbf{I}(\boldsymbol{\theta}_j)} (\overline{\lambda}(\eta))^{-d} \varepsilon^{-d}, \tag{8.15}$$

where the symbol $\sim$ signifies that the limit of the ratio of the two sides is 1 as $\varepsilon \to 0$. As all the metrics $\underline{H}_j$ and $\overline{H}_j$, $j = 1, \ldots, J$, arise from elliptic norms, it can be easily concluded by making a suitable linear transformation that $\tau_j = \tau_j' = \tau$ (say) for all $j = 1, \ldots, J$. Thus we obtain from (8.10)–(8.15) that

$$\limsup_{\varepsilon \to 0} \frac{D(\varepsilon, Q; H)}{D(\varepsilon, K; H)} \leq \frac{\sum_{j=1}^{J} \sqrt{\det \mathbf{I}(\boldsymbol{\theta}_j)} \mathrm{vol}(Q_j)}{\sum_{j=1}^{J} \sqrt{\det \mathbf{I}(\boldsymbol{\theta}_j)} \mathrm{vol}(K_j')} \left(\frac{\overline{\lambda}(\eta)}{\underline{\lambda}(\eta)} \right)^d \tag{8.16}$$

and

$$\liminf_{\varepsilon \to 0} \frac{D(\varepsilon, Q; H)}{D(\varepsilon, K; H)} \leq \frac{\sum_{j=1}^{J} \sqrt{\det \mathbf{I}(\boldsymbol{\theta}_j)} \mathrm{vol}(Q_j')}{\sum_{j=1}^{J} \sqrt{\det \mathbf{I}(\boldsymbol{\theta}_j)} \mathrm{vol}(K_j)} \left(\frac{\underline{\lambda}(\eta)}{\overline{\lambda}(\eta)} \right)^d. \tag{8.17}$$

Now let $\eta \to 0$. By the convergence of Riemann sums, $\sum_{j=1}^{J} \sqrt{\det \mathbf{I}(\boldsymbol{\theta}_j)} \mathrm{vol}(Q_j) \to \int_Q \sqrt{\mathbf{I}(\boldsymbol{\theta})} d\boldsymbol{\theta}$ and $\sum_{j=1}^{J} \sqrt{\det \mathbf{I}(\boldsymbol{\theta}_j)} \mathrm{vol}(Q_j') \to \int_Q \sqrt{\mathbf{I}(\boldsymbol{\theta})} d\boldsymbol{\theta}$ and similarly for the sums involving K_j's and K_j''s. Also, $\underline{\lambda}(\eta) \to 1$ and $\overline{\lambda}(\eta) \to 1$, so the desired result follows. ∎

Remark 8.3.1 During a discussion Prof. Hartigan remarked that Jeffreys had envisaged constructing non-informative priors by approximating Θ with Kullback-Liebler neighborhoods, and asked us if the construction in the last section can be carried out using the K-L neighborhoods. Since the K-L divergence $K(\theta, \theta') = \int f_\theta \log \frac{f_\theta}{f_{\theta'}}$ is not a metric there would be obvious difficulties in formalizing the notion of an ϵ-net. However, if the family of densities $\{f_\theta : \theta \in \Theta\}$ have well-behaved tails such that , for any $\theta, \theta', K(\theta, \theta') \leq \phi(H(\theta, \theta'))$, where $\phi(\epsilon)$ goes to 0 as ϵ goes to 0, then any ϵ-net $\{\theta_1, \theta_2 \cdots \theta_k\}$ in the Hellinger metric can be thought of as a K-L net in the sense that

1. $K(\theta_i, \theta_j) > \epsilon$ for $i, j = 1, 2, \ldots, k$,

2. for any θ there exists an i such that $K(\theta_i, \theta) < \phi(\epsilon)$.

In such situations, the above theorem allows us to view the Jeffreys' prior as a limit of uniform distributions arising out of K-L neighborhoods. Wong and Shen (1995) show that suitable tail behavior for all $\theta, \theta', \int_{(f_\theta/f_{\theta'} \geq \exp \frac{1}{\delta})} f_\theta (\frac{f_\theta}{f_{\theta'}})^\delta < M$.

We now consider the case when there is a nuisance parameter. Let θ be the parameter of interest and φ be the nuisance parameter, and we assume for simplicity that both are real-valued. We can write the information matrix as

$$\begin{pmatrix} I_{11}(\theta,\varphi) & I_{12}(\theta,\varphi) \\ I_{21}(\theta,\varphi) & I_{22}(\theta,\varphi) \end{pmatrix}, \tag{8.18}$$

where $I_{12} = I_{21}$. In view of Theorem 8.3.1, it is natural to put the prior $\pi(\varphi|\theta) = \sqrt{I_{22}(\theta,\varphi)}$ for φ given θ. So we need to construct a non-informative marginal prior for θ. First let us assume that the parameter space is compact. With n i.i.d. observations, the joint density of the observations given θ only is given by

$$g(\mathbf{x}^n;\theta) = (c(\theta))^{-1} \int \prod_{i=1}^{n} f(x_i;\theta,\varphi)\sqrt{I_{22}(\theta,\varphi)}d\varphi, \tag{8.19}$$

where $c(\theta) = \int \sqrt{I_{22}(\theta,\varphi)}d\varphi$ is the constant of normalization. Let $I_n(\theta;g)$ denote the information for the family $\{g(\mathbf{x}^n;\theta) : \theta \in \Theta\}$. Under adequate regularity conditions, it can be shown that the information per observation $I_n(\theta;g)/n$ satisfies

$$\lim_{n\to\infty} I_n(\theta;g)/n = (c(\theta))^{-1} \int I_{11.2}(\theta,\varphi)\sqrt{I_{22}(\theta,\varphi)}d\varphi = J(\theta) \text{ (say)}, \tag{8.20}$$

where $I_{11.2} = I_{11} - I_{12}^2/I_{22}$. Let $H_n(\theta,\theta+h)$ be the Hellinger distance between $g(\mathbf{x}^n;\theta)$ and $g(\mathbf{x}^n;\theta+h)$. Locally as $h \to 0$, $H_n^2(\theta,\theta+h)$ behaves like $I_n(\theta;g)h^2$. Hence by Theorem 8.3.1, the non-informative (marginal) prior for θ would be proportional to $\sqrt{I_n(\theta;g)}$. In view of (8.20), passing to the limit as $n \to \infty$, the (sample size independent) marginal non-informative prior for θ should be taken to be proportional to $(J(\theta))^{1/2}$, and so the prior for (θ,φ) is proportional to $J(\theta)\pi(\varphi|\theta)$. Generally, for a noncompact parameter space, we proceed like Berger and Bernardo (1989). Fix a sequence of compact sets Λ_l increasing to the whole parameter space. Put $\Phi_l(\theta) = \{\varphi : (\theta,\varphi) \in \Lambda_l\}$ and normalize $\pi(\varphi|\theta)$ on $\Phi_l(\theta)$ as

$$p_l(\varphi|\theta) = (c_l(\theta))^{-1}\pi(\varphi|\theta)I\{\varphi \in \Phi_l(\theta)\}, \tag{8.21}$$

where $c_l(\theta) = \int_{\Phi_l(\theta)} \sqrt{I_{22}(\theta,\varphi)}d\varphi$ is the constant of normalization, as before. The marginal non-informative prior for θ at stage l is then defined as

$$\pi_l(\theta) = \sqrt{\int I_{11.2}(\theta,\varphi)p_l(\varphi|\theta)d\varphi}. \tag{8.22}$$

Let θ_0 be a fixed value of θ. The (joint) non-informative prior is finally defined as

$$\lim_{l\to\infty} \frac{(c_l(\theta))^{-1}\pi_l(\theta)}{(c_l(\theta_0))^{-1}\pi_l(\theta_0)} \frac{\pi(\varphi|\theta)}{\pi(\varphi|\theta_0)}, \tag{8.23}$$

assuming that the above limit exists for all θ. Informally, the prior for θ is obtained by taking the average of $I_{11.2}(\theta, \varphi)$ (with respect to $\sqrt{I_{22}(\theta, \varphi)}$) and then taking the square-root. The reference prior of Berger and Bernardo or the probability matching prior takes averages of other functions of $\sqrt{I_{11.2}(\theta, \varphi)}$ and then transforms back. So they all have common structure and are worth comparing through examples. In the examples of Datta and Ghosh (1995b), we believe that they reduce to the same prior.

8.4   An Infinite Dimensional Example

In this section, we show that in a certain class of infinite dimensional families, the third approach mentioned in the introduction leads to consistent posterior.

Theorem 8.4.1 *Let $\mathcal{P}$ be a family of densities where $\mathcal{P}$, metrized by the Hellinger distance, is compact. Let ε_n be a positive sequence satisfying $\sum_{n=1}^{\infty} n^{1/2} \varepsilon_n < \infty$. Let $\mathcal{F}_n$ be an ε_n-net in $\mathcal{P}$, μ_n be the uniform distribution on $\mathcal{F}_n$ and μ be the probability on $\mathcal{P}$ defined by $\mu = \sum_{n=1}^{\infty} \lambda_n \mu_n$, where λ_n's are positive numbers adding upto unity. If for any $\beta > 0$*

$$\lim_{n \to \infty} e^{\beta n} \frac{\lambda_n}{D(\varepsilon_n, \mathcal{F}_n)} = \infty, \tag{8.24}$$

then the posterior distribution based on the prior μ and i.i.d. observations $X_1, X_2, \ldots$ is consistent at every $p_0 \in \mathcal{P}$.

PROOF OF THEOREM 8.4.1. Fix a $p_0 \in \mathcal{P}$ and a neighborhood U of p_0. Let P_0 stand for the probability corresponding to the density p_0. We need to show that

$$\frac{\int_{U^c} \prod_{i=1}^{n} \frac{p(X_i)}{p_0(X_i)} \mu(dp)}{\int_{\mathcal{P}} \prod_{i=1}^{n} \frac{p(X_i)}{p_0(X_i)} \mu(dp)} \to 0 \text{ a.s.}[P_0]. \tag{8.25}$$

Since $\mathcal{P}$ is compact under the Hellinger metric, the weak topology and the Hellinger topology coincide on $\mathcal{P}$. Thus U is also a weak neighborhood of p_0 and since there is a uniformly consistent test for $p = p_0$ against $p \notin U$, it follows from the arguments of Schwartz (1965) and Barron (1986), that for some $\beta > 0$,

$$e^{n\beta} \int_{U^c} \prod_{i=1}^{n} \frac{p(X_i)}{p_0(X_i)} \mu(dp) \to 0 \text{ a.s. } [P_0]. \tag{8.26}$$

To establish (8.25), it suffices to show that for every $\beta > 0$,

$$e^{n\beta} \int_{\mathcal{P}} \prod_{i=1}^{n} \frac{p(X_i)}{p_0(X_i)} \mu(dp) \to \infty \text{ a.s. } [P_0]. \tag{8.27}$$

Let p_n and p_{0n}, respectively, stand for the joint densities $\prod_{i=1}^{n} p(X_i)$ and $\prod_{i=1}^{n} p_0(X_i)$. Let $\|\cdot\|_r$ and $H(\cdot,\cdot)$ denote the L^r-norms, $r = 1, 2$, and Hellinger distance respectively. Then

$$
\begin{aligned}
E_{P_0}\left|\frac{p_n}{p_{0n}} - 1\right| &= \|p_n - p_{0n}\|_1 \\
&= \|(p_n^{1/2} + p_{0n}^{1/2})(p_n^{1/2} - p_{0n}^{1/2})\|_1 \\
&\leq \|p_n^{1/2} + p_{0n}^{1/2}\|_2 H(p_n, p_{0n}) \\
&\leq 2\sqrt{1 - \int p_n p_{0n}} \\
&= 2\sqrt{1 - \left(\int p p_0\right)^n} \\
&= 2\sqrt{1 - \left(1 - \frac{1}{2}H^2(p, p_0)\right)^n}.
\end{aligned}
$$

Using the elementary inequality $1 - x^n \leq n(1-x)$ for $0 \leq x \leq 1$, we observe that if $p \in B(p_0, \varepsilon_n) = \{p : H(p_0, p) < \varepsilon_n\}$, then

$$
E_{P_0}\left|\frac{p_n}{p_{0n}} - 1\right| \leq \sqrt{2n}\varepsilon_n,
$$

and so

$$
\begin{aligned}
E_{P_0}\left|\int_{B(p_0,\varepsilon_n)}\left(\frac{p_n}{p_{0n}} - 1\right)\mu(dp)\right| &\leq \int_{B(p_0,\varepsilon_n)} E_{P_0}\left|\frac{p_n}{p_{0n}} - 1\right|\mu(dp) \\
&\leq \sqrt{2n}\varepsilon_n \mu(B(p_0, \varepsilon_n)).
\end{aligned}
$$

Hence

$$
\begin{aligned}
P_0&\left\{\int_{\mathcal{P}}\left(\frac{p_n}{p_{0n}}\right)\mu(dp) \leq \frac{1}{2}\mu(B(p_0, \varepsilon_n))\right\} \\
&\leq P_0\left\{\int_{(B(p_0,\varepsilon_n)}\left(\frac{p_n}{p_{0n}}\right)\mu(dp) \leq \frac{1}{2}\mu(B(p_0, \varepsilon_n^*))\right\} \\
&\leq P_0\left\{\left|\int_{B(p_0,\varepsilon_n)}\left(\frac{p_n}{p_{0n}} - 1\right)\mu(dp)\right| > \frac{1}{2}\mu(B(p_0, \varepsilon_n))\right\} \\
&\leq \frac{2E_{P_0}\left|\int_{B(p_0,\varepsilon_n)}\left(\frac{p_n}{p_{0n}} - 1\right)\mu(dp)\right|}{\mu(B(p_0, \varepsilon_n))} \\
&\leq \sqrt{8n}\varepsilon_n.
\end{aligned}
$$

By the construction, $\exp[n\beta]\mu(B(p_0, \varepsilon_n)) \geq \exp[n\beta]\lambda_n/D(\varepsilon_n, \mathcal{P})$, which goes to infinity in view of Assumption (8.24). An application of the Borel-Cantelli Lemma yields (8.27). ∎

Remark 8.4.1 Consistency is obtained in the last theorem by requiring (8.24) for sieves whose width ε_n was chosen carefully. However it is clear from the proof that consistency would follow for sieves with width $\varepsilon_n \downarrow 0$ by imposing (8.24) for a carefully chosen subsequence.

Precisely, if $\varepsilon_n \downarrow 0$, $\mathcal{F}_n$ is an ε_n-net, μ is the probability on $\mathcal{P}$ defined by $\mu = \sum_{n=1}^{\infty} \lambda_n \mu_n$ and δ_n is a positive summable sequence, then by choosing $j(n)$ with

$$\varepsilon_{j(n)} \leq \sqrt{\frac{2}{n}} \delta_n, \tag{8.28}$$

the posterior is consistent, if

$$\exp[n\beta] \frac{\lambda_{j(n)}}{D(\varepsilon_{j(n)}, \mathcal{P})} \to \infty. \tag{8.29}$$

A useful case corresponds to

$$D(\varepsilon, \mathcal{P}) \leq A \exp[c\varepsilon^{-\alpha}], \tag{8.30}$$

where $0 < \alpha < 2/3$ and A and c are positive constants, $\delta_n = n^{-\gamma}$ for some $\gamma > 1$. If in this case, $j(n)$ is the smallest integer satisfying (8.28), then (8.29) becomes

$$\exp[n\beta - c\varepsilon_{j(n)}^{-\alpha}]\lambda_{j(n)} \to \infty. \tag{8.31}$$

If $\varepsilon_n = \varepsilon/2^n$ for some $\varepsilon > 0$ and λ_n decays no faster than n^{-s} for some $s > 0$, then (8.31) holds. Moreover, the condition $0 < \alpha < 2$ in (8.30) is enough for posterior consistency in probability.

An example of this kind is the following class of densities considered in density estimation [see, e.g., Wong and Shen (1995)]:

$$\mathcal{P} = \{f = g^2 : g \in C^r[0,1], \int g^2(x)dx = 1, \|g^{(j)}\|_{\sup} \leq L_j, j = 1, \ldots, r,$$

$$|g^{(r)}(x_1) - g^{(r)}(x_2)| \leq L_{r+1}|x_1 - x_2|^m\},$$

where r is a positive integer, $0 \leq m \leq 1$ and L_j's are fixed constants. By Theorem XV of Kolmogorov and Tihomirov (1961), $D(\varepsilon, \mathcal{P}) \leq \exp[c\varepsilon^{-1/(r+m)}]$. Hence the hierarchical prior constructed in Theorem 8.4.1 leads to consistent posterior.

Acknowledgements. Research of the first author was supported by the National Board of Higher Mathematics, Department of Atomic Energy, Bombay, India. Research of the second author was supported by NSF grant number 9307727. Research of the third author was supported by NIH grant number 1R01 GM49374.

References

1. Barron, A. R. (1986). Discussion of "On the consistency of Bayes estimates" by P. Diaconis and D. Freedman, *Annals of Statistics*, **14**, 26–30.

2. Barron, A. R., Schervish, M. and Wasserman, L. (1996). The consistency of posterior distributions in non parametric problems, *Preprint*.

3. Basu, D. (1975). Statistical information and likelihood, *Sankhyā, Series A*, **37** 1–71.

4. Berger, J. O. and Bernardo, J. M. (1989). Estimating a product of means: Bayesian analysis with reference priors, *Journal of the American Statistical Association*, **84**, 200–207.

5. Berger, J. O. and Bernardo, J. M. (1992). On the development of reference priors (with discussions), In *Bayesian Statistics V* (Eds., J. M. Bernardo *et al.*), pp. 35–60.

6. Berger, J. O. and Pericchi, L. (1994). Intrinsic Bayes factor for model selection and prediction in general linear model, *Preprint*.

7. Berger, J. O. and Pericchi, L. (1996). The intrinsic Bayes factor for model selection and prediction, *Journal of the American Statistical Association*, **91**, 109–121.

8. Bernardo, J. M. (1979). Reference posterior distributions for Bayesian inference (with discussions), *Journal of the Royal Statistical Society, Series B*, **41**, 113–147.

9. Datta, G. S. and Ghosh, J. K. (1995a). On priors providing frequentist validity for Bayesian inference, *Biometrika*, **82**, 37–46.

10. Datta, G. S. and Ghosh, J. K. (1995b). Noninformative priors for maximal invariant in group models, *Test*, **4**, 95–114.

11. Datta, G. S. and Ghosh, M. (1995). Some remarks on noninformative priors, *Journal of the American Statistical Association*, **90**, 1357–1363.

12. Datta, G. S. and Ghosh, M. (1996). On the invariance of noninformative priors, *Annals of Statistics*, **24** (to appear).

13. Dembski, W. A. (1990). Uniform probability, *Journal of Theoretical Probability*, **3**, 611–626.

14. Gasparini, M. (1992). Bayes nonparametrics for biased sampling and density estimation, *Ph.D thesis*, University of Michigan.

15. Ghosh, J. K. (1994). *Higher Order Asymptotics*, NSF-CBMS Regional Conference Series in Probability and Statistics **4**, IMS, Hayward, CA.

16. Ghosh, J. K. and Mukerjee, R. (1992). Non-informative priors (with discussions), In *Bayesian Statistics*, **4** (Eds., J. M. Bernardo *et al.*), pp. 195–210.

17. Ghosh, J. K. and Ramamoorthi R. V. (1997). *Lecture notes on Bayesian asymptotics*, Under preparation.

18. Heath, D. and Sudderth, W. (1978). On finitely additive priors, coherence and extended admissibility, *Annals of Statistics*, **6**, 333–345.

19. Ibragimov, I. A. and Has'minskii, R. Z. (1981). *Statistical Estimation: Asymptotic Theory*, New York: Springer-Verlag.

20. Kass, R. E. and Wasserman, L. (1992). A reference Bayesian test for nested hypotheses with large samples, *Technical Report*, #567, Carnegie Mellon University.

21. Kass, R. E. and Wasserman, L. (1996). The selection of prior distribution by formal rules, *Journal of the American Statistical Association*, **96**, 1343–1370.

22. Kolmogorov, A. N. and Tihomirov. V. M. (1961). ε-entropy and ε-capacity of sets in function spaces, *American Mathematics Society Transl. Ser.* 2, **17**, 277–364. [Translated from Russian: *Uspekhi Mat. Nauk*, **14**, 3–86, (1959).]

23. Lo, A. Y. (1984). On a class of Bayesian nonparametric estimates, *Annals of Statistics*, **12**, 351–357.

24. O'Hagan, A. (1995). Fractional Bayes factors for model comparisons, *Journal of the Royal Statistical Society, Series B*, **57**, 99–138.

25. Schwartz, L. (1965). On Bayes procedures, *Z. Wahrsch. Verw. Gebiete*, **4**, 10–26.

26. Spiegelhalter, D. J. and Smith, A. F. M. (1982). Bayes factors for linear and log-linear models with vague prior information, *Journal of the Royal Statistical Society, Series B*, **44**, 377–387.

27. Tibshirani, R. (1989). Noninformative priors for one parameter of many, *Biometrika*, **76**, 604–608.

28. Wong, W. H. and Shen, X. (1995). Probability inequalities for likelihood ratios and convergence rates of sieve MLEs, *Annals of Statistics*, **23**, 339–362.

29. West, M. (1992). Modelling with mixtures, In *Bayesian Statistics*, **4**, (Eds., J. M. Bernardo *et al.*), pp. 503-524.

30. West, M., Muller, P. and Escobar, M. D. (1994). Hierarchical priors and mixture models, with applications in regression and density estimation, In *Aspects of uncertainty: A Tribute to D. V. Lindley*, pp. 363-386.

31. Zellner, A. (1990). Bayesian methods and entropy in economics and econometrics, In 10 *International MaxEnt Workshop*, University of Wyoming.

PART III

POINT AND INTERVAL ESTIMATION—CLASSICAL APPROACH

9

From Neyman's Frequentism to the Frequency Validity in the Conditional Inference

J. T. Gene Hwang

Cornell University, Ithaca, NY

Abstract: Assume that C_X is a confidence set with coverage probability $1-\alpha$. The quantity $1-\alpha$ is called the *confidence* of C_X, since in the long run the proportion of cases in which C_X would cover the true parameter in repeated experiments is $1-\alpha$.

We consider the generalization of this justification, called the long term frequency validity, for data dependent confidence reports. The long term frequency validity is related to Berger's (1983, 1986) validity definition involving expectations, called the *expectation validity*. In one of the models, it is shown that long term frequency validity is equivalent to the expectation validity and in the other model the former is weaker. We also characterize situations under which frequency valid reports are not too conservative. Such reports are called *frequency appropriate*. Frequency appropriateness is shown to be related to Casella's criterion.

Keywords and phrases: Expectation validity, frequency appropriateness, admissibility

---

9.1 Introduction

For a confidence set C_X with coverage probability $1-\alpha$, Neyman (1937) called $1-\alpha$ the confidence of C_X. Why does $1-\alpha$ deserve such a name? The major reason is that if C_X is used repeatedly, the proportion of times in which C_X will cover θ is $1-\alpha$. Here, θ is the parameter and the distribution of X depends on θ. Therefore, if $1-\alpha$ is high, say 0.99, the frequentist is confident that C_X will cover θ in 99% of the experiments.

Under what scenario, however, should the experiment be repeated? Does the experiment have to be conducted at the same θ? The answer is no. Indeed a more realistic case would be:

Model 1. Similar experiments are conducted at possibly different values of θ.

It would be even more satisfactory to consider:

Model 2. Different experiments with, of course, different $\theta_i's$.

Model 2 is the case in Neyman's formulation. Neyman (1937, p. 346) wrote: "it is not necessary that the problem of estimation should be the same in all cases." Also the steps in calculating confidence intervals are "different in details of sampling and arithmetic." The relative frequency will still converge to $1 - \alpha$, by the application of the law of large numbers to i.i.d. Bernoulli trials. Berger (1983) also pointed this out.

In general, it is not always the case that C_X has a constant coverage probability. Therefore, the frequentist's promise should be one–sided, i.e., in the long run C_X will cover θ in at least $(1 - \alpha)$ proportion of cases. In such cases, $1 - \alpha$ shall be called a long term frequency valid confidence for C_X. For a precise definition, see (9.3). Under Model 1, it can then be shown that $1 - \alpha$ is long term frequency valid if and only if

$$P_\theta(\theta \in C_X) \geq 1 - \alpha \text{ for all } \theta. \tag{9.1}$$

When (9.1) holds, $1 - \alpha$ is said to be expectation valid for C_X.

Although Neyman's theory is based on a constant confidence, there are reasons to consider a data dependent confidence, $\gamma(X)$. One primary reason is that in some situations, the report of a conditional coverage probability, is more appropriate. Cox's (1958) example [see also the confidence interval version of Bondar (1988)] points out a need to condition on an ancillary statistic. The conditional coverage probability is itself a data dependent report. The idea of conditioning on an ancillary statistic dates back to Fisher (1925). Kiefer (1977a,b) and Brownie and Kiefer (1977) represent the first major attempts in treating the data dependent report, often by considering the conditional probability report. Of particular interest to the readers of this volume is that Kiefer (1977a) also studied the fundamental issue of the ranking and selection procedures of Professor Shanti Gupta (1956).

Kiefer (1976, 1977a, 1977b) and Brownie and Kiefer (1977) apparently were the first series of papers which deal with the validity and optimality of conditional reports. Some lucid comments about the work can be found in Berger (1986). Recently, the subject of conditional inference has regained much attention. See Johnstone (1988), Rukhin (1988a,b), Lu and Berger (1989a,b), Robert and Casella (1990), Brown and Hwang (1991), Hwang and Brown (1991), Hwang et al. (1992), McCullagh (1991), Casella (1992), and a review paper, Goutis and Casella (1995).

An important question then is how to define the long term (frequency) validity and the expectation validity for the data dependent confidence. Discussions among some Cornell statisticians about the definition of the expectation validity when the coverage probability is non-constant have highlighted some confusion. However, defining the long term frequency validity as laid out in Berger (1983) seems intuitively obvious. The purpose of this article is to derive carefully a necessary and sufficient condition equivalent to the long term validity. For Model 1, this turns out to be what Berger proposed, i.e.,

$$P_\theta(\theta \in C_X) \geq E_\theta \gamma(X) \text{ for every } \theta,$$

which is a condition that is, in many situations, easier to check than the long term frequency validity. Indeed, the fact that expectation validity implies long term validity have been stated in Kiefer (1976, p. 839; 1977b, p. 801), and in Berger and Wolpert (1988, p. 16). Here we show that the reverse is also true. By doing so, we hope that we have clarified the confusion. One of the new interesting implications is that an asymptotic confidence level may not be long term frequency valid. However, for Model 2, the expectation validity is weaker than the long term validity, to which an equivalent condition is derived in Theorem 9.3.4.

In this paper, another major goal is to discuss some refined ideas such as frequency appropriateness, referring to frequency valid reports which are not excessively conservative. Frequency appropriateness is implied by a criterion proposed by Casella (1988). Also a theorem (Theorem 9.2.3) is established that shows that any frequency inappropriate (i.e., not frequency appropriate) estimator can be improved by a frequency appropriate estimator. Section 9.3 contains some discussion about Model 2. Results about estimating a loss function are established in Theorems 9.3.1 and 9.3.2.

9.2 Frequency Validity under Model 1

In this section, we focus on one confidence set C_X, and hence we shall consider Model 1. Specifically, let

θ_i be an arbitrary sequence of possibly different parameters
and $X_i \sim P_{\theta_i}$ be the corresponding random variables. $\qquad$ (9.2)

Since the experiments are similar, $P.$ does not depend on i. Let $I(\cdot)$ denote the indicator function, i.e.,

$$I(\theta_i \in C_{X_i}) = \begin{cases} 1 & \text{if } \theta_i \in C_{X_i} \\ 0 & \text{otherwise.} \end{cases}$$

A confidence $\gamma = 1 - \alpha$ is defined as long–term frequency valid if, with probability one,

$$\liminf \frac{1}{n} \sum I(\theta_i \in C_{X_i}) \geq \gamma \qquad (9.3)$$

where θ_i and X_i are any sequences defined in (9.2). This is Neyman's justification for the term "confidence". Therefore, according to (9.3), if an experiment characterized by θ is repeated at possibly different θ's, the long term relative frequency for C_X to cover θ is at least γ. However, the confidence γ should not be underestimated at all times. We shall call γ to be long term frequency appropriate if in addition to (9.3) there exist $\theta_1^*, \ldots, \theta_n^*, \ldots$ such that

$$\lim_{n \to \infty} \frac{1}{n} \sum I(\theta_i^* \in C_{X_i}) = \gamma, \qquad (9.4)$$

i.e., in some situations the long term relative frequency is exactly γ.

Technically, if γ is long term frequency appropriate, then it is the ess inf of the left hand side of (9.3) over $\theta_1, \ldots, \theta_n, \ldots$. For the definition of ess inf, see for example, Brown (1978). The definition of frequency appropriateness is stronger than the assertion that γ is the ess inf, since it asserts also that the ess inf can be achieved for some $\theta_1^*, \ldots, \theta_n^*, \ldots$. A similar comment applies to Definition 9.2.1 below and its related concepts where the essential infimum is zero.

As indicated in the introduction, it is important to be able to deal with a data dependent report $\gamma(X)$. The inequality in (9.3) is equivalent to

$$\liminf_{n \to \infty} \frac{1}{n} \sum \{I(\theta_i \in C_{X_i}) - \gamma\} \geq 0.$$

Therefore, we make the following definition.

Definition 9.2.1 The data dependent report $\gamma(X)$ is said to be long term frequency valid if, with probability one,

$$\liminf_{n \to \infty} \frac{1}{n} \sum \{I(\theta_i \in C_{X_i}) - \gamma(X_i)\} \geq 0 \qquad (9.5)$$

for any θ_i and X_i defined in (9.2).

Similarly, $\gamma(X)$ is said to be long term frequency appropriate if in addition to (9.5), there exist $\theta_1^*, \ldots, \theta_n^*, \ldots$ such that

$$\lim_{n \to \infty} \frac{1}{n} \sum \{I(\theta_i^* \in C_{X_i}) - \gamma(X_i)\} = 0, \qquad (9.6)$$

where $X_i \sim P_{\theta_i^*}$.

The main result of this paper relates long term frequency validity to the expectation frequency validity, as defined below.

Definition 9.2.2 A data dependent report $\gamma(X)$ is said to be expectation frequency valid if for every θ

$$P_\theta(\theta \in C_X) \geq E_\theta \gamma(X). \tag{9.7}$$

It is called expectation frequency appropriate if

$$\inf_\theta \{P_\theta(\theta \in C_X) - E_\theta \gamma(X)\} = 0. \tag{9.8}$$

The main purpose of this section is to establish that Definitions 9.2.1 and 9.2.2 are equivalent under no regularity conditions. We shall use the following lemma.

Lemma 9.2.1 *[Law of large numbers] Assume that Y_i's are uncorrelated random variables and the variance of Y_i is bounded above for all i by a finite number. Then*

$$\frac{1}{n}\sum(Y_i - EY_i) \to 0 \text{ almost surely.}$$

The theorem can be found in most standard probability text books. See for example, Theorem 5.1.2 on p. 103 of Chung (1974). Using this lemma, we establish an useful identity below.

Lemma 9.2.2 *Under the assumption that X_i's are independent, and $0 \leq \gamma(X_i) \leq 1$*

$$\liminf \frac{1}{n}\sum [I(\theta_i \in C_{X_i}) - \gamma(X_i)] = \liminf \frac{1}{n}\sum [P_{\theta_i}(\theta_i \in C_{X_i}) - E_{\theta_i}\gamma(X_i)] \tag{9.9}$$

Similarly, (9.9) holds if $\liminf$ *is replaced by* $\limsup$.

PROOF. Write

$$\frac{1}{n}\sum \{I(\theta_i \in C_{X_i}) - \gamma(X_i)\}$$
$$= \frac{1}{n}\sum \{I(\theta_i \in C_{X_i}) - \gamma(X_i) - E_{\theta_i}[I(\theta_i \in C_{X_i}) - \gamma(X_i)]\}$$
$$+ \frac{1}{n}\sum E_{\theta_i}[I(\theta_i \in C_{X_i}) - \gamma(X_i)]. \tag{9.10}$$

Since $I(\theta_i \in C_{X_i}) - \gamma(X_i)$ has an absolute value bounded by one, its variance is bounded by one satisfying the assumption of Lemma 9.2.1. Lemma 9.2.1 then implies that the first summation on the right hand side of (9.10) converges to zero, establishing (9.9). The above argument applies to $\limsup$ as well. ∎

Now we are ready for the main result. Throughout the paper, we shall always assume that a confidence report is bounded between 0 and 1 as in Lemma 9.2.2, which, of course, is a reasonable assumption. The "if" part of the following theorem has been stated in Kiefer (1976, p. 839; 1977b, p. 801) and Berger and Wolpert (1988, p. 16).

Theorem 9.2.1 *A data dependent report $\gamma(X)$ is long term frequency valid if and only if it is expectation frequency valid.*

PROOF. We deal with the "if" statement first. The inequality (9.7) implies that the right hand side of (9.9) is greater than or equal to zero. Eq. (9.9) therefore implies the long term frequency validity of $\gamma(X)$. Conversely, if $\gamma(X)$ is long term frequency valid, then (9.9) is greater than or equal to zero for all $\theta_1, \ldots, \theta_n, \ldots$. Setting $\theta_i = \theta$ reduces it to

$$P_\theta(\theta \in C_X) - E_\theta \gamma(X) \geq 0$$

which implies (9.7) and the expectation frequency validity.   ∎

Since long term frequency validity is equivalent to expectation frequency validity, we shall from now on refer to them as frequency validity.

If the coverage probability $P_\theta(\theta \in C_X)$ equals a constant $1-\alpha$, the frequency validity is then

$$1 - \alpha \geq E_\theta \gamma(X) \text{ for every } \theta.$$

In this simpler case, the definition of validity is less controversial. There have been some inadmissibility and optimality results concerning data dependent confidence reports restricted to this class. See Lu and Berger (1989a,b), Hwang and Brown (1991), and Hsieh and Hwang (1993). However, opinions differ about the validity definition for the nonconstant coverage probability case. The natural generalization of Neyman's formulation specifically requires (9.7).

One of the interesting implications of Theorem 9.2.1 is that asymptotic confidence levels may not have frequency validity. There are situations in which a confidence set C_X has an asymptotic coverage probability $1 - \alpha \geq 0$ and yet its minimum finite sample coverage probability is zero, i.e.,

$$\inf_\theta P_\theta(\theta \in C_X) = 0. \tag{9.11}$$

See Gleser and Hwang (1987). In such a case, $1 - \alpha$ is not frequency valid. Theorem 9.2.2 then implies that there exists a sequence of $\theta_1^*, \ldots, \theta_n^*, \ldots$ and $X_i \sim P_{\theta_i^*}$, such that

$$\lim_{n \to \infty} \frac{1}{n} \sum I(\theta_i^* \in C_{X_i}) = 0.$$

Therefore, in the worst case, there is no guarantee.

Next, we shall prove equivalence of the two definitions of frequency appropriateness.

Theorem 9.2.2 *A data dependent report $\gamma(X)$ is long term frequency appropriate if and only if it is expectation frequency appropriate.*

PROOF. "Only if" part: Assume that $\gamma(X)$ is long term frequency appropriate. Hence, $\gamma(X)$ is frequency valid which, by (9.7), implies

$$\inf_{\theta}[P_\theta(\theta \in C_X) - E_\theta\gamma(X)] \geq 0. \tag{9.12}$$

To complete the proof, we need to establish (9.8). It then suffices to show that the nonstrict inequality in (9.12) can be reversed. By the definition of long term frequency appropriateness, there is a sequence $\theta_1^*, \ldots, \theta_n^*, \ldots$ such that (9.6) holds, which implies

$$\lim \frac{1}{n} \sum [P_{\theta_i^*}(\theta_i^* \in C_{X_i}) - E_{\theta_i^*}\gamma(X_i)] = 0. \tag{9.13}$$

However, the left hand side of (9.12) is less than or equal to the left hand side of (9.13). This implies that the reverse nonstrict inequality in (9.12) holds.

Now the "if" part: Note that (9.8) implies (9.7), the frequency validity, and hence (9.5) holds. To show that it is long term frequency appropriate, we now find $\theta_1^*, \ldots, \theta_n^*, \ldots$ such that (9.6) holds. To do so, let θ_i^* be such that

$$P_{\theta_i^*}(\theta_i^* \in C_{X_i}) - E_{\theta_i^*}\gamma(X_i) < \frac{1}{i^2}.$$

Such θ_i^* exists by (9.8). Here, $1/i^2$ can be replaced by any sequence of i approaching zero as $i \to \infty$. Now

$$\limsup_{n \to \infty} \frac{1}{n} \sum \{P_{\theta_i^*}(\theta_i^* \in C_{X_i}) - E_{\theta_i^*}\gamma(X_i)\}$$
$$\leq \lim_{n \to \infty} \frac{1}{n} \sum \frac{1}{i^2} = 0.$$

Thus, the Lemma 9.2.2 and (9.5) imply (9.6), establishing the long term frequency appropriateness. ∎

The expectation validity is typically much easier to check than the long term frequency validity.

By using these theorems, we provide below a decision theoretic justification of our definition of the long term frequency appropriateness. The following theorem depends on a comparison criterion involving two confidence estimators $\gamma_1(X)$ and $\gamma_2(X)$. A confidence estimator $\gamma(X)$ is considered to be an estimate of the indicator function $I(\theta \in C_X)$. Therefore, we evaluate it with respect to a score function such as the squared score function

$$\{I(\theta \in C_X) - \gamma(X)\}^2.$$

The estimator $\gamma_1(X)$ is said to be better than $\gamma_2(X)$ if

$$E\{I(\theta \in C_X) - \gamma_1(X)\}^2 \leq E\{I(\theta \in C_X) - \gamma_2(X)\}^2$$

for every θ and with strict inequality holding for some θ. The criterion was first proposed in Sandved (1968) who proved that the conditional probability given an ancillary statistic performs better than the unconditional probability. Later on, Robinson (1979a) studied this criterion and, among other results, related it to existence of relevant subsets. Recently, this criterion is becoming a popular one and has been used in Brown and Hwang (1991), Goutis and Casella (1992), Lu and Berger (1989a), Hwang and Brown (1991), Robert and Casella (1990), and Robinson (1979a,b). Similar criteria have been used by Hwang *et al.* (1992) for the hypothesis testing and by Lu and Berger (1989b), Hsieh and Hwang (1993), and Johnstone (1988) for the point estimation setting. In the latter two frameworks, the only difference between their criteria and the one presented here is the objective function. Returning to long term frequency appropriateness, we have the following theorem.

Theorem 9.2.3 *Let $\gamma(X)$ be a frequency valid estimator which is not frequency appropriate. Then $\gamma_*(X) = \gamma(X) + c$ is better than $\gamma(X)$ where c is a positive number defined in (9.15). Indeed*

$$E\{I(\theta \in C_X) - \gamma_*(X)\}^2 \leq E\{I(\theta \in C_X) - \gamma(X)\}^2 - c^2. \tag{9.14}$$

PROOF. Let

$$c = \inf_\theta P_\theta(\theta \in C_X) - E_\theta(\gamma(X)). \tag{9.15}$$

Since $\gamma(X)$ is valid and inappropriate (i.e., not frequency appropriate), c is positive. Now

$$E\{I(\theta \in C_X) - \gamma_*(X)\}^2 = E\{I(\theta \in C_X) - \gamma(X)\}^2 + c^2 - 2cE\{I(\theta \in C_X) - \gamma(X)\}.$$

Employing the lower bound

$$E\{I(\theta \in C_X) - \gamma(X)\} \geq c$$

which follows directly from (9.15) establishes (9.14).   ∎

The rationale behind the theorem is that $\gamma(X)$ has a negative "bias" at least c in estimating $I(\theta \in C_X)$ and $\gamma(X) + c$ will reduce the bias without affecting the variance and hence is obviously better.

The frequency appropriateness also relates to a criterion proposed in Casella (1988), which asserts that $\gamma(X)$ should satisfy

$$\inf_\theta P_\theta(\theta \in C_X) \leq E_\theta \gamma(X) \leq P_\theta(\theta \in C_X) \text{ for every } \theta. \tag{9.16}$$

The upper bound is the frequency validity requirement. This, together with the lower bound, implies frequency appropriateness as we shall show now.

Theorem 9.2.4 *Casella's criterion (9.16) implies that $\gamma(X)$ is frequency appropriate.*

PROOF. Let $\{\theta_n\}$ be a sequence of θ such that

$$P_{\theta_n}(\theta_n \in C_X) \to \inf_\theta P_\theta(\theta \in C_X).$$

Therefore by (9.16)

$$\lim_{n \to \infty} P_{\theta_n}(\theta_n \in C_X) \le E_{\theta_n}\gamma(x) \le P_{\theta_n}(\theta_n \in C_X).$$

Hence

$$\lim_{n \to \infty} E_{\theta_n}\gamma(X) = \lim_{n \to \infty} P_{\theta_n}(\theta_n \in C_X).$$

Now

$$
\begin{aligned}
0 &\le \inf_\theta \{P_\theta(\theta \in C_X) - E_\theta\gamma(X)\} \\
&\le \lim_{n \to \infty} \{P_{\theta_n}(\theta_n \in C_X) - E_{\theta_n}\gamma(X)\} = 0,
\end{aligned}
$$

implying the frequency appropriateness. ∎

Casella's criterion implies that the θ that minimizes $P_\theta(\theta \in C_X)$ also equates $P_\theta(\theta \in C_X) - E_\theta\gamma(X)$ to the minimum value zero. The frequency appropriateness criterion, however, does not specify the θ that minimizes $P_\theta(\theta \in C_X) - E_\theta\gamma(X)$ as long as its minimum value is zero. It is easy to see that Casella's criterion is not equivalent to (but is stronger than) the frequency appropriateness criterion. The following picture illustrates a counterexample to this equivalence. The estimator $\gamma(x)$ is frequency appropriate and does not satisfy Casella's criterion.

9.3 Generalizations

9.3.1 Other statistical procedures

In general, we may consider a statistical procedure $\delta(X)$ and a loss function $\widehat{L}(\theta, \delta(X))$. We may address the similar question as to what loss estimator $\widehat{L}(X)$ shall be long term frequency valid, which can be similarly defined as

$$\limsup_{n \to \infty} \frac{1}{n} \sum \{L(\theta_i, \delta(X_i)) - \widehat{L}(X_i)\} \le 0 \tag{9.17}$$

for θ_i's and X_i's given in (9.2), where Model 1 is assumed. Therefore, we focus on the case where similar experiments are performed although possibly at different θ_i values. Note that a conservatism of frequency theory requires that we should

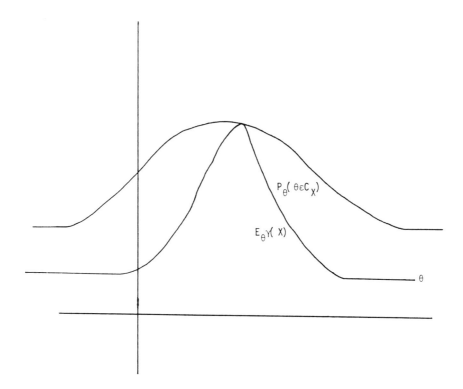

Figure 9.1: The estimator $\gamma(X)$ is valid, frequency appropriate and yet does not satisfy Casella's criterion

not underestimate the loss. Hence the inequality is "$\leq$", rather than "$\geq$". This general setting reduces to the setting of Section 9.2 if the loss is taken to be $L(\theta, C_X) = 1 - I(\theta \in C_X)$ and its estimator $\widehat{L} = 1 - \gamma(X)$. In such a case, (9.17) reduces to (9.5). The formulation, however, is more general, since it applies to the loss of a point estimator as well. We could write down a theorem similar to Theorems 9.2.1 and 9.2.2, provided that L is uniformly bounded in θ and X and its estimator $\widehat{L}(X)$ is uniformly bounded in X. This assumption is satisfied for the confidence problem. For the point estimation problem, the condition is normally satisfied if the parameter space is contained in a known interval.

In general though, we provide an alternative sufficient condition as below.

Theorem 9.3.1 *Assume that for every θ,*

$$Var(L(\theta, \delta(X)) - \widehat{L}(X)) < M$$

where M is a finite number independent of θ. Then, the long term validity of $\widehat{L}(X)$ is equivalent to the expectation validity of $\widehat{L}$, i.e.,

$$E_\theta L(\theta, \delta(X)) \leq E_\theta \widehat{L}(X) \ \ \forall \ \theta.$$

PROOF. The theorem can be proved similarly by the arguments that were used to establish Theorem 9.2.1. ∎

We could also establish a theorem similar to Theorem 9.2.2 regarding the long term frequency appropriateness of $\widehat{L}$, i.e., it satisfies (9.17) and there exists $\theta_1^*, \ldots, \theta_n^*, \ldots$ such that

$$\limsup_{n \to \infty} \frac{1}{n} \sum_{i=1}^{n} \{ L(\theta_i^*, \delta(X_i)) - \widehat{L}(X_i) \} = 0.$$

Theorem 9.3.2 *Under the assumption of Theorem 9.3.1, $\widehat{L}$ is long term frequency appropriate if and only if*

$$\sup_{\theta} E_\theta L(\theta, \delta(X)) - E_\theta \widehat{L}(X) = 0. \tag{9.18}$$

9.3.2 Frequency validity under Model 2 involving different experiments

In more realistic situations, a statistician works on a different problem each time. It is important to consider Model 2 to evaluate whether, in the long run, something interesting can be determined. Therefore, in this setting, X_i has a probability distribution $P_{\theta_i}^i$, where the distributions and θ_i vary as i varies. For instance, the dimensionality of θ_i may even be different. More precisely, we consider

an arbitrary sequence of possibly different parameters θ_i's

and their corresponding random variables $X_i \sim P_{\theta_i}^i$. (9.19)

We focus on confidence set problems first. Naturally, the confidence set for θ_i depends on i also and is denoted as $C_{X_i}^i$. Similar to Definition 9.2.1, we shall say that the sequence of confidence reports $\gamma^i(X_i)$ is long term frequency valid if

$$\liminf \frac{1}{n} \sum_{i=1}^{n} \{ I(\theta_i \in C_{X_i}^i) - \gamma^i(X_i) \} \geq 0 \tag{9.20}$$

for θ_i and X_i defined in (9.19).

Using arguments similar to those in Section 9.2, we establish the following theorem.

Theorem 9.3.3 *Assume that the report $\gamma^i(X_i)$ is bounded between 0 and 1.*

(i) If $\gamma^i(X_i)$ is expectation valid, i.e.,

$$P_{\theta_i}^i(\theta_i \in C_{X_i}^i) \geq E_{\theta_i}^i \gamma^i(X_i) \text{ for every } \theta_i, \tag{9.21}$$

then the sequence $\gamma^i(X_i)$ is long term frequency valid.

(ii) If, furthermore, for each i,

$$\inf_{\theta_i}[P^i_{\theta_i}(\theta_i \in C^i_{X_i}) - E^i_{\theta_i}\gamma^i(X_i)] = 0 \qquad (9.22)$$

then there exists $\theta^*_1, \ldots, \theta^*_n, \ldots$ so that

$$\lim_{n \to \infty} \frac{1}{n} \sum_{i=1}^{n} \{I(\theta^*_i \in C^i_{X_i}) - \gamma^i(X_i)\} = 0.$$

PROOF. Use arguments similar to those establishing Theorems 9.2.1 and 9.2.2. In proving this theorem, the fact that the confidence sets and the confidence reports depend on i does not make the argument more difficult. ∎

This seems comforting to the frequentist since the very fact that he has worked throughout his life on different problems does not prevent him from enjoying the fruit of the long term guarantee (9.20) as long as his confidence reports are expectation valid for every problem.

Unlike in the situations of Theorems 9.2.1 and 9.2.2, the reverse of Theorem 9.3.3 is, however, not true in general. The following theorem gives a necessary and sufficient condition.

Theorem 9.3.4 *Under the setting of Theorem 9.3.3, $\gamma^i(X_i)$ is long term frequency valid if and only if*

$$\liminf \frac{1}{n} \sum \{P^i_{\theta_i}(\theta_i \in C^i_{X_i}) - E^i_{\theta_i}\gamma^i(X_i)\} \geq 0. \qquad (9.23)$$

Similar to Lemma 9.2.2, the above theorem can be established assuming that the experiments are performed in a fixed order. Conclusion (9.23) does not say anything about, for instance, $i = 1$. In fact, even if (9.21) fails for $i = 1$, as long as it stands for all other i's, (9.23) and hence long term validity still hold. This shows that a procedure does not have to be expectation valid every time to be long term frequency valid. Long term frequency holds as long as the number n_v of problems in which he uses expectation valid reports grows as fast as n (i.e., $n_v/n \to 1$). [In another scenario, the frequentist could use confidence estimators, half of which are valid for 0.85 and the other half are valid for 0.95 and in the end of his career, he could still claim validity of 0.90 due to (9.23).] This seems somewhat surprising, especially when compared with Theorems 9.2.1 and 9.2.2, since the frequentist can do less and claim more in that he needs not be concerned with a few problems regarding the expectation validity and still enjoy long term validity under a less stringent Model 2. The reason there is no logical contradiction is because Model 2 does not include Model 1 as a subset. Model 2, however, is more realistic and hence the fact that long term frequentist validity is weaker than the expectation validity for every i is an interesting result. When one procedure is involved, however, Model 1 seems more appropriate.

Finally, we may also deal with the general problem of estimating a loss of a statistical procedure under Model 2. Under a setting similar to Theorem 9.3.1, the following theorem is proved.

Theorem 9.3.5 *Assume that for every i*

$$Var(L^i(\theta_i, \delta^i(X_i)) - \widehat{L}^i(X_i)) \le M.$$

(i) If $\widehat{L}^i$ is expectation valid, i.e.,

$$E^i_{\theta_i} L(\theta_i, \delta^i(X)) \le E^i_{\theta_i} \widehat{L}^i(X_i) \text{ for every } \theta_i \text{ and every } i,$$

then the sequence of estimators $\widehat{L}^i(X_i)$ for $L^i(\theta_i, \delta^i(X_i))$ is long term frequency valid, i.e.,

$$\limsup_{n \to \infty} \frac{1}{n} \sum_{i=1}^{n} \{L^i(\theta_i, \delta^i(X_i)) - \widehat{L}^i(X_i)\} \le 0 \text{ for all } \theta_1, \ldots, \theta_n, \ldots.$$

(ii) If, furthermore, for every i,

$$\sup_{\theta_i} E^i_{\theta_i} L^i(\theta_i, \delta^i(X_i)) - E^i_{\theta_i} \widehat{L}^i(X_i) = 0,$$

then there exists $\theta_1^*, \ldots, \theta_n^*, \ldots$ such that

$$\lim_{n \to \infty} \frac{1}{n} \sum \{L^i(\theta_i^*, \delta^i(X_i)) - \widehat{L}^i(X_i)\} = 0.$$

A theorem parallel to Theorem 9.3.4 can also be established.

9.4   Conclusions

In this paper, the Neyman's formulation of frequency justification (or validity) is generalized to the data dependent reports. The generalization appears to be quite natural. This natural generalization is shown to be equivalent (in one important case, viz., Model 1) to Berger's validity definition without any regularity conditions. A frequency appropriate confidence estimator (which are not overly conservative) is defined. The definition is shown to be related to (but weaker than) Casella's criterion. The derivation is quite elementary and pedagogical. However, through the derivation, it is hoped that the subject of frequency criteria has become more clear.

Acknowledgements. The writing of this paper is primarily motivated by the discussion of the conditional inference research group formed at Tsing Hua University in Spring 1993, during a visit by the author. The author thanks the participants of the research group, and in particular, Professor Wun–Yi Shu for his insightful comments relating to this paper.

The author is also very grateful to Professors Lawrence D. Brown and George Casella at Cornell University, with whom the author had numerous discussions concerning the subject. The editorial comments of Dr. Lauress Ackman helped to improve the flow of the paper.

References

1. Berger, J. O. (1983). The frequentist viewpoint and conditioning, In *Proceedings of the Berkeley Conference in Honor of Jerzy Neyman and Jack Kiefer* (Eds., L. LeCam and R. Olshen), Volume 1, pp. 15–55, Monterey: Wadsworth.

2. Berger, J. O. (1986). Commentary on Kiefer's papers, In *Jack Kiefer Collected Papers*, Supplementary Volume (Eds., L. D. Brown, I. Olkin, J. Sacks and H. Wynn), New York, Springer-Verlag.

3. Berger, J. O. and Wolpert, R. L. (1988). *The Likelihood Principle*, Second edition, Institute of Mathematical Statistics, Hayward, CA.

4. Bondar, J. V. (1988). Discussion of Casella's paper, *Statistical Decision Theory and Related Topics–IV* (Eds., S. Gupta and J. Berger), Volume 1, pp. 91–93, New York: Academic Press.

5. Brown, L. D. (1978). A contribution to Kiefer's theory of conditional confidence procedures, *Annals of Statistics*, **6**, 59–71.

6. Brown, L. D. and Hwang, J. T. (1991). Admissibility of confidence estimators, In *Proceedings of the 1990 Taipei Symposium in Statistics, June 28–30* (Eds., M. T. Chao and P. E. Chang), pp. 1–10, Institute of Statistical Science, Academia Sinica, Taiwan.

7. Brownie, C. and Kiefer, J. (1977). The ideas of conditional confidence in the simplest setting, *Communications in Statistics—Theory and Methods*, **6**, 691–751.

8. Casella, G. (1988). Conditional acceptable frequentist solutions info,In *Statistical Decision Theory and Related Topics–IV* (Eds., S. Gupta and J. Berger), Volume 1, Reply to discussion, p. 111, New York: Academic Press.

9. Casella, G. (1992). Conditional inference, *Lecture notes, Monograph Series* (Eds., A. D. Barbour, J. A. Rice, R. J. Serfling and W. E. Strawderman), pp. 1–12 , Hayward, California: Institute of Mathematical Statistics.

10. Chung, K. L. (1974). *A Course in Probability Theory*, Second edition, New York: Academic Press.

11. Cox, D. R. (1958). Some problems connected with statistical inference, *Annals of Mathematical Statistics*, **29**, 357–372.

12. Fisher, R. A. (1925). Theory of statistical estimation, *Proceedings of the Cambridge Philosophical Society*, **22**, 700–725.

13. Gleser, L. J. and Hwang, J. T. (1987). The nonexistence of $100(1-\alpha)\%$ confidence sets of finite expected diameter in errors-in-variables and related models, *Annals of Statistics*, **15**, 1351–1362.

14. Goutis, C. and Casella, G. (1992). Increasing the confidence in Student's t-interval, *Annals of Statistics*, **20**, 1501–1513.

15. Goutis, C. and Casella, G. (1995). Frequentist post-data inference, *International Statistical Review*, **63**, 325–344.

16. Gupta, S. (1956). On a decision rule for a problem in ranking means, *Ph.D. Thesis*, University of North Carolina, Chapel Hill, NC.

17. Hsieh, F. and Hwang, J. T. (1993). Admissibility under the validity constraint in estimating the loss of the least squares estimator, *Journal of Multivariate Analysis*, **44**, 279–285.

18. Hwang, J. T. and Brown, L. D. (1991). The estimated confidence approach under the validity constraint criterion, *Annals of Statistics*, **19**, 1964–1977.

19. Hwang, J. T., Casella, G., Robert, C., Wells, M. T. and Farrell, R. H. (1992). Estimation of accuracy in testing, *Annals of Statistics*, **20**, 490–509.

20. Hwang, J. T. and Pemantle, R. (1990). Evaluation of estimators of statistical significance under a class of proper loss functions, *Statistics and Decisions*.

21. Johnstone, I. (1988). On the inadmissibility of some unbiased estimates of loss, In *Statistical Decision Theory and Related Topics–IV* (Eds., S. S. Gupta and J. O. Berger), pp. 361–379, New York: Springer-Verlag.

22. Kiefer, J. (1976). Admissibility of conditional confidence procedures, *Annals of Statistics*, **4**, 836–865.

23. Kiefer, J. (1977a). Conditional confidence and estimated confidence in multidecision problems (with applications to selection and ranking), In *Multivariate Analysis–IV, Proceedings of the Fourth International Symposium on Multivariate Analysis* (Ed., P. R. Krishnaiah), pp. 143–158, Amsterdam: North-Holland.

24. Kiefer, J. (1977b). Conditional confidence statements and confidence estimators (with discussion), *Journal of the American Statistical Association*, **72**, 789–827.

25. Lu, K. L. and Berger, J. O. (1989a). Estimated confidence procedures for multivariate normal means, *Journal of Statistical Planning and Inference*,

26. Lu, K. L. and Berger, J. O. (1989b). Estimation of normal means: frequentist estimation loss, *Annals of Statistics*, **17**, 890–906.

27. McCullagh P. (1991). Conditional inference and Cauchy models, *Technical Report*, Department of Statistics, University of Chicago, Chicago, Illinois.

28. Neyman, J. (1937). Outline of a theory of statistical estimation based on the classical theory of probability, *Philosophical Transactions of the Royal Society of London, Series A*, **236**, 333–380.

29. Robert, C. and Casella, G. (1990). Improved confidence statements for the usual multivariate normal confidence set, *Technical Report*, Statistics Center, Cornell University, Ithaca, New York.

30. Robinson, G. K. (1979a). Conditional properties of statistical procedures, *Annals of Statistics*, **7**, 742–755.

31. Robinson, G. K. (1979b). Conditional properties of statistical procedures for location and scale families, *Annals of Statistics*, **7**, 756–771.

32. Rukhin, A. (1988a). Estimated loss and admissible loss, In *Statistical Decision Theory and Related Topics–IV* (Eds., S. Gupta and J. Berger), Vol. 1, pp. 409–416, New York: Academic Press.

33. Rukhin, A. (1988b). Loss functions for loss estimation, *Annals of Statistics*, **16**, 1262–1269.

34. Sandved, E. (1968). Ancillary statistics and estimation of the loss in estimation problem, *Annals of Mathematical Statistics*, **39**, 1756–1758.

10

Asymptotic Theory for the Simex Estimator in Measurement Error Models

Raymond J. Carroll and L. A. Stefanski

Texas A& M University, College Station, TX
North Carolina State University, Raleigh, NC

Abstract: Carroll *et al.* (1996) make a conjecture about the asymptotic distribution of the SIMEX estimator [Cook and Stefanski (1994)], a promising method in the analysis of measurement error models. Here we prove their conjecture under a set of technical conditions.

Keywords and phrases: Asymptotics, bootstrap, computationally intensive methods, estimating equations, linear regression, logistic regression, measurement error models, regression calibration

10.1 Introduction

We consider regression problems where some of the predictors are measured with additive error. The response is denoted by Y, the predictors are (Z, X), but X cannot be observed. Instead we can observe $W = X + \sigma U$, where U has mean zero and variance 1. We will consider the structural case, so that (Y_i, Z_i, X_i, W_i) for $i = 1, ..., n$ are independent and identically distributed. For linear regression, this is the classical additive structural measurement error model described in detail by Fuller (1987).

In linear regression, as well as nonlinear models such as the generalized linear models, it is well known that the *naive estimator* which ignores measurement error leads to inconsistent regression parameter estimates. Correction for this bias in linear regression has a long history, but the analysis of nonlinear measurement error models is of more recent vintage. One of the more useful general methods is what we call *regression calibration*, wherein X is replaced by an estimate of $E(X|Z, W)$ and the standard analysis then performed; see Rosner, Willett and Spiegelman (1989), Rosner, Spiegelman and Willett (1990),

Carroll and Stefanski (1990), Gleser (1990) and Carroll, Ruppert and Stefanski (1995). The regression calibration estimates are usually easy to compute, have straightforward standard error estimates [Carroll and Stefanski (1990, 1994)] and are amenable to bootstrap analysis.

An alternative general method has recently been proposed by Cook and Stefanski (1994), hereafter denoted by CS. Their idea, called SIMEX for *simulation* and *extrapolation*, relies on computer simulation to generate parameter estimates. SIMEX has essentially all the same properties outlined above for regression calibration, with two exceptions: (i) it is as easy to program but far more computationally intensive; and (ii) the small measurement error bias of SIMEX is of order $O(\sigma^6)$ instead of oder $O(\sigma^4)$, suggesting that it might prove superior in some highly nonlinear models.

Carroll *et al.* (1996) investigate the asymptotic distribution of the SIMEX estimator, and give a sketch of a proof. The purpose of this paper is to expand their sketch to give a proof under explicit regularity conditions.

This paper consists of three sections. Section 10.2 reviews the SIMEX method and the sketch of Carroll *et al.* (1996). In Section 10.3, we provide general conditions under which SIMEX estimators based on different location functionals all have the same asymptotic distribution, in the case of a scalar parameter θ. The result is applied to logistic regression, under strong regularity conditions.

The general result relies on the continuity of location functionals, which is not guaranteed for the mean. Section 10.4 gives an explicit proof for the mean functional in simple linear regression.

10.2 The SIMEX Method

We provide here a brief description of the SIMEX algorithm. CS and Carroll *et al.* (1995) should be consulted for more details and motivation. Stefanski and Cook (1995) should also be consulted.

The algorithm consists of a *simulation* step, followed by an *extrapolation* step. We review each step in turn.

10.2.1 Simulation step

Suppose one has an unknown vector parameter Θ_0. If one could observe the X's, then we suppose one could estimate Θ_0 by solving an estimating equation

$$\mathbf{0} = \sum_{i=1}^{n} \psi(Y_i, Z_i, X_i, \Theta). \tag{10.1}$$

Estimating equations and their theory have a long history in statistics and include as special cases most of the forms of regression currently used in practice.

As used by CS, SIMEX works as follows. Fix $B > 0$ (they use $B = 50, 100$) and for $b = 1, ..., B$, generate via computer independent standard normal random variables $\{\epsilon_{ib}\}$. In place of W_i, "add–on" pseudo–errors of the form $\sigma \lambda^{1/2} \epsilon_{ib}$. Then the variance of $W_i + \sigma \lambda^{1/2} \epsilon_{ib}$ given X_i is $\sigma^2(1+\lambda)$. For each b, define $\widehat{\Theta}_{b,\sigma^2(1+\lambda)}$ as the solution to

$$0 = \sum_{i=1}^{n} \psi(Y_i, Z_i, W_i + \sigma \lambda^{1/2} \epsilon_{ib}, \Theta). \tag{10.2}$$

Now form the average of the $\widehat{\Theta}_{b,\sigma^2(1+\lambda)}$'s, namely

$$\widehat{\Theta}_{S,\sigma^2(1+\lambda)} = B^{-1} \sum_{b=1}^{B} \widehat{\Theta}_{b,\sigma^2(1+\lambda)}, \tag{10.3}$$

the subscript S emphasizing the simulation nature of the estimator. Of course, when $\lambda = 0$ the simulation process is vacuous and $\widehat{\Theta}_{S,0}$ denotes the naive estimator of Θ_0. By standard estimating equation theory, under sufficient regularity conditions, $\widehat{\Theta}_{S,\sigma^2(1+\lambda)}$ converges in probability to $\Theta_{\sigma^2(1+\lambda)}$, the solution in Θ of

$$0 = E\left\{\psi(Y, Z, W + \sigma \lambda^{1/2} \epsilon, \Theta)\right\}. \tag{10.4}$$

CS also consider the limit for infinite B, namely

$$\widehat{\Theta}_{\sigma^2(1+\lambda)} = E\left\{\widehat{\Theta}_{b,\sigma^2(1+\lambda)} | (Y_i, Z_i, W_i), i = 1, ..., n\right\}. \tag{10.5}$$

It is the asymptotic distribution of $\widehat{\Theta}_{\sigma^2(1+\lambda)}$ which is of interest here.

For later use, we note that for any fixed b, standard asymptotic theory yields the expansion

$$n^{1/2}\left\{\widehat{\Theta}_{b,\sigma^2(1+\lambda)} - \Theta_{\sigma^2(1+\lambda)}\right\} = -\mathcal{A}^{-1}(\sigma^2, \lambda, \Theta_{\sigma^2(1+\lambda)})$$

$$\times \; n^{-1/2} \sum_{i=1}^{n} \psi(Y_i, Z_i, W_i + \sigma \lambda^{1/2} \epsilon_{ib}, \Theta_{\sigma^2(1+\lambda)}) + o_p(1), \tag{10.6}$$

where

$$\mathcal{A}(\sigma^2, \lambda, \Theta_{\sigma^2(1+\lambda)}) = E\left\{(\partial/\partial\Theta^t)\psi(Y, Z, W + \sigma \lambda^{1/2} \epsilon, \Theta_{\sigma^2(1+\lambda)})\right\}. \tag{10.7}$$

10.2.2   Extrapolation step

CS suggest that one compute $\widehat{\Theta}_{S,\sigma^2(1+\lambda)}$ on a fixed grid $\Lambda = (\lambda_1, ..., \lambda_M)$, thus yielding an understanding of the behavior of the estimators for different amounts of measurement error. Then, as motivated in the introduction, they suggest that one fit a parametric model $\mathcal{G}(\Gamma, \lambda)$ in a vector parameter Γ to the $\widehat{\Theta}_{S,\sigma^2(1+\lambda)}$'s as a function of the λ's, thus resulting in estimates $\widehat{\Gamma}$. Finally, the SIMEX estimator of Θ_0 is

$$\widehat{\Theta}_{\text{SIMEX}} = \mathcal{G}(\widehat{\Gamma}, -1). \tag{10.8}$$

Various parametric models are possible, CS suggesting linear, quadratic and, with $\Gamma = (\gamma_0, \gamma_1, \gamma_2)^t$, the nonlinear model $\mathcal{G}(\lambda, \Gamma) = \gamma_0 + \gamma_1(\lambda + \gamma_2)^{-1}$. They fit the models by ordinary unweighted least squares.

10.2.3   Sketch of asymptotics when $B = \infty$

There are three types of asymptotics which can be investigated: (i) $n \to \infty$ for B fixed; (ii) $n \to \infty$ and $B \to \infty$ simultaneously; and (iii) $B = \infty$ and $n \to \infty$. Carroll *et al.* (1996) derive the distribution of the SIMEX estimator in the first two cases.

We have proposed the estimator (10.3), based on taking the sample mean of the terms $\widehat{\Theta}_{b,\sigma^2(1+\lambda)}$ for $b = 1, ..., B$. Other location functions can be used, e.g., the sample median, trimmed mean, etc. We will distinguish between a sample location functional, such as the sample mean or median, and a population location functional. Let $\mathcal{T}$ be any population location functional of a random variable V, such as the population mean or median, satisfying $\mathcal{T}(a + bV) = a + b\mathcal{T}(V)$. The population location functional also must satisfy the condition that if V is normally distributed with mean zero, then $\mathcal{T}(V) = 0$.

It is easiest to understand the behavior of the SIMEX estimator for large B if one first fixes B, lets $n \to \infty$ and then lets $B \to \infty$. For fixed B, Carroll *et al.* (1996) have shown that the random variables $\left\{ \widehat{\Theta}_{b,\sigma^2(1+\lambda)} \right\}_{b=1}^{B}$ form a set of correlated normal random variables as $n \to \infty$, with common variance and correlation. They thus satisfy approximately the one-way random effects model

$$n^{1/2} \left(\widehat{\Theta}_{b,\sigma^2(1+\lambda)} - \Theta_{\sigma^2(1+\lambda)} \right) \approx \xi + \eta_b, \tag{10.9}$$

say, where ξ is normally distributed and independent of $(\eta_1, \ldots, \eta_B)$, which are themselves independent and identically distributed normal random variables with mean zero. As $B \to \infty$, subject to further regularity conditions, all sample location functionals of the left hand side of (10.9), including the mean and the median, converges to ξ, and hence they all have the same limit distribution.

A more involved analysis occurs if asymptotics are done for $B = \infty$. Here is the sketch given by Carroll *et al.* (1996).

As indicated by CS and in (10.5), if $B = \infty$ then for any given λ, the SIMEX building blocks are the terms

$$\mathcal{T}\left\{ \widehat{\Theta}_{b,\sigma^2(1+\lambda)} | (Y_i, Z_i, W_i), i = 1, ..., n \right\}. \tag{10.10}$$

It is important to emphasize that b in the above expression is any *single, fixed* b. Equation (10.5) arises when $\mathcal{T}$ is the expectation functional, while when using the median, $\mathcal{T}$ is the median of the indicated distribution.

At the risk of repetition, (10.10) is a functional of the conditional distribution of a random variable. If we can understand the behavior of the random variable for large sample sizes as a function of the data, then under sufficient regularity conditions we can compute its conditional distribution.

Our analysis then requires two steps.

Step 1: For any fixed b, find an expansion for $n^{1/2}\left(\widehat{\Theta}_{b,\sigma^2(1+\lambda)} - \Theta_{\sigma^2(1+\lambda)}\right)$.

This describes the unconditional distribution of $\widehat{\Theta}_{b,\sigma^2(1+\lambda)}$ as a function of the data (Y_i, Z_i, W_i) for $i = 1, ..., n$.

Step 2: Compute the conditional distribution of the expansion.

Here is Step 1. For any fixed b (remember, $B = \infty$, and we are not letting $B \to \infty$), from (10.6), as $n \to \infty$,

$$n^{1/2}\left\{\widehat{\Theta}_{b,\sigma^2(1+\lambda)} - \Theta_{\sigma^2(1+\lambda)}\right\}$$
$$= -n^{-1/2}\sum_{i=1}^{n} A^{-1}(\cdot)\psi\left\{Y_i, Z_i, W_i + \sigma\lambda^{1/2}\varepsilon_{ib}, \Theta_{\sigma^2(1+\lambda)}\right\} + o_p(1),$$

where $A(\cdot)$ is defined in (10.7) and where the $(\varepsilon_{ib})_{i=1}^{n}$ are independent and identically distributed standard normal random variables.

Let $G(Y, Z, W, \Theta)$ be the conditional mean of $-A^{-1}(\cdot)\psi(Y, Z, W + \sigma\lambda^{1/2}\varepsilon, \Theta)$ given (Y, Z, W); note that it has unconditional mean zero. It then follows that

$$n^{1/2}\left\{\widehat{\Theta}_{b,\sigma^2(1+\lambda)} - \Theta_{\sigma^2(1+\lambda)}\right\}$$
$$= n^{-1/2}\sum_{i=1}^{n} G\left\{Y_i, Z_i, W_i, \Theta_{\sigma^2(1+\lambda)}\right\}$$
$$+ n^{-1/2}\sum_{i=1}^{n}\left[A^{-1}(\cdot)\psi\left\{Y_i, W_i + \lambda^{1/2}\sigma\varepsilon_{ib}, \Theta_{\sigma^2(1+\lambda)}\right\}\right.$$
$$\left. - G\left\{Y_i, Z_i, W_i, \Theta_{\sigma^2(1+\lambda)}\right\}\right] + o_p(1). \quad (10.11)$$

The two terms on the right hand side of (10.11) are uncorrelated. Further, given $(Y_i, Z_i, W_i)_{i=1}^{n}$, the next to last term in (10.11), which we write as $\mathcal{G}_n$, converges to a normal random variable, say $\mathcal{G}$, which has mean zero and variance $\Omega = \Omega\left\{(Y_i, Z_i, W_i)_{i=1}^{\infty}\right\}$. Thus,

$$n^{1/2}\left\{\widehat{\Theta}_{b,\sigma^2(1+\lambda)} - \Theta_{\sigma^2(1+\lambda)}\right\}$$
$$= \mathcal{G}_n + n^{-1/2}\sum_{i=1}^{n} G\left\{Y_i, Z_i, W_i, \Theta_{\sigma^2(1+\lambda)}\right\} + o_p(1). \quad (10.12)$$

The two terms on the right hand side of (10.12) are uncorrelated, asymptotically normally distributed and hence asymptotically independent.

Here is Step 2. Condition on the terms $(Y_i, Z_i, W_i)_{i=1}^n$. Under sufficient regularity conditions (we later give one such set), it follows that

$$n^{1/2} \left\{ \mathcal{T}(\widehat{\Theta}_{b,\sigma^2(1+\lambda)}) - \mathcal{T}(\Theta_{\sigma^2(1+\lambda)}) \right\}$$

$$= \mathcal{T} \left\{ \mathcal{G}_n + n^{-1/2} \sum_{i=1}^n G\left(Y_i, W_i, \Theta_{\sigma^2(1+\lambda)}\right) + o_p(1) \right\}$$

$$= n^{-1/2} \sum_{i=1}^n G\left(Y_i, W_i, \Theta_{\sigma^2(1+\lambda)}\right) + \mathcal{T}\left\{\mathcal{G}_n + o_p(1)\right\}$$

$$= n^{-1/2} \sum_{i=1}^n G\left(Y_i, W_i, \Theta_{\sigma^2(1+\lambda)}\right) + o_p(1), \qquad (10.13)$$

the last step following from the fact that $\mathcal{G}_n$ is conditionally normally distributed with mean zero, and $\mathcal{T}(V) = 0$ when V is normally distributed with mean zero.

We have thus shown heuristically that, subject to regularity conditions in going from Step 1 to Step 2, if $B = \infty$, then as $n \to \infty$ SIMEX estimators based on different location functionals are all asymptotically equivalent. In the next two sections, we give details justifying the heuristics in special situations.

Finally, a remark on regularity conditions. The $o_p(1)$ term in Eq. (10.12) is an unconditional one, while the $o_p(1)$ term in (10.13) is conditional. Justifying this passage can be extraordinarily difficult and technical as can be seen below.

10.3   General Result and Logistic Regression

10.3.1   The general result

We now provide a formal proof under explicit regularity conditions in the case of a scalar parameter θ. We will consider the random variables R and ϵ, and further define $\tilde{R} = (R_1, R_2, ...)$. In the SIMEX application, $R = (Y, W)$. While the statement of the result is general, in applications of it both ϵ and R must be in a compact set.

We first remember that a *location functional* $\mathcal{T}$ of the distribution of a random variable Z is such that $\mathcal{T}(Z) = 0$ if Z is symmetrically distributed about zero, and $\mathcal{T}(a + Z) = a + \mathcal{T}(Z)$ for any constant a. We will assume that $\mathcal{T}$ is bounded and continuous, and in particular that for two sequences of continuous random variables Z_n and V_n, if $Z_n = V_n + o_p(1)$, and if V_n is converging in distribution to a continuous random variable, then $\mathcal{T}(Z_n) - \mathcal{T}(V_n) \to 0$. The median satisfies these conditions, as do essentially all robust measures of location. The mean satisfies the convergence condition only under stronger assumptions.

We assume that $\theta \in \mathcal{C}_1$, a compact set, and we write $(\epsilon, R) \in \mathcal{C}_2$. Although not used in the Theorem itself, in the application we will require that ϵ be a bounded random variable.

Let $\widehat{\theta}$ be the solution to the equation

$$0 = n^{-1} \sum_{i=1}^{n} \psi(\epsilon_i, R_i, \theta), \tag{10.14}$$

and define θ_0 to be the (assumed unique) solution to the equation

$$0 = E\left\{\psi(\epsilon_i, R_i, \theta)\right\}.$$

Make the further definitions

$$
\begin{aligned}
\xi(\epsilon, R, \theta) &= -\frac{\partial}{\partial \theta}\psi(\epsilon, R, \theta); \\
A &= E\left\{\xi(\epsilon, R, \theta_0)\right\}; \\
G(R) &= A^{-1}E\left\{\psi(\epsilon, R, \theta_0)|R\right\}; \\
\Lambda(\epsilon, R) &= A^{-1}\psi(\epsilon, R, \theta_0) - G(R).
\end{aligned}
$$

Consider the following assumptions:

A.1: Conditional on $\tilde{R}$, for almost all such subsequences $\tilde{R}$, a solution to (10.14) exists in $\mathcal{C}_1$ with probability approaching 1.

A.2: The function $\psi(\cdot)$ is continuously twice differentiable.

A.3: There is a constant $b_1 < \infty$ such that for $\theta \in \mathcal{C}_1$, $(\epsilon, R) \in \mathcal{C}_2$, $|\xi(\epsilon, R, \theta)| \leq b_1$.

A.4: There is a constant $b_2 < \infty$ such that for $(\theta_1, \theta_2) \in \mathcal{C}_1 \times \mathcal{C}_1$, $(\epsilon, R) \in \mathcal{C}_2$, $|\xi(\epsilon, R, \theta_2) - \xi(\epsilon, R, \theta_1)| \leq b_2|\theta_2 - \theta_1|$.

A.5: For $b_4 < \infty$ such that for $\theta \in \mathcal{C}_1$, $(\epsilon, R) \in \mathcal{C}_2$, $E\left\{\psi^2(\epsilon, R, \theta)|R\right\} \leq b_4$.

A.6: For some $c > 0$, conditional on $\tilde{R}$, for almost all such subsequences $\tilde{R}$, $n^c(\widehat{\theta} - \theta_0) \rightarrow 0$ in probability.

A.7: There is a constant $b_3 > 0$ such that for $\theta \in \mathcal{C}_1$, $(\epsilon, R) \in \mathcal{C}_2$, $|\xi(\epsilon, R, \theta)| \geq b_3$.

Theorem 10.3.1 *Under* **A.1**–**A.5** *and either* **A.6** *or* **A.7**, *conditional on* $\tilde{R}$, *for almost all such subsequences* $\tilde{R}$,

$$n^{1/2}(\widehat{\theta} - \theta_0) = n^{-1/2}\sum_{i=1}^{n} G(R_i) + n^{-1/2}\sum_{i=1}^{n} \Lambda(\epsilon_i, R_i) + o_p(1). \tag{10.15}$$

If $\mathcal{T}(\cdot|\tilde{R})$ is a location functional given $\tilde{R}$, this means that for almost all subsequences $\tilde{R}$,

$$n^{1/2}\left\{\mathcal{T}(\widehat{\theta}|\tilde{R}) - \theta_0\right\} = n^{-1/2}\sum_{i=1}^{n}G(R_i) + \mathcal{T}\left\{n^{-1/2}\sum_{i=1}^{n}\Lambda(\epsilon_i, R_i)|\tilde{R}\right\} + o_p(1)$$

$$= n^{-1/2}\sum_{i=1}^{n}G(R_i) + o_p(1), \qquad (10.16)$$

as claimed.

PROOF OF THE THEOREM. We only prove (10.15), since (10.16) follows from our conditions on the functional, and the fact that given $\tilde{R}$, $n^{-1/2}\sum_{i=1}^{n}\Lambda(\epsilon_i, R_i)$ is asymptotically normally distributed

By a Taylor series expansion, we have that

$$n^{1/2}\left(\widehat{\theta} - \theta_0\right) = \frac{n^{-1/2}\sum_{i=1}^{n}\psi(\epsilon_i, R_i, \theta_0)}{n^{-1}\sum_{i=1}^{n}\xi(\epsilon_i, R_i, \theta_*)}, \qquad (10.17)$$

where θ_* is between $\widehat{\theta}$ and θ_0.

We first show that **A.1–A.5** and **A.7** imply **A.6**. Using (10.17) and **A.7**, for $c > 0$,

$$n^c|\widehat{\theta} - \theta_0| \le b_3^{-1}|n^{-1+c}\sum_{i=1}^{n}\psi(\epsilon_i, R_i, \theta_0)|.$$

But

$$n^{-1+c}\sum_{i=1}^{n}\psi(\epsilon_i, R_i, \theta_0) = n^{-1+c}\sum_{i=1}^{n}[\psi(\epsilon_i, R_i, \theta_0) - E\left\{\psi(\epsilon, R, \theta_0)|R_i\right\}]$$

$$+ n^{-1+c}\sum_{i=1}^{n}E\left\{\psi(\epsilon, R, \theta_0)|R_i\right\}$$

$$= S_{n1} + S_{n2}. \qquad (10.18)$$

Now, for almost all sequences $\tilde{R}$, $S_{n2} \to 0$ by the law of the iterated logarithm (LIL), and $S_{n1} = o_p(1)$ given $\tilde{R}$ because of **A.5** and Chebychev's inequality. This proves **A.6**.

We now prove (10.15), using **A.1–A.6**. Invoke (10.17) again, and rewrite

$$n^{-1}\sum_{i=1}^{n}\xi(\epsilon_i, R_i, \theta_*) = T_{n1} + T_{n2} + T_{n3} + A,$$

where

$$T_{n1} = n^{-1}\sum_{i=1}^{n}\left\{\xi(\epsilon_i, R_i, \theta_*) - \xi(\epsilon_i, R_i, \theta_0)\right\},$$

$$T_{n2} = n^{-1} \sum_{i=1}^{n} \left[\xi(\epsilon_i, R_i, \theta_0) - E\left\{ \xi(\epsilon, R, \theta_0) | R_i \right\} \right],$$

$$T_{n3} = n^{-1} \sum_{i=1}^{n} \left[E\left\{ \xi(\epsilon, R, \theta_0) | R_i \right\} - A \right].$$

Now $T_{n3} = o_{a.s.}(n^{-c})$ for some $c > 0$ by the LIL and **A.3**. Also, given $\tilde{R}$, $T_{n2} = o_p(n^{-c})$ for some $c > 0$ by an appeal to the Chebychev inequality and **A.3**. Finally, given $\tilde{R}$, $T_{n1} = o_p(n^{-c})$ by **A.4–A.5**.

We have thus shown that for almost all sequences $\tilde{R}$, given $\tilde{R}$, for some $c > 0$,

$$n^{1/2}(\widehat{\theta} - \theta_0) = \left\{ A^{-1} n^{-1/2} \sum_{i=1}^{n} \psi(\epsilon_i, R_i, \theta_0) \right\} \left\{ 1 + o_p(n^{-c}) \right\}.$$

The same argument used in (10.18) now proves the Theorem.   ∎

10.3.2   Logistic regression

We show here that the conditions of the theorem apply to logistic regression, under strong assumptions and when the measurement error variance σ^2 is known. In the theorem, the random variables $R = (Y, W)$, where Y is the binary response and $W = X + U$ is the observed predictor, where X is the unobserved true predictor and U is the measurement error. Strict application of the conditions requires that W be bounded, and hence that each of X and U be bounded. While from a theoretical point of view the conditions are too strong, as a practical matter we simply need to show that there are conditions under which the SIMEX estimator with $B = \infty$ is asymptotically normally distributed, with distribution independent of the location functional used. Weakening the conditions would be a matter of some theoretical interest.

Logistic regression based on the observed data fits into our framework by setting

$$\psi(\epsilon, R, \theta) = (W + \sigma\lambda^{1/2}\epsilon) \left[Y - H\left\{ \theta(W + \sigma\lambda^{1/2}\epsilon) \right\} \right],$$

$$\xi(\epsilon, R, \theta) = (W + \sigma\lambda^{1/2}\epsilon)^2 H^{(1)} \left\{ \theta(W + \sigma\lambda^{1/2}\epsilon) \right\},$$

where $H^{(1)}(v) = H(v)\left\{ 1 - H(v) \right\}$.

Assumptions **A.1**, **A.3–A.5** hold if W is bounded. Assumption **A.7** holds if, in addition, $|\epsilon| \leq d$ while simultaneously $|W| > d$. Basically, these conditions require that the true values of X be bounded away from zero, and both the measurement error U and the pseudo-errors ϵ be bounded. These conditions are also sufficient to verify **A.2**, for a sufficiently large set $\mathcal{C}_1$.

10.4   SIMEX Estimation in Simple Linear Regression

The previous argument relied on the continuity of the location functional $\mathcal{T}$, something which need not hold when the functional is the mean, i.e., two distributions can have the same limit distribution without their means even existing, much less converging.

In this section, our intent is to provide an explicit argument for the mean functional. The details are complex even for the simple linear regression case considered here.

The regression model under investigation is $Y_i = \alpha + \theta X_i + \eta_i$, $i = 1, \ldots, n$, where the equation errors $\{\eta_i\}_1^n$ are identically distributed with $E\{\eta_1\} = 0$ and $\mathrm{Var}\{\eta_1\} = \sigma_\eta^2$, mutually independent, and are independent of the measurement errors. We assume the functional version of this model, i.e., that $X_1, \ldots, X_n$ are nonrandom constants. We add the regularity condition that the sample variance of $\{X_i\}_1^n$ converges to $\sigma_x^2 > 0$, as $n \to \infty$. Finally, the results assume that we observe $W = X + U$, where U has a normal distribution with known variance σ^2.

We do not address asymptotic normality per se, because this follows from the easily established asymptotic equivalence of the SIMEX estimator and the much-studied method-of-moments estimator [Fuller (1987, pp. 15-17)].

Define $S_{yx} = n^{-1} \sum (Y_i - \overline{Y})(X_i - \overline{X})$ and $S_{xx} = n^{-1} \sum (X_i - \overline{X})^2$, etc. We proceed under the minimal assumptions that: (1) $S_{yy} = \theta^2 \sigma_x^2 + \sigma_\eta^2 + o_p(1)$; (2) $\sqrt{n}(S_{yw} - S_{xx}\theta) = O_p(1)$; and (3) $\sqrt{n}\{S_{ww} - S_{xx} - (n-1)\sigma^2/n\} = O_p(1)$. Note that (3) implies that $S_{ww} = \sigma_x^2 + \sigma^2 + o_p(1)$. We consider estimating θ.

For this problem

$$\widehat{\theta}_b(\lambda) = \frac{\widehat{N}_b(\lambda)}{\widehat{D}_b(\lambda)},$$

where $\widehat{N}_b(\lambda) = n^{-1} \sum_{i=1}^n (Y_i - \overline{Y})(W_i + \sigma\sqrt{\lambda}\,\epsilon_{b,i})$ and $\widehat{D}_b(\lambda) = n^{-1} \sum_{i=1}^n \{(W_i - \overline{W}) + \sigma\sqrt{\lambda}\,(\epsilon_{b,i} - \overline{\epsilon}_b)\}^2$. Define

$$
\begin{aligned}
\widehat{N}(\lambda) &= E\{\widehat{N}_b(\lambda) \mid \{Y_i, W_i\}_1^n\} = S_{yw}, \\
N(\lambda) &= E\{S_{yw}\} = S_{xx}\theta, \\
\widehat{D}(\lambda) &= E\{\widehat{D}_b(\lambda) \mid \{Y_i, W_i\}_1^n\} = S_{ww} + \left(\frac{n-1}{n}\right)\lambda\sigma^2, \\
D(\lambda) &= E\{S_{ww} + \left(\frac{n-1}{n}\right)\lambda\sigma^2\} = S_{xx} + \left(\frac{n-1}{n}\right)(\lambda+1)\sigma^2.
\end{aligned}
$$

The Cauchy-Schwartz inequality is used to show that

$$\widehat{\theta}_b^2(\lambda) \leq \frac{nS_{yy}}{\sum_{i=1}^n \{W_i - \overline{W} + \sqrt{\lambda}\,\sigma(\epsilon_{b,i} - \overline{\epsilon}_b)\}^2}. \tag{10.19}$$

Conditioned on $\{Y_i, W_i\}_1^n$, the right hand side of (10.19) is proportional to the reciprocal of a noncentral chi-squared random variable with $n-1$ degrees of freedom and thus has its first k moments finite provided $n > 1 + 2k$ [see Johnson, Kotz and Balakrishnan (1995, p. 449)]. So for $k > 0$, $\widehat{\theta}_b(\lambda)$ possesses a finite $2k$ conditional moment, provided $n > 1 + 2k$. Thus, the random quantity $\widehat{\theta}(\lambda)$ is well defined for $n > 3$.

Define

$$\tilde{\theta}_b(\lambda) = \frac{N(\lambda)}{D(\lambda)} + \frac{\widehat{N}_b(\lambda) - N(\lambda)\widehat{D}_b(\lambda)/D(\lambda)}{D(\lambda)}.$$

Then $\widehat{\theta}_b(\lambda) = \tilde{\theta}_b(\lambda) + \tilde{R}_b(\lambda)$ where

$$\tilde{R}_b(\lambda) = \left(\frac{\widehat{D}_b(\lambda) - D(\lambda)}{\widehat{D}_b(\lambda)}\right)\left(\frac{\widehat{N}_b(\lambda) - N(\lambda)\widehat{D}_b(\lambda)/D(\lambda)}{D(\lambda)}\right).$$

Letting $\tilde{\theta}(\lambda) = E\{\tilde{\theta}_b(\lambda) \mid \{Y_i, W_i\}_1^n\}$ and $\tilde{R}(\lambda) = E\{\tilde{R}_b(\lambda) \mid \{Y_i, W_i\}_1^n\}$, we have that $\widehat{\theta}(\lambda) = E\{\widehat{\theta}_b(\lambda) \mid \{Y_i, W_i\}_1^n\} = \tilde{\theta}(\lambda) + \tilde{R}(\lambda)$ where

$$\tilde{\theta}(\lambda) = \frac{N(\lambda)}{D(\lambda)} + \frac{\widehat{N}(\lambda) - N(\lambda)\widehat{D}(\lambda)/D(\lambda)}{D(\lambda)}.$$

We now show that the remainder terms, $\tilde{R}_b(\lambda)$ and $\tilde{R}(\lambda)$ can be ignored asymptotically. It is not sufficient to show that these are $o_p(n^{-1/2})$. It must be shown that $E\{n[\tilde{R}_b(\lambda) - \tilde{R}(\lambda)]^2 \mid \{Y_i, W_i\}_1^n\} = o_p(1)$. But, because $E\{n[\tilde{R}_b(\lambda) - \tilde{R}(\lambda)]^2 \mid \{Y_i, W_i\}_1^n\} = E\{n\tilde{R}_b^2(\lambda) \mid \{Y_i, W_i\}_1^n\} - n\tilde{R}^2(\lambda)$ and $n\tilde{R}^2(\lambda) \le E\{n\tilde{R}_b^2(\lambda) \mid \{Y_i, W_i\}_1^n\}$ almost surely, it is sufficient to show that $E\{n\tilde{R}_b^2(\lambda) \mid \{Y_i, W_i\}_1^n\} = o_p(1)$.

For the remainder of this section $E\{\cdot\}$ denotes conditional expectation, $E\{\cdot \mid \{Y_i, W_i\}_1^n\}$, and all equalities and inequalities hold almost surely.

The Cauchy-Schwartz inequality and the inequality $(a+b)^4 \le 8(a^4 + b^4)$ show that $E\{n\tilde{R}_b^2(\lambda)\}$ is bounded above by

$$8E\left\{\left(\frac{\widehat{D}_b(\lambda) - D(\lambda)}{\widehat{D}_b(\lambda)}\right)^4\right\}$$

$$\times E\left\{\left[\frac{n^{1/2}[\widehat{N}_b(\lambda) - N(\lambda)]}{D(\lambda)}\right]^4 + \left[\frac{N(\lambda)}{D(\lambda)}\frac{n^{1/2}[\widehat{D}_b(\lambda) - D(\lambda)]}{D(\lambda)}\right]^4\right\}.$$

Thus we must show that $E\{(\sqrt{n}[\widehat{N}_b(\lambda) - N(\lambda)])^4\} = O_p(1)$, $E\{(\sqrt{n}[\widehat{D}_b(\lambda) - D(\lambda)])^4\} = O_p(1)$, and

$$E\left\{\left(\frac{\widehat{D}_b(\lambda) - D(\lambda)}{\widehat{D}_b(\lambda)}\right)^4\right\} = o_p(1).$$

Conditioned on $\{Y_i, W_i\}_1^n$, $\sqrt{n}[\widehat{N}_b(\lambda) - N(\lambda)]$ is normally distributed with mean $M_C = \sqrt{n}(S_{yw} - S_{xx}\theta)$ and variance $V_C = \lambda\sigma^2 S_{yy}$. Its fourth (conditional) moment is thus bounded by $8(M_C^4 + 3V_C^2)$ which is $O_p(1)$ provided $\sqrt{n}(S_{yw} - S_{xx}\theta)$ and S_{yy} are $O_p(1)$.

Define $\tau_n = nS_{ww}/(2\lambda\sigma^2)$. Conditioned on $\{Y_i, W_i\}_1^n$, $\sqrt{n}[\widehat{D}_b(\lambda) - D(\lambda)]$ is equal in distribution to

$$A_n = \sqrt{n}\left\{\frac{\lambda\sigma^2}{n}\left[\chi^2_{(n-1)}(\tau_n) - 2\tau_n - (n-1)\right] + \left(S_{ww} - S_{xx} - \left(\frac{n-1}{n}\right)\sigma^2\right)\right\},$$

where $\chi^2_{(n-1)}(\tau_n)$ denotes a noncentral chi-squared random variable with non-centrality parameter τ_n.

The inequality $(a+b)^4 \leq 8(a^4 + b^4)$, and evaluation of the fourth central moment of a noncentral chi-squared distribution show that $E\{(\sqrt{n}[\widehat{D}_b(\lambda) - D(\lambda)])^4\} \leq 8(B_n + C_n)$, where

$$B_n = \frac{\sigma^8\lambda^4}{n^2}\{48(8\tau_n + n - 1) + 3[8\tau_n + 2(n-1)]^2\},$$
$$C_n = (\sqrt{n}[S_{ww} - S_{xx} - (n-1)\sigma^2/n])^4.$$

Thus, $E\{(\sqrt{n}[\widehat{D}_b(\lambda) - D(\lambda)])^4\} = O_p(1)$ provided C_n and τ_n/n are $O_p(1)$.

Note that

$$\left[\frac{\widehat{D}_b(\lambda) - D(\lambda)}{\widehat{D}_b(\lambda)}\right]^4 = \left[1 - \frac{D(\lambda)}{\widehat{D}_b(\lambda)}\right]^4,$$

and thus to show that

$$E\left\{\left(\frac{\widehat{D}_b(\lambda) - D(\lambda)}{\widehat{D}_b(\lambda)}\right)^4\right\} = o_p(1),$$

it is sufficient to show that $E\{[D(\lambda)/\widehat{D}_b(\lambda)]^j\} = 1 + o_p(1)$ for $j = 1, \ldots, 4$. We present the proof for $j = 4$ only. Define τ_n and $\chi^2_{(n-1)}(\tau_n)$ as above. We must show that

$$\left[S_{xx} + \left(\frac{n-1}{n}\right)(\lambda+1)\sigma^2\right]^4 E\left\{n^4\left[\lambda\sigma^2\chi^2_{(n-1)}(\tau_n)\right]^{-4}\right\} = 1 + o_p(1). \quad (10.20)$$

When $n > 9$ the indicated expectation exists. Furthermore,

$$E\left\{n^4\left[\chi^2_{(n-1)}(\tau_n)\right]^{-4}\right\} = n^4\sum_{j=1}^{\infty}\frac{\exp(-\tau_n)\tau_n^j}{j!}\frac{1}{(n+2j-3)\times\cdots\times(n+2j-9)}.$$

Let a_k be the coefficient of t^{4-k} in the expansion of $(1-t)^3/48$, $k = 1, \ldots, 4$. Then

$$\sum_{k=1}^{4}\frac{a_k}{n+2j-2k-1} = \frac{1}{(n+2j-3)\times\cdots\times(n+2j-9)}.$$

For $0 \leq s \leq 1$, define the generating function

$$g_n(s) = n^4 \sum_{j=1}^{\infty} \frac{\exp(-\tau_n)\tau_n^j}{j!} \sum_{k=1}^{4} \frac{a_k s^{n+2j-2k-1}}{n+2j-2k-1}.$$

Note that $g_n(1) = E\left\{n^4 \left[\chi^2_{(n-1)}(\tau_n)\right]^{-4}\right\}$. The derivative of $g_n(s)$ with respect to s, $g'_n(s)$, exists (almost surely) and furthermore

$$
\begin{aligned}
g'_n(s) &= n^4 s^{n-10} \sum_{j=1}^{\infty} \frac{\exp(-\tau_n)(\tau_n s^2)^j}{j!} \sum_{k=1}^{4} a_k (s^2)^{4-k} \\
&= \frac{n^4 s^{n-10}(1-s^2)^3}{48} \exp\left\{\tau_n(s^2 - 1)\right\}.
\end{aligned}
$$

Let '$\rightarrow$' denote convergence in probability. Integrating, making the change-of-variables $y = -\tau_n(s^2 - 1)$, and appealing to the Lebesgue Dominated Convergence Theorem using the fact that $\tau_n/n \rightarrow (\sigma_x^2 + \sigma^2)/(2\sigma^2\lambda)$ in probability, shows that

$$
\begin{aligned}
g_n(1) &= n^4 \int_0^1 \frac{n^4 s^{n-10}(1-s^2)^3}{48} \exp\left\{\tau_n(s^2 - 1)\right\} ds \\
&= \frac{n^4}{96\tau_n^4} \int_0^{\tau_n} \left(1 - \frac{y}{\tau_n}\right)^{(n-11)/2} y^3 \exp(-y)\, dy \\
&\rightarrow \frac{1}{96}\left(\frac{2\lambda\sigma^2}{\sigma_x^2 + \sigma^2}\right)^4 \int_0^{\infty} y^3 \exp\left(-y - \frac{y2\sigma^2\lambda}{2(\sigma_x^2 + \sigma^2)}\right) dy \\
&= \frac{(\lambda\sigma^2)^4}{[\sigma_x^2 + \sigma^2(\lambda + 1)]^4}.
\end{aligned}
$$

In light of (10.20), it follows that $E\{[D(\lambda)/\widehat{D}_b(\lambda)]^4\} = 1 + o_p(1)$.

Acknowledgements. Carroll's research was supported by a grant from the National Cancer Institute (CA–57030). Stefanski's research was supported by grants from the National Science Foundation (DMS–92009) and the Environmental Protection Agency.

References

1. Carroll, R. J., Küchenhoff, H., Lombard, F. and Stefanski, L. A. (1996). Asymptotics for the SIMEX estimator in nonlinear measurement error models, *Journal of the American Statistical Association*, **91**, 242–250.

2. Carroll, R. J., Ruppert, D. and Stefanski, L. A. (1995). *Measurement Error in Nonlinear Models*, London: Chapman and Hall.

3. Carroll, R. J. and Stefanski, L. A. (1990). Approximate quasilikelihood estimation in models with surrogate predictors, *Journal of the American Statistical Association*, **85**, 652–663.

4. Carroll, R. J. and Stefanski, L. A. (1994). Measurement error, instrumental variables and corrections for attenuation with applications to meta-analysis, *Statistics in Medicine*, **13**, 1265–1282.

5. Cook, J. R. and Stefanski, L. A. (1994). Simulation–extrapolation estimation in parametric measurement error models, *Journal of the American Statistical Association*, **89**, 1314–1328.

6. Fuller, W. A. (1987). *Measurement Error Models*, New York: John Wiley & Sons.

7. Gleser, L. J. (1990). Improvements of the naive approach to estimation in nonlinear errors-in-variables regression models, In *Statistical Analysis of Measurement Error Models and Application* (Eds., P. J. Brown and W. A. Fuller), pp. 99–114, Providence, RI: American Mathematical Society.

8. Johnson, N. L., Kotz, S. and Balakrishnan, N. (1995). *Continuous Univariate Distributions–Vol. 2*, Second edition, New York: John Wiley & Sons.

9. Rosner, B., Spiegelman, D. and Willett, W. C. (1990). Correction of logistic regression relative risk estimates and confidence intervals for measurement error: the case of multiple covariates measured with error, *American Journal of Epidemiology*, **132**, 734–745.

10. Rosner, B., Willett, W. C. and Spiegelman, D. (1989). Correction of logistic regression relative risk estimates and confidence intervals for systematic within-person measurement error, *Statistics in Medicine*, **8**, 1051–1070.

11. Stefanski, L. A. and Cook, J. R. (1995). Simulation-extrapolation: the measurement error jackknife, *Journal of the American Statistical Association*, **90**, 1247–1256.

11

A Change Point Problem for Some Conditional Functionals

Pranab K. Sen and Zhenwei Zhou

University of North Carolina at Chapel Hill, NC

Abstract: A change point problem for certain conditional sample functionals is considered. This type of functionals includes the strength of a bundle of parallel filaments as a special case. The consistency along with a first order representation of the proposed procedure is established under appropriate regularity conditions.

Keywords and phrases: Boundary detection, bundle strength, change point, conditional distribution, conditional sample extremal functional, first order representation, induced order statistics

11.1   Introduction

We motivate the statistical problem by reference to the strength of a bundle of n parallel filaments, clamped at each end and subjected to increasing load which they share together. Generally, the individual filament strengths are subject to stochastic variation, and as such the bundle would break under a certain (possibly stochastic) load. The minimum load B_n beyond which all the filaments of the bundle give away is defined to be the *strength of the bundle*. In this setup, often, the filaments are not homogeneous with respect to their cross-sections, length or other characteristics having good impact on their strength. However, assuming that these filaments are statistically identical copies, which is permissible when the auxiliary variables are concomitant ones in a statistical sense, we denote their individual strengths by $X_1, \ldots, X_n$ and let $X_{n1} \leq \ldots \leq X_{nn}$ be the corresponding n order statistics; all of these are nonnegative random variables. If we assume that the force of a free load on the bundle is distributed equally on each filament and the strength of an individual filament is indepen-

165

dent of the number of filaments in a bundle, then, if a bundle breaks under a load L, the inequalities $nX_{n1} \leq L$, $(n-1)X_{n2} \leq L, \ldots, X_{nn} \leq L$ must be simultaneously satisfied. Motivated by this observation, Daniels (1945) defined the bundle strength as:

$$B_n = \max\{nX_{n1}, (n-1)X_{n2}, \ldots, X_{nn}\}.$$

Note that if $F_n(x)$ is the empirical distribution function for $X_1, \ldots, X_n$, then

$$B_n^* \equiv n^{-1}B_n = \sup_{x \geq 0} x[1 - F_n(x)]; \qquad (11.1)$$

see Suh, Bhattacharyya and Grandage (1970). Daniels (1945) investigated the probability distribution of (B_n) and established its asymptotic normality by an elaborate analysis. Later on, Sen, Bhattacharyya and Suh (1973) used the above representation as a sample extremal functional, and provided a vastly simplified proof of its asymptotic normality and related properties, even in a more general case of a class of extremal functionals and allowing dependence of observations to a certain extent. Some other workers studied some extensions of Daniels' model under alternative formulations, and obtained similar results. Note that the strength of a bundle of fibers generally depends on the material, length, cross-section, and other auxiliary variables. Hence, it is natural to study the effect of concomitant variations on the bundle strength. The current study focuses on this aspect through the formulation of some general extremal conditional functionals.

Let $(X_1, \mathbf{Y}_1), \ldots, (X_n, \mathbf{Y}_n)$ be n independent and identically distributed (i.i.d.) random vectors, where the primary variate $X \in \mathcal{R}$, and the concomitant variate $\mathbf{Y} \in \mathcal{R}^q$, for some $q \geq 1$. Let $G(x|\mathbf{y})$ be the conditional distribution function of X_1, given $\mathbf{Y}_1 = \mathbf{y}$. As in Sen, Bhattacharyya and Suh (1973), we may consider a general functional of the form $\sup_x \Psi(x, F(x))$, where F is the (marginal) distribution function of X, and note that this includes the Daniels model as a particular case where $\psi(x, F(x))) = x\{1 - F(x)\}$. However, to accommodate concomitant variation, we replace the marginal distribution function F by the conditional distribution function $G(x|\mathbf{y})$, and consider a conditional extremal functional of the form $\theta(\mathbf{y}) = \sup_{x \in \mathcal{R}} \Psi(x, G(x|\mathbf{y}))$, $\mathbf{y} \in \mathcal{C}$ (*compact*) $\subset \mathcal{R}^q$. Note that $\theta(\mathbf{y})$ has a natural estimator

$$\hat{\theta}_n(\mathbf{y}) = \sup_{x \geq 0} \Psi(x, G_n(x|\mathbf{y})), \quad \mathbf{y} \in \mathcal{C}, \qquad (11.2)$$

where $G_n(x|\mathbf{y})$ is an estimator of the conditional distribution $G(x|\mathbf{y})$.

Note the structural difference between the two statistics in (11.1) and (11.2). The empirical distribution function F_n is expressible in terms of an average of i.i.d. random variables, and it is a reversed martingale. On the other hand, an estimator of $G(x|\mathbf{y})$, as will be posed later on, may not have this reversed martingale property, not to speak about it being an average of i.i.d. random variables. As a matter of fact, the rate of convergence of such an estimator is typically slower than $n^{1/2}$, and more noticably when the dimension of $\mathbf{y}$ is ≥ 2. Therefore, studies of asymptotic properties rest on alternative approaches wherein weak convergence of empirical probability measures plays a basic role. Basically, we consider the following stochastic process

$$V_n(\mathbf{y}) = k_n^{1/2}[\hat{\theta}_n(\mathbf{y}) - \theta(\mathbf{y})], \ \mathbf{y} \in \mathcal{C}, \tag{11.3}$$

where $\{k_n\}$ is a sequence of positive integers such that $k_n = o(n)$ and it goes to ∞ as $n \to \infty$. Basically, our goal is to incorporate the asymptotic properties of the above stochastic process in deriving parallel results for the functional estimator. As has been pointed out earlier, $\theta(\mathbf{y})$ can be related to the strength of a bundle of n parallel filaments when we allow stochastic variation in the length, cross-section or other characteristics of the filaments that may influence their individual strengths. This is usually the case with fibres of coir, cotton and other natural materials, and even in an engineering setup, such variations are common, though more within controlable limits. In this way, we can conceive of a conditional functional that depends on the conditional distribution $G(\cdot|\mathbf{y})$. It is also possible to conceive of a threshold model wherein a reduction of the cross-section of a filament below a lower bound (or increasing the length beyond a threshold value) can drastically affect its strength, while above (or below) that mark, the change may be smooth enough to be parameterizable in a simple linear or quadratic (surface) form. Generally, the strength of a bundle of filaments may change considerably beyond a compact zone of the covariate specification, and hence, a boundary detection (change point) model is quite relevant in this conditional framework.

There are two basic types of models for change point (boundary detection) problems. The first type refers to a possible jump-discontnuity of a function at an unknown boundary, and well posed statistical problems relate to the detection of the nature or magnitude of the probable change as well as the boundary where it might have occurred. For example, considering an image as a regression function with possible jump discontinuity between an object and its background, and given the (white) noisy observations on the image, we may want to estimate the discontinuity curve. This problem has been studied by Tsybakov (1989, 1994), Korostelev and Tsybakov (1993), Rudemo and Stryhn (1991), Carlstein and Krishnamoorthy (1992), Müller and Song (1992), and others. The other type relates to a function that is assumed to be continuous but may have a jump discontinuity in its gradient (first derivative in the one dimension case), so that the basic problem is to detect the point at which such a gradient change occurs, along with the nature of the change. This problem has been studied by Müller and Song (1992), Eubank and Speckman (1994), among others. In our situation, since $\theta(\mathbf{y})$ is continuous inside the compact domain C, so a possible change should relate to its gradient vector. Therefore, we consider the second change point model. For simplicity of presentation, in Section 11.2 we consider the case when $y \in [0, 1]$. The general situation of $\mathbf{y} \in C \in \mathcal{R}^q$ is then briefly treated in Section 11.3.

11.2   Regularity Conditions and the Main Result

In order to study the behavior of $\theta(y)$, through its plug-in estimator, first, we need to incorporate an estimator $G_{nk}(x|y)$ of $G(x|y)$. Suppose that we have a collection of i.i.d. observations $\{(X_i, Y_i), i \geq 1\}$, and for a given y, we set $Z_i = |Y_i - y|$, $i = 1, \cdots, n$. Based on this, we have the data transform:

$$(X_i, Y_i) \to (X_i, Z_i), i = 1, \cdots, n.$$

Let $0 \leq Z_{n1} \leq Z_{n2} \leq \cdots \leq Z_{nn}$ be the order statistics corresponding to $Z_1, Z_2, \cdots, Z_n$, and let $X_{n1}, X_{n2}, \cdots, X_{nn}$ be the *induced (concomitants of) order statistics*, i.e.,

$$X_{ni} = X_j \text{ if } Z_{ni} = Z_j \text{ for } i, j = 1, \cdots, n.$$

Defining $\{k_n = o(n)\}$ as before, we formulate the k_n nearest neighbor (k_n-NN) empirical distribution estimator of $G(x|y)$ as

$$G_{nk_n}(x|y) = k_n^{-1} \sum_{i=1}^{k_n} I(X_{ni} \leq x), \quad x \in \mathcal{R},$$

where $I(A)$ stands for the indicator function of the set A.

The following result, due to Bhattacharya (1974), exhibits a basic property of induced order statistics; here, we let $G^*(\cdot|z)$ be the conditional distribution of X given $Z = z$.

Proposition 11.2.1 *For every n, the induced order statistics $X_{n1}, X_{n2}, \cdots, X_{nn}$ are conditionally independent given $Z_1, Z_2, \cdots, Z_n$ with conditional distributions $G^*(\cdot|Z_{n1}), G^*(\cdot|Z_{n2}), \cdots, G^*(\cdot|Z_{nn})$, respectively.*

We may remark that as $n^{-1}k_n \to 0$ with $n \to \infty$, for any two distinct points y_1, y_2, the intersection of the corresponding induced order statistics sets becomes asymptotically a null set [by the Bahadur (1966) Theorem]. From Proposition 11.2.1, we obtain that the two estimators $G_{nk}(x|y_1)$ and $G_{nk}(x|y_2)$ are conditionally asymptotically independent.

Once we have the estimator $G_{nk}(x|y)$, we plug it into $\theta(y)$, and obtain a natural estimator $\hat{\theta}_n(y)$. Then proceeding as in Sen, Bhattacharyya and Suh (1973), it follows that for a given y, one can obtain a first order asymptotic representation for $\hat{\theta}_n(y) - \theta(y)$ with a leading term

$$\Psi_{01}(x^0(y), G(x^0(y)|y))[G_{nk_n}(x^0(y)|y) - G(x^0(y)|y)], \qquad (11.4)$$

where $x^0(y)$ is the point at which $\Psi(x, G(x|y))$ attains its (unique) maximum, and Ψ_{01} is the first order partial derivative of $\Psi(x, z)$ with respect to z. In this representation, the remainder term is $o_p(\|G_{nk_n}(\cdot|y) - G(\cdot|y)\|)$, so that the order of stochastic convergence of the empirical conditional distribution function G_{nk_n} provides the rate of stochastic convergence of $\hat{\theta}_n(y)$ for a given y. Since we treat here the concomitant variate as stochastic, we need to have a parallel result that holds uniformly in y in a compact interval. This calls for some regularity assumptions on the distribution functions $G(\cdot|y)$, $F(x)$, their density functions $g(\cdot|y)$ and $f(y)$, as well as the function Ψ; these are presented below:

A1: (a) $f(y_0) > 0$.

(b) $f''(y)$ exists in a neighborhood of y_0, and there exist $\epsilon > 0$ and a constant $A^* < \infty$ such that $|y - y_0| < \epsilon$ implies

$$|f''(y) - f''(y_0)| \le A^*|y - y_0|. \tag{11.5}$$

A2: (a) $g(\cdot|y_0) > 0$.

(b) The partial derivatives $g_{yy}(x|y)$ of $g(x|y)$ and $G_{yy}(x|y)$ of $G(x|y)$ exist in a neighborhood of y_0, and there exist $\epsilon > 0$ and Lebesgue measurable functions $u_1^*(x)$, $u_2^*(x)$ and $u_3^*(x)$ such that for $|y - y_0| < \epsilon$,

$$|g_y(x|y)| \le u_1^*(x), \quad |g_{yy}(x|y)| \le u_2^*(x),$$
$$|g_{yy}(x|y) - g_{yy}(x|y_0)| \le u_3^*(x)|y - y_0|. \tag{11.6}$$

A3: (a) Define $x^0(y)$ as in (11.4). Then, $x^0(y)$ is a continuous function of y, and for every $\delta > 0$, there exists an $\eta > 0$, such that

$$\Psi(x, G(x|y)) \le \Psi(x^0(y), G(x^0(y)|y)) - \delta \tag{11.7}$$

$\forall x : |x - x^0(y)| > \eta$, uniformly in $y \in \mathcal{C}$.

(b) There exists a compact $\mathcal{C}^0$, such that

$$x^0(y) \in \mathcal{C}^0 \quad \text{whenever} \quad y \in \mathcal{C}. \tag{11.8}$$

(c) There exists an $\epsilon > 0$, such that $\forall y \in \mathcal{C}$, and $\forall x$,

$$|\Psi(x, H(x, y)) - \Psi(x, G(x|y))| \le q(x)|H(x, y) - G(x|y)| \tag{11.9}$$

$\forall H : \|H - G\| \le \epsilon$, where $q(x)$ is nonnegative, and

$$E\{[q(X_i)]^r\} < \infty, \quad \text{for some} \quad r \ge 3. \tag{11.10}$$

Recall that $F_Z(z)$, the distribution function of $Z_i = |Y_i - y_0|$, has a density function

$$f_Z(z) = f(y_0 + z) + f(y_0 - z), \ z \geq 0. \tag{11.11}$$

Thus from **A1** and **A2**, we conclude that **A1** holds for the density f_Z too. Let $G^*(x|z)$ and $g^*(x|z)$ be, respectively, the conditional distribution function and density function of X, given $Z = z$. Then we have

$$G^*(x|z) = \{f(y_0 + z)G(x|y_0 + z) + f(y_0 - z)G(x|y_0 - z)\}/f_Z(z), \tag{11.12}$$

$$g^*(x|z) = \{f(y_0 + z)g(x|y_0 + z) + f(y_0 - z)g(x|y_0 - z)\}/f_Z(z). \tag{11.13}$$

Lemma 11.2.1 *Under conditions **A1** and **A2**, the following expansions hold for the conditional density function and distribution function:*

$$g^*(x|z) = g(x) + \frac{1}{2}q(x)z^2 + r(x, z)z^3, \tag{11.14}$$

$$G^*(x|z) = G(x) + \frac{1}{2}Q(x)z^2 + R(x, z)z^3, \tag{11.15}$$

where

$$\begin{aligned} g(x) &= g(x|y_0), \ \ G(x) = G(x|y_0), \\ q(x) &= g_{yy}(x|y_0) + 2f'(y_0)g_y(x|y_0)/f(y_0), \\ Q(x) &= G_{yy}(x|y_0) + 2f'(y_0)G_y(x|y_0)/f(y_0), \end{aligned} \tag{11.16}$$

and there exist $\epsilon > 0$ and $M < \infty$ such that $|q(x)|, |Q(x)|, |r(x, z)|$ and $|R(x, z)|$ are all bounded by M for $0 < z < \epsilon$.

For proof, we refer to Bhattacharya and Gangopadhyay (1990) and Sen (1993).

Proposition 11.2.2 *Under the conditions **A1** and **A2**, for any $0 < \gamma < 1/2$,*

$$\sup_x \sup_{0 \leq y \leq 1} |G_{nk_n}(x|y) - G(x|y)| = o_p(k_n^{-\gamma}). \tag{11.17}$$

PROOF. Let

$$S_n(x, y) = k_n^\gamma(G_{nk_n}(x|y) - G(x|y)). \tag{11.18}$$

We show that for any fixed y, x and given $\epsilon > 0$,

$$P(|S_n((x, y)| > \epsilon) \leq e^{-c_1 k_n^{c_2}} \tag{11.19}$$

for some constants $c_1 > 0, c_2 > 0$. For this, note that

$$S_n(x, y) = k_n^{\gamma-1} \sum [I_{X_{ni} \leq x} - G(x|y)], \tag{11.20}$$

and that from Lemma 11.2.1

$$ES_n(x, y) = O(n^{\gamma - \frac{7}{10}}). \tag{11.21}$$

Therefore, the proof of (11.19) follows by using the Hoeffding (1963) inequality.

We return to the proof of Proposition 11.2.2. Without any loss of generality, we assume that both x and y are defined on the unit interval $[0, 1]$ (as otherwise use the marginal probability integral transformations to do so). We then divide interval $[0, 1]$ equally into m_{n_1} subintervals, where $m_{n_1} = O(n^{\frac{4}{5}})$, and denote the corresponding points as $y_j = j/m_{n_1}$, $j = 1, 2, \cdots, m_{n_1}$. Then, for $y \in [y_j, y_{j+1})$

$$
\begin{aligned}
&S_n(x(y), y) - S_n(x'(y_j), y_j) \\
&= k_n^{\gamma - 1} \left\{ \underbrace{\sum [I_{X_{ni}^y \leq x(y)} - G(x(y)|y)]}_{upper} - \underbrace{\sum [I_{X_{ni}^{y_j} \leq x'(y_j)} - G(x'(y_j)|y_j)]}_{lower} \right\} \\
&+ k_n^{\gamma - 1} \underbrace{\sum [I_{\min(x(y), x'(y_j)) \leq X_{ni} \leq \max(x(y), x'(y_j))}]}_{common},
\end{aligned}
\tag{11.22}
$$

where $x(y)$ and $x'(y_j)$ satisfy $G(x(y)|y) = G(x'(y_j)|y_j)$. Again, through some standard steps and using the Hoeffding (1963) inequality, it can be shown that there exist constants $c_3 > 0$ and $c_4 > 0$ such that for any $\epsilon > 0$

$$P(|S_n(x(y), y) - S_n(x'(y_j), y_j)| > \epsilon) \leq e^{-c_3 k_n^{c_4}}. \tag{11.23}$$

Now, we divide the x-axis equally into m_{n_2} subintervals, where $m_{n_2} = O(n^{\frac{3}{4}})$, and denote the corresponding points by $x_j = j/m_{n_2}$, $j = 1, \ldots, m_{n_2}$. Then for $x \in [x_j, x_{j+1})$, by the Bahadur (1966) theorem, there exist constants c_5 and c_6 such that

$$|S_n(x, y) - S_n(x_j, y)| \leq c_5 k_n^{-c_6}. \tag{11.24}$$

Finally, when n is large enough, we have

$$
\begin{aligned}
&P(\sup_{x \in [0,1]} \sup_{y \in [0,1]} |S_n(x, y)| > \epsilon) \\
&\leq P(\max_{i,j} \sup_{x \in [x_i, x_{i+1})} \sup_{y \in [y_j, y_{j+1})} |S_n(x, y) - S_n(x_i, y_j)| > \epsilon/2) \\
&\quad + P(\max_{i,j} |S_n(x_i, y_j)| > \epsilon/2) \\
&\leq P(\max_{i,j} \sup_{x \in [x_i, x_{i+1})} \sup_{y \in [y_j, y_{j+1})} |S_n(x, y) - S_n(x_i, y)| > \epsilon/4) \\
&\quad + P(\max_{i,j} \sup_{x \in [x_i, x_{i+1})} \sup_{y \in [y_j, y_{j+1})} |S_n(x_i, y) - S_n(x_i, y_j)| > \epsilon/4) \\
&\quad + P(\max_{i,j} |S_n(x_i, y_j)| > \epsilon/2) \\
&\leq 2 m_{n_1} m_{n_2} (e^{-c_1 k_n^{c_2}} + e^{-c_1 k_n^{c_2}}), \tag{11.25}
\end{aligned}
$$

and the right hand side of (11.25) converges to 0 as $n \to \infty$. ∎

Next, we note that by the moment condition in **A3**(c),

$$\max_{1\leq i\leq n}q(X_i) = o(n^{1/r}) \quad \text{a.s.} \quad \text{as} \quad n \to \infty; \tag{11.26}$$

so noting that $\hat{\theta}_n(y) = \Psi(\hat{x}_n(y), G_{nk_n}(\hat{x}_n(y)|y))$, at a sample point $\hat{x}_n(y)$, by Proposition 11.2.2 and **A3**(a)–(c), we conclude that with probability converging to 1 (as $n \to \infty$),

$$\hat{x}_n(y) \in \mathcal{C}_0 \quad \forall\, y \in \mathcal{C}. \tag{11.27}$$

Next, we obtain the following two inequalities directly from the definitions of $\theta(y)$ and $\hat{\theta}_n(y)$: $\forall y$,

$$\hat{\theta}_n(y) - \theta(y) \geq \Psi(x^0(y), G_{nk_n}(x^0(y)|y)) - \Psi(x^0(y), G(x^0(y)|y)), \tag{11.28}$$

$$\hat{\theta}_n(y) - \theta(y) \leq \Psi(\hat{x}_n(y), G_{nk_n}(\hat{x}_n(y)|y)) - \Psi(\hat{x}_n(y), G(\hat{x}_n(y)|y)). \tag{11.29}$$

Therefore, using again Proposition 11.2.2 along with **A3** and the above results, we obtain that

$$\sup_{y\in\mathcal{C}} |\hat{\theta}_n(y) - \theta(y)| = O_p\|G_{nk_n} - G\| = o_p(k_n^{-\gamma}), \tag{11.30}$$

where $\gamma(< 1/2)$ can be taken arbitrary close to $1/2$.

While this result relates to the equivalence of the stochastic order of convergence of both G_{nk_n} and $\hat{\theta}_n(y)$, in order to obtain our main result, an asymptotic first order representation for $\hat{\theta}_n(y)$, $y \in \mathcal{C}$, we make the following additional assumption:

A4: $\Psi_{01}(x, G(x|y))$ is continuous in a neighborhood of $(x^0(y), G(x^0(y)|y))$, uniformly in $y \in \mathcal{C}$.

By virtue of the uniform stochastic order in (11.30), Proposition 11.2.2, and Assumption **A4**, it follows by some simple argument that there exists a $\delta : 0 < \delta \leq 1/2$, such that

$$\sup_{y\in\mathcal{C}} \{|\hat{x}_n(y) - x^0(y)|\} = O_p(k_n^{-\delta}) \text{ as } n \to \infty. \tag{11.31}$$

Moreover, repeating the proof of Proposition 11.2.2, we obtain the following result yielding a better order of stochastic convergence in a shrinking neighborhood. For every $K : 0 < K < \infty$, as $n \to \infty$,

$$\sup_{y\in\mathcal{C}} \sup_{x: k_n^\delta |x-x^0(y)|\leq K} |[G_{nk_n}(x|y) - G(x|y)] - [G_{nk_n}(x^0(y)|y) - G(x^0(y)|y)]|$$

$$= o_p(k_n^{-1/2}). \tag{11.32}$$

From the above results, we arrive at the following.

Theorem 11.2.1 *Under the conditions* **A1**, **A2**, **A3** *and* **A4**, *we have*

$$
\begin{aligned}
\hat{\theta}_n(y) &= \theta(y) + \Psi_{01}(x^0(y), G(x^0(y)|y))[G_{nk_n}(x^0(y)|y) - G(x^0(y)|y)] \\
&\quad + o_p(\|G_{nk_n} - G\|), \ uniformly \ in \ y \in \mathcal{C}. \tag{11.33}
\end{aligned}
$$

We may remark that under the assumed regularity conditions, for any fixed $y_0 \in \mathcal{C}$, when we consider a shrinking neighborhood $I_n(y_0) = \{y \in \mathcal{C} : \|y - y_0\| \leq Kk_n^{-1/2}\}$, the two dimensional time parameter stochastic process

$$
V_n(x, t; y_0) = k_n^{1/2}[G_{nk_n}(x|y_0 + k_n^{-1/2}t) - G(x|y_0 + k_n^{-1/2}t)], \ x \in \mathcal{R}, \ |t| \leq K, \tag{11.34}
$$

converges weakly to a two-dimensional time parameter Gaussian function; for some related studies, we refer to Sen (1993) where other references are cited. On the other hand, for any two distinct points, y_0 and y_0', the two related processes are asymptotically independent. Therefore, by virtue of the first order asymptotic representation for the estimators $\hat{\theta}_n(y)$, $y \in \mathcal{C}$, prsented in the theorem, and the uniform stochastic convergence of $\hat{x}_n(y)$ to $x^0(y)$, $y \in \mathcal{C}$, we may conclude that the stochastic process

$$
\{k_n^{1/2}[\hat{\theta}_n(y) - \theta(y)] : y \in \mathcal{C}\} \tag{11.35}
$$

converges weakly to a Gaussian function on $\mathcal{C}$. This result in turn makes it easier to study local behavior of the estimators when there is a change point τ_0 in the interior of $\mathcal{C}$. In this sense, the standard asymptotics for change point models related to Gaussian functions can be used for this model too.

We conclude this section with a remark that the regularity assumptions **A1**, **A2**, **A3**, **A4** are all sufficient but not necessary conditions. Nevertheless, they are very mild in nature, and they all pertain to the particular case of bundle strength. Some further discussion of these regularity conditions are made in the concluding section.

11.3 Some Discussions

In the one dimensional case, a change-point model relating to a possible jump discontinuity in the gradient at an unknown time point, can be described, in a canonical form, by means of a set of equispaced time-points on the unit interval $[0, 1]$ (i.e., $t_{nr} = r/n$, $r = 0, 1, \ldots, n$) and the realization of the response variables at these points, namely, $Y_{n0}, \ldots, Y_{nn}$, in the form

$$
Y_{nr} = \theta(t_{nr}) + \epsilon_{nr}, \ r = 0, 1, \ldots, n, \tag{11.36}
$$

where the ϵ_{nr} are the error components, while the unknown regression function θ is assumed to be continuous on $[0, 1]$ and to be twice continuously differentiable on $(0, \tau_0)$ and $(\tau_0, 1)$ for some point $\tau_0 \in (0, 1)$. At τ_0, θ is assumed to have a jump discontinuity in its first derivative, and therefore, it may be taken to have the following form:

$$\theta(t) = \gamma_0 \phi_{\tau_0}(t) + f(t), \tag{11.37}$$

where

$$\phi_\tau(t) = \begin{cases} 0, & t < \tau, \\ t - \tau, & t \geq \tau, \end{cases} \tag{11.38}$$

so that τ_0 is the unknown change point, γ_0 is the size of the jump in the first derivative of θ at τ_0,; in this setup, $f(t)$ is an unknown continuously differentiable function. The situation is somewhat different in the current context, as here the auxiliary variable Y is stochastic in nature. In this case, an analogous design would have been the set of (r/n)-quantiles, $r = 0, 1, \ldots, n$, for the marginal distribution function $F(y)$. However, since F is not totally specified these quantiles are not known in advance, and hence, it is difficult to adopt such a design. Moreover, in the current situation, $\theta(y)$ is not a conventional regression function (of X on y) even when we allow y to be subject to errors, but is an extremal functional of the conditional distribution function $G(\cdot|y)$. For these reasons, the above formulation for the conventional gradient change-point model is not totally adoptable in the current setup.

If we look into this extremal type conditional functional in a regular case (i.e., when there is no change point) where $\mathbf{Y}$ is generally multivariate, though there are complications in performing exact statistical analyses, the scheme suggested by Sen and Zhou (1996) for the bundle strength problem, works out well in an asymptotic setup. This provides some clues for handling the multidimensional case with possible change-points in a reasonable manner. Note that in a general multivariate case where $\mathbf{y}$ is allowed to be a point in the unit cube $[0, 1]^q$, for some $q \geq 1$, one may consider the model

$$\theta(\mathbf{y}) = \boldsymbol{\gamma}' \boldsymbol{\phi}_{\boldsymbol{\tau}_0} + f(\mathbf{y}), \tag{11.39}$$

where $\mathbf{y}' = (y_1, \ldots, y_q)$, $\boldsymbol{\gamma}' = (\gamma_1, \ldots, \gamma_q)$, $\boldsymbol{\phi}'_{\boldsymbol{\tau}} = (\phi_{1\tau_1}, \ldots, \phi_{q\tau_q})$ and

$$\phi_{j\tau_j} = \begin{cases} 0, & y_j < \tau_j, \\ y_j - \tau_j, & y_j \geq \tau_j. \end{cases} \tag{11.40}$$

As in the one-dimensional case, in view of the stochastic nature of $\mathbf{Y}$, and the extremal nature of the functional $\theta(\mathbf{y})$, such a model may be difficult to adopt. The change-point concept extends to possible changes on a boundary which is a lower dimensional space. Thus, it may be generally difficult to identify such boundaries in general, and to detect jump discontinuities in the gradient vector occurring on a general boundary instead of a single point. To illustrate this point, we assume that $\theta(\mathbf{y})$ has the following simple form:

$$\theta(y) = (a^{(1)} + \mathbf{b}^{(1)'}\mathbf{y} + \mathbf{y}'\boldsymbol{\Sigma}_1 \mathbf{y})I(\mathbf{y} \in \mathcal{C}_1)$$

$$+ (a^{(2)} + \mathbf{b}^{(2)'}\mathbf{y} + \mathbf{y}'\Sigma_2\mathbf{y})I(\mathbf{y} \in C_2)$$
$$\equiv \theta_1(\mathbf{y})I(\mathbf{y} \in C_1) + \theta_2(\mathbf{y})I(\mathbf{y} \in C_2), \tag{11.41}$$

where

$$\Sigma_j = (l_{ik}^{(j)})_{q \times q}, \quad \mathbf{y}' = (y_1 \ldots y_q), \quad j = 1, 2, \tag{11.42}$$

$I(\cdot)$ is the indicator function, and C_1, C_2 are two regions of C such that $C_1 \cap C_2 = C_0$, boundary, and $C_1 \cup C_2 = C$. With these notations, we may state that $\theta(\mathbf{y})$ has no jump at a boundary point is equivalent to

$$\theta_1(\mathbf{y}) = \theta_2(\mathbf{y}) \text{ for } \mathbf{y} \in C_0. \tag{11.43}$$

In general, it is quite difficult to identify such a boundary and cosider appropriate statistical tools for its detection. Toward this end, consider the more specific case where the boundary has the form:

$$B \equiv \{(y_1, \ldots, y_q) : y_q = \alpha + \boldsymbol{\beta}'\mathbf{y}_{(q-1)}\}, \tag{11.44}$$

where α is a constant, $\boldsymbol{\beta}' = (\beta_1, \ldots, \beta_{q-1})$, and $\mathbf{y}'_{(q-1)} = (y_1, \ldots, y_{q-1})$. The above constraint then becomes

$$a^{(1)} + \mathbf{b}'^{(1)} \left\{ \begin{array}{c} y_1 \\ \vdots \\ y_{q-1} \\ \alpha + \boldsymbol{\beta}'\mathbf{y}_{(q-1)} \end{array} \right\}$$

$$+ (y_1, \cdots, y_{q-1}, \alpha + \boldsymbol{\beta}'\mathbf{y}_{(q-1)})\Sigma_1 \left\{ \begin{array}{c} y_1 \\ \vdots \\ y_{q-1} \\ \alpha + \boldsymbol{\beta}'\mathbf{y}_{(q-1)} \end{array} \right\}$$

$$= a^{(2)} + \mathbf{b}'^{(2)} \left\{ \begin{array}{c} y_1 \\ \vdots \\ y_{q-1} \\ \alpha + \boldsymbol{\beta}'\mathbf{y}_{(q-1)} \end{array} \right\}$$

$$+ (y_1, \cdots, y_{q-1}, \alpha + \boldsymbol{\beta}'\mathbf{y}_{(q-1)})\Sigma_2 \left\{ \begin{array}{c} y_1 \\ \vdots \\ y_{q-1} \\ \alpha + \boldsymbol{\beta}'\mathbf{y}_{(q-1)} \end{array} \right\}. \tag{11.45}$$

After some manipulations, we can show that the above equation is equivalent to

$$a^{(1)} + b^{(1)}\alpha + l_{qq}^{(1)}\alpha^2 + \sum_{j=1}^{q-1}(b_j^{(1)} + \beta_j b_q^{(1)} + \alpha_{jq}^{(1)} + \alpha l_{qj}^{(1)} + 2\alpha\beta_j l_{qq}^{(1)})y_j$$

$$+ \sum_{i,j \leq q-1}(l_{ij}^{(1)} + \beta_i l_{qj}^{(1)} + \beta_j l_{iq}^{(1)} + \beta_i\beta_j l_{qq}^{(1)})y_iy_j$$

$$= a^{(2)} + b^{(2)}\alpha + l_{qq}^{(2)}\alpha^2 + \sum_{j=1}^{q-1}(b_j^{(2)} + \beta_j b_q^{(2)} + \alpha_{jq}^{(2)} + \alpha l_{qj}^{(2)}$$

$$+ 2\alpha\beta_j l_{qq}^{(2)})y_j + \sum_{i,j \leq q-1}(l_{ij}^{(2)} + \beta_i l_{qj}^{(2)} + \beta_j l_{iq}^{(2)} + \beta_i\beta_j l_{qq}^{(2)})y_iy_j.$$

$$(11.46)$$

From the above equation, we actually have

$$l_{ij}^{(1)} + \beta_i l_{qj}^{(1)} + \beta_j l_{iq}^{(1)} + \beta_i\beta_j l_{qq}^{(1)} = l_{ij}^{(2)} + \beta_i l_{qj}^{(2)} + \beta_j l_{iq}^{(2)} + \beta_i\beta_j l_{qq}^{(2)},$$
$$1 \leq i,j \leq q-1.$$

Though this simple case is somehow manageable, for $q > 1$, with $\theta(\mathbf{y})$ of non-parametric nature, such a simplification may not be always possible. In approximating such boundaries by linear subspaces, based on the concept of projection persuits, we may sacrifice a lot of model flexibility, and hence lose some generality of the model. For this reason, local smoothness conditions on $\theta(\mathbf{y})$ are generally needed in a nonparametric approach, and in view of the slower rate of convergence for higher dimensional concomitant variates, statistical modeling and analysis may presumably be more complex. Nevertheless, the weak invariance principle for the normalized $\hat{\theta}_n(y)$, developed here for the one dimensional case, goes through for the q-dimensional case as well, and hence, we have a workable approach to this boundary detection problem. We intend to cover such a general situation in a future communication.

References

1. Bahadur, R. R. (1966). A note on quantiles for large samples, *Annals of Mathematical Statistics*, **37**, 577–580.

2. Bhattacharya, P. K. (1974). Convergence of sample paths of normalized sums of induced order statistics, *Annals of Statistics*, **2**, 1034–1039.

3. Bhattacharya, P. K. and Gangopadhyay, A. K. (1990). Kernel and nearest neighbor estimation of a conditional quantile, *Annals of Statistics*, **18**, 1400–1415.

4. Bhattacharya, P. K. and Mack, Y. P. (1987). Weak convergence of k-NN density and regression estimators with varying k and applications, *Annals of Statistics*, **15**, 976–994.

5. Carlstein, E. and Krishnamoorthy, C. (1992). Boundary estimation, *Journal of the American Statistical Association*, **87**, 430–438.

6. Daniels, H. E. (1945). The statistical theory of the strength of bundles of threads, *Proceedings of the Royal Society of London, Series A*, **183**, 405–435.

7. Eubank, R. L. and Speckman, P. L. (1994). Nonparametric estimation of functions with jump discontinuities, *IMS Lecture Notes—Monograph Series*, **23**, 130–144.

8. Hoeffding, W. (1963). Probability inequalities for sums of bounded random variables, *Journal of the American Statistical Association*, **58**, 13–30.

9. Korostelev, A. P. (1991). Minimax reconstruction of two-dimensional images, *Theory of Probability and its Applications*, **36**, 153–159.

10. Korostelev, A. P. and Tsybakov, A. B. (1993). *Minimax Theory of Image Reconstruction*, Lecture Notes in Statistics—**82**, New York: Springer-Verlag.

11. Müller, H. and Song, K. (1992). On estimation of multidimensional boundaries, *Technical Report*, Division of Statistics, University of California, Davis, CA.

12. Rudemo, I. and Stryhn, H. (1991). Approximating the distribution of maximum likelihood contour estimators in two-region images, *Technical Report 91-2*, Department of Mathematics and Physics, Royal Veterinary and Agricultural University, Thorvaldsensvej 40, 1871 Frederiksberg C, Denmark.

13. Sen, P. K. (1981). *Sequential Nonparametrics: Invariance Principles and Statistical Inference*, New York: John Wiley & Sons.

14. Sen, P. K. (1993). Perspectives in Multivariate nonparametrics: Conditional functionals and ANOCOVA models, *Sankhyā, Series A*, **55**, 516–532.

15. Sen, P. K., Bhattacharyya, B. B. and Suh, M. W. (1973). Limiting behavior of the extremum of certain sample functions, *Annals of Probability*, **1**, 297–311.

16. Sen, P. K. and Zhou, Z. (1996). Stochastic approximation and selection of a conditional extremal function in continuum, *Sequential Analysis* (Submitted).

17. Suh, M. W., Bhattacharyya, B. B. and Grandage, A. (1970). On the distribution of the strength of a bundle of filaments, *Journal of Applied Probability*, **7**, 712–720.

18. Tsybakov, A. B. (1989). Optimal estimation accuracy of nonsmooth images, *Problems of Information and Transmission*, **25**, 180–191.

19. Tsybakov, A. B. (1994). Multidimensional change point problems, *IMS Lecture Notes—Monograph Series*, **23**, 317–329.

20. Zhou, Z. (1996). Limiting behavior of extrema of certain conditional sample functions and some applications, *Ph.D. Dissertation*, Department of Statistics, University of North Carolina, Chapel Hill, NC.

12

On Bias Reduction Methods in Nonparametric Regression Estimation

W. C. Kim, B. U. Park and K. H. Kang

Seoul National University, Korea

Abstract: Standard nonparametric regression estimators such as Nadaraya-Watson, Priestly-Chao, Gasser-Müller, and local linear smoothing have convergence of order $O(n^{-2/5})$ when the kernel weight functions used are of second order. We discuss here two recently proposed techniques which improve the convergence rate of any given nonparametric regression estimator. When they are applied to the basic $O(n^{-2/5})$ methods, the convergence rate reduces to $O(n^{-4/9})$. In this paper, we focus on the cases when these two methods are applied to the local linear smoothing. It is demonstrated by means of a Monte Carlo study that the asymptotic improvements are noticeable even for moderate sample sizes.

Keywords and phrases: Bias reduction, nonparametric regression, transformation, multiplicative correction, local linear smoothing

12.1 Introduction

Given data (x_i, Y_i), $i = 1, 2, \ldots, n$, there exist a number of nonparametric techniques for estimating the regression mean function m. Four of the leading kernel prescriptions are the Nadaraya-Watson estimator [Nadaraya (1964) and Watson (1964)], the two so called "convolution type" estimators due to Priestley and Chao (1972), Gasser and Müller (1979, 1984), and the local linear smoother [Cleveland (1979), Fan (1992) and Hastie and Loader (1993)]. These all have the convergence of order $O(n^{-2/5})$ to the true regression function. This rate of convergence comes from the fact that, as the sample size $n \to \infty$ and the smoothing parameter (or bandwidth) $h = h(n) \to 0$, the bias is $O(h^2)$ with the variance of order $O(n^{-1}h^{-1})$.

Recent years have seen two general techniques to improve the bias of any given regression estimators. One of them is the multiplicative correction method by Jones, Linton and Nielsen (1995). The basic idea is to multiply a given initial estimator $\tilde{m}$ by an estimator of the correction factor $m/\tilde{m}$. The other technique proposed by Park *et al.* (1997) is based on transformation of the design points. Given an intial estimator $\tilde{m}$, it simply regresses $\{Y_i\}$ on $\{\tilde{m}(x_i)\}$ and then transforms the resulting estimator back to the original design scale. A nice feature of these two methods is that they can be applied to any regression estimator in order to improve the bias. For instance, when they are applied to the four aforementioned basic estimators, the bias reduces from $O(h^2)$ to $O(h^4)$, sufficient smoothness of m permitting, and this ensures the faster $O(n^{-4/9})$ rate of convergence. This is in contrast to the variable bandwidth approach, known to be another way of reducing the bias, which fails for the local linear fitting [see Jones (1995)].

In this paper, we investigate by means of a Monte Carlo study how these techniques are successful for finite sample sizes. The main interest here is to see to what extent the bias reduction methods improve the finite sample performance of a given nonparametric regression estimator. Although the two techniques can be applied to any regression estimator, we focus only on the local linear fitting. We compare the integrated squared biases and variances of the three estimators : the local linear fitting, and the two resulting estimators when each of the multiplicative correction and the transformation technique is applied to it. One of the observations made is that the asymptotic improvements are noticeable even for moderate sample sizes. One might expect that these techniques would increase the variance of the estimator due to the extra estimation stage. Contrary to this initial expectation, however, it turns out that they produce stabilized variances.

In the next section, we briefly introduce the local linear fitting, and the two bias reduction schemes applied to it along with their asymptotic behaviour. We present in Section 12.3 the results of the Monte Carlo study.

12.2   The Bias Reduction Methods

The kernel weighted local linear fitting is known as an example of locally weighted regression which has been popularized by Cleveland (1979) after initial suggestions by Stone (1977). The basic idea of locally weighted regression is to fit a polynomial locally at each x. In the case of assigning the local weights by a kernel function K, and denoting by p the degree of the polynomial being fitted, it is given by the value of a_0 when $a_0, \ldots, a_p$ are chosen to minimize the local sum of squares

$$\sum_{i=1}^{n} K_h(x - x_i)\{Y_i - a_0 - \cdots - a_p(x - x_i)^p\}^2$$

where $K_h(\cdot) = K(\cdot/h)/h$ is the kernel function scaled by the bandwidth h.

The linear fit $(p = 1)$ gives the kernel weighted local linear fitting estimator which can be written explicitly by

$$\widehat{m}_h^{LL}(x) = n^{-1} \sum_{i=1}^{n} \{w_2(x)K_h(x - x_i) - w_1(x)(P_1K)_h(x - x_i)\}Y_i$$

where $w_\ell(x) = s_\ell(x)/\{s_0(x)s_2(x) - s_1^2(x)\}$, $s_\ell(x) = n^{-1}\sum_{i=1}^{n}(P_\ell K)_h(x - x_i)$ and $(P_\ell K)_h(z) = h^{-\ell-1}z^\ell K(z/h)$. When the kernel function is of second order (e.g., symmetric probability density functions), the bias and variance of $\widehat{m}_h^{LL}(x)$ are, under certain conditions, given by

$$\begin{aligned} \text{bias}(\widehat{m}_h^{LL}(x)) &= h^2 m''(x)\mu_2/2 + o(h^2), \\ \text{var}(\widehat{m}_h^{LL}(x)) &= n^{-1}h^{-1}\sigma^2(x)\int K^2(u)du/f(x) + o(n^{-1}h^{-1}), \end{aligned}$$

where $f(\cdot)$ and $\sigma^2(\cdot)$ are the design density and the variance function respectively. Here and in the sequel, μ_r denotes $\int u^r K(u)du$. Interested readers may refer to Fan (1992, 1993), Fan and Gijbels (1992), Jones, Davies and Park (1994) for discussions on other properties of this estimator.

Now we introduce the two bias reduction techniques. As mentioned earlier, we apply these methods to $\widehat{m}_h^{LL}$. First, for the multiplicative correction method, note that $m(x) = \widehat{m}_h^{LL}(x)r(x)$ where $r(x) = m(x)/\widehat{m}_h^{LL}(x)$. The basic idea is to multiply $\widehat{m}_h^{LL}(x)$ by an estimator of the correction factor $r(x)$. For estimation of $r(x)$, one can repeat the local linear fitting on $\{x_i, Y_i/\widehat{m}_h^{LL}(x_i)\}$. This yields the estimator

$$\widehat{m}_h^{JLN}(x) = \widehat{m}_h^{LL}(x)\widehat{r}_0(x)$$

where $\widehat{r}_0(x)$ is given by the value of a_0 when a_0 and a_1 are chosen to minimize

$$\sum_{i=1}^{n} K_h(x - x_i)\{Y_i/\widehat{m}_h^{LL}(x_i) - a_0 - a_1(x - x_i)\}^2.$$

Jones, Linton and Nielsen (1995) have shown that the bias and variance of $\widehat{m}_h^{JLN}(x)$ are, under certain conditions, given by

$$\begin{aligned} \text{bias}(\widehat{m}_h^{JLN}(x)) &= h^4 m(x)\{m''(x)/m(x)\}''\mu_2^2/4 + o(h^4), \\ \text{var}(\widehat{m}_h^{JLN}(x)) &= n^{-1}h^{-1}\sigma^2(x)\int (2K(u) - K*K(u))^2 du/f(x) + o(n^{-1}h^{-1}) \end{aligned}$$

where $K*L$ denotes the convolution of K and L. Note that the bias is improved from $O(h^2)$ to $O(h^4)$.

The idea of the transformation techniques stems from noting that, if m were available, a suitable nonparametric regression of $\{Y_i\}$ on $\{m(x_i)\}$ would not introduce any bias because of the linear relationship. The function m is, of course, not available, but one can use an initial estimator of it. Hence, starting with $\widehat{m}_h^{LL}$, one repeats the local linear fitting on $\{Y_i, \widehat{m}_h^{LL}(x_i)\}$ and then transforms the resulting estimator back to the original design scale. Specifically, the estimator is given by the value of a_0 when a_0 and a_1 are chosen to minimize

$$\sum_{i=1}^{n} K_h(x - x_i)\{Y_i - a_0 - a_1(\widehat{m}_h^{LL}(x) - \widehat{m}_h^{LL}(x_i))\}^2.$$

Denote this by $\widehat{m}_h^{TR}(x)$. Park *et al.* (1997) showed that the bias and variance of $\widehat{m}_h^{TR}(x)$ are, under certain conditions, given by

$$\text{bias}(\widehat{m}_h^{TR}(x)) = \frac{h^4}{m'(x)}\{m''(x)m'''(x) - m''''m'(x)\}\mu_2^2/4 + o(h^4),$$

$$\text{var}(\widehat{m}_h^{TR}(x)) = n^{-1}h^{-1}\sigma^2(x)\int(2K(u) - K * K(u))^2 du/f(x) + o(n^{-1}h^{-1}).$$

Again, note that the bias is reduced to $O(h^4)$.

From the above bias formula, one can expect that $\widehat{m}_h^{JLN}$ and $\widehat{m}_h^{TR}$ exhibit somewhat strange and unappealing behaviour as $m(x) \to 0$ and $m'(x) \to 0$, respectively. Our previous experience is that this is not a real problem for $\widehat{m}_h^{TR}$, but it is for $\widehat{m}_h^{JLN}$. In fact, for the simulation study in the next section, we modified $\widehat{m}_h^{JLN}$ by first doing regression with $\{Y_i + \delta\}$ and then subtracting δ from the resulting estimator. The value of δ can be chosen to ensure that $m(x) + \delta > 0$ for all x, and in fact we chose $\delta = 5$. A similar modification could be tried on $\widehat{m}_h^{TR}$, although we did not do it in the next section, by first doing regression with $\{Y_i + \delta x_i\}$ and then subtracting δx from the resulting estimator.

Before concluding this section, we want to point out that the above two bias reduction methods can be applied to more sophisticated kernel estimators. For example, when they are applied to the local cubic fitting which already has $O(h^4)$ bias, the bias is further reduced to $O(h^8)$.

12.3   A Monte Carlo Study

We undertook a simulation study in order to see to what extent the two bias reduction methods improve the finite sample performance of the local linear smoothing estimator. We used three different target functions: (a) a sine function; (b) a parabolic function; and (c) a linear function with a Gaussian peak,

$$m(x) = 2 - 5x + 5\exp\{-400(x - 0.5)^2\}, \quad x \in [0, 1].$$

These functions are illustrated in Figure 12.1.

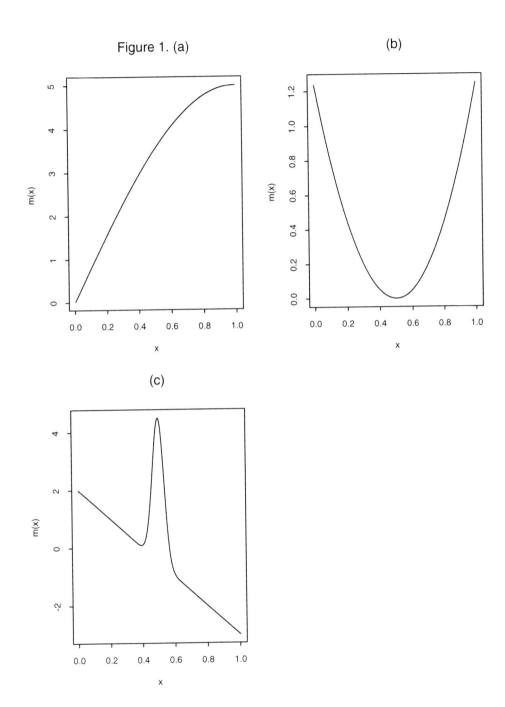

Figure 12.1: The target functions (a) a sine function, (b) a parabolic function, and (c) a linear function with Gaussian peak

The design points x_i selected are the quantiles of the uniform and truncated normal distributions. Specifically, for the fixed uniform design, $x_i = (2i - 1)/(2n)$, and for the fixed truncated normal design, $x_i = G^{-1}\{(2i - 1)/(2n)\}$, $i = 1, \ldots, n$, where $G(t) = \{\Phi(t - 0.5) - \Phi(-0.5)\}/\{\Phi(0.5) - \Phi(-0.5)\}$ and $\Phi(t)$ is the cumulative distribution function of $N(0, 1)$. We investigated two noise levels, $\sigma^2 = 0.05$ and $\sigma^2 = 0.5$, and two sample sizes, $n = 100$ and 400. As the kernel function, we used the standard normal probability density function.

The values of mean integrated squared error (MISE) were calculated by averaging over 500 simulations. Figure 12.2 depicts MISE as a function of bandwidth in the case of the linear function with Gaussian peak, uniform design, high noise, and $n = 100$. The grid of bandwidths consists of 15 logarithmically equispaced values. As one would expect, the minimum MISEs of the estimators with reduced bias are achieved at larger bandwidths. The figures for the other cases reiterate the same point and hence are not presented here.

To assess the efficiency of the two bias reduction methods, we also calculated the minimal values of the MISE curves. This was done by quadratic interpolation around the value of the h-grid which achieve the smallest MISE value. Furthermore, for the values of the bandwidth which achieve the minimal MISEs, we computed the integrated squared biases and the integrated variances. These are summarized in Table 12.1. Here, we present the results only for the case of the uniform design and high noise level. Those for the case of the truncated normal design or low noise level were of similar nature and hence are omitted.

The first observation from Table 12.1 is that, for the sine function, both bias reduction techniques are successful even for the sample size $n = 100$. In particular, the multiplicative correction method is more efficient than the transformation method in reducing the bias. Note that this function is almost linear and so the local linear smoothing is already doing quite well.

Looking at the results for the second (parabolic) and the third (linear with a Gaussian peak) functions, one can find that, for $n = 100$, the multiplicative correction method is doing well for the second function, but fails to reduce the bias of the local linear smoothing for the third function. On the other hand, the transformation is still successful in reducing the bias for the two functions, but for the second function the inflated variance cancels this effect resulting in an overall worse performance. Nevertheless, the two bias reduction methods outperform the local linear smoothing for $n = 400$.

Finally, one can observe that, except in the case of the parabolic function and $n = 100$, the bias reduction techniques do not inflate the variance. This is contrary to the initial expectation that the extra estimation would increase variability of the resulting estimators.

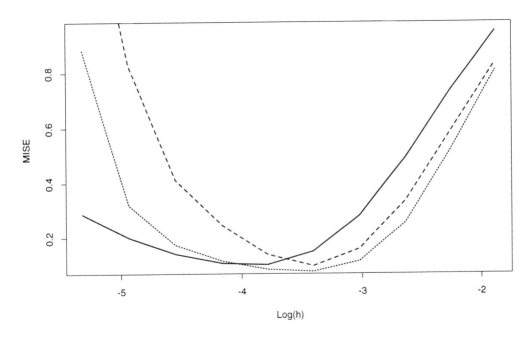

Figure 12.2: Mean integrated squared error as a function of bandwidth on a logarithmic scale, for the local linear smoothing (solid), the multiplicative correction (dashed), and the transformation (dotted)

Table 12.1: Integrated squared biases, integrated variances and mean integrated squared errors (the latter with standard errors in brackets) of the local linear smoothing (LL), the multiplicative correction (JLN), and the transformation (TR) estimators averaged over 500 runs from the three functions, (a) the sine function, (b) the parabolic function, and (c) the linear function with a Gaussian peak. All entries are multiplied by 10^2

regression function	estimator	$n = 100$ Int. Sq. Bias	Int. Variance	MISE (s.e.)	$n = 400$ Int. Sq. Bias	Int. Variance	MISE (s.e.)
(a)	LL	0.3363	1.7139	2.0502 (0.0621)	0.0750	0.5904	0.6654 (0.0182)
	JLN	0.1928	1.5561	1.7489 (0.0563)	0.0261	0.5019	0.5280 (0.0157)
	TR	0.0931	1.6738	1.7669 (0.0632)	0.0239	0.4995	0.5234 (0.0158)
(b)	LL	0.3396	1.8282	2.1678 (0.0597)	0.0876	0.6206	0.7082 (0.0194)
	JLN	0.1821	2.1092	2.2913 (0.0768)	0.0117	0.4801	0.4918 (0.0159)
	TR	0.0306	1.7654	1.7960 (0.0758)	0.0239	0.5090	0.5329 (0.0161)
(c)	LL	2.2047	7.5855	9.7902 (0.1212)	0.5322	2.7823	3.3145 (0.0384)
	JLN	1.0530	6.3749	7.4279 (0.1147)	0.1759	2.2025	2.3784 (0.0336)
	TR	2.3588	7.2035	9.5623 (0.1412)	0.1801	2.7066	2.8867 (0.0358)

Acknowledgements. This research was supported by Korea Science and Engineering Foundation grant 931-0100-006-2.

References

1. Cleveland, W. S. (1979). Robust locally weighted regression and smoothing scatterplots, *Journal of the American Statistical Association*, **74**, 829–836.

2. Fan, J. (1992). Design-adaptive nonparametric regression, *Journal of the American Statistical Association*, **87**, 998–1004.

3. Fan, J. (1993). Local linear regression smoothers and their minimax efficiencies, *Annals of Statistics*, **21**, 196–216.

4. Fan, J. and Gijbels, I. (1992). Variable bandwidth and local linear regression smoothers, *Annals of Statistics*, **20**, 2008–2036.

5. Gasser, T. and Müller, H. G. (1979). Kernel estimation of regression functions, In *Smoothing Techniques for Curve Estimation* (Eds., Th. Gasser and M. Rosenblatt), pp. 23–68, *Lecture Notes in Mathematics*—**757**, Heidelberg: Springer-Verlag.

6. Gasser, T. and Müller, H. G. (1984). Estimating regression functions and their derivatives by the kernel method, *Scandinavian Journal of Statistics*, **11**, 171–185.

7. Hastie, T. J. and Loader, C. (1993). Local regression; automatic kernel carpentry (with comments), *Statistical Science*, **8**, 120–143.

8. Jones, M. C. (1995). Local and variable bandwidths and local linear regression, *Statistics*, **26**, 65–71.

9. Jones, M. C., Davies, S. J. and Park, B. U. (1994). Versions of kernel-type regression estimators, *Journal of the American Statistical Association*, **89**, 825–832.

10. Jones, M. C., Linton, O. and Nielsen, J. P. (1995). A simple bias reduction method for density estimation, *Biometrika*, **82**, 327–338.

11. Nadaraya, E. A. (1964). On estimating regression, *Theory of Probability and its Applications*, **9**, 141–142.

12. Park, B. U., Kim, W. C., Ruppert, D., Jones, M. C., Signorini, D. F. and Kohn, R. (1997). Simple transformation techniques for improved nonparametric regression, *Scandinavian Journal of Statistics* (to appear).

13. Priestley, M. B. and Chao, M. T. (1972). Non-parametric function fitting, *Journal of the Royal Statistical Society, Series B*, **34**, 385–392.

14. Stone, C. J. (1977). Consistent nonparametric regression (with discussion), *Annals of Statistics*, **5**, 595–645.

15. Watson, G. S. (1964). Smooth regression analysis, *Sankhyā, Series A*, **26**, 359–372.

13

Multiple Comparisons With the Mean

Kathleen S. Fritsch and Jason C. Hsu

The Ohio State University, Columbus, OH

Abstract: Multiple Comparisons with the Mean (MCM), also known as Analysis of Means (ANOM), compares within-treatment means to the grand mean so as to identify treatments with means significantly different from the grand mean. Due to the relative difficulty of exact MCM probabilistic computations, MCM has not been applied much in practice except in the quality assurance setting in which controlled experiments with variance-balanced designs can be conducted. Recently, the health industry has shown an interest in MCM. Since data from health care providers tend to have simple but unbalanced designs, we investigate the feasibility of performing essentially exact MCM probabilistic computations for such designs in this chapter. We find that, for small number of treatments, an algorithm involving complex variable integration is required. However, for moderate to large number of treatments, two simple approximations seem adequate.

Keywords and phrases: Multiple comparisons, analysis of means, weighted mean, unweighted mean

13.1   Different Types of Multiple Comparisons

Multiple Comparisons with the Mean (MCM), also known in its graphical form as Analysis of Means (ANOM), is a multiple comparisons method which compares within-treatment means to the grand mean so as to identify treatments with means significantly different from the grand mean.

To discuss the relationship of MCM with other multiple comparison methods, consider the one-way model

$$X_{ij} = \mu_i + \varepsilon_{ij}, \tag{13.1}$$

189

where

$$i = 1, \ldots, k,$$
$$j = 1, \ldots, n_i,$$
$$k = \text{the number of treatments,}$$
$$n_i = \text{sample size of the } i\text{-th treatment.}$$

We assume that $\varepsilon_{11}, \ldots, \varepsilon_{kn_k}$ are i.i.d. $N(0, \sigma^2)$, and we define $N = \sum_{i=1}^{k} n_i$.

Different types of multiple comparisons differ in the parameters each considers primary, as follows:

- MCA (all-pairwise multiple comparisons) considers $\mu_i - \mu_j$ for all $i \neq j$ to be of primary interest;

- MCB (multiple comparisons with the best) considers $\mu_i - \max_{j \neq i} \mu_j$ for $i = 1, \ldots, k$ to be of primary interest;

- MCC (multiple comparisons with a control) considers $\mu_i - \mu_k$ for $i = 1, \ldots, k-1$ to be of primary interest;

- MCM (multiple comparisons with the mean) considers $\mu_i - \bar{\mu}$ or $\mu_i - \bar{\bar{\mu}}$ for $i = 1, \ldots, k$ to be of primary interest, where $\bar{\mu}$ and $\bar{\bar{\mu}}$ are the unweighted and the weighted means of the μ_i's.

MCM was first proposed by Halperin et al. (1955). Tukey (1992) recommends MCM over MCA for large k, because the result of k comparisons in MCM would be easier to comprehend than the result of $k(k-1)/2$ comparisons in MCA when k is large. This advantage is shared by MCB and MCC, which make k and $k-1$ comparisons, respectively. Hsu (1996) suggests choosing a multiple comparison type according to the parameters of primary interest, not only to maintain comprehension, but also to obtain the sharpest possible inference.

In contrast to MCA, MCB, and MCC, exact probabilities associated with MCM are relatively difficult to compute for unbalanced designs. Therefore, MCM has not been applied much in practice in most disciplines. An exception is industrial quality control, in which controlled experiments with variance-balanced designs can often be conducted. In the quality control setting, MCM is usually known as Analysis of Means (ANOM). Ott (1967) originally developed ANOM to analyze the balanced, one-way model with k treatments. He presented it graphically, using decision lines similar to those in Shewart control charts to indicate which treatments are different from the overall mean. Nelson (1993) discusses other designs, such as Latin Squares and Balanced Incomplete Block Designs, which are so-called variance-balanced designs. With such designs, estimates of the main effects have the same correlation structure as in a balanced one-way model. Thus, with the appropriate substitution of parameter estimates, variance-balanced designs can be analyzed with procedures developed for the balanced one-way model.

Recently, the health maintenance industry has shown an interest in MCM. Since data from health care providers tend to be observational with a one-way design, we investigate the feasibility of performing essentially exact MCM probabilistic computations for such designs in this chapter.

Remark. In early literature, MCM was also discussed in the context of outlier detection, since ordinary residuals from a single normal population have the same joint distribution as deviations of treatment means from the grand mean in the balanced case; see, for example, Barnett and Lewis (1993). However, since modern regression diagnostics are based on deleted residuals instead of ordinary residuals, we do not discuss MCM in this context.

We estimate the treatment means, μ_i, with

$$\hat{\mu}_i = \overline{X}_i = \frac{1}{n_i} \sum_{j=1}^{n_i} X_{ij}, \quad i = 1, \ldots, k.$$

Two definitions of the "overall mean," the reference point from which we compare the individual treatment means, are considered. The *unweighted* mean is defined as

$$\overline{\mu} = \frac{1}{k} \sum_{i=1}^{k} \mu_i \tag{13.2}$$

and the *weighted* mean is defined as

$$\overline{\overline{\mu}} = \frac{1}{N} \sum_{i=1}^{k} n_i \mu_i . \tag{13.3}$$

We estimate (13.2) with

$$\overline{X} = \frac{1}{k} \sum_{i=1}^{k} \overline{X}_i$$

and (13.3) with

$$\overline{\overline{X}} = \frac{1}{N} \sum_{i=1}^{k} n_i \overline{X}_i .$$

We also define the mean squared error as

$$\hat{\sigma}^2 = \sum_{i=1}^{k} \sum_{j=1}^{n_i} (X_{ij} - \overline{X}_i)^2 / \sum_{i=1}^{k} (n_i - 1) .$$

The quantities of interest in ANOM and MCM are either the deviations from the unweighted mean $\mu_i - \bar{\mu}$, estimated by $\overline{X}_i - \overline{X}$, or the deviations from the weighted mean $\mu_i - \bar{\bar{\mu}}$, estimated by $\overline{X}_i - \overline{\overline{X}}$. If the model is balanced, i.e. $n_i = n$ for $i = 1, \ldots, k$, then $\mu_i - \bar{\mu}$ and $\mu_i - \bar{\bar{\mu}}$ are identical.

13.2 Balanced MCM

For balanced designs $\overline{X} = \overline{\overline{X}}$, and we can formulate our comparisons in terms of either the weighted or unweighted mean. The variances and covariances of $\overline{X}_i - \overline{\overline{X}}$ are

$$\text{Var}\,(\overline{X}_i - \overline{\overline{X}}) = \sigma^2 \frac{(k-1)}{N}$$

$$\text{Cov}\,(\overline{X}_i - \overline{\overline{X}},\ \overline{X}_j - \overline{\overline{X}}) = -\frac{\sigma^2}{N}\,.$$

The ANOM procedure computes decision lines which are two-sided confidence intervals for the $\mu_i - \bar{\bar{\mu}}$. These intervals are of the form

$$\overline{X}_i - \overline{\overline{X}} \pm h_{\alpha,k,\nu}\,\hat{\sigma}\,\sqrt{(k-1)/N},\quad i = 1, \ldots, k$$

where $h_{\alpha,k,\nu}$ is the $(1-\alpha)$th quantile of the distribution of the statistic

$$\max_{1 \leq i \leq k} \frac{|\overline{X}_i - \overline{\overline{X}}|}{\hat{\sigma}\sqrt{(k-1)/N}}\,.$$

[Tukey (1992) calls this statistic the MAD, for Maximum Absolute Deviation.] Treatment means are considered to be significantly different from the overall mean if the associated confidence interval for $\mu_i - \bar{\mu}$ does not cover 0.

In the balanced case, Nelson (1982, 1993) has tabulated the critical values $h_{\alpha,k,\nu}$ for various combinations of α, k, and error degrees of freedom $\nu = \sum_{i=1}^{k}(n_i - 1) = N - k$. The computation needed to calculate these critical values is involved, requiring double integration of a complex integrand. Before exact tables of critical values were available, the literature on MCM and ANOM recommended using conservative Bonferroni t quantiles, $t_{\alpha/2k,\nu}$, instead of $h_{\alpha,k,\nu}$. Since exact critical values are difficult to compute, we investigate the adequacy of this approximation. We also investigate the adequacy of two other approximations: one based on the Studentized Maximum Modulus quantile, and the other based on the Hunter-Worsley inequality.

The Studentized Maximum Modulus (SMM) method is a method designed to give simultaneous confidence intervals of k means which can be estimated independently. In this setting, the only source of dependence among the statistics $|\overline{X}_i|/(\hat{\sigma}/\sqrt{n_i})$ arises from the fact that each one is studentized by the common pooled standard deviation estimate $\hat{\sigma}$. Of course, in our setting, the statistics in MCM and ANOM, $\overline{X}_i - \overline{\overline{X}}$, are not independent, but are negatively correlated with

$$\text{Corr}(\overline{X}_i - \overline{\overline{X}}, \ \overline{X}_j - \overline{\overline{X}}) = -\frac{1}{k-1}$$

in the balanced case. Thus, intuitively, the SMM quantile may serve as an adequate approximation when k is large.

We can use Šidák's inequality [Šidák (1968)] to show that the SMM quantile is conservative. Define

$$E_i = \{|\overline{X}_i - \overline{\overline{X}}|/\hat{\sigma}\sqrt{(k-1)/N} < h\} .$$

Then, Šidák's inequality states

$$P(\bigcap_{i=1}^{k} E_i \mid \hat{\sigma}) \geq \prod_{i=1}^{k} P(E_i \mid \hat{\sigma}) . \tag{13.4}$$

The $(1-\alpha)$th SMM quantile, $|m|$, can be computed as the solution to the equation

$$\int_0^\infty [\Phi(|m|s) - \Phi(-|m|s)]^k d\Gamma(s) = 1 - \alpha,$$

where Φ is the standard normal distribution function and Γ is the distribution function of $\hat{\sigma}/\sigma$. Thus, we see that the SMM quantile $|m|$ is a conservative approximation to $h_{\alpha,k,\nu}$.

We can also take equation (13.4) one step further by integrating over $\hat{\sigma}$, applying Kimball's (1951) inequality (also known as Chebyshev's other inequality) to get the unconditional inequality

$$P(\bigcap_{i=1}^{k} E_i) \geq \prod_{i=1}^{k} P(E_i) . \tag{13.5}$$

From Eq. (13.5), we see that another conservative approximation to the exact quantile is the Šidák t quantile, $t_{[1-(1-\alpha)^{1/k}]/2,\nu}$. Since the Šidák t quantile is always more conservative than the SMM quantile [significantly so when the error degrees of freedom is small; see, for example, Hsu (1996, p. 11)], the Šidák t quantile approximation is not included in our study.

The Hunter-Worsley inequality is a refinement of the Bonferroni inequality. It is as follows.

Theorem 13.2.1 (The Hunter–Worsley inequality) *If $\mathcal{T}$ is a spanning tree of the vertices $\{1,\ldots,k\}$, and $\{i,j\} \in \mathcal{T}$ denotes i and j are adjacent, then*

$$P(\bigcup_{i=1}^{k} E_i^c) \leq \sum_{i=1}^{k} P(E_i^c) - \sum_{\{i,j\}\in\mathcal{T}} P(E_i^c \bigcap E_j^c). \tag{13.6}$$

Thus, a conservative approximation h to $h_{\alpha,k,\nu}$ results if one solves the equation

$$E_{\hat\sigma}\left[\sum_{i=1}^{k}P(E_i^c\mid\hat\sigma)-\sum_{\{i,j\}\in\mathcal{T}}P(E_i^c\cap E_j^c\mid\hat\sigma)\right]=1-\alpha.$$

We investigate the adequacy of the sharpest Hunter-Worsley bound, achieved when $\mathcal{T}$ is a spanning tree that maximizes the second term.

Table 13.1 presents exact, Hunter-Worsley, Studentized Maximum Modulus, and Bonferroni quantiles for selected numbers of treatments and sample sizes. Comparing the Hunter-Worsley and Studentized Maximum Modulus approximations, we find that except when $k=3$ the SMM approximation is better than the Hunter-Worsley approximation. (The Hunter-Worsley approximation also gets a slight edge when $k=4$ and the degrees of freedom is large.) When k is small, however, even for large degrees of freedom, neither approximation is sufficiently close to the exact value. When both k and the degrees of freedom are sufficiently large, though, the SMM quantile is essentially exact to two decimal places. Figure 13.1 displays the region where the SMM quantile is essentially exact. For very small k, the approximation never gets sufficiently close to the exact values, even when the degrees of freedom is very large. For slightly larger k, the approximation may be good enough when the degrees of freedom are large enough so as to be essentially infinite. For even larger k, then, the approximation is sufficient when the degrees of freedom exceeds a finite threshold.

13.3 Unbalanced MCM

When the design is unbalanced, we must distinguish whether we wish to compare the treatment means μ_i with the weighted mean $\bar{\bar{\mu}}$, or the unweighted mean $\bar{\mu}$. Using the weighted mean makes sense if the unequal sample sizes arise from a setting such as where the sample sizes are proportional to population sizes. Using the unweighted mean may make sense if the unequal sample sizes are due to missing values which were unintended in the original design.

13.3.1 Comparisons with the weighted mean

For comparisons with the weighted mean in the unbalanced case, we can compute the variances and covariances of the deviations from the mean as

$$\text{Var}\left(\overline{X}_i-\overline{\overline{X}}\right)=\sigma^2\left[\frac{1}{n_i}-\frac{1}{N}\right],\tag{13.7}$$

$$\text{Cov}\left(\overline{X}_i-\overline{\overline{X}},\ \overline{X}_j-\overline{\overline{X}}\right)=-\frac{\sigma^2}{N}$$

Table 13.1: Comparison of exact, Hunter-Worsley, Studentized Maximum Modulus, and Bonferroni approximations for MCM, balanced case, where $k =$ number of treatments and $n =$ sample size for each treatment (one-way model)

$$\alpha = 0.10$$

	$k = 3$					$k = 4$				
n	Exact	H-W	SMM	Bonf.	df	Exact	H-W	SMM	Bonf.	df
2	3.16	3.34	3.37	3.74	3	3.09	3.27	3.20	3.50	4
5	2.27	2.32	2.36	2.40	12	2.38	2.43	2.43	2.47	16
10	2.15	2.18	2.22	2.24	27	2.27	2.31	2.31	2.34	36
∞	2.05	2.08	2.11	2.13	∞	2.19	2.22	2.23	2.24	∞

	$k = 10$					$k = 20$				
n	Exact	H-W	SMM	Bonf.	df	Exact	H-W	SMM	Bonf.	df
2	3.02	3.13	3.03	3.17	10	3.07	3.14	3.07	3.15	20
5	2.66	2.70	2.67	2.70	40	2.86	2.88	2.86	2.89	80
10	2.60	2.63	2.61	2.63	90	2.82	2.84	2.82	2.84	180
∞	2.55	2.57	2.56	2.58	∞	2.79	2.81	2.79	2.81	∞

$$\alpha = 0.05$$

	$k = 3$					$k = 4$				
n	Exact	H-W	SMM	Bonf.	df	Exact	H-W	SMM	Bonf.	df
2	4.18	4.39	4.43	4.86	3	3.89	4.08	4.00	4.31	4
5	2.67	2.71	2.75	2.78	12	2.74	2.78	2.78	2.81	16
10	2.48	2.51	2.54	2.55	27	2.58	2.61	2.62	2.63	36
∞	2.34	2.36	2.39	2.39	∞	2.47	2.48	2.49	2.50	∞

	$k = 10$					$k = 20$				
n	Exact	H-W	SMM	Bonf.	df	Exact	H-W	SMM	Bonf.	df
2	3.45	3.55	3.47	3.58	10	3.40	3.45	3.40	3.46	20
5	2.95	2.97	2.95	2.97	40	3.11	3.12	3.11	3.12	80
10	2.86	2.87	2.87	2.88	90	3.06	3.07	3.06	3.07	180
∞	2.80	2.80	2.80	2.81	∞	3.02	3.02	3.02	3.02	∞

$$\alpha = 0.01$$

	$k = 3$					$k = 4$				
n	Exact	H-W	SMM	Bonf.	df	Exact	H-W	SMM	Bonf.	df
2	7.51	7.84	7.92	8.58	3	6.20	6.45	6.36	6.76	4
5	3.57	3.60	3.63	3.65	12	3.54	3.56	3.57	3.58	16
10	3.18	3.19	3.21	3.22	27	3.23	3.24	3.25	3.25	36
∞	2.91	2.92	2.93	2.94	∞	3.01	3.02	3.02	3.03	∞

	$k = 10$					$k = 20$				
n	Exact	H-W	SMM	Bonf.	df	Exact	H-W	SMM	Bonf.	df
2	4.49	4.56	4.50	4.59	10	4.12	4.14	4.12	4.15	20
5	3.54	3.55	3.54	3.55	40	3.63	3.63	3.63	3.63	80
10	3.40	3.40	3.40	3.40	90	3.55	3.55	3.55	3.55	180
∞	3.29	3.29	3.29	3.29	∞	3.48	3.48	3.48	3.49	∞

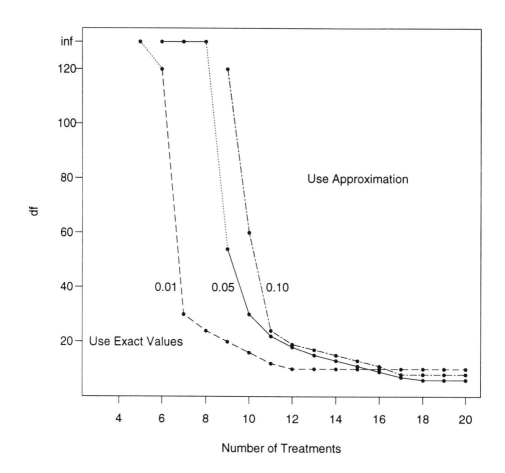

Figure 13.1: Region where SMM approximation is essentially exact for
$\alpha = 0.01,\ 0.05,\ 0.10$

Minitab has attempted to incorporate the ANOM procedure into its Release 10 version, using the weighted definition of the overall mean in the unbalanced case. Unfortunately, Minitab has incorporated it incorrectly. Minitab computes

$$\text{Var}\,(\overline{X}_i - \overline{\overline{X}}) = \sigma^2\,(N-1)/(Nn_i)$$

instead of

$$\text{Var}\,(\overline{X}_i - \overline{\overline{X}}) = \sigma^2\,(N-n_i)/(Nn_i)$$

which is the correct rearrangement of (13.7). This mistake creates confidence intervals which are wider than they should be. One other thing to note about this implementation in Minitab is that it uses the conservative Bonferroni t quantile, $t_{\alpha/2k,\nu}$, instead of the exact $h_{\alpha,k,\nu}$.

Discussions in the literature about how to deal with unbalanced designs in the Analysis of Means has been limited to two articles by Nelson (1974, 1983). Nelson's first article recommends using the weighted definition of the overall mean in the unbalanced case and conservative Bonferroni t quantiles, $t_{\alpha/2k,\nu}$. In his second article, he refines his recommended quantiles to the slightly less conservative Šidák quantiles, $t_{[1-(1-\alpha)^{1/k}]/2,\nu}$.

Exact quantiles for comparisons with the weighted mean in unbalanced designs can be computed analogously to those for balanced designs because of the similarity in correlation structure. Our statistics of interest,

$$T_j = \frac{\overline{X}_j - \overline{\overline{X}}}{\hat{\sigma}\,\sqrt{(1/n_j) - (1/N)}}\,,$$

have multivariate t distribution with correlation matrix of the form $\mathbf{R} = \mathbf{D} - \boldsymbol{\lambda}\boldsymbol{\lambda}'$ (negative product structure) where $\lambda_j = \sqrt{n_j/(N-n_j)}$, and $\mathbf{D} = \text{diag}(1+\lambda_j^2)$. It is known that when $\mathbf{R}$ has the *positive* product structure, $\mathbf{D} + \boldsymbol{\lambda}\boldsymbol{\lambda}'$,

$$P(|T_1| \le h, \ldots, |T_k| \le h) \tag{13.8}$$

can be expressed as

$$\int_0^\infty \int_{-\infty}^\infty \prod_{j=1}^k \left[\Phi\left(\frac{sh - \lambda_j x}{\sqrt{1-\lambda_j^2}}\right) - \Phi\left(\frac{-sh - \lambda_j x}{\sqrt{1-\lambda_j^2}}\right) \right] d\Phi(x) d\Gamma(s)$$

where Φ is the standard normal distribution and Γ is the distribution of $\hat{\sigma}/\sigma$. [The corresponding expression for normal random variables was given in Dunnett and Sobel (1955). The expression for t variables is derived by a conditioning argument.]

We do not have the required positive product structure, but we can achieve it by making the substitution $\lambda_j = \sqrt{-1}\eta_j$. If we define the standard normal distribution with complex argument as

$$\Phi(x+iy) = \frac{1}{\sqrt{2\pi}}\exp(y^2/2)\int_{-\infty}^x \exp(-ity - t^2/2)dt$$

where $i = \sqrt{-1}$, then we can compute the probability (13.8) as

$$\int_0^\infty \int_{-\infty}^\infty \prod_{j=1}^k \left[\Phi\left(\frac{sh - i\eta_j x}{\sqrt{1 + \eta_j^2}}\right) - \Phi\left(\frac{-sh - i\eta_j x}{\sqrt{1 + \eta_j^2}}\right) \right] d\Phi(x) d\Gamma(s). \qquad (13.9)$$

The proof of the validity of (13.9) as an expression for (13.8) is presented in Soong and Hsu (1996). Thus, exact quantiles can be computed, but the computation is involved, requiring adaptive Romberg integration of the complex integrand.

Since computations for exact quantiles for unbalanced designs are time consuming, again the adequacy of the SMM approximation, which continues to be conservative by Šidák's inequality, is of interest. Since the adequacy of this approximation in the balanced case has already been studied in the last section, we study the difference between the MCM critical values for balanced and unbalanced designs with the same number of treatments and error degrees of freedom. This study has an additional motivation, as follows.

When $k = 3$, the correlation matrix for MCM is equivalent to the correlation matrix for all pairwise multiple comparisons (MCA). In MCA the correlation matrix is also of the form $\mathbf{R} = \mathbf{D} - \boldsymbol{\lambda}\boldsymbol{\lambda}'$, where $\lambda_1 = \sqrt{n_3/(n_1 + n_2)}$, $\lambda_2 = \sqrt{n_2/(n_1 + n_3)}$, $\lambda_3 = \sqrt{n_1/(n_2 + n_3)}$, and $\mathbf{D}$ is as before. Thus, it is a permutation of the MCM correlation matrix. [See Soong and Hsu (1996) for derivations of the matrices.] Since Hayter (1984) proved that using the quantile from the balanced setting is conservative in the unbalanced MCA setting, his proof also shows that the balanced quantile in MCM is conservative for unbalanced designs when $k = 3$. Therefore, there is the intriguing possibility that, for all k, MCM critical values for balanced designs turn out to be conservative for unbalanced designs with the same number of treatments and error degrees of freedom.

For several selected sample size configurations, Table 13.2 lists simulated quantiles computed using the technique of Edwards and Berry (1987), as well as exact quantiles from Nelson (1993) for the corresponding balanced case with the same number of treatments and error degrees of freedom. It appears that the quantiles for balanced designs are conservative in the unbalanced case, and that the quantiles are essentially identical to two decimal places when $k \geq 4$.

For an example implementing MCM, consider the data of Campbell and McCabe (1984) who reported on the Scholastic Aptitude Test (SAT) mathematics scores of first year computer science majors and their choice of majors in their second year. The sample sizes and sample means are given in Table 13.3, and the pooled sample standard deviation is $\hat{\sigma} = 82.52$. We use $\alpha = 0.10$.

We compare an average student in each group with a student who is average when the three groups are combined. We assume that the sample sizes 103, 31, 122 are proportional to the population sizes of the three groups. Because the populations are assumed to be of different sizes, it makes the most sense to work with the weighted mean.

Table 13.2: Comparison of simulated quantiles for selected sample size configurations with quantiles for balanced designs with the same error degrees of freedom and $\alpha = 0.05$

$k = 3$

df	10	30	100
sample size configuration	(2,4,7)	(2,11,20)	(2,34,67)
weighted simulated quantile	2.73	2.44	2.32
unweighted simulated quantile	2.73	2.40	2.24
quantile for balanced design	2.74	2.47	2.38

$k = 4$

df	10	30	100
sample size configuration	(2,3,4,5)	(2,7,10,15)	(2, 20,32,50)
weighted simulated quantile	2.92	2.61	2.50
unweighted simulated quantile	2.92	2.59	2.43
quantile for balanced design	2.93	2.61	2.50

$k = 5$

df	10	30	100
sample size configuration	(2,2,3,4,4)	(2,5,7,9,12)	(2,10,20,30,43)
weighted simulated quantile	3.06	2.71	2.60
unweighted simulated quantile	3.06	2.70	2.57
quantile for balanced design	3.07	2.71	2.60

$k = 6$

df	10	30	100
sample size configuration	(1,2,2,3,4,4)	(2,3,6,6,9,10)	(2,4,15,20,30,35)
weighted simulated quantile	3.17	2.79	2.67
unweighted simulated quantile	3.17	2.79	2.65
quantile for balanced design	3.17	2.79	2.67

Table 13.3: SAT mathematics scores

Second-year major	Label	Sample size	Sample mean	Sample standard deviation
Computer science	1	103	619	86
Engineering	2	31	629	67
Other	3	122	575	83

The exact critical value for $\alpha = 0.10$, $k = 3$ treatments, and $\nu = 253$ error degrees of freedom, and sample size configuration $n_1 = 103, n_2 = 31, n_3 = 122$, is $h^w_{.10,(103,31,122)} = 2.047$. In this example, $\overline{\overline{X}} = 599.24$. Therefore, 90% multiple comparisons with the weighted mean confidence intervals $\overline{X}_i - \overline{\overline{X}} \pm h^w_{\alpha,\, \mathbf{n}} \hat{\sigma} \sqrt{(1/n_i) - (1/N)}$ are

$$
\begin{aligned}
6.89 &< \mu_1 - \overline{\overline{\mu}} < 32.63, \\
1.32 &< \mu_2 - \overline{\overline{\mu}} < 58.20, \\
-35.30 &< \mu_3 - \overline{\overline{\mu}} < -13.18.
\end{aligned}
$$

Thus, we conclude that those students who remain computer science majors or switch to engineering had SAT math scores above the overall mean on average, and those who switch to other majors had scores below the overall mean on average.

13.3.2   Comparisons with the unweighted mean

For comparisons with the unweighted mean when the design is unbalanced, the structure of the variances and covariances of the deviations from the mean is much more complicated and are given by

$$
\mathrm{Var}\,(\overline{X}_i - \overline{X}) = \frac{\sigma^2}{k}\left[\frac{k-2}{n_i} + \frac{1}{k}\sum_{l=1}^{k}\frac{1}{n_l}\right],
$$

$$
\mathrm{Cov}\,(\overline{X}_i - \overline{X},\, \overline{X}_j - \overline{X}) = \frac{\sigma^2}{k}\left[-\left(\frac{1}{n_i} + \frac{1}{n_j}\right) + \frac{1}{k}\sum_{l=1}^{k}\frac{1}{n_l}\right]
$$

In this setting, we no longer have product correlation structure. Therefore, the algorithm in Soong and Hsu (1996) is no longer applicable. So the only options are either to use the SMM quantile, which continues to be conservative by Šidák's inequality, or to use the quantile for the balanced design with the same number of treatments and error degrees of freedom. Table 13.2 also lists simulated critical values for the unweighted mean setting. Again, the quantile for the balanced setting with the same number of treatments and error degrees of freedom seems to be conservative for the unbalanced design. We note, however, that the degree of conservatism of the approximation seems to be greater for comparisons with the unweighted mean than for comparisons with the weighted mean.

As an example where comparisons with the unweighted mean may be of interest, consider the data presented in Fleiss (1986). This data consists of measurements of four different examiners on the total number of decayed, missing, and filled surfaces on each of 10 patients' teeth. The original data come from a balanced design with no missing values. To illustrate multiple comparisons with the unweighted mean in an unbalanced design, four observations were randomly deleted from the original data set. The data are presented in Table 13.4.

Table 13.4: Decayed, missing, and filled permanent teeth surfaces

Patient	Examiner 1	2	3	4
1	8	–	11	7
2	13	11	15	–
3	0	0	2	1
4	3	6	9	6
5	13	13	17	10
6	19	23	27	18
7	0	–	1	0
8	2	0	4	5
9	–	20	22	16
10	5	3	8	3
Mean	7.00	9.50	11.60	7.33

In this example, we wish to compare the average effect of an examiner to the average effect over all examiners to identify examiners with unusually high or low effects. We don't want to weigh one examiner's average differently from other examiners' averages just because examiners observed different numbers of patients. We want each examiner's average to contribute equally, so we use the unweighted average. In this data there is an obvious "patient effect," different patients have different numbers of decayed surfaces. A more complete model for this data would include a term to model the patient effect, but for purposes of demonstrating multiple comparisons with the unweighted mean, we will ignore the patient effect. For the one-way model, we have $\overline{X} = 8.86$, $\hat{\sigma} = 7.678$, $k = 4$ treatments, and $\nu = 32$ degrees of freedom. We now compute confidence intervals for $\overline{X}_i - \overline{X}$. We consider $\alpha = .05$. Using $h_{.05, 4, 32} = 2.60$ from the balanced setting to approximate $h_{\alpha,\mathbf{n}}^u$, we compute the confidence intervals

$$\overline{X}_i - \overline{X} \pm h_{\alpha, k, \nu} \, \hat{\sigma} \sqrt{\frac{1}{k}\left[\frac{k-2}{n_i} + \frac{1}{k}\sum_{l=1}^{k}\frac{1}{n_l}\right]} \text{ as}$$

$$\begin{aligned} -7.63 &< \mu_1 - \overline{\mu} < 3.91, \\ -5.36 &< \mu_2 - \overline{\mu} < 6.64, \\ -2.83 &< \mu_3 - \overline{\mu} < 8.31, \\ -7.30 &< \mu_4 - \overline{\mu} < 4.24. \end{aligned}$$

Since this data was fit with a one-way model, the standard deviation estimate must absorb the patient-to-patient variability. Thus, the large estimate overwhelms the examiner-to-examiner variability, and we cannot conclude that any examiner is different from the overall mean since all confidence intervals contain 0.

13.4   Recommendations

Multiple comparisons with the mean (MCM) is a useful multiple comparison method. To compute probabilities associated with MCM, we recommend the following:

- For balanced designs with small number of treatments, or moderate number of treatments and inadequate error degrees of freedom, the numerical algorithm discussed in Soong and Hsu (1996) is available.

- For balanced designs with large number of treatments, or moderate number of treatments and adequate error degrees of freedom, the Studentized Maximum Modulus quantile is an adequate approximation.

- For multiple comparisons with the weighted mean with small number of treatments, or moderate number of treatments and inadequate error degrees of freedom, the algorithm discussed in Soong and Hsu (1996) is available.

- For multiple comparisons with the weighted mean with large number of treatments, or moderate number of treatments and adequate error degrees of freedom, one can use either the SMM quantile (which is guaranteed to be conservative) or the quantile from the balanced design with the same number of treatments and error degrees of freedom (which is smaller but still appears to be conservative).

- For multiple comparisons with the unweighted mean in an unbalanced design, one can use either the SMM quantile (which is guaranteed to be conservative) or the quantile from the balanced design with the same number of treatments and error degrees of freedom (which is smaller and appears to be conservative).

References

1. Barnett, V. and Lewis, T. (1993). *Outliers in Statistical Data*, Third edition, Chichester, England: John Wiley & Sons.

2. Campbell, P. F. and McCabe, G. P. (1984). Predicting the success of freshmen in a computer science major, *Communications of the ACM*, **27**, 1108–1113.

3. Dunnett, C. W. and Sobel, M. (1955). Approximations to the probability integral and certain percentage points of a multivariate analogue of Student's t-distribution, *Biometrika*, **42**, 258–260.

4. Edwards, D. G. and Berry, J. J. (1987). The efficiency of simulation-based multiple comparisons, *Biometrics*, **43**, 913–928.

5. Fleiss, J. L. (1986) *The Design and Analysis of Clinical Experiments*, New York: John Wiley & Sons.

6. Halperin, M., Greenhouse, S. W., Cornfield, J. and Zalokar, J. (1955). Tables of percentage points for the studentized maximum absolute deviate in normal samples, *Journal of the American Statistical Association*, **50**, 185–195.

7. Hayter, A. J. (1984). A proof of the conjecture that the Tukey-Kramer multiple comparisons procedure is conservative, *Annals of Statistics*, **12**, 61–75.

8. Hsu, J. C. (1996). *Multiple Comparisons: Theory and Methods*, London: Chapman & Hall.

9. Kimball, A. W. (1951). On dependent tests of significance in analysis of variance, *Annals of Mathematical Statistics*, **22**, 600–602.

10. Nelson, L. S. (1974). Factors for the analysis of means, *Journal of Quality Technology*, **6**, 175–181.

11. Nelson, L. S. (1983). Exact critical values for use with the analysis of means, *Journal of Quality Technology*, **15**, 40–44.

12. Nelson, P. R. (1982). Exact critical points for the analysis of means, *Communications in Statistics—Theory and Methods*, **11**, 699–709.

13. Nelson, P. R. (1993). Additional uses for the analysis of means and extended tables of critical values, *Technometrics*, **35**, 61–71.

14. Ott, E. R. (1967). Analysis of means—A graphical procedure, *Industrial Quality Control*, **24**, 101–109.

15. Šidák, Z. (1968). On multivariate normal probabilities of rectangles: Their dependence on correlations, *Annals of Mathematical Statistics*, **39**, 1425–1434.

16. Soong, W. C. and Hsu, J. C. (1996). Using complex integration to compute multivariate normal probabilities, *Submitted for publication.*

17. Tukey, J. W. (1992). Where should multiple comparisons go next?, In *Multiple Comparisons, Selection, and Applications in Biometry: A Festschrift in Honor of Charles W. Dunnett* (Ed., F. M. Hoppe), pp. 187–208, New York: Marcel Dekker.

PART IV

Tests of Hypotheses

14

Properties of Unified Bayesian-Frequentist Tests

J. O. Berger, B. Boukai and Y. Wang

Purdue University, West-Lafayette, IN
Indiana University—Purdue University, Indianapolis, IN
Indiana University—Purdue University, Indianapolis, IN

Abstract: The modified Bayesian-Frequentist test of Berger, Brown and Wolpert (1994) is considered here in the context of normal hypothesis testing. We focus attention on the testing of a precise null hypothesis versus a composite alternative, either the one-sided or the two-sided type. We study the properties of the corresponding modified Bayesian-Frequentist test and in particular the large-sample behavior of its *no decision region* under two different classes of prior distributions, viz., the shifted conjugate class and a domain-restricted noninformative class. It is shown that under these prior classes, the size of the no-decision region of the test is rather small, compared to the relevant sample size. A lower bound on the conditional probability of the type I error is also provided.

Keywords and phrases: Bayes factor, likelihood ratio, composite hypothesis, conditional test, error probabilities

14.1 Introduction

Over the past twenty years, Shanti S. Gupta has been a major contributor to efforts to integrate Bayesian and Frequentist concepts in statistical inference. An area in which such integration has, until recently, been lacking is hypothesis testing. Indeed, testing statistical hypotheses has been one of the focal points for disagreement between Bayesians and Frequentists.

Given the data, one has to decide whether the underlying statistical model is as described by the null hypothesis H_0 or as described by the alternative model-hypothesis H_1. The classical Frequentist approach is to construct a *critical rejection* region and report corresponding error probabilities. The incorrect rejection of H_0 (the Type I error), has a designated probability α and the incorrect acceptance of H_0 (the Type II error), has a designated probability

β. However, the traditional (α, β)-Frequentist approach has been greatly criticized for reporting these error probabilities independently of the given data. In practice, a common alternative is to compute the test's P-value and report it as a data-dependent measure of the strength of the evidence against the null hypothesis H_0. However, the P-value can be highly misleading as a measure of the evidence provided by the data against the null hypothesis. This was demonstrated by Edwards, Lindman and Savage (1963), Berger and Sellke (1987), Berger and Delampady (1987) and Delampady and Berger (1990) who have reviewed the practicality of the P-value and explored the dramatic conflict between the P-value and other data dependent measures of evidence.

There have been many attempts to rectify these deficiencies and to modify the classical Frequentist approach by incorporating data-dependent procedures which are based on conditioning. Earlier works in this direction are summarized in Kiefer (1977) and in Berger and Wolpert (1988). In a seminal series of papers, Kiefer (1975, 1976, 1977) and Brownie and Kiefer (1977), the Conditional Frequentist approach was formalized. The basic idea behind this approach is to condition on a statistic measuring the evidential strength of the data, and then to provide error probabilities conditional on the observed value of this statistic. Unfortunately, the approach never achieved substantial popularity, in part because of the difficulty of choosing the statistic upon which to condition [cf., the Discussion of Kiefer (1977)].

The Bayesian approach to testing is based on consideration of the most extreme form of conditioning, namely conditioning on the given data. Suppose we observe the realization, x, of the random variable $X \in \mathcal{X}$ from a density $f(x|\theta)$, with θ being an unknown element of the parameter space Θ. In the sequel, we let $P_\theta(\cdot)$ denote conditional probability given $\theta \in \Theta$. Consider the problem of testing simple versus composite hypotheses as given by

$$H_0 : \theta = \theta_0 \text{ versus } H_1 : \theta \in \Theta_1, \tag{14.1}$$

where $\theta_0 \notin \Theta_1 \subset \Theta$. We will consider here primarily the two-sided alternative case with $\Theta_1 = \{\theta \in \Theta : \theta \neq \theta_0\}$.

Within the Bayesian framework for testing the above hypotheses, one usually specifies the *prior probabilities*, π_0 for H_0 being true and $1 - \pi_0$ for H_1 being true, while assigning to Θ_1 the prior density $(1 - \pi_0)\pi(\theta)$, where $\pi \in \Pi$, some class of proper density functions over Θ_1. To simplify the presentation, we assume throughout this paper the default prior probability of $\pi_0 = 1/2$ for the simple hypothesis $H_0 : \theta = \theta_0$. Note that H_0 should typically be considered an approximation to the more realistic hypothesis $H_0 : |\theta - \theta_0| < \epsilon$ for some small ϵ [see Berger and Delampady (1987)].

Let $m_\pi(x)$ and $m_0(x)$ denote the marginal densities of X conditional on H_1 and H_0 being true, respectively; then $m_0(x) = f(x|\theta_0)$ and

$$m_\pi(x) = \int_{\Theta_1} f(x|\theta)\pi(\theta)d\theta. \tag{14.2}$$

For the Classical Frequentist, π might be thought of as a weight function which allows computation of an average likelihood for H_1 [as given by $m_\pi(x)$ in (14.2)]. However, for a Bayesian, π is the prior density for θ conditional on H_1 being true. It follows that, for the Bayesian, the test of (14.1) can be reduced to an equivalent test of two *"simple"* hypotheses concerning the marginal distribution of X, namely

$$H_0: \; X \sim m_0(x) \;\; \text{versus} \;\; H_1: \; X \sim m_\pi(x). \tag{14.3}$$

We denote by

$$B_\pi(x) \equiv \frac{m_0(x)}{m_\pi(x)} \tag{14.4}$$

the *likelihood ratio of H_0 to H_1* (or, equivalently, the *Bayes factor in favor of H_0*). With the default choice of prior probability $\pi_0 = 1/2$ of H_0 being true, the posterior probability, given the data, of H_0 being true is

$$\Pr[H_0|x \equiv \alpha^*(B_\pi(x))] = \frac{B_\pi(x)}{1 + B_\pi(x)} \tag{14.5}$$

and the posterior probability, given the data, of H_1 being true is

$$\Pr[H_1|x \equiv \beta^*(B_\pi(x))] = \frac{1}{1 + B_\pi(x)}. \tag{14.6}$$

For a Bayesian, $B_\pi(x)$ in (14.4) is the Bayes factor in favor of H_0; for the default choice of $\pi_0 = 1/2$, it is also the posterior odds in favor of H_0. Clearly, the decision to either *reject* or *accept* H_0 will depend on the observed value of $B_\pi(x)$, where *small* values of $B_\pi(x)$ correspond to rejection of H_0. In this situation, the standard and optimal Bayesian test, for a Bayesian who assumes the loss to be zero for correct decision and ℓ_i for an incorrect decision when H_i is true $(i = 0, 1)$, can be written as (defining $\ell = \ell_1/\ell_0$)

$$\mathbf{T}_\ell : \begin{cases} \text{if } B_\pi(x) \leq \ell, \text{ reject } H_0 \text{ and report the posterior risk } \ell_0 \, \alpha^*(B_\pi(x)); \\ \text{if } B_\pi(x) > \ell, \text{ accept } H_0 \text{ and report the posterior risk } \ell_1 \, \beta^*(B_\pi(x)). \end{cases}$$

For inference without specific losses, one can take in $\mathbf{T}_\ell$ the default values of $\ell_0 = \ell_1 \equiv 1$. With these default losses, the optimal Bayes test becomes

$$\mathbf{T}_1 : \begin{cases} \text{if } B_\pi(x) \leq 1, \text{ reject } H_0 \text{ and report the posterior probability } \alpha^*(B_\pi(x)); \\ \text{if } B_\pi(x) > 1, \text{ accept } H_0 \text{ and report the posterior probability } \beta^*(B_\pi(x)). \end{cases}$$

$$\tag{14.7}$$

14.2   The New Bayesian-Frequentist Test

The testing of simple versus simple hypotheses was considered by Berger, Brown and Wolpert (1994) who proposed a modification of the Bayesian test above, which includes a structured *no decision region*. The modified Bayesian test was shown to also be a valid Conditional Frequentist test, with error probabilities computed conditionally upon a certain statistic $S(X)$. The surprising aspect of this result is not that both the Bayesian and the Conditional Frequentist might have the same decision rule for rejecting or accepting the null hypothesis (this is not so uncommon), but rather that they will report the same conditional error probabilities (CEP) upon rejecting or accepting. That is, the error probabilities reported by the conditional Frequentist using the proposed conditioning strategy are the same as the posterior probabilities of the relevant errors reported by the Bayesian.

Recently, Berger, Boukai and Wang (1996a) adapted the modified Bayesian test to the composite alternative case as given in (14.1). They have shown that, even in the composite alternative case, the modified test retains its valid Conditional Frequentist interpretations.

Let $\mathcal{B}$ denote the range of $B_\pi(x)$, as x varies over $\mathcal{X}$. We will restrict attention here to the case where $\mathcal{B}$ is an interval that contains 1. Let F_0 and F_1 be the cumulative distributions of $B_\pi(X)$ under m_0 and m_π, respectively. For simplicity, we assume in the following that their inverses, F_0^{-1} and F_1^{-1}, exist over the range $\mathcal{B}$ of $B_\pi(x)$. For any $b \in \mathcal{B}$, define

$$\psi(b) = F_0^{-1}(1 - F_1(b)), \tag{14.8}$$

so that $\psi^{-1}(b) = F_1^{-1}(1 - F_0(b))$, and set

$$r = \min(1, \psi^{-1}(1)) \qquad \text{and} \qquad a = \max(1, \psi(1)). \tag{14.9}$$

The new modified Bayesian test of the hypothesis H_0 versus H_1 in (14.1) [or equivalently (14.3)] is given by

$$\mathbf{T}_1^* : \begin{cases} \text{if } B_\pi(x) \leq r, & \text{reject } H_0 \text{ and report the CEP } \alpha^*(B_\pi(x)); \\ \text{if } r < B_\pi(x) < a, & \text{make no-decision;} \\ \text{if } B_\pi(x) > a, & \text{accept } H_0 \text{ and report the CEP } \beta^*(B_\pi(x)), \end{cases} \tag{14.10}$$

where $\alpha^*(B_\pi)$ and $\beta^*(B_\pi)$ are as given in (14.5) and (14.6), respectively.

The "surprise" already observed in Berger, Brown and Wolpert (1994) and in Berger, Boukai and Wang (1996a) is that $\mathbf{T}_1^*$ is also a conditional Frequentist test, arising from use of the conditioning statistic

$$S(X) = \min\{B_\pi(X), \ \psi^{-1}(B_\pi(X))\}, \tag{14.11}$$

over the domain $\mathcal{X}^* = \{x \in \mathcal{X} : 0 \leq S(x) \leq r\}$ (whose complement is the no-decision region in the test $\mathbf{T}_1^*$).

Thus, the Conditional Frequentist who uses the acceptance and rejection regions in $\mathbf{T}_1^*$, along with the conditioning statistic in (14.11), will report conditional error probabilities upon accepting or rejecting which are in complete agreement with the Bayesian posterior probabilities.

To be more precise, let

$$\alpha(s) \equiv P_{\theta_0}(rejecting\ H_0\ |S(X) = s) \tag{14.12}$$

and

$$\beta(\theta|s) \equiv P_\theta(accepting\ H_0\ |S(X) = s) \tag{14.13}$$

denote the conditional probabilities of Type I and Type II error as reported by the Conditional Frequentist who utilizes $\mathbf{T}_1^*$ along with the conditioning statistic (14.11) in order to test $H_0 : \theta = \theta_0$ versus $H_1 : \theta \in \Theta_1$, as in (14.1).

Theorem 14.2.1 [Berger, Boukai and Wang (1996a)] *For the test $\mathbf{T}_1^*$ of H_0 versus H_1 above, and the conditioning statistic $S(X)$ as given in (14.11), we have:* $\alpha(s) \equiv \alpha^*(B_\pi)$ *and*

$$E^{\pi_1(\theta|s)}[\beta(\theta|s)] \equiv \beta^*(B_\pi), \tag{14.14}$$

where $\pi_1(\theta|s)$ denotes the posterior density of θ conditional on H_1 being true and on the observed value s of $S(X)$.

For a further discussion of this result and the Conditional Frequentist interpretation of this modified Bayesian test, see Berger, Boukai and Wang (1996a).

It is clear from the definitions of a and r that the *no decision region* in $\mathbf{T}_1^*$ disappears whenever $F_0(1) = 1 - F_1(1)$, in which case $r = a = 1$. This can happen in cases with *Likelihood Ratio Symmetry* [for a definition, see Berger, Brown and Wolpert (1994)]. Without the presence of the *no decision region* in $\mathbf{T}_1^*$, it would be the optimal Bayes test $\mathbf{T}_1$ [as presented in (14.7)] for a Bayesian who assumes equal prior probabilities of the hypotheses as well as equal losses for the two incorrect decisions (as we do in this paper). For such a Bayesian, the report of the conditional error probabilities $\alpha^*(B_\pi)$ and $\beta^*(B_\pi)$ (being the posterior probabilities of H_0 and H_1, respectively) has valid Bayesian interpretations. The inclusion of the *no decision region* in the test is primarily intended to eliminate possible conflict between the intuitive interpretation of the conditional error probabilities and the respective decision to either reject or accept the null hypothesis [see Berger, Brown and Wolpert (1994) for discussion]. It should be viewed as the *price* that must be paid to obtain a conditional Frequentist interpretation for the optimal Bayes test.

The notation in (14.2) and (14.4) emphasizes the dependence of the testing procedure $\mathbf{T}_1^*$, and especially of the constants r and a defined in (14.9) (through the marginal density m_π), on the particular choice of a prior density $\pi \in \Pi$ for θ over Θ_1. Moreover, these constants are also greatly affected by the sample size n. These effects, combined, will determine the "size" of the *no decision region* on which the Conditional Frequentist and the Bayesian might report differing error probabilities. The test would be impractical if the size of the "no decision region" is too large.

This Chapter has three goals. One is to study the size of the *no-decision region*, both numerically (for small sample sizes) and asymptotically. It will be shown that the *no decision region* does appear to be largely ignorable. A second goal is to show that, in the studied problems, $r = 1$. Then, $\mathbf{T}_1^*$ simplifies considerably in that the *no decision region* is never involved in rejection of H_0 (assuming that one would not want to consider rejection when $B_\pi > 1$). Since practitioners are (historically) mainly interested in rejection of H_0, this means that the *no decision region* can typically be completely ignored. The final goal of the paper is to show how results from robust Bayesian testing can be employed to yield lower bounds on conditional Frequentist error probabilities over wide classes of π [or, equivalently, over wide classes of conditioning statistics $S(X)$]. This can greatly ease concerns about appropriateness of any particular π [or $S(X)$]. For reasons of space limitations, proofs of the main results presented here are omitted. They can be found in Berger, Boukai and Wang (1996b).

14.3   Normal Testing

Suppose that $X_1, X_2, \ldots, X_n$ are n i.i.d. random variables from a normal distribution having unknown mean θ and known variance σ^2, i.e., $\mathcal{N}(\theta, \sigma^2)$ distribution. We denote by $\bar{X}_n = \sum X_i/n$ their average and recall that $\bar{X}_n \sim \mathcal{N}(\theta, \frac{\sigma^2}{n})$. We denote by $\phi(z)$ and $\Phi(z)$ the density function and the cumulative distribution function, respectively, of the standard normal distribution $\mathcal{N}(0, 1)$, so that

$$\phi(z) = \frac{1}{\sqrt{2\pi}} \exp\left\{-\frac{z^2}{2}\right\} \quad \text{and} \quad \Phi(z) = \int_{-\infty}^{z} \phi(x)\, dx.$$

Given the observed value $\bar{x}_n$ of $\bar{X}_n$, we wish to test, as in (14.1), the hypotheses

$$H_0 : \theta = \theta_0 \quad \text{versus} \quad H_1 : \theta \in \Theta_1, \tag{14.15}$$

where $\theta_0 \notin \Theta_1 \subset \Theta$. We assume that, conditional on H_1 being true, the prior distribution of θ has a continuous density $\pi(\theta)$ over the set Θ_1 ($\pi \in \Pi$).

Let

$$t \equiv \sqrt{n}(\bar{x}_n - \theta_0)/\sigma$$

denote the "standard" statistic for many of the normal testing problems. It is easy to verify that the marginal densities of t, as calculated under H_0 and H_1, are given (respectively) by

$$m_0(t) = \phi(t) \equiv \frac{1}{\sqrt{2\pi}} \exp\left\{-\frac{t^2}{2}\right\} \tag{14.16}$$

and

$$m_\pi(t) = \int_{\Theta_1} \frac{1}{\sqrt{2\pi}} \exp\left\{-\frac{1}{2}(t - \sqrt{n}(\theta - \theta_0)/\sigma)^2\right\} \pi(\theta)d\theta. \tag{14.17}$$

It follows from (14.16) and (14.17) that the problem of testing the hypotheses (14.15) is reduced to that of testing the hypotheses

$$H_0: \ t \sim m_0(t) \quad \text{versus} \quad H_1: \ t \sim m_\pi(t), \tag{14.18}$$

which, according to (14.10), is based on the Bayes factor

$$B_\pi(t) = \frac{\phi(t)}{\int_{\Theta_1} \phi(t - \sqrt{n}(\theta - \theta_0)/\sigma)\pi(\theta)d\theta}. \tag{14.19}$$

The corresponding modified Bayesian testing procedure $\mathbf{T}_1^*$ is given in (14.10) with critical testing values r, a obtained from (14.9).

Ideally, the prior π in (14.16) should be specified, whenever possible, in a completely subjective fashion. However, when this cannot be, it is useful to consider classes of priors that depend on what could be termed *minimal Bayesian input*. Here, we consider (for the normal testing problem at hand) two possible choices for the class of conditional priors π. These classes are the (shifted) conjugate normal class and a uniform class.

To be specific, when $\Theta_1 = \{\theta \neq \theta_0\}$ (as corresponds to the "two-sided" normal test), these classes are

$$\Pi_c = \left\{\pi : \pi(\theta) = \frac{1}{\tau} \phi(\frac{\theta - \theta_1}{\tau}) \text{ for } \theta \neq \theta_0, -\infty < \theta_1 < \infty, \ 0 < \tau < \infty\right\},$$

$$\Pi_u = \{\pi : \pi(\theta) = \mathcal{U}(\theta_0 - \tau, \ \theta_0 + \tau), \ 0 < \tau < \infty\}.$$

The class Π_c of normal priors allows a mean other than θ_0, although the choice $\theta_1 = \theta_0$ is the common choice. The class Π_u of symmetric (about θ_0) uniform priors is of interest not only because they are appealing in their simplicity, but also because they form the extreme points of the class of all symmetric (about θ_0) unimodal distributions. When $\Theta_1 = \{\theta > \theta_0\}$ (as corresponds to the "one-sided" normal test), we will consider the class of one-sided normal priors given by

$$\Pi_+ = \left\{\pi : \pi(\theta) = \frac{2}{\tau} \phi(\frac{\theta - \theta_0}{\tau}) \text{ for } \theta > \theta_0, \ 0 < \tau < \infty\right\}.$$

In all these cases, the prior density π depends (heavily) on the value of τ which has to be specified; this is the required *minimal prior input*. Intuitively, this minimal prior input corresponds to the degree of departure from θ_0 that is anticipated, conditional on H_1 being true. (For Π_c, one must also specify θ_1 but, since the default choice of $\theta_1 = \theta_0$ can always be made, specification of θ_1 is not absolutely necessary.)

To clarify (and thus also to emphasize) the choice of the relevant class of prior densities, we will write throughout this chapter, m_c, B_c, F_c and m_u, B_u, F_u for the marginal density m_π, the Bayes factor B_π and its cumulative distribution F_1 (under H_1), corresponding to the classes Π_c and Π_u, respectively. Similar notation is used for the case of Π_+.

14.4 The "Two-Sided" Normal Test With Shifted Conjugate Prior

We consider the problem of testing the hypotheses (14.15) assuming the shifted conjugate prior for θ. That is, $\pi \in \Pi_c$ over $\Theta_1 \equiv \{\theta \neq \theta_0\}$. Thus, conditional on H_1 being true, θ has the $\mathcal{N}(\theta_1, \tau^2)$ prior distribution. Without loss of generality, we assume throughout that $\theta_0 > \theta_1$.

Since $\pi(\theta)$ is now specified, $m_\pi(t)$ in (14.17) can be computed as

$$m_c(t) = \frac{1}{\sqrt{2\pi}\sqrt{1+\lambda}} \, \exp\left\{\frac{-(t+\sqrt{\lambda}\Delta)^2}{2(1+\lambda)}\right\}, \qquad (14.20)$$

where $\Delta = (\theta_0 - \theta_1)/\tau$ and $\lambda = n\tau^2/\sigma^2$. Thus, with $\pi \in \Pi_c$, $t \sim \mathcal{N}(-\sqrt{\lambda}\Delta, 1+\lambda)$, under H_1. The parameter θ_1 is the conditional prior mean of θ, given H_1 is true. This allows, under H_1, a measurable *shift* of the conditional prior density of θ away from H_0. When $\Delta = 0$, the prior density is symmetric about θ_0. This choice of Δ is often considered as the default choice for applications, and is used in Example 14.4.1 below. Also in Example 14.4.1, the default choice of $\tau^2 = 2\sigma^2$ was made; the resulting $\mathcal{N}(0, 2\sigma^2)$ prior is similar to the Cauchy$(0, \sigma^2)$ default prior recommended by Jeffreys (1961).

Combining (14.20) and (14.15), in (14.19), we obtain a simple expression for the Bayes factor $B_c(t)$ as

$$B_c(t) = \sqrt{1+\lambda} \, \exp\left\{-\frac{\lambda}{2(1+\lambda)}\left(t - \frac{\Delta}{\sqrt{\lambda}}\right)^2 + \frac{\Delta^2}{2}\right\}. \qquad (14.21)$$

Observe that, for fixed $\lambda > 0$ and $\Delta \in \mathbf{R}$, $B_c(t)$ attains its maximum value at $t^* = \Delta/\sqrt{\lambda}$, at which point

$$B_c(t^*) \equiv B_c^* \equiv \sqrt{1+\lambda} \, \exp\left\{\frac{\Delta^2}{2}\right\} > 1.$$

For any $b \in \mathcal{B} = (0, B_c^*)$, let $t_b^{\pm}$ be the two solutions of the equation $B_c(t) = b$. Then, it follows from expression (14.21) for $B_c(t)$ that

$$t_b^{\pm} = \frac{\Delta}{\sqrt{\lambda}} \pm \sqrt{\frac{1+\lambda}{\lambda}} \sqrt{\Delta^2 + \log\left(\frac{1+\lambda}{b^2}\right)}. \qquad (14.22)$$

It is straightforward to verify that, for all $b \in \mathcal{B}$, the marginal cumulative distributions of B_c under m_0 and m_c are given (respectively) by

$$F_0(b) = \Phi(t_b^-) + \Phi(-t_b^+) \qquad (14.23)$$

and

$$F_c(b) = \Phi\left(\frac{t_b^- + \sqrt{\lambda}\Delta}{\sqrt{1+\lambda}}\right) + \Phi\left(\frac{-t_b^+ - \sqrt{\lambda}\Delta}{\sqrt{1+\lambda}}\right). \qquad (14.24)$$

Lemma 14.4.1 *For any $\lambda > 0$, $\Delta \in \mathbf{R}$ and $\pi \in \Pi_c$, we have $F_0(1) \leq 1 - F_c(1)$, so that $\psi(1) > 1$ in (14.9).*

The proof of this result can be found in Berger, Boukai and Wang (1996b); we omit the details here for reasons of space.

By Lemma 14.4.1, $\psi(1) > 1$ so that $r = 1$ and $a = \psi(1)$ in (14.10). Accordingly, when $\pi \in \Pi_c$, the modified Bayesian test of the hypotheses $H_0 : \theta = \theta_0$ versus $H_1 : \theta \neq \theta_0$ becomes

$$\mathbf{T}_1^* : \begin{cases} \text{if } B_c(t) \leq 1, & \text{reject } H_0 \text{ and report the CEP } \alpha^*(B_c(t)); \\ \text{if } 1 < B_c(t) < a, & \text{make no decision}; \\ \text{if } B_c(t) \geq a, & \text{accept } H_0 \text{ and report the CEP } \beta^*(B_c(t)). \end{cases} \qquad (14.25)$$

This is an important simplification because it is natural to even consider rejection only when $B_c \leq 1$ (at least for a default Bayesian), and then the *no decision region* never comes into play.

Unfortunately, there is no explicit expression for a in this case. But it can be computed numerically as the solution of the equation $F_0(a) = 1 - F_c(1)$, which, in view of (14.23) and (14.24), may be written as

$$\Phi(t_a^-) + \Phi(-t_a^+) = \Phi\left(\frac{t_1^+ + \sqrt{\lambda}\Delta}{\sqrt{1+\lambda}}\right) - \Phi\left(\frac{t_1^- + \sqrt{\lambda}\Delta}{\sqrt{1+\lambda}}\right), \qquad (14.26)$$

where $t_a^{\pm}$ and $t_1^{\pm}$ are as defined by (14.22) with $b \equiv a$ and $b \equiv 1$, respectively. It is clear that this solution $a \equiv a(\lambda, \Delta)$ depends on both Δ and on τ and the sample size n, through λ. Table 14.1 below lists several values of a evaluated numerically for different choices of λ and Δ.

Example 14.4.1 Fisher and Van Belle (1993) provide the birth weights in grams of $n = 15$ cases of SIDS (Sudden Infant Death Syndrome) born in King County in 1977:

$$2013 \ 3827 \ 3090 \ 3260 \ 4309$$
$$3374 \ 3544 \ 2835 \ 3487 \ 3289$$
$$3714 \ 2240 \ 2041 \ 3629 \ 3345$$

With the standing assumption of normality and a supposed known standard deviation of $\sigma = 800$ g, we consider the test of

$$H_0 : \theta = 3300 \quad \text{versus} \quad H_1 : \theta \neq 3300;$$

here, 3300 g is the overall average birth weight in King County in 1977 (which can effectively be considered to be known), so that H_0 would correspond to the (believable) hypothesis that SIDS is not related to birth weight. We apply the test (14.25) with $\theta_1 = \theta_0$ (i.e. $\Delta = 0$) and the default choice of $\tau^2 = 2\sigma^2$. From Table 14.1, we find that $a \equiv a(30,0) = 3.051$, and simple calculations yield $t = 0.485$ and $B_c(t) = 4.968$, so that $B_c(t) > a$. Thus, according to $\mathbf{T}_1^*$, we accept H_0 and report the CEP $\beta^*(B_c) = 1/(1 + 4.968) = 0.168$.

Table 14.1: Selected values of $a(\lambda, \Delta)$ for the two-sided test with the conjugate prior

λ	$\mid\Delta\mid=0$	1	2	3	4	5
1	1.317	1.655	1.777	1.793	1.780	1.802
5	1.932	2.506	3.306	3.449	3.483	3.500
10	2.321	2.966	4.206	4.617	4.683	4.710
15	2.576	3.256	4.744	5.442	5.559	5.593
20	2.768	3.471	5.121	6.085	6.272	6.314
25	2.922	3.642	5.407	6.608	6.882	6.936
30	3.051	3.783	5.637	7.046	7.421	7.490
40	3.260	4.010	5.990	7.749	8.343	8.455
50	3.425	4.188	6.257	8.293	9.116	9.287
60	3.563	4.336	6.470	8.732	9.781	10.026
70	3.681	4.462	6.647	9.096	10.362	10.694
80	3.784	4.571	6.798	9.404	10.878	11.305
90	3.876	4.668	6.929	9.671	11.338	11.868
100	3.958	4.756	7.045	9.903	11.754	12.390

One can, alternatively, write the test $\mathbf{T}_1^*$ in terms of the standard statistic, t, as follows:

$$\mathbf{T}_1^* : \begin{cases} \text{if } t \leq t_1^- \text{ or } t \geq t_1^+, & \text{reject } H_0 \text{ and report the CEP } \alpha^*(B_c(t)); \\ \text{if } t_1^- < t < t_a^- \text{ or } t_a^+ < t < t_1^+, & \text{make no decision;} \\ \text{if } t_a^- \leq t \leq t_a^+, & \text{accept } H_0 \text{ and report the CEP } \beta^*(B_c(t)). \end{cases}$$

Figure 14.1 below illustrates the effects of the *"shift"* parameter Δ and of the sample size n (through λ) on the *no decision region* corresponding to the test $\mathbf{T}_1^*$ as expressed in terms of the standard statistic t. Note the symmetry of the regions when $\Delta = 0$ and that the size of the *no decision region* decreases as Δ increases.

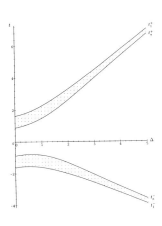
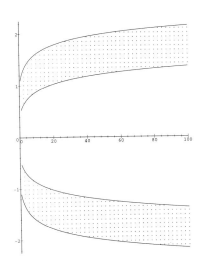

Figure 14.1: Illustration of the no-decision region
(a) with $\Delta = 0$ and (b) with $\lambda = 10$ (fixed)

As we can see in (14.25) and in these figures, the magnitude of $a \equiv \psi(1)$ determines the "size" of the *no decision region*. If this "size" is too large, the provided procedure will not be useful in practice. We first show that, for the testing problem at hand, a grows as $\log n$ for fixed Δ, and is bounded for fixed λ. Recall that $\lambda = n\tau^2/\sigma^2$. Here and henceforth, we will use the notation $f(y) \sim g(y)$, as $y \to \infty$ to mean $\lim_{y\to\infty} f(y)/g(y) = 1$.

Theorem 14.4.1 *Assume $a(\lambda, \Delta)$ satisfies the equation $F_0(a) = 1 - F_c(1)$. Then*

(a) *for each fixed Δ,* $\quad a(\lambda, \Delta) \sim \log \lambda$ *as* $\lambda \to \infty$;

(b) *for each fixed λ,* $\quad \lim_{\Delta \to \infty} a(\lambda, \Delta) = (1 + \lambda)^{\frac{1}{2} \frac{\lambda}{1+\lambda - \sqrt{1+\lambda}}}.$

See Berger, Boukai and Wang (1996b) for a detailed proof. A similar result on the asymptotic behavior of $t_a^{\pm}$ and $t_1^{\pm}$ can be obtained directly from Theorem 14.4.1.

Corollary 14.4.1 *For $\pi \in \Pi_c$ and with fixed Δ, $|t_1|^{\pm} \sim \sqrt{\log(n)}$ and $|t_a|^{\pm} \sim \sqrt{\log(n)}$ as $n \to \infty$.*

While the value of a is growing with the sample size n, so that the no-decision region is growing in size, a much more relevant concern in practice is the probability (under H_0 and H_1) of falling in the no-decision region. The following immediate corollary shows that this probability is vanishingly small asymptotically.

Corollary 14.4.2 *For $\pi \in \Pi_c$ and with fixed Δ,*

$$P_i(\textit{no-decision}) = O(n^{-\frac{1}{2}}(\log(n))^{\frac{1}{2}}), \quad i = 0, 1, \quad as \ n \to \infty.$$

14.5   The "Two-Sided" Test With Uniform Prior

We consider here the same two-sided normal testing problem as in Section 14.4 [i.e. with $\Theta_1 = \{\theta \neq \theta_0\}$ in (14.15)], but now with the uniform prior over Θ_1, so that $\pi \in \Pi_u$. It is straightforward to verify that, with $\pi \in \Pi_u$, the marginal density $m_\pi(t)$ in (14.17) can be written as

$$m_\pi(t) = \frac{1}{\lambda} \left[\Phi(t + \lambda) - \Phi(t - \lambda) \right], \tag{14.27}$$

where we have set $\lambda = \sqrt{n}\tau/\sigma$. From (14.16) and (14.20), the corresponding expression for the Bayes factor B_u, from (14.19), is

$$B_u(t) = 2\lambda \, \phi(t)/[\Phi(t + \lambda) - \Phi(t - \lambda)], \tag{14.28}$$

which has a maximum value $B_u^* \equiv B_u(0) = \sqrt{2/\pi}\lambda \,/[2\Phi(\lambda) - 1]$. The situation here is somewhat more complicated than that of the previous section, since the Bayes factor $B_u(t)$ in (14.28) involves the normal distribution function $\Phi(\cdot)$. To arrive at similar results to those of Section 14.4, we present the following series of Lemmas regarding the Bayes factor B_u. Their proofs are provided in Berger, Boukai and Wang (1996b).

Lemma 14.5.1 *Let $\pi \in \Pi_u$, and $0 < b < B_u^*$ be fixed. Then, there is only one positive solution t_b to the equation $B_u(t_b) = b$, and furthermore, this positive solution t_b satisfies $t_b \sim \sqrt{2 \log \lambda}$ as $\lambda \to \infty$.*

In practice, for all $b > 0$, the value of t_b can be determined numerically as the solution of $B_u(t_b) = b$, which by (14.28) can be written as

$$2\lambda \exp\{-\frac{t_b^2}{2}\} = b \int_{t_b-\lambda}^{t_b+\lambda} \exp\left\{-\frac{s^2}{2}\right\} ds. \tag{14.29}$$

This identity is used repeatedly in the sequel. It is also used to obtain expressions for $F_0(1)$ and $F_u(1)$ as:

$$F_0(1) = 2\Phi(-t_1) \tag{14.30}$$

and

$$F_u(1) = \frac{1}{\lambda} \int_{t_1}^{\infty} [\Phi(t + \lambda) - \Phi(t - \lambda)]dt. \tag{14.31}$$

With the result of Lemma 14.5.1, the two distribution functions (14.30) and (14.31) can be evaluated asymptotically, as summarized in Lemma 14.5.2 below.

Lemma 14.5.2 *If $\pi \in \Pi_u$, then $F_0(1) \sim \frac{1}{\lambda\sqrt{2\log\lambda}}$ and $1 - F_u(1) \sim \frac{\sqrt{2\log\lambda}}{\lambda}$ as $\lambda \to \infty$. Hence, for sufficiently large λ, $F_0(1) < 1 - F_u(1)$, so that $\psi(1) > 1$ in (14.9).*

Consequently, the form of the modified Bayesian test $\mathbf{T}_1^*$, in this case, is exactly as given in (14.25), with the Bayes factor $B_c(t)$ replaced by $B_u(t)$ from (14.28), and with $a \equiv \psi(1)$, as determined by the equation $F_0(a) = 1 - F_u(1)$. In fact, it follows from (14.29)–(14.31) that the numerical values of t_1, t_a and of a (see Table 14.2 below), for each fixed λ, must satisfy the following set of equations:

$$\begin{cases} 2\lambda \exp\{-\frac{t_1^2}{2}\} &= \int_{t_1-\lambda}^{t_1+\lambda} \exp\{-\frac{s^2}{2}\}ds; \\ 2\lambda\Phi(t_a) &= \int_{t_1}^{\infty} [\Phi(t+\lambda) - \Phi(t-\lambda)]dt; \\ a &= \frac{2\lambda \exp\{-\frac{t_a^2}{2}\}}{\int_{t_a-\lambda}^{t_a+\lambda} \exp\{-\frac{s^2}{2}\}ds}. \end{cases}$$

In similarity to the conjugate prior case discussed in Section 14.4, a still has a nice asymptotic behavior, as presented in Theorem 14.5.1 [see Berger, Boukai and Wang (1996b) for a detailed proof.]

Theorem 14.5.1 *For $\pi \in \Pi_u$, we have $a \sim 2\log\lambda$ and*

$$P_i(no\text{-}decision) = O(\lambda^{-1}(2\log(\lambda))^{\frac{1}{2}}), \quad i = 0, 1, \quad as \ \lambda \to \infty.$$

With $\lambda = \sqrt{n}\tau/\sigma$, the results of this theorem should be compared to those of Theorem 14.4.1(a) and Corollary 14.4.2.

Table 14.2: Selected values of a, t_a and t_1 for the two-sided test with a uniform prior

λ	a	t_a	t_1
1	1.131	0.478	1.049
2	1.442	0.623	1.184
3	1.814	0.762	1.360
4	2.178	0.875	1.528
5	2.496	0.969	1.664
10	3.559	1.271	2.038

λ	a	t_a	t_1
20	4.693	1.565	2.354
30	5.379	1.728	2.520
40	5.874	1.840	2.632
50	6.262	1.924	2.715
70	6.853	2.048	2.836
100	7.486	2.175	2.960

14.6   Lower Bound on α^*

A lower bound on the conditional error probability, $\alpha^*(B_\pi)$, over a class Π of priors is clearly of great interest here. It provides a bound on the amount of evidence in support of the null hypothesis (through the posterior probability of H_0) obtained in the modified Bayesian test $\mathbf{T}_1^*$. While the conditional Frequentist might be reluctant to employ a particular prior $\pi \in \Pi$, the lower bound on the error probability is quite compelling.

Theorem 14.6.1 *Let $\pi \in \Pi_c$ and let $B_c(t)$ be as defined by (14.21). Then*

$$\alpha^*(B_c(t)) \equiv \frac{B_c(t)}{1 + B_c(t)} \geq [1 + e^{\frac{t^2}{2}}]^{-1} \quad \forall\, t \in \mathbf{R}.$$

Unfortunately, there is no closed form expression for the lower bound in the uniform prior case. However, it can be computed numerically. In Table 14.3, we provide selected numerical values of the lower bound of α^* over the two classes of priors considered here, for selected values of t along with the corresponding $P-$values. The implications are quite surprising. For instance, a $P-$value of 0.05 corresponds to a conditional Frequentist error-probability of *at least* 0.290 for any $\pi \in \Pi_u$ (and indeed any unimodal prior symmetric about θ_0). Thus, a $P-$value of 0.05 does not seem to be substantial evidence against H_0 from almost any reasonable conditional Frequentist measure.

Table 14.3: Lower bound on $\alpha^*(B_c)$ and $\alpha^*(B_u)$

$P-$Value	t	$\pi \in \Pi_c$	$\pi \in \Pi_u$
0.10	1.645	0.205	0.389
0.05	1.960	0.128	0.290
0.01	2.576	0.035	0.109
0.001	3.291	0.004	0.018

These results on the lower bound are simply "borrowed" from the robust Bayesian literature [cf., Edwards, Lindman and Savage (1963) and Berger and Sellke (1987).] One of the strengths of the dual Bayesian-Frequentist interpretation of $\mathbf{T}_1^*$ is that known Bayesian results can directly be carried over and given a conditional Frequentist interpretation.

14.7 The "One-Sided" Test

We continue with the same testing problem as in Section 14.4, but now with $\Theta_1 = \{\theta > \theta_0\}$ and a prior density $\pi \in \Pi_+$. With this prior density, the marginal density (14.17) of t, conditional on H_1, becomes

$$m_+(t) = \frac{2}{\sqrt{1+\lambda}} \phi\left(\frac{t}{\sqrt{(1+\lambda)}}\right) \Phi\left(\frac{\sqrt{\lambda}t}{\sqrt{1+\lambda}}\right),$$

where, as in Section 14.4, $\lambda = n\tau^2/\sigma^2$. Hence, the corresponding Bayes factor may be written as

$$B_+(t) = \frac{\sqrt{1+\lambda}}{2} \exp\left\{\frac{-\lambda t^2}{2(1+\lambda)}\right\} \left(\Phi\left(\frac{\sqrt{\lambda}t}{\sqrt{1+\lambda}}\right)\right)^{-1}.$$

It can be verified that, in this case also, the *no decision region* is of the form $(1, a)$, where a is determined numerically by the following set of equations:

$$\begin{cases} B_+(t_1) &= 1; \\ 1 - \Phi(t_a) &= 2\int_{-\infty}^{t_1/\sqrt{1+\lambda}} \Phi(\lambda t)\phi(t)dt; \\ a &= B_+(t_a). \end{cases}$$

Thus, the corresponding test $\mathbf{T}_1^*$ (as presented in terms of the standard test statistic t) is

$$\mathbf{T}_1^* : \begin{cases} \text{if } t \geq t_1, & \text{reject } H_0 \text{ and report the CEP } \alpha^*(B_+(t)); \\ \text{if } t_a < t < t_1, & \text{make no decision;} \\ \text{if } t \leq t_a, & \text{accept } H_0 \text{ and report the CEP } \beta^*(B_+(t)). \end{cases}$$

Table 14.4 below presents values of a, t_a and t_1 as were evaluated numerically for selected choices of λ. Note that the *no decision region* is somewhat smaller than that in the two-sided test, unless λ is very small.

Table 14.4: Selected values of a, t_a and t_1 for the one-sided test

λ	a	t_a	t_1
1	1.271	0.183	0.560
2	1.448	0.262	0.731
3	1.580	0.320	0.841
4	1.858	0.367	0.923
5	1.774	0.406	0.987
10	2.084	0.541	1.190

λ	a	t_a	t_1
20	2.436	0.690	1.390
30	2.659	0.781	1.505
40	2.825	0.847	1.584
50	2.956	0.898	1.645
70	3.161	0.976	1.734
100	3.385	1.057	1.825

Acknowledgements. Research was supported, in part, by the National Science Foundation under grant DMS-9303556.

References

1. Berger, J. O., Boukai, B. and Wang, Y. (1996a). Unified frequentist and Bayesian testing of a precise hypothesis, *Statistical Science* (to appear).

2. Berger, J. O., Boukai, B. and Wang, Y. (1996b). Asymptotic properties of modified Bayesian-Frequentist tests, *Preprint Series in Mathematics and Statistics, #PR-96-17*, Department of Mathematical Sciences, Indiana University Purdue University, Indianapolis, IN.

3. Berger, J. O., Brown, L. D. and Wolpert, R. L. (1994). A unified conditional frequentist and Bayesian test for fixed and sequential simple hypothesis testing, *Annals of Statistics*, **22**, 1787–1807.

4. Berger, J. O. and Delampady, M. (1987). Testing precise hypotheses, *Statistical Science*, **3**, 317–352.

5. Berger, J. O. and Sellke, T. (1987). Testing a point null hypothesis: the irreconcilability of *P*-values and evidence, *Journal of the American Statistical Association*, **82**, 112–122.

6. Berger, J. O. and Wolpert, R. L. (1988). *The Likelihood Principle*, Second edition, Institute of Mathematical Statistics, Hayward, CA.

7. Brownie, C. and Kiefer, J. (1977). The ideas of conditional confidence in the simplest setting, *Communications in Statistics—Theory and Methods*, **6**, 691–751.

8. Delampady, M. and Berger, J. O. (1990). Lower bounds on Bayes factors for the multinomial distribution, with application to chi-squared tests of fit, *Annals of Statistics*, **18**, 1295–1316.

9. Edwards, W., Lindman, H. and Savage, L. J. (1963). Bayesian statistical inference for psychological research, *Psychology Review*, **70**, 193–242.

10. Fisher, L. D. and Van Belle, G. (1993). *Biostatistics: A Methodology for the Health Sciences*, New York: John Wiley & Sons.

11. Jeffreys, H. (1961). *Theory of Probability*, London, England: Oxford University Press.

12. Kiefer, J. (1975). Conditional confidence approach in multi decision problems, In *Multivariate Analysis IV* (Ed., P. R. Krishnaiah), New York: Academic Press.

13. Kiefer, J. (1976). Admissibility of conditional confidence procedures, *Annals of Statistics*, **4**, 836–865.

14. Kiefer, J. (1977). Conditional confidence statements and confidence estimators (with discussion), *Journal of the American Statistical Association*, **72**, 789–827.

15. Wolpert, R. L. (1995). Testing simple hypotheses, In *Studies in Classification, Data Analysis, and Knowledge Organization*, Vol. 7 (Eds., H. H. Bock and W. Polasek), pp. 289-297, Heidelberg, Germany: Springer-Verlag.

15

Likelihood Ratio Tests and Intersection-Union Tests

Roger L. Berger

North Carolina State University, Raleigh, NC

Abstract: The likelihood ratio test (LRT) method is a commonly used method of hypothesis test construction. The intersection-union test (IUT) method is a less commonly used method. We will explore some relationships between these two methods. We show that, under some conditions, both methods yield the same test. But, we also describe conditions under which the size-α IUT is uniformly more powerful than the size-α LRT. We illustrate these relationships by considering the problem of testing $H_0 : \min\{|\mu_1|, |\mu_2|\} = 0$ versus $H_a : \min\{|\mu_1|, |\mu_2|\} > 0$, where μ_1 and μ_2 are means of two normal populations.

Keywords and phrases: Likelihood ratio test, intersection-union test, size, power, normal mean, sample size

15.1 Introduction and Notation

The likelihood ratio test (LRT) method is probably the most commonly used method of hypothesis test construction. Another method, which is appropriate when the null hypothesis is expressed as a union of sets, is the intersection-union test (IUT) method. We will explore some relationships between tests that result from these two methods. We will give conditions under which both methods yield the same test. But, we will also give conditions under which the size-α IUT is uniformly more powerful than the size-α LRT.

Let $\boldsymbol{X}$ denote the random vector of data values. Suppose the probability distribution of $\boldsymbol{X}$ depends on an unknown parameter θ. The set of possible values for θ will be denoted by Θ. $L(\theta|\boldsymbol{x})$ will denote the likelihood function for the observed value $\boldsymbol{X} = \boldsymbol{x}$. We will consider the problem of testing the null hypothesis $H_0 : \theta \in \Theta_0$ versus the alternative hypothesis $H_a : \theta \in \Theta_0^c$, where Θ_0 is a specified subset of Θ and Θ_0^c is its complement.

The likelihood ratio test statistic for this problem is defined to be

$$\lambda(\boldsymbol{x}) = \frac{\sup_{\theta \in \Theta_0} L(\theta|\boldsymbol{x})}{\sup_{\theta \in \Theta} L(\theta|\boldsymbol{x})}.$$

A LRT rejects H_0 for small values of $\lambda(\boldsymbol{x})$. That is, the rejection region of a LRT is a set of the form $\{\boldsymbol{x} : \lambda(\boldsymbol{x}) < c\}$, where c is a chosen constant. Typically, c is chosen so that the test is a size-α test. That is, $c = c_\alpha$ is chosen to satisfy

$$\sup_{\theta \in \Theta_0} P_\theta(\lambda(\boldsymbol{X}) < c_\alpha) = \alpha, \tag{15.1}$$

where α is the Type-I error probability chosen by the experimenter.

We will consider problems in which the null hypothesis set is conveniently expressed as a union of k other sets, i.e., $\Theta_0 = \cup_{i=1}^k \Theta_i$. (We will consider only finite unions, although arbitrary unions can also be considered.) Then the hypotheses to be tested can be stated as

$$H_0 : \theta \in \bigcup_{i=1}^k \Theta_i \quad \text{versus} \quad H_a : \theta \in \bigcap_{i=1}^k \Theta_i^c. \tag{15.2}$$

The IUT method is a natural method for constructing a hypothesis test for this kind of problem. Let $R_i, i = 1, \ldots, k$, denote a rejection region for a test of $H_{i0} : \theta \in \Theta_i$ versus $H_{ia} : \theta \in \Theta_i^c$. Then the IUT of H_0 versus H_a, based on $R_1, \ldots, R_k$, is the test with rejection region $R = \cap_{i=1}^k R_i$. The rationale behind an IUT is simple. The overall null hypothesis, $H_0 : \theta \in \cup_{i=1}^k \Theta_i$, can be rejected only if each of the individual hypotheses, $H_{i0} : \theta \in \Theta_i$, can be rejected.

An IUT was described as early as 1952 by Lehmann. Gleser (1973) coined the term IUT. Berger (1982) proposed IUTs for acceptance sampling problems, and Cohen, Gatsonis and Marden (1983a) proposed IUTs for some contingency table problems. Since then, many authors have proposed IUTs for a variety of problems. The IUT method is the reverse of Roy's (1953) well-known union-intersection method, which is useful when the null hypothesis is expressed as an intersection.

Berger (1982) proved the following two theorems about IUTs.

Theorem 15.1.1 *If R_i is a level-α test of H_{0i}, for $i = 1, \ldots, k$, then the IUT with rejection region $R = \cap_{i=1}^k R_i$ is a level-α test of H_0 versus H_a in (15.2).*

An important feature in Theorem 15.1.1 is that each of the individual tests is performed at level-α. But the overall test also has the same level α. There is no need for an adjustment (like Bonferroni) for performing multiple tests. The reason there is no need for such a correction is the special way the individual tests are combined. H_0 is rejected only if every one of the individual hypotheses, H_{0i}, is rejected.

Theorem 15.1.1 asserts that the IUT is level-α. That is, its size is at most α. In fact, a test constructed by the IUT method can be quite conservative. Its size can be much less than the specified value α. But, Theorem 15.1.2 [a generalization of Theorem 2 in Berger (1982)] provides conditions under which the IUT is not conservative; its size is exactly equal to the specified α.

Theorem 15.1.2 *For some $i = 1, \ldots, k$, suppose R_i is a size-α rejection region for testing H_{0i} versus H_{ai}. For every $j = 1, \ldots, k$, $j \neq i$, suppose R_j is a level-α rejection region for testing H_{0j} versus H_{aj}. Suppose there exists a sequence of parameter points $\theta_l, l = 1, 2, \ldots$, in Θ_i such that*

$$\lim_{l \to \infty} P_{\theta_l}(\boldsymbol{X} \in R_i) = \alpha,$$

and, for every $j = 1, \ldots, k$, $j \neq i$,

$$\lim_{l \to \infty} P_{\theta_l}(\boldsymbol{X} \in R_j) = 1.$$

Then, the IUT *with rejection region $R = \bigcap_{i=1}^{k} R_i$ is a size-α test of H_0 versus H_a.*

Note that in Theorem 15.1.2, the one test defined by R_i has size exactly α. The other tests defined by R_j, $j = 1, \ldots, k$, $j \neq i$, are level-α tests. That is, their sizes may be less than α. The conclusion is the IUT has size α. Thus, if rejection regions $R_1, \ldots, R_k$ with sizes $\alpha_1, \ldots, \alpha_k$, respectively, are combined in an IUT and Theorem 15.1.2 is applicable, then the IUT will have size equal to $\max_i\{\alpha_i\}$.

15.2   Relationships Between LRTs and IUTs

For a hypothesis testing problem of the form (15.2), the LRT statistic can be written as

$$\lambda(\boldsymbol{x}) = \frac{\sup_{\theta \in \Theta_0} L(\theta|\boldsymbol{x})}{\sup_{\theta \in \Theta} L(\theta|\boldsymbol{x})} = \frac{\max_{1 \leq i \leq k} \sup_{\theta \in \Theta_i} L(\theta|\boldsymbol{x})}{\sup_{\theta \in \Theta} L(\theta|\boldsymbol{x})} = \max_{1 \leq i \leq k} \frac{\sup_{\theta \in \Theta_i} L(\theta|\boldsymbol{x})}{\sup_{\theta \in \Theta} L(\theta|\boldsymbol{x})}.$$

But,

$$\lambda_i(\boldsymbol{x}) = \frac{\sup_{\theta \in \Theta_i} L(\theta|\boldsymbol{x})}{\sup_{\theta \in \Theta} L(\theta|\boldsymbol{x})}$$

is the LRT statistic for testing $H_{i0} : \theta \in \Theta_i$ versus $H_{ia} : \theta \in \Theta_i^c$. Thus, the LRT statistic for testing H_0 versus H_a is

$$\lambda(\boldsymbol{x}) = \max_{1 \leq i \leq k} \lambda_i(\boldsymbol{x}). \tag{15.3}$$

The LRT of H_0 is a combination of tests for the individual hypotheses, $H_{10}, \ldots,$ H_{k0}. In the LRT, the individual LRT statistics are first combined via (15.3). Then, the critical value, c_α, that yields a size-α test is determined by (15.1).

Another way to combine the individual LRTs is to use the IUT method. For each $i = 1, \ldots, k$, the critical value that defines a size-α LRT of H_{i0} is the value $c_{i\alpha}$ that satisfies

$$\sup_{\theta \in \Theta_{i0}} P_\theta(\lambda_i(\boldsymbol{X}) < c_{i\alpha}) = \alpha. \tag{15.4}$$

Then, $R_i = \{\boldsymbol{x} : \lambda_i(\boldsymbol{x}) < c_{i\alpha}\}$ is the rejection region of the size-α LRT of H_{i0}, and, by Theorem 15.1.1, $R = \cap_{i=1}^k R_i$ is the rejection region of a level-α test of H_0. If the conditions of Theorem 15.1.2 are satisfied, this IUT has size α.

In general, the two methods of combining $\lambda_1(\boldsymbol{x}), \ldots, \lambda_k(\boldsymbol{x})$ need not yield the same test. But, the following theorem gives a common situation in which the two methods do yield the same test. Theorems 15.2.1 and 15.2.2 are similar to Theorems 5 and 6 in Davis (1989).

Theorem 15.2.1 *If the constants $c_{1\alpha}, \ldots, c_{k\alpha}$ defined in (15.4) are all equal and the conditions of Theorem 15.1.2 are satisfied, then the size-α LRT of H_0 is the same as the IUT formed from the individual size-α LRTs of $H_{10}, \ldots, H_{k0}$.*

PROOF. Let $c = c_{1\alpha} = \cdots = c_{k\alpha}$. The rejection region of the IUT is given by

$$R = \bigcap_{i=1}^k \{\boldsymbol{x} : \lambda_i(\boldsymbol{x}) < c_{i\alpha}\} = \bigcap_{i=1}^k \{\boldsymbol{x} : \lambda_i(\boldsymbol{x}) < c\}$$
$$= \{\boldsymbol{x} : \max_{1 \le i \le k} \lambda_i(\boldsymbol{x}) < c\} = \{\boldsymbol{x} : \lambda(\boldsymbol{x}) < c\}.$$

Therefore, R has the form of a LRT rejection region. Because each of the individual LRTs has size-α and the conditions of Theorem 15.1.2 are satisfied, R is the size-α LRT. ∎

Theorem 15.2.1 is particularly useful in situations in which the individual LRT statistics (or a transformation of them) have simple known distributions. In this case, the determination of the critical values, $c_{1\alpha}, \ldots, c_{k\alpha}$, is easy. But the distribution of $\lambda(\boldsymbol{X}) = \max_{1 \le i \le k} \lambda_i(\boldsymbol{X})$ may be difficult, and the determination of its critical value, c_α, from (15.1) may be difficult. Examples of this kind of analysis may be found in Sasabuchi (1980, 1988a,b). In these papers about normal mean vectors, the alternative hypothesis is a polyhedral cone. The individual LRTs are expressed in terms of t-tests, each one representing the LRT corresponding to one face of the cone. All of the t-tests are based on the same degrees of freedom, so all the critical values are equal. Assumptions are made that ensure that the conditions of Theorem 15.1.2 are satisfied, and, in this way, the LRT is expressed as an intersection of t-tests. Sasabuchi does not use the IUT terminology, but it is clear that this is the argument that is used.

Theorem 15.2.1 gives conditions under which, if $c_{1\alpha} = \cdots = c_{k\alpha}$, the size-$\alpha$ LRT and size-α IUT are the same test. But, if the $c_{i\alpha}$s are not all equal, these two tests are not the same, and, often, the IUT is the uniformly more powerful test. Theorem 15.2.2 gives conditions under which this is true.

Theorem 15.2.2 *Let $c_{1\alpha}, \ldots, c_{k\alpha}$ denote the critical values defined in (15.4). Suppose that for some i with $c_{i\alpha} = \min_{1 \leq j \leq k}\{c_{j\alpha}\}$, there exists a sequence of parameter points $\theta_l, l = 1, 2, \ldots$, in Θ_i such that the following three conditions are true:*

(i) $\lim_{l \to \infty} P_{\theta_l}(\lambda_i(\boldsymbol{X}) < c_{i\alpha}) = \alpha$,

(ii) *For any $j \neq i$, $\lim_{l \to \infty} P_{\theta_l}(\lambda_j(\boldsymbol{X}) < c_{i\alpha}) = 1$.*

Then, the following are true:

(a) *The critical value for the size-α LRT is $c_\alpha = c_{i\alpha}$.*

(b) *The IUT with rejection region $R = \cap_{j=1}^k\{\boldsymbol{x} : \lambda_j(\boldsymbol{x}) < c_{j\alpha}\}$ is a size-α test.*

(c) *The IUT in (b) is uniformly more powerful than the size-α LRT.*

PROOF. To prove (a), recall that the LRT rejection region using critical value $c_{i\alpha}$ is

$$\{\boldsymbol{x} : \lambda(\boldsymbol{x}) < c_{i\alpha}\} = \bigcap_{j=1}^k \{\boldsymbol{x} : \lambda_j(\boldsymbol{x}) < c_{i\alpha}\}. \tag{15.5}$$

For each $j = 1, \ldots, k$, because $c_{i\alpha} = \min_{1 \leq j \leq k}\{c_{j\alpha}\}$ and $\{\boldsymbol{x} : \lambda_j(\boldsymbol{x}) < c_{j\alpha}\}$ is a size-α rejection region for testing H_{j0} versus H_{ja}, $\{\boldsymbol{x} : \lambda_j(\boldsymbol{x}) < c_{i\alpha}\}$ is a level-α rejection region for testing H_{j0} versus H_{ja}. Thus, by Theorem 15.1.1, the LRT rejection region in (15.5) is level-α. But, in fact, this LRT rejection region is size-α because

$$\sup_{\theta \in \Theta_0} P_\theta(\lambda(\boldsymbol{X}) < c_{i\alpha}) \geq \lim_{l \to \infty} P_{\theta_l}(\lambda(\boldsymbol{X}) < c_{i\alpha})$$

$$= \lim_{l \to \infty} P_{\theta_l}\left(\bigcap_{j=1}^k \{\lambda_j(\boldsymbol{X}) < c_{i\alpha}\}\right)$$

$$= 1 - \lim_{l \to \infty} P_{\theta_l}\left(\bigcup_{j=1}^k \{\lambda_j(\boldsymbol{X}) < c_{i\alpha}\}^c\right)$$

$$\geq 1 - \lim_{l \to \infty} \sum_{j=1}^k P_{\theta_l}\left(\{\lambda_j(\boldsymbol{X}) < c_{i\alpha}\}^c\right)$$

$$= 1 - (1 - \alpha) = \alpha.$$

The last inequality follows from (i) and (ii).

For each $j = 1, \ldots, k$, $\{\boldsymbol{x} : \lambda_j(\boldsymbol{x}) < c_{j\alpha}\}$ is a level-α rejection region for testing H_{j0} versus H_{ja}. Thus, Theorem 15.1.2, (i) and (iii) allow us to conclude that Part (b) is true.

Because $c_{i\alpha} = \min_{1 \le j \le k}\{c_{j\alpha}\}$, for any $\theta \in \Theta$,

$$P_\theta(\lambda(\boldsymbol{X}) < c_{i\alpha}) = P_\theta\left(\bigcap_{j=1}^k \{\lambda_j(\boldsymbol{X}) < c_{i\alpha}\}\right) \le P_\theta\left(\bigcap_{j=1}^k \{\lambda_j(\boldsymbol{X}) < c_{j\alpha}\}\right).$$

(15.6)

The first probability in (15.6) is the power of the size-α LRT, and the last probability in (15.6) is the power of the IUT. Thus, the IUT is uniformly more powerful. ∎

In Part (c) of Theorem 15.2.2, all that is proved is that the power of the IUT is no less than the power of the LRT. However, if all the $c_{j\alpha}$s are not equal, the rejection region of the LRT is a proper subset of the rejection region of the IUT, and, typically, the IUT is strictly more powerful than the LRT. An example in which the critical values are unequal and the IUT is more powerful than the LRT is discussed in Berger and Sinclair (1984). They consider the problem of testing a null hypothesis that is the union of linear subspaces in a linear model. If the dimensions of the subspaces are unequal, then the critical values from an F-distribution have different degrees of freedom and are unequal.

15.3   Testing $H_0 : \min\{|\mu_1|, |\mu_2|\} = 0$

In this section, we consider an example that illustrates the previous results. We find that the size-α IUT is uniformly more powerful than the size-α LRT. We then describe a different IUT that is much more powerful than both of the preceding tests. This kind of improved power, that can be obtained by judicious use of the IUT method, has been described for other problems by Berger (1989) and Liu and Berger (1995). Saikali (1996) found tests more powerful than the LRT for a one-sided version of the problem we consider in this section.

Let $X_{11}, \ldots, X_{1n_1}$ denote a random sample from a normal population with mean μ_1 and variance σ_1^2. Let $X_{21}, \ldots, X_{2n_2}$ denote an independent random sample from a normal population with mean μ_2 and variance σ_2^2. All four parameters, μ_1, μ_2, σ_1^2, and σ_2^2, are unknown. We will consider the problem of testing the hypotheses

$$H_0 : \mu_1 = 0 \text{ or } \mu_2 = 0 \quad \text{versus} \quad H_a : \mu_1 \ne 0 \text{ and } \mu_2 \ne 0. \tag{15.7}$$

Another way to express these hypotheses is

$$H_0 : \min\{|\mu_1|, |\mu_2|\} = 0 \quad \text{versus} \quad H_a : \min\{|\mu_1|, |\mu_2|\} > 0.$$

The parameters μ_1 and μ_2 could represent the effects of two different treatments. Then, H_0 states that at least one treatment has no effect, and H_a states that both treatments have an effect.

Cohen, Gatsonis and Marden (1983b) considered tests of (15.7) in the variance known case. They proved an optimality property of the LRT in a class of monotone, symmetric tests.

15.3.1 Comparison of LRT and IUT

Standard computations yield that, for $i = 1$ and 2, the LRT statistic for testing $H_{i0} : \mu_i = 0$ is

$$\lambda_i(x_{11}, \ldots, x_{1n_1}, x_{21}, \ldots, x_{2n_2}) = \left(1 + \frac{t_i^2}{n_i - 1}\right)^{-n_i/2},$$

where $\overline{x}_i$ and s_i^2 are the sample mean and variance from the ith sample and

$$t_i = \frac{\overline{x}_i}{s_i/\sqrt{n_i}} \tag{15.8}$$

is the usual t-statistic for testing H_{i0}. Note that λ_i is computed from both samples. But, because the likelihood factors into two parts, one depending only on μ_1, σ_1^2, $\overline{x}_1$ and s_1^2 and the other depending only on μ_2, σ_2^2, $\overline{x}_2$ and s_2^2, the part of the likelihood for the sample not associated with the mean in H_{i0} drops out of the LRT statistic.

Under H_{i0}, t_i has a Student's t distribution. Therefore, the critical value that yields a size-α LRT of H_{i0} is

$$c_{i\alpha} = \left(1 + \frac{t_{\alpha/2,n_i-1}^2}{n_i - 1}\right)^{-n_i/2},$$

where $t_{\alpha/2,n_i-1}$ is the upper $100\alpha/2$ percentile of a Student's t distribution with $n_i - 1$ degrees of freedom. The rejection region of the IUT is the set of sample points for which $\lambda_1(\boldsymbol{x}) < c_{1\alpha}$ and $\lambda_2(\boldsymbol{x}) < c_{2\alpha}$. This is more simply stated as reject H_0 if and only if

$$|t_1| > t_{\alpha/2,n_1-1} \quad \text{and} \quad |t_2| > t_{\alpha/2,n_2-1}. \tag{15.9}$$

Theorem 15.1.2 can be used to verify that the IUT formed from these individual size-α LRTs is a size-α test of H_0. To verify the conditions of Theorem 15.1.2, consider a sequence of parameter points with σ_1^2 and σ_2^2 fixed at any positive values, $\mu_1 = 0$, and let $\mu_2 \to \infty$. Then, $P(\lambda_1(\boldsymbol{x}) < c_{1\alpha}) = P(|t_1| > t_{\alpha/2,n_1-1}) = \alpha$, for any such parameter point. However, $P(\lambda_2(\boldsymbol{x}) < c_{2\alpha}) = P(|t_2| > t_{\alpha/2,n_2-1}) \to 1$ for such a sequence because the power of the t-test converges to 1 as the noncentrality parameter goes to infinity.

If $n_1 = n_2$, then $c_{1\alpha} = c_{2\alpha}$, and, by Theorem 15.2.1, this IUT formed from the individual LRTs is the LRT of H_0.

If the sample sizes are unequal, the constants $c_{1\alpha}$ and $c_{2\alpha}$ will be unequal, and the IUT will not be the LRT. In this case, let $c = \min\{c_{1\alpha}, c_{2\alpha}\}$. By Theorem 15.2.2, c is the critical value that defines a size-α LRT of H_0. The same sequence as in the preceding paragraph can be used to verify the conditions of Theorem 15.2.2, if $c_{1\alpha} < c_{2\alpha}$. If $c_{1\alpha} > c_{2\alpha}$, a sequence with $\mu_2 = 0$ and $\mu_1 \to \infty$ can be used.

If $c = c_{1\alpha} < c_{2\alpha}$, then the LRT rejection region, $\lambda(\boldsymbol{x}) < c$, can be expressed as

$$|t_1| > t_{\alpha/2, n_1 - 1} \quad \text{and} \quad |t_2| > \left\{ \left[\left(1 + \frac{t_{\alpha/2, n_1 - 1}^2}{n_1 - 1} \right)^{n_1/n_2} - 1 \right] (n_2 - 1) \right\}^{1/2}.$$

(15.10)

The cutoff value for $|t_2|$ is larger than $t_{\alpha/2, n_2 - 1}$, because this rejection region is a subset of the IUT rejection region.

The critical values $c_{i\alpha}$ were computed for the three common choices of $\alpha = 0.10, 0.05$, and 0.01, and for all sample sizes $n_i = 2, \ldots, 100$. On this range, it was found that $c_{i\alpha}$ is increasing in n_i. So, at least on this range, $c = \min\{c_{1\alpha}, c_{2\alpha}\}$ is the critical value corresponding to the smaller sample size. This same property was observed by Saikali (1996).

15.3.2   More powerful test

In this section, we describe a test that is uniformly more powerful than both the LRT and the IUT. This test is similar and may be unbiased. The description of this test is similar to tests described by Wang and McDermott (1996).

The more powerful test will be defined in terms of a set, S, a subset of the unit square. S is the union of three sets, S_1, S_2, and S_3, where

$$
\begin{aligned}
S_1 \;=\; & \{(u_1, u_2) : 1 - \alpha/2 < u_1 \le 1, 1 - \alpha/2 < u_2 \le 1\} \\
& \bigcup \{(u_1, u_2) : 0 \le u_1 < \alpha/2, 1 - \alpha/2 < u_2 \le 1\} \\
& \bigcup \{(u_1, u_2) : 1 - \alpha/2 < u_1 \le 1, 0 \le u_2 < \alpha/2\} \\
& \bigcup \{(u_1, u_2) : 0 \le u_1 < \alpha/2, 0 \le u_2 < \alpha/2\},
\end{aligned}
$$

$$
\begin{aligned}
S_2 \;=\; & \{(u_1, u_2) : \alpha/2 \le u_1 \le 1 - \alpha/2, \alpha/2 \le u_2 \le 1 - \alpha/2\} \\
& \bigcap \Big(\{(u_1, u_2) : u_1 - \alpha/4 \le u_2 \le u_1 + \alpha/4\} \\
& \qquad \bigcup \{(u_1, u_2) : 1 - u_1 - \alpha/4 \le u_2 \le 1 - u_1 + \alpha/4\} \Big),
\end{aligned}
$$

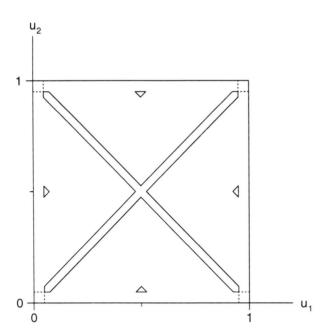

Figure 15.1: The set S for $\alpha = 0.10$. Solid lines are in S, dotted lines are not

and

$$
\begin{aligned}
S_3 \;=\; & \{(u_1,u_2) : \alpha/2 \le u_1 \le 1-\alpha/2, \alpha/2 \le u_2 \le 1-\alpha/2 \\
& \cap \Big(\{(u_1,u_2) : |u_1 - 1/2| + 1 - 3\alpha/4 \le u_2\} \\
& \cup \{(u_1,u_2) : -|u_1 - 1/2| + 3\alpha/4 \ge u_2\} \\
& \cup \{(u_1,u_2) : |u_2 - 1/2| + 1 - 3\alpha/4 \le u_1\} \\
& \cup \{(u_1,u_2) : -|u_2 - 1/2| + 3\alpha/4 \ge u_1\} \Big).
\end{aligned}
$$

The set S for $\alpha = 0.10$ is shown in Figure 15.1. S_1 consists of the four squares in the corners. S_2 is the middle, X-shaped region. S_3 consists of the four small triangles.

The set S has this property. Consider any horizontal or vertical line in the unit square. Then, the total length of all the segments of this line that intersect with S is α. This property implies the following theorem.

Theorem 15.3.1 *Let U_1 and U_2 be independent random variables. Suppose the supports of U_1 and U_2 are both contained in the interval $[0,1]$. If either U_1 or U_2 has a uniform$(0,1)$ distribution, then $P((U_1,U_2) \in S) = \alpha$.*

PROOF. Suppose $U_1 \sim$ uniform$(0,1)$. Let G_2 denote the cumulative distribution function of U_2. Let $S(u_2) = \{u_1 : (u_1,u_2) \in S\}$, for each $0 \le u_2 \le 1$.

Then,

$$P((U_1, U_2) \in S) = \int_0^1 \int_{S(u_2)} 1 \, du_1 \, dG_2(u_2) = \int_0^1 \alpha \, dG_2(u_2) = \alpha.$$

The second equality follows from the property of S mentioned before the theorem.

If $U_2 \sim \text{uniform}(0, 1)$, the result is proved similarly.   ∎

Our new test, which we will call the S-test, of the hypotheses (15.7) can be described as follows. Let F_i, $i = 1, 2$, denote the distribution function of a central t distribution with $n_i - 1$ degrees of freedom. Let $U_i = F_i(t_i)$, $i = 1, 2$, where t_i is the t statistic defined in (15.8). Then, the S-test rejects H_0 if and only if $(U_1, U_2) \in S$.

U_1 and U_2 are independent because t_1 and t_2 are independent. If μ_1 $(\mu_2) = 0$, then $F_1(t_1)$ $(F_2(t_2)) \sim \text{uniform}(0, 1)$, and, by Theorem 15.3.1, $P((U_1, U_2) \in S) = \alpha$. That is, the S-test is a size-α test of H_0. The event $(U_1, U_2) \in S_1$ is the same as the event in (15.9). So, the rejection region of the S-test contains the rejection region of the IUT from the previous section, and the S-test is a size-α test that is uniformly more powerful than the size-α IUT.

We have seen that the IUT is uniformly more powerful than the LRT, and the S-test is uniformly more powerful than the IUT. Table 15.1 gives an example of the differences in power for these three tests. This example is for $n_1 = 5$, $n_2 = 30$ and $\alpha = 0.05$. The table gives the rejection probabilities for some parameter points of the form $(\mu_1, \mu_2) = (r \cos(\theta), r \sin(\theta))$, where $r = 0(0.25)2$ and $\theta = 0(\pi/8)\pi/2$. These are equally spaced points on five lines emanating from the origin in the first quadrant. In Table 15.1, $\sigma_1^2 = \sigma_2^2 = 1$.

The $\theta = 0$ and $\theta = \pi/2$ entries in Table 15.1 are on the μ_1 and μ_2 axes, respectively. For the S-test, the rejection probability is equal to α for all such points. But, the other two tests are biased and their rejection probabilities are much smaller than α for (μ_1, μ_2) close to $(0, 0)$. For the IUT, the power converges to α as the parameter goes to infinity along either axis. For the LRT, this is also true along the μ_2 axis. But, as is suggested by the table, for the LRT

$$\lim_{\mu \to \pm\infty} P(\text{reject } H_0 | \mu_1 = \mu, \mu_2 = 0) = P(|T_{29}| > 2.384) = .024,$$

where T_{29} has a central t distribution with 29 degrees of freedom and 2.384 is the critical value for t_2 from (15.10). The power of the IUT along the μ_i axis is proportional to the power of a univariate, two-sided, size-α t-test of $H_{0i} : \mu_i = 0$. Because the test of H_{01} is based on 4 degrees of freedom while the test of H_{02} is based on 29 degrees of freedom, the power increases more rapidly along the μ_2 axis.

The sections of Table 15.1 for $\theta = \pi/8$, $\pi/4$ and $3\pi/8$ (except for $r = 0$) correspond to points in the alternative hypothesis. There, it can be seen that the

Table 15.1: Powers of S-test, IUT and LRT for $n_1 = 5$, $n_2 = 30$ and $\alpha = 0.05$. Power at parameters of form $(\mu_1, \mu_2) = (r\cos(\theta), r\sin(\theta))$ with $\sigma_1 = \sigma_2 = 1$

	\multicolumn{9}{c}{r}								
	0.00	0.25	0.50	0.75	1.00	1.25	1.50	1.75	2.00
\multicolumn{10}{c}{$\theta = 0$}									
S-test	.050	.050	.050	.050	.050	.050	.050	.050	.050
IUT	.002	.004	.007	.013	.020	.028	.036	.041	.045
LRT	.001	.002	.003	.006	.010	.013	.017	.020	.022
\multicolumn{10}{c}{$\theta = \pi/8$}									
S-test	.050	.051	.066	.117	.224	.384	.567	.731	.849
IUT	.002	.006	.022	.074	.186	.359	.555	.727	.848
LRT	.001	.003	.013	.050	.141	.299	.499	.688	.829
\multicolumn{10}{c}{$\theta = \pi/4$}									
S-test	.050	.052	.076	.137	.227	.329	.440	.554	.663
IUT	.002	.009	.044	.122	.223	.329	.440	.554	.663
LRT	.001	.006	.032	.106	.214	.327	.440	.554	.663
\multicolumn{10}{c}{$\theta = 3\pi/8$}									
S-test	.050	.051	.060	.079	.103	.133	.170	.213	.262
IUT	.002	.012	.043	.076	.103	.133	.170	.213	.262
LRT	.001	.008	.036	.073	.102	.133	.170	.213	.262
\multicolumn{10}{c}{$\theta = \pi/2$}									
S-test	.050	.050	.050	.050	.050	.050	.050	.050	.050
IUT	.002	.013	.038	.049	.050	.050	.050	.050	.050
LRT	.001	.009	.032	.048	.050	.050	.050	.050	.050

S-test has much higher power than the other two tests, especially for parameters close to $(0,0)$. The IUT, which is very intuitive and easy to describe, offers some power improvement over the LRT.

15.4 Conclusion

For a null hypothesis expressed as a union, as in (15.2), the IUT method is a simple, intuitive method of constructing a level-α test. We have described situations in which the IUT defined by size-α LRTs of the individual hypotheses is a uniformly more powerful test than the size-α LRT of the overall hypothesis. And, we have illustrated in an example how even more powerful tests might be found by careful consideration of the specific problem at hand.

Acknowledgements. I thank Yining Wang for helpful suggestions about Theorem 15.2.2.

References

1. Berger, R. L. (1982). Multiparameter hypothesis testing and acceptance sampling, *Technometrics*, **24**, 295–300.

2. Berger, R. L. (1989). Uniformly more powerful tests for hypotheses concerning linear inequalities and normal means, *Journal of the American Statistical Association*, **84**, 192–199.

3. Berger, R. L. and Sinclair, D. F. (1984). Testing hypotheses concerning unions of linear subspaces, *Journal of the American Statistical Association*, **79**, 158–163.

4. Cohen, A., Gatsonis, C. and Marden, J. I. (1983a). Hypothesis testing for marginal probabilities in a $2 \times 2 \times 2$ contingency table with conditional independence, *Journal of the American Statistical Association*, **78**, 920–929.

5. Cohen, A., Gatsonis, C. and Marden, J. I. (1983b). Hypothesis tests and optimality properties in discrete multivariate analysis, In *Studies in Econometrics, Time Series, and Multivariate Statistics* (Eds., S. Karlin, T. Amemiya and L. A. Goodman), pp. 379–405, New York: Academic Press.

6. Davis, L. J. (1989). Intersection union tests for strict collapsibility in three-dimensional contingency tables, *Annals of Statistics*, **17**, 1693–1708.

7. Gleser, L. J. (1973). On a theory of intersection-union tests, *Institute of Mathematical Statistics Bulletin*, **2**, 233 (Abstract).

8. Lehmann, E. L. (1952). Testing multiparameter hypotheses, *Annals of Mathematical Statistics*, **23**, 541–552.

9. Liu, H. and Berger, R. L. (1995). Uniformly more powerful, one-sided tests for hypotheses about linear inequalities, *Annals of Statistics*, **23**, 55–72.

10. Roy, S. N. (1953). On a heuristic method of test construction and its use in multivariate analysis, *Annals of Mathematical Statistics*, **24**, 220–238.

11. Saikali, K. G. (1996). Uniformly more powerful tests for linear inequalities, *Ph.D. Dissertation*, North Carolina State University, Raleigh, NC.

12. Sasabuchi, S. (1980). A test of a multivariate normal mean with composite hypotheses determined by linear inequalities, *Biometrika*, **67**, 429–439.

13. Sasabuchi, S. (1988a). A multivariate test with composite hypotheses determined by linear inequalities when the covariance matrix has an unknown scale factor, *Memoirs of the Faculty of Science, Kyushu University, Series A*, **42**, 9–19.

14. Sasabuchi, S. (1988b). A multivariate one-sided test with composite hypotheses when the covariance matrix is completely unknown, *Memoirs of the Faculty of Science, Kyushu University, Series A*, **42**, 37–46.

15. Wang, Y. and McDermott, M. P. (1996). Construction of uniformly more powerful tests for hypotheses about linear inequalities, *Technical Report 96/05*, Departments of Statistics and Biostatistics, University of Rochester, Rochester, NY.

16

The Large Deviation Principle for Common Statistical Tests Against a Contaminated Normal

N. Rao Chaganty and J. Sethuraman

Old Dominion University, Norfolk, VA
Florida State University, Tallahassee, FL

Abstract: We examine the performance of the standard tests—the mean test, the t-test, the Wilcoxon test and the sign test—for testing that the measure of central tendency of a distribution is zero. We do this by comparing the Bahadur slopes in a contaminated normal model. We first establish the large deviation principle (LDP) and then calculate the Bahadur slopes for the standard test statistics when the observations come from a contaminated normal distribution. An examination of tables of Bahadur efficiencies reveals that the Wilcoxon test outperforms other tests in a neighborhood of the null hypothesis, even in the presence of moderate contamination, but not uniformly over the whole alternative hypothesis.

Keywords and phrases: Bahadur slope, large deviations, Pitman efficiency, robustness, Tukey model, Wilcoxon test

16.1 Introduction

One of the most common testing problems encountered in statistics is testing

$$H_0 : \theta = 0 \qquad vs. \qquad H_1 : \theta > 0$$

where θ is a measure of central tendency. For simplicity, one makes the assumption that the sample forms an i.i.d. sample from a normal distribution with unknown variance. In this case, the t-test is known to be the uniformly most powerful unbiased test. Other tests that have been proposed include the mean test, the sign test, and the Wilcoxon test. Examining the robustness of these tests against departures from this model has been the subject of a large number of papers; see Staudte (1980) and the books by Andrews *et al.* (1972),

Huber (1981) and Tiku, Tan and Balakrishnan (1986), and more recently by
DasGupta (1994). It has been the standard practice to examine the robustness
of these tests in the famous Tukey model [see Tukey (1960)], which models a
certain form of departure from normality. Under the Tukey model, the sample
consists of i.i.d. observations from the density

$$f_{\theta,\epsilon,\sigma}(x) = (1 - \epsilon)\,\phi(x;\theta,1) + \epsilon\,\phi(x;\theta,\sigma). \tag{16.1}$$

Here $\phi(x;\theta,\sigma)$ denotes the probability density function of a normal random
variable with mean θ and standard deviation σ, and $\epsilon \in (0,1)$ represents the
level of contamination.

Two measures which are commonly used to compare the large sample performance of tests are Pitman efficiency and Bahadur efficiency. In Andrews
et al. (1972), Huber (1981) and Lehmann (1983, Chapter 5), robustness was
measured by Pitman efficiency, which is obtainable by comparing asymptotic
efficacies of tests. In this paper, we measure the robustness of these tests by Bahadur efficiency, which is obtainable by comparing Bahadur slopes. We present
some tables showing the Bahadur efficiencies of the Wilcoxon test relative to
other three tests. From an examination of these tables, it appears that the
Wilcoxon test is the best performer in a neighborhood of the null hypothesis, even under the presence of moderate contamination, but is not the best
performer uniformly over the whole region of the alternative hypothesis.

The concept of Bahadur slope can be briefly described as follows. Let
$X_1, \ldots, X_n$ be i.i.d., whose distribution depends on a parameter λ taking values
in a set Λ. The parameter λ can be a vector like $(\theta, \epsilon, \sigma)$ as occurs in our problem. Consider the problem of testing the hypothesis that λ lies in a subset Λ_0 of
Λ. For each n, let T_n be a real valued function of the sample $\{X_1, X_2, \ldots, X_n\}$,
such that large values of T_n are significant for testing the null hypothesis. For
any λ and t, let

$$F_n(t, \lambda) = P_\lambda(T_n < t) \tag{16.2}$$

and

$$G_n(t) = \inf\{F_n(t, \lambda) : \lambda \in \Lambda_0\}. \tag{16.3}$$

If Λ_0 were a singleton, then $F_n(t, \lambda)$ and $G_n(t)$ are equal; otherwise, the significance probability of a test based on T_n is obtained from $G_n(t)$. In fact, the
level attained by T_n is

$$L_n(T_n) = 1 - G_n(T_n). \tag{16.4}$$

The rate at which L_n tends to zero when a non-null λ obtains is a measure of the
discriminating power of the sequence of test statistics $\{T_n\}$ in discriminating

that λ; see Bahadur (1960, 1967, 1971). The sequence of test statistics $\{T_n\}$ is said to have exact slope $c(\lambda)$ when λ obtains if

$$\lim_{n \to \infty} \frac{1}{n} \log L_n(T_n) = -\frac{1}{2} c(\lambda) \quad \text{a.s.} \quad [P_\lambda]. \tag{16.5}$$

It is only the values of $c(\lambda)$ for $\lambda \in \Lambda \setminus \Lambda_0$ that are of interest, with larger values indicating that the alternative hypothesis λ is discriminated better. In general, it is a nontrivial matter to determine the exact slope of a given sequence $\{T_n\}$. A convenient and frequently used method to obtain Bahadur slopes is due to Bahadur (1967) and can be stated in the form of the following theorem.

Theorem 16.1.1 *Suppose that for each* $\lambda \in \Lambda \setminus \Lambda_0$,

$$\lim_{n \to \infty} T_n = b(\lambda) \quad \text{a.s.} \quad [P_\lambda] \tag{16.6}$$

where $-\infty < b(\lambda) < \infty$. *Suppose that for* $\lambda \in \Lambda_0$

$$\lim_{n \to \infty} \frac{1}{n} \log L_n(s) = -I(s) \quad \text{a.s.} \quad [P_\lambda] \tag{16.7}$$

for each s in an open interval which includes $b(\lambda)$, and $I(s)$ is a positive continuous function on that interval. Then, the exact slope of $\{T_n\}$ exists for each $\lambda \in \Lambda \setminus \Lambda_0$ and equals $c(\lambda) = 2I(b(\lambda))$.

In practice, verification of condition (16.6) is easy and usually follows from a strong law of large numbers. However, a large deviation theorem is needed to establish (16.7) and is usually the difficult part. For sums of i.i.d. random variables, one can use Cramér's or Chernoff's theorem [see Theorem 3.1 of Bahadur (1971), for instance] to establish (16.7). For statistics with completely general structure, it is more convenient to use the main theorem of Ellis (1984) to establish (16.7). We will use both these methods in this paper.

We will now briefly give the definition of the large deviation principle and state the main theorem of Ellis (1984).

A function $I(s) : \mathcal{R}^k \to [0, \infty]$ is said to be a *rate function* if it is lower semicontinuous. For any subset A, we write $I(A) = \inf\{I(s) : s \in A\}$. Let $\{\mu_n\}$ be a sequence of probability measures on $(\mathcal{R}^k, \mathcal{B})$. We say that $\{\mu_n\}$ obeys the *large deviation principle* (LDP) with rate function $I(s)$ [see Varadhan (1984)] if the following conditions are satisfied:

$$\limsup_n \frac{1}{n} \log \mu_n(C) \leq -I(C) \tag{16.8}$$

$$\liminf_n \frac{1}{n} \log \mu_n(G) \geq -I(G) \tag{16.9}$$

for all closed sets C and for all open sets G, respectively, of $\mathcal{R}^k$. The rate function $I(s)$ is said to be a *proper rate function* if it further satisfies the condition that the level set $\{s : I(s) \leq L\}$ is a compact subset of $\mathcal{R}^k$, for each $L \geq 0$.

Let $\{T_n\}$ be a sequence of $\mathcal{R}^k$ valued random variables with distribution given by the sequence of probability measures $\{\mu_n\}$. Define

$$c_n(t) = \frac{1}{n} \log \, E[\exp(< t, \, n\,T_n >)]. \qquad (16.10)$$

Suppose that $\lim_n c_n(t)$ exists and is equal to $c(t)$, for all $t \in \mathcal{R}^k$, where we allow both $c_n(t)$ and $c(t)$ to take the value $+\infty$. Let $\mathcal{D}(c) = \{t \in \mathcal{R}^k : c(t) < \infty\}$. The function $c(t) : \mathcal{R}^k \to \mathcal{R}$ is said to be closed if $\{t : c(t) \leq \alpha\}$ is closed for each real α. This is equivalent to $c(t)$ being lower semi-continuous. If $c(t)$ is differentiable on the interior of $\mathcal{D}(c)$, then we call $c(t)$ steep if $\| \, \mathrm{grad}(t_n)) \, \| \to \infty$ for any sequence $\{t_n\} \subset \mathrm{int}\,(\mathcal{D}(c))$ which tends to a boundary point of $\mathcal{D}(c)$. Let

$$I(s) = \sup_{t \in \mathcal{R}^k} \, [< t, \, s > \, - \, c(t)], \qquad (16.11)$$

for $s \in \mathcal{R}^k$ be the Legendre-Fenchel transform of $c(t)$. The main theorem of Ellis (1984) can then be stated as follows.

Theorem 16.1.2 (Ellis): *If $\mathcal{D}(c)$ has a nonempty interior containing the point t=0 and $c(t)$ is a closed convex function of $\mathcal{R}^k$, then the function $I(s)$ defined in (16.11) is a proper rate function on $\mathcal{R}^k$ and the sequence of probability measures $\{\mu_n\}$ satisfies the upper bound (16.8) for all closed sets C of $\mathcal{R}^k$ with proper rate function $I(s)$. Furthermore, if $c(t)$ is differentiable on all of interior of $\mathcal{D}(c)$ and is steep, then the sequence of probability measures $\{\mu_n\}$ satisfies the lower bound (16.9) for all open sets G of $\mathcal{R}^k$ with proper rate function $I(s)$.*

16.2   LDP for Common Statistical Tests

Let $X_1, X_2, \ldots, X_n$ be a random sample from the distribution (16.1). In this section, we will first establish the LDP for the commonly used test statistics for testing the hypothesis $H_0 : \theta = 0$ *vs.* $H_1 : \theta > 0$. The LDP results for the Wilcoxon and t-statistics are new. We do this even though the full force of the LDP is not required to calculate Bahadur slopes.

We will consider four test statistics—the mean test, the t-test, the Wilcoxon test, and the sign test—the last two of which are nonparametric tests.

Mean test.   The test statistic (under the assumption that the population variance is known) for the mean test is $T_{1n} = \overline{X}_n = \frac{1}{n} \sum_{i=1}^{n} X_i$. Under the null hypothesis $H_0 : \theta = 0$, we have

$$c_{1n}(t) = \frac{1}{n} \log E[\exp(n\,t\,T_{1n})]$$

$$= \frac{t^2}{2} + \log \left[(1-\epsilon) + \epsilon \exp\left(\frac{t^2(\sigma^2 - 1)}{2} \right) \right]$$

$$= c_1(t) \quad (\text{say}), \quad -\infty < t < \infty, \tag{16.12}$$

is independent of n. It is easy to verify that the function $c_1(t)$ is a closed convex function on the real line, and satisfies the hypothesis of Theorem 16.1.2. Therefore, T_{1n} obeys the LDP with proper rate function

$$I_1(s) = \sup_{-\infty < t < \infty} [s\,t - c_1(t)]$$

$$= s\,t_s - c_1(t_s) \tag{16.13}$$

where t_s satisfies the equation $s = c_1'(t_s)$, which simplifies to

$$e^{t_s^2/2} (1-\epsilon)(s - t_s) + \epsilon\, e^{\sigma^2\, t_s^2/2} (s - t_s \sigma^2) = 0. \tag{16.14}$$

The above equation (16.14) can be solved numerically using the Newton-Raphson method.

Sign test. Let $Y_i = 1$ if $X_i > 0$ and $Y_i = 0$ if $X_i \leq 0$. The nonparametric sign test is based on the statistic $T_{2n} = \frac{1}{n} \sum_{i=1}^{n} Y_i$. Note that the random variables Y_i's are i.i.d. Bernoulli with mean $1/2$ under the null hypothesis $H_0 : \theta = 0$. It is well known, and can also be alternatively derived from Theorem 16.1.2, that T_{2n} obeys the LDP with proper rate function given by

$$I_2(s) = \begin{cases} \log(2) + s\log(s) + (1-s)\log(1-s) & \text{if } 0 \leq s \leq 1 \\ \infty & \text{otherwise.} \end{cases} \tag{16.15}$$

Wilcoxon test. Arrange $|X_1|, \ldots, |X_n|$ in increasing order and assign ranks. Let U_i be the sign of X_j where $|X_j|$ has rank i. The Wilcoxon statistic is equivalent to

$$T_{3n} = \frac{1}{n(n+1)} \sum_{i=1}^{n} i\, U_i. \tag{16.16}$$

The following theorem generalizes a result of Klotz (1965) and gives the LDP for the Wilcoxon statistic.

Theorem 16.2.1 *Let E_{ni} denote the expected value of the ith smallest order statistic from a sample of n observations with distribution function G on $(0, \infty)$ satisfying $\int_0^\infty x \, dG(x) < \infty$. Let U_i be independent such that $P(U_i = \pm) = 1/2$ for $i = 1, 2, \ldots, n$ and let $S_n = \dfrac{1}{n} \sum_{i=1}^n E_{ni} U_i$. Then, $\{S_n\}$ obeys the LDP with proper rate function given by*

$$I(s) = \sup_t \left[st - \int_0^\infty \log(\cosh(xt)) \, dG(x) \right]. \tag{16.17}$$

PROOF. From Theorem 1 of Hoeffding (1953), we have for each $t \in \mathcal{R}$

$$
\begin{aligned}
c_{3n}(t) &= \frac{1}{n} \log \, E[\exp(n \, t \, S_n)] \\
&= \frac{1}{n} \sum_{i=1}^n \log[\cosh(t \, E_{ni})] \\
&\to \int_0^\infty \log(\cosh(t \, x)) \, dG(x) = c_3(t) \quad \text{(say)} \tag{16.18}
\end{aligned}
$$

as $n \to \infty$. It is easy to check that the function $c_3(t)$ satisfies the conditions of Theorem 16.1.2. Theorem 16.2.1 now follows from Theorem 16.1.2. ∎

Note that under the null hypothesis $\theta = 0$, the random variables $U_1, \ldots, U_n$ in (16.16) are i.i.d. symmetric Bernoulli. Also in (16.16), $E_{ni} = \frac{i}{n+1}$ is the expected value of the ith smallest order statistic from a random sample of n observations distributed uniformly on $(0, 1)$. Therefore, conditions of Theorem 16.2.1 apply with G as the *uniform cdf* on $(0, 1)$. Let

$$c_3(t) = \int_0^1 \log[\cosh(t \, x)] \, dx, \qquad -\infty < t < \infty. \tag{16.19}$$

Using Theorem 16.2.1, we can conclude that T_{3n} obeys the LDP with proper rate function

$$
\begin{aligned}
I_3(s) &= \sup_t \, [s \, t - c_3(t)] \\
&= s \, t_s - c_3(t_s) \tag{16.20}
\end{aligned}
$$

where t_s is the solution of the equation

$$
\begin{aligned}
s &= c_3'(t) \\
&= \int_0^1 x \, \tanh(t \, x) \, dx \\
&= \frac{1}{2} - \frac{\pi^2}{24 \, t^2} + \frac{\log(1 + \exp(-2t))}{t} + \frac{1}{2t^2} \sum_{k=1}^\infty (-1)^{k+1} \frac{\exp(-2tk)}{k^2}.
\end{aligned}
$$

$$\tag{16.21}$$

Equation (16.21) can be solved numerically using the Newton-Raphson method. By substituting an alternate expression for $c_3(t)$ obtainable from (16.19) using integration by parts, we can rewrite (16.20) as

$$I_3(s) = 2s\, t_s - \log(\cosh(t_s)) \tag{16.22}$$

where t_s is the solution of the equation (16.21).

t-test. Let $\overline{X}_n$ and $S_n^2 = \dfrac{1}{n}\sum_{i=1}^{n}(X_i - \overline{X}_n)^2$ be the mean and variance of the sample. The t-statistic is simply defined as $T_{4n} = \overline{X}_n / S_n$. The LDP for the t-statistic does not follow from Theorem 16.1.2. However, we can establish the LDP for the t-statistic using a recent large deviation theorem of Chaganty (1997). Let K_n be distributed as Binomial with parameters n and $(1-\epsilon)$. Let $Z_1, Z_2, \ldots$ be i.i.d. $N(0,1)$ and $Y_1, Y_2, \ldots$ be i.i.d. $N(0,\sigma)$, independent of K_n. Let $\overline{Z}_n$ and $\overline{Y}_n$ be the sample means and let S_{zn}^2 and S_{yn}^2 be the sample variances of a sample of n observations from Z and Y, respectively. Note that T_{4n} is equal in distribution to the statistic

$$\frac{P_n\, \overline{Z}_{nP_n} + (1 - P_n)\, \overline{Y}_{n(1-P_n)}}{\sqrt{P_n\, S_{z\,nP_n}^2 + (1 - P_n)\, S_{y\,n(1-P_n)}^2 + P_n\,(1 - P_n)(\overline{Z}_{nP_n} - \overline{Y}_{n(1-P_n)})^2}}$$

where $P_n = K_n/n$. It is well known that $\overline{Z}_n$ obeys the LDP with proper rate function $h_1(z) = z^2/2$ and $\overline{Y}_n$ obeys the LDP with proper rate function $h_2(y) = y^2/2\sigma^2$. Also, S_{zn}^2 and S_{yn}^2 obey the LDP with proper rate functions $h_3(u) = [u - 1 - \log(u)]/2$ and $h_4(v) = [v/\sigma^2 - 1 - \log(v/\sigma^2)]/2$, respectively. Conditional on $P_n = p$, the variables $\{\overline{Z}_{nP_n}, \overline{Y}_{n(1-P_n)}, S_{z\,nP_n}^2, S_{y\,n(1-P_n)}^2\}$ are all independent. Using Corollary 2.9 in Lynch and Sethuraman (1987) and Example 3.11 in Chaganty (1997), we can see that this conditional joint distribution obeys the LDP continuity condition in p with proper rate function given by $h_1(z) + h_2(y) + h_3(u) + h_4(v)$. See Chaganty (1997) for the definition of the LDP continuity condition and the contraction principle in that connection. It follows from that contraction principle that the conditional distribution of T_{4n} given $P_n = p$ also satisfies the LDP continuity condition in p with proper rate function given by

$$J(p,\, s) = \inf_{(z,y,u,v)\,:\, s = \frac{pz+(1-p)y}{\sqrt{pu+qv+p(1-p)(z-y)^2}}} [h_1(z) + h_2(y) + h_3(u) + h_4(v)]. \tag{16.23}$$

From the LDP for binomial distributions, P_n obeys the LDP with proper rate function given by $h_5(p) = p\log(p/(1-\epsilon)) + (1-p)\log((1-p)/\epsilon)$. It then follows from Theorem 2.3 of Chaganty (1997) that T_{4n} obeys the LDP with proper rate function

$$I_4(s) = \inf_{0 < p < 1} [J(p,\, s) + h_5(p)] \tag{16.24}$$

where $J(p, s)$ is given by (16.23). The above expression (16.24) for $I_4(s)$ is not convenient for computational purposes. However, using a different approach, Chaganty and Sethuraman (1997) have derived the following equivalent form for the rate function

$$I_4(s) = \inf_{0<p<1} \frac{1}{2} \left[p \log(1 + 2at^*) + (1 - p) \log(1 + 2a\sigma^2 t^*) + 2h_5(p) \right],$$
(16.25)

where

$$a = s^2/(1 + s^2)$$

$$t^* = \frac{\sigma^2(p + a - 1) + (a - p) + \sqrt{(\sigma^2(p + a - 1) + a - p)^2 + 4\sigma^2 a(1 - a)}}{4\sigma^2 a(1 - a)}.$$

We use the expression in (16.25) in our calculations.

16.3   Bahadur Slopes and Efficiencies

We now derive the Bahadur slopes of the common test statistics for testing $H_0 : \theta = 0$ vs. $H_1 : \theta > 0$ in the Tukey model, using the results of Section 16.2. These slopes will depend on the alternative hypothesis, i.e., on the vector $\lambda = (\theta, \epsilon, \sigma)$ with $\theta > 0$.

1. The Bahadur slope of the mean test is

$$c_m(\lambda) = 2[\theta\, t_\lambda - c_1(t_\lambda)]$$
(16.26)

   where $c_1(t)$ is defined in (16.12) and t_λ satisfies the equation

$$e^{t_\lambda^2/2}\,(1 - \epsilon)(\theta - t_\lambda) + \epsilon\, e^{\sigma^2 t_\lambda^2/2}\,(\theta - t_\lambda \sigma^2) = 0.$$
(16.27)

2. The Bahadur slope of the sign test is

$$c_s(\lambda) = 2[\log 2 + p_\lambda \log(p_\lambda) + q_\lambda \log(q_\lambda)]$$
(16.28)

   where $p_\lambda = (1 - \epsilon)\, \Phi(\theta) + \epsilon\, \Phi(\theta/\sigma)$, $q_\lambda = 1 - p_\lambda$ and Φ is the cdf of the standard normal distribution.

3. The Bahadur slope of the Wilcoxon test is

$$c_w(\lambda) = 2\,[2\,b(\lambda)\, t_\lambda - \log(\cosh(t_\lambda))]$$
(16.29)

where

$$b(\lambda) = \epsilon^2 \Phi\left(\frac{\sqrt{2}\,\theta}{\sigma}\right) + 2\epsilon(1-\epsilon)\Phi\left(\frac{2\theta}{\sqrt{1+\sigma^2}}\right)$$
$$+ (1-\epsilon)^2 \Phi(\sqrt{2}\theta) - \frac{1}{2} \tag{16.30}$$

and t_λ is the solution of the equation (16.21) with $s = b(\lambda)$.

4. The Bahadur slope of the t-statistic is

$$c_t(\lambda) = 2\, I_4(b(\lambda)) \tag{16.31}$$

where $b(\lambda) = \theta/\sqrt{(1-\epsilon) + \epsilon\sigma^2}$ and $I_4(s)$ is given by (16.24).

The Bahadur efficiencies of the mean test, t-test, and the sign test with respect to the Wilcoxon test are defined as the ratio of the slopes and they are given by $e_\lambda(m, w) = c_m(\lambda)/c_w(\lambda)$, $e_\lambda(t, w) = c_t(\lambda)/c_w(\lambda)$ and $e_\lambda(s, w) = c_s(\lambda)/c_w(\lambda)$, respectively. The Pitman efficiencies of these test statistics can be obtained from more general formulas given in Serfling (1980, p. 321); see also Hodges and Lehmann (1956, 1961). In our problem, the Pitman efficiencies of the mean test and the t-test with respect to the Wilcoxon test are equal and the common value is given by

$$e_{p\lambda}(m, w) = e_{p\lambda}(t, w)$$
$$= \frac{\pi}{3}[(1-\epsilon) + \epsilon\sigma^2]^{-1}\left[(1-\epsilon)^2 + \frac{2\sqrt{2}\epsilon(1-\epsilon)}{\sqrt{1+\sigma^2}} + \frac{\epsilon^2}{\sigma}\right]^{-2} \tag{16.32}$$

whereas the Pitman efficiency of the sign test with respect to the Wilcoxon test is

$$e_{p\lambda}(s, w) = \frac{2}{3}[(1-\epsilon) + \epsilon/\sigma]^2\left[(1-\epsilon)^2 + \frac{2\sqrt{2}\epsilon(1-\epsilon)}{\sqrt{1+\sigma^2}} + \frac{\epsilon^2}{\sigma}\right]^{-2}. \tag{16.33}$$

Following the convention set in Andrews *et al.* (1972), we have set $\sigma = 3$ and computed the Bahadur efficiencies $e_\lambda(m, w)$, $e_\lambda(t, w)$ and $e_\lambda(s, w)$. These efficiencies will depend on the alternative θ, and the level of contamination ϵ. For simplicity in notation, we shall drop the subscript λ and denote them simply as $e(m, w)$, $e(t, w)$ and $e(s, w)$.

Figures 16.1, 16.2 and 16.3 give the surface of the Bahadur efficiencies $e(m, w)$, $e(t, w)$ and $e(s, w)$ as a function of θ and ϵ. Tables 16.1, 16.2 and 16.3 can be used to view the same information by looking at the performances of the mean test, the t-test and the sign test, simultaneously, with respect to the Wilcoxon test for a fixed level of contamination and varying values of θ. From a numerical point of view, the corresponding Pitman efficiencies are given by the restriction of these surfaces to the plane $\theta = 0$. The fact that the limiting Bahadur efficiency as $\theta \to 0$ yields the Pitman efficiency has been established in great generality in Wieand (1976), and we conjecture that it is true in this case also. It is clear that the Bahadur efficiencies of the mean test, t-test and the sign test with respect to the Wilcoxon test is less than 1 in a neighborhood of $\theta = 0$ and $\epsilon = 0$, but not on the whole region of alternatives. This leads us to the conclusion that the Wilcoxon test outperforms the remaining tests in a neighborhood of the null hypothesis even under the presence of moderate contamination.

Acknowledgements. Research partially supported by the U.S. Army research office grant numbers DAAH04-96-1-0070 (first author) and DAAH04-93-G-0201 (second author). The United States Government is authorized to reproduce and distribute reprints for Governmental purposes notwithstanding any copyright notation thereon.

References

1. Andrews, D. F., Bickel, P. J., Hampel, F. R., Huber, P. J., Rogers, W. H. and Tukey, J. W. (1972). *Robust Estimates of Location: Survey and Advances*, Princeton: Princeton University Press.

2. Bahadur, R. R. (1960). Stochastic comparison of tests, *Annals of Mathematical Statistics*, **31**, 276–295.

3. Bahadur, R. R. (1967). Rates of convergence of estimates and test statistics, *Annals of Mathematical Statistics*, **38**, 303–324.

4. Bahadur, R. R. (1971). *Some Limit Theorems in Statistics*, CBMS/NSF Regional Conference in Applied Mathematics–**4**, Philadelphia: SIAM.

5. Chaganty N. R. (1997). Large deviations for joint distributions and statistical applications, *Sankhyā, Series A* (to appear).

6. Chaganty, N. R. and Sethuraman, J. (1997). Bahadur slope of the t-statistic for a contaminated normal, *Statistics & Probability Letters* (to appear).

7. DasGupta, A. (1994). Bounds on asymptotic relative efficiencies of robust estimates of locations for random contaminations, *Journal of Statistical Planning and Inference*, **41**, 73–93.

8. Ellis, R. S. (1984). Large deviations for a general class of random vectors, *Annals of Probability*, **12**, 1–12.

9. Hodges, J. L., Jr. and Lehmann, E. L. (1956). The efficiency of some non-parametric competitors of the *t*-test, *Annals of Mathematical Statistics*, **27**, 324–335.

10. Hodges, J. L., Jr. and Lehmann, E. L. (1961). Comparison of the normal scores and Wilcoxon tests, *Proceedings of the 4th Berkeley Symposium on Mathematical Statistics and Probability*, **1**, 307–317.

11. Hoeffding, W. (1953). On the distribution of the expected values of the order statistics, *Annals of Mathematical Statistics*, **24**, 93–100.

12. Huber, P. J. (1981). *Robust Statistics*, New York: John Wiley & Sons.

13. Klotz, J. (1965). Alternative efficiencies for the signed rank tests, *Annals of Mathematical Statistics*, **36**, 1759–1766.

14. Lehmann, E. L. (1983). *Theory of Point Estimation*, New York: John Wiley & Sons.

15. Lynch, J. and Sethuraman, J. (1987). Large deviations for processes with independent increments, *Annals of Probability,* **15**, 610–627.

16. Serfling, R. J. (1980). *Approximation Theorems of Mathematical Statistics*, New York: John Wiley & Sons.

17. Staudte, R. G., Jr. (1980). Robust estimation, *Queen's Papers in Pure and Applied Mathematics*, No. 53, Queen's University, Kingston, Ontario.

18. Tiku, M. L., Tan, W. Y. and Balakrishnan, N. (1986). *Robust Inference*, New York: Marcel Dekker.

19. Tukey, J. W. (1960). A survey of sampling from contaminated distributions, In *Contributions to Probability and Statistics* (Ed., I. Olkin), Stanford, CA: Stanford University Press.

20. Varadhan, S. R. S. (1984). *Large Deviations and Applications*, CBMS/NSF Regional Conference in Applied Mathematics–**46**, Philadelphia: SIAM.

21. Wieand, H. S. (1976). A condition under which the Pitman and Bahadur approaches to efficiency coincide, *Annals of Statistics*, **4**, 1003–1011.

Table 16.1: Bahadur efficiencies of the mean test, t-test and the sign test with respect to the Wilcoxon test when the level of contamination is 5%

ϵ	θ	$e(m, w)$	$e(t, w)$	$e(s, w)$
0.05	0.000	0.83615	0.83615	0.69638
	0.250	0.84208	0.86082	0.70317
	0.500	0.86294	0.89467	0.72332
	1.000	0.97967	0.94991	0.79873
	1.500	1.23220	1.07476	0.90046
	2.000	1.61877	1.26326	0.98765
	2.500	2.08962	1.46462	1.02746
	3.000	2.59458	1.64439	1.02776

Table 16.2: Bahadur efficiencies of the mean test, t-test and the sign test with respect to the Wilcoxon test when the level of contamination is 10%

ϵ	θ	$e(m, w)$	$e(t, w)$	$e(s, w)$
0.10	0.000	0.72819	0.72819	0.72689
	0.250	0.73488	0.75380	0.73400
	0.500	0.75811	0.81043	0.75505
	1.000	0.87895	0.91614	0.83311
	1.500	1.12328	1.05802	0.93433
	2.000	1.47700	1.24380	1.01025
	2.500	1.67873	1.33772	1.02781
	3.000	2.32384	1.58321	1.02866

Table 16.3: Bahadur efficiencies of the mean test, t-test and the sign test with respect to the Wilcoxon test when the level of contamination is 25%

ϵ	θ	$e(m, w)$	$e(t, w)$	$e(s, w)$
0.25	0.000	0.61885	0.61885	0.82077
	0.250	0.62890	0.63528	0.82785
	0.500	0.65991	0.68437	0.84847
	1.000	0.79132	0.86165	0.91906
	1.500	1.01506	1.06753	0.99199
	2.000	1.30210	1.24270	1.02229
	2.500	1.61679	1.38110	1.00918
	3.000	1.94835	1.49314	0.98289

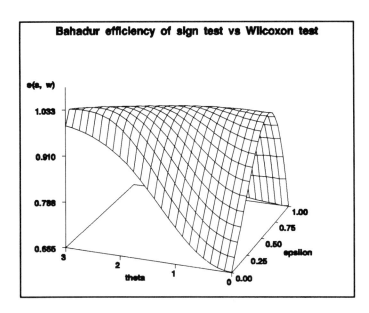

Figure 16.1: Bahadur efficiency of the mean test with respect to the Wilcoxon test

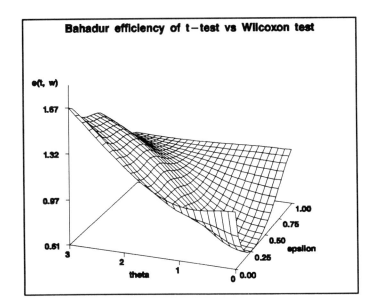

Figure 16.2: Bahadur efficiency of the *t*-test with respect to the Wilcoxon test

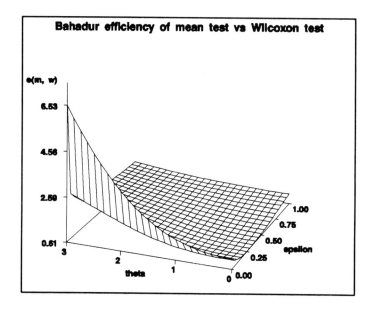

Figure 16.3: Bahadur efficiency of the sign test with respect to the Wilcoxon
test

17

Multiple Decision Procedures for Testing Homogeneity of Normal Means With Unequal Unknown Variances

Deng-Yuan Huang and Ching-Ching Lin

Fu Jen Catholic University, Taipei
National Taiwan Normal University, Taipei

Abstract: In this paper, we propose a multiple decision procedure for testing the homogeneity of normal means when the sample sizes and unknown variances are unequal. When there is a substantial departure from the null hypothesis we reject the null hypothesis and also identify the levels that contributed most towards the departure from homogeneity.

Keywords and phrases: Behrens-Fisher problem, gamma distribution, noncentral chi-square, noncentral F

17.1 Introduction

Gupta, Huang and Panchapakesan (1995) proposed a multiple decision procedure for testing the homogeneity of treatment means (factor levels) in the ANOVA model with the usual assumption that the unknown error variances are equal. In this paper, we consider testing the equality of several normal means when the unknown variances can possibly be unequal. In the case of two normal means, this is the well-known Behrens-Fisher problem.

Let $X_{i1}, \ldots, X_{in_i}$, $i = 1, \ldots, k$, be k independent random samples from the normal populations $N(\mu_i, \sigma_i^2)$, $i = 1, \ldots, k$. We want to test the hypothesis $H_0 : \mu_1 = \cdots = \mu_k$. Let $\theta_i = \mu_i - \frac{1}{k}\sum_{j=1}^{k}\mu_j$ and $\tau_i = \theta_i^2$, $i = 1, \ldots, k$. Obviously, $\theta_1 + \cdots + \theta_k = 0$ and H_0 is equivalent to $\theta_1 = \cdots = \theta_k = 0$. The θ_i are the treatment effects subject to the restriction that $\theta_1 + \cdots + \theta_k = 0$. We move away from homogeneity when two or more θ_i's move away from zero in either direction and such treatments are said to be significant. When H_0 is rejected by our test procedure, we want to identify treatments that contribute substantially to departure from homogeneity. The treatment associated with the largest τ_i is the most significant in the sense that its effect is the farthest from the average of all treatment effects which is zero. Motivated by these considerations, we wish to test $H_0 : \tau_1 = \cdots = \tau_k = 0$ (which is equivalent to $\theta_1 = \cdots = \theta_k = 0$) at significance level α. When H_0 is false, we want to reject H_0 and select a non-empty subset (preferably of a small size) of the k treatments which includes the treatment associated with the largest τ_i; if this happens we say that a *correct decision* (CD) occurs.

To state our formulation of the problem precisely, let $\boldsymbol{\tau} = (\tau_1, \ldots, \tau_k)$ and denote the ordered τ_i by $\tau_{[1]} \leq \cdots \leq \tau_{[k]}$. Define, for $\boldsymbol{\sigma} = (\sigma_1, \ldots, \sigma_k)$,

$$A_i(\sigma_1, \ldots, \sigma_k) = \frac{(k-1)\sigma_i^2}{n_i k^2} + \frac{1}{k^2} \sum_{i \neq j=1}^{k} \frac{\sigma_j^2}{n_j},$$

$$i = 1, \ldots, k,$$

$$A_{i0}(\sigma_1, \ldots, \sigma_k) = \frac{(k-1)\sigma_i^2}{2k^2} + \frac{1}{k^2} \sum_{i \neq j=1}^{k} \frac{\sigma_j^2}{2}, \tag{17.1}$$

and, for $\Delta > 0$,

$$\Omega_\Delta = \{(\boldsymbol{\mu}, \boldsymbol{\sigma}) \mid \tau_{[k]} \geq \Delta A_{i0}(\sigma_1, \ldots, \sigma_k) \text{ for } i = 1, \ldots, k\}. \tag{17.2}$$

For any valid test procedure R, we require that

$$\Pr\{\text{Reject } H_0 \mid \tau_1 = \cdots = \tau_k = 0\} \leq \alpha \tag{17.3}$$

and

$$\Pr\{\text{CD} \mid R\} \geq P^* \text{ whenever } (\boldsymbol{\mu}, \boldsymbol{\sigma}) \in \Omega_\Delta. \tag{17.4}$$

Here Δ and P^* are specified in advance of the experiment, and for a meaningful problem, $\frac{1}{k} < P^* < 1$. The requirement (17.3) is to satisfy the bound on the probability of a type I error. The requirement (17.4) will be referred to as the *P*-condition* and it controls $\Pr\{\text{CD} \mid R\}$ as an operating characteristic in a specified part of the alternative region.

In Section 17.2, we consider some preliminary results on related distribution theory. Section 17.3 proposes a test procedure R and obtains the critical value for the test and sample sizes necessary to satisfy the requirements (17.3) and (17.4). An illustrative example is discussed in Section 17.4.

17.2 Some Preliminary Results

Let $\overline{X}_i$ and s_i^2 be the sample mean and sample variance (divisor $n_i - 1$), respectively, for the sample of size n_i from $N(\mu_i, \sigma_i^2)$, $i = 1, \ldots, k$. Then $\hat{\theta}_i = \overline{X}_i - \frac{1}{k}\sum_{k=1}^{k} \overline{X}_j$ is an unbiased estimator of θ_i and $\mathrm{Var}(\hat{\theta}_i) = A_i(\sigma_1, \ldots, \sigma_k)$, which is defined in (17.1). An unbiased estimator of $\mathrm{Var}(\hat{\theta}_i)$ is given by $\widehat{\mathrm{Var}}(\hat{\theta}_i) = A_i(s_1, \ldots, s_k)$.

Define $Y_i = \frac{\hat{\theta}_i^2}{\mathrm{Var}(\hat{\theta}_i)}$ and $Z_i = \frac{\widehat{\mathrm{Var}}(\hat{\theta}_i)}{\mathrm{Var}(\hat{\theta}_i)}$, $i = 1, \ldots, k$. Since $\hat{\theta}_i$ is $N(\theta_i, \mathrm{Var}(\hat{\theta}_i))$, then Y_i has a noncentral chi-square distribution χ^2_{1,λ_i} with 1 degree of freedom and noncentrality parameter $\lambda_i = \theta_i^2/A_i(\sigma_1, \ldots, \sigma_k)$. As regards the distribution of Z_i, we note that Z_i can be written in the form: $Z_i = \sum_{j=1}^{k} a_{ij} W_j$, where, for $i = 1, \ldots, k$,

$$a_{ij} = \frac{\sigma_j^2}{n_j} \bigg/ \left[\frac{(k-1)^2 \sigma_i^2}{n_i} + \sum_{i \neq j=1}^{k} \frac{\sigma_j^2}{n_j} \right], \quad j \neq i,$$

$$a_{ii} = \frac{(k-1)\sigma_i^2}{n_i} \bigg/ \left[\frac{(k-1)^2 \sigma_i^2}{n_i} + \sum_{i \neq j=1}^{k} \frac{\sigma_j^2}{n_j} \right],$$

$$W_i = \frac{s_i^2}{\sigma_i^2}.$$

Now, letting $f_i = n_i - 1$, $(i = 1, \ldots, k)$, W_i has the gamma distribution, $\mathrm{Gamma}\left(\frac{f_i}{2}, \frac{f_i}{2}\right)$, where $\mathrm{Gamma}(\alpha, \nu)$ denotes the gamma distribution with scale parameter α^{-1} and shape parameter ν. Thus Z_i is a linear combination of independent gamma random variables. We approximate the distribution of Z_i by a gamma distribution with same first and second moments. Towards this end, we use the following lemmas.

Lemma 17.2.1 *Let W_i, $i = 1, \ldots, k$, be independent $\mathrm{Gamma}\left(\frac{f_i}{2}, \frac{f_i}{2}\right)$ variables. Let $c_1, \ldots, c_k$ be positive constants such that $c_1 + \cdots + c_k = 1$. Further, let $\frac{1}{f^*} = \frac{c_1^2}{f_1} + \cdots + \frac{c_k^2}{f_k}$. Then $c_1 W_1 + \cdots + c_k W_k$ has the same first and second moments as W which has $\mathrm{Gamma}\left(\frac{f^*}{2}, \frac{f^*}{2}\right)$ distribution.*

The proof of this lemma is straightforward and is omitted.

Lemma 17.2.2 *Given $f_1, \ldots, f_k$ and $c_1, \ldots, c_k$, sets of positive numbers, such that $c_1 + \cdots + c_k = 1$ and $\frac{1}{f^*} = \frac{c_1^2}{f_1} + \cdots + \frac{c_k^2}{f_k}$, the minimum and the maximum values of f^* are, respectively, $f_{\min} = \min\{f_1, \ldots, f_k\}$ and $f_{\mathrm{tot}} = f_1 + \cdots + f_k$.*

PROOF. By Cauchy's Inequality, $\left(\sum_{i=1}^{k} c_i\right)^2 \leq \left(\sum_{i=1}^{k} \frac{c_i^2}{f_i}\right)\left(\sum_{i=1}^{k} f_i\right)$. This gives $f^* \leq f_{\text{tot}}$. Since $\frac{1}{f^*} \leq \frac{(c_1+\cdots+c_k)^2}{f_{\min}} = \frac{1}{f_{\min}}$, we have $f^* \geq f_{\min}$.

By using Lemmas 17.2.1 and 17.2.2, we approximate the distribution of Z_i by $\text{Gamma}\left(\frac{f_i^*}{2}, \frac{f_i^*}{2}\right)$, where $\frac{1}{f_i^*} = \frac{a_{i1}^2}{f_1} + \cdots + \frac{a_{ik}^2}{f_k}$ and $f_{\min} \leq f_i^* \leq f_{\text{tot}}$. ∎

We now state as a lemma a well-known theorem of Basu (1955). This will be used to prove the independence of Y_i and Z_i, $i = 1, \ldots, k$.

Lemma 17.2.3 *(Basu, 1955). Let X be a random vector whose distribution depends on θ. Suppose that T, based on X, is a complete sufficient statistic for θ and that Y, based on X has a distribution which does not depend on θ. Then T and Y are stochastically independent.*

Since $\hat{\theta}_i$ is a complete sufficient statistic for θ_i and Z_i has a distribution which does not depend on θ, then by Lemma 2.3, Y_i and Z_i are independent. Since Y_i has a χ^2_{1,λ_i} distribution and Z_i is approximately distributed as $\text{Gamma}\left(\frac{f_i^*}{2}, \frac{f_i^*}{2}\right)$ which is (central) $\chi^2_{f_i^*}/f_i^*$, we conclude that Y_i/Z_i is approximately distributed as a noncentral F variable with 1 and f_i^* degrees of freedom and noncentrality parameter λ_i denoted by F_{1,f_i^*,λ_i}. Summarizing the above discussion, we state the following lemma.

Lemma 17.2.4 $Y_i = \hat{\theta}_i^2/Var(\hat{\theta}_i)$ *and* $Z_i = \widehat{Var}(\hat{\theta}_i)/Var(\hat{\theta}_i)$ *are independent. The distribution of Y_i/Z_i is approximately noncentral F with 1 and f_i^* degrees of freedom and noncentrality parameter λ_i, where*

$\lambda_i = \theta_i^2 \Big/ \left[\frac{(k-1)\sigma_i^2}{k^2 n_i} + \frac{1}{k^2}\sum_{i\neq j=1}^{k} \frac{\sigma_j^2}{n_j}\right]$, $f_i^* \in [f_{\min}, f_{\text{tot}}]$, $f_{\min} = \min\{f_1, \ldots, f_k\}$, $f_{\text{tot}} = f_1 + \cdots + f_k$, $f_i = n_i - 1$, $i = 1, \ldots, k$.

Remark 17.2.1 When $\frac{a_{i1}}{f_1} = \cdots = \frac{a_{ik}}{f_k}$, the distribution of Z_i is approximated by $\text{Gamma}\left(\frac{f_1+\cdots+f_k}{2}, \frac{f_1+\cdots+f_k}{2}\right)$. For $f_i \geq 4$, $i = 1, \ldots, k$; $k = 2, 3, 4$, based on simulation with $10,000$ repetitions using SAS RANGAM, it was found that the maximum absolute error is about 0.029. It should also be pointed out that when $k = 2$ and $\frac{a_{i1}}{f_1} = \frac{a_{i2}}{f_2}$ $\left(\text{i.e. } \frac{\sigma_1^2}{n_1(n_1-1)} = \frac{\sigma_2^2}{n_2(n_2-1)}\right)$, we get $\frac{\hat{\theta}_1^2}{\widehat{Var}(\hat{\theta}_1)} = \frac{\hat{\theta}_2^2}{\widehat{Var}(\hat{\theta}_2)} = \frac{(\overline{X}_1-\overline{X}_2)^2}{(s_1^2/n_1)+(s_2^2/n_2)}$ which has exactly $F_{1,f_1+f_2,\lambda}$ distribution, where $\lambda = \lambda_1 = \lambda_2$. This special case is Welch t-test [see Welch (1937, 1947) and Lee (1992)].

Finally, we state two lemmas, one regarding an approximation for noncentral F distribution and another regarding the stochastic ordering for central and noncentral F. Lemma 17.2.5 is due to Huang (1996) and, for Lemma 17.2.6, see Graybill (1976, p. 130).

Lemma 17.2.5 *If U has the noncentral F distribution $F_{u,v,\Delta}$, then*

$$Pr\{U \leq y\} \approx [1 + \exp(-1.794148x)]^{-1}$$

where

$$x = \frac{\left(1 - \frac{2}{9v}\right)\left(\frac{uy}{u+\Delta}\right)^{1/3} - 1 + \frac{2(u+2\Delta)}{9(u+\Delta)^2}}{\sqrt{\frac{2(u+2\Delta)}{9(u+\Delta)^2} + \frac{2}{9v}\left(\frac{uy}{u+\Delta}\right)^{2/3}}}.$$

When $u \geq 1$ and $v \geq 5$, the maximum absolute error is 0.05.

Lemma 17.2.6

1. *The distribution function $F_{1,f,0}(x)$ is increasing in f for a fixed x.*

2. *The distribution function $F_{1,f,\lambda}(x)$ is decreasing in λ for fixed f and x.*

17.3   Decision Rule R

First, we define $SN_i = 10 \log_{10}[\hat{\theta}_i^2/\widehat{\mathrm{Var}}(\hat{\theta}_i)]$. For testing $H_0 : \theta_1 = \ldots = \theta_k = 0$ as formulated in Section 1, we propose the following procedure R.

R: If $SN_i \geq C$ for some $i \in \{1, \ldots, k\}$, reject H_0 and include in the selected subset of significant populations all those for which $SN_i \geq C$.

The constant C and the sample sizes $n_1, \ldots, n_k$ must be chosen so that the conditions (17.3) and (17.4) are satisfied.

17.3.1   Pr{Type I error}

Recalling that $\tau_i = \theta_i^2$ and $\tau_{[1]} \leq \ldots \leq \tau_{[k]}$ are the ordered τ_i,

Pr{Type I error}

$$= Pr\{\max_{1 \leq i \leq k} SN_i \geq C \mid \tau_{[k]} = 0\}$$

$$= Pr\{\max_{1 \leq i \leq k} \frac{\hat{\theta}_i^2}{\widehat{\mathrm{Var}}(\hat{\theta}_i)} \geq C' \mid \tau_{[k]} = 0\}, \text{ where } C' = 10^{C/10}$$

$$\begin{cases} \leq \sum_{i=1}^k Pr\left\{\frac{\hat{\theta}_i^2}{\widehat{\mathrm{Var}}(\hat{\theta}_i)} \geq C' \mid \theta_i^2 = 0\right\} & \text{for } k \geq 3 \\ = Pr\left\{\frac{\hat{\theta}_i^2}{\widehat{\mathrm{Var}}(\hat{\theta}_1)} \geq C' \mid \theta_1^2 = 0\right\} & \text{for } k = 2 \text{ (since } SN_1 = SN_2) \end{cases}$$

$$\approx \begin{cases} \sum_{i=1}^{k} \Pr\left\{F_{1,f_i^*,0} \geq C'\right\} & \text{for } k \geq 3 \\ \\ \Pr\left\{F_{1,f_1^*,0} \geq C'\right\} & \text{for } k = 2 \end{cases} \quad \text{by Lemma 17.2.4}$$

$$\leq \begin{cases} k \Pr\{F_{1,f_{\min},0} \geq C'\} & \text{for } k \geq 3 \\ \Pr\{F_{1,f_{\min},0} \geq C'\} & \text{for } k = 2 \end{cases}$$

by Lemma 17.2.6 [part (1)]. Thus, condition (17.3) is satisfied if

$$\Pr\{F_{1,f_{\min},0} \geq C'\} = \alpha^*, \tag{17.5}$$

where $\alpha^* = \frac{\alpha}{k}$ for $k \geq 3$ and $\alpha^* = \alpha$ for $k = 2$. By using the approximation for the noncentral F distribution given in Lemma 17.2.5 with $\Delta = 0$, we get an approximate solution for C', given by

$$1 - \{1 + e^{-bx}\}^{-1} = \alpha^*, \tag{17.6}$$

where $b = 1.794148$ and

$$x = \frac{\left(1 - \frac{2}{9f_{\min}}\right) C'^{1/3} - \frac{7}{9}}{\sqrt{\frac{2}{9}\left(1 + \frac{C'^{2/3}}{f_{\min}}\right)}}. \tag{17.7}$$

From (17.6), we get

$$x = \frac{1}{b} \ln\left(\frac{1}{\alpha^*} - 1\right) = A, \text{ say.} \tag{17.8}$$

Now, equating the right-hand sides of (17.7) and (17.8), we can obtain a quadratic equation in $C'^{1/3}$, the positive square-root of which yields

$$C' = \left\{ \frac{\frac{7}{9}\left(1 - \frac{2}{9f_{\min}}\right) + A\sqrt{\frac{2}{9}\left(1 - \frac{2}{9f_{\min}}\right)^2 + \frac{98}{729f_{\min}} - \frac{4A^2}{81f_{\min}}}}{\left(1 - \frac{2}{9f_{\min}}\right)^2 - \frac{2A^2}{9f_{\min}}} \right\}^3. \tag{17.9}$$

17.3.2   Pr{ CD| R} and its infimum over Ω_Δ

Let $\hat{\theta}_{(k)}^2$ be the statistic associated with $\tau_{[k]}$. Let $\sigma_{(k)}^2$ denote the variance of the corresponding population and let its estimator be denoted by $s_{(k)}^2$. Then, by Lemma 2.4, $\hat{\theta}_{(k)}^2/\widehat{\text{Var}}(\hat{\theta}_{(k)})$ has approximately $F_{1,f_{(k)}^*,\lambda_{(k)}}$ distribution, where $f_{(k)}^*$ and $\lambda_{(k)}$ are the f_i^* and λ_i associated with $\hat{\theta}_{(k)}^2$. Obviously, $f_{(k)}^* \in [f_{\min}, f_{\text{tot}}]$. Finally, let $SN_{(k)}$ denote the SN_i associated with $\hat{\theta}_{(k)}^2$. Then

$$\begin{aligned} \Pr\{CD \mid R\} &= \Pr\{SN_{(k)} \geq C\} \\ &= \Pr\{\hat{\theta}_{(k)}^2/\widehat{\text{Var}}(\hat{\theta}_{(k)}) \geq C'\}, \text{ where } C' = 10^{C/10} \\ &\approx \Pr\{F_{1,f_{(k)}^*,\lambda_{(k)}} \geq C'\}. \end{aligned}$$

In order to satisfy (17.4), we need to evaluate the infimum of $\Pr\{F_{1,f^*_{(k)},\lambda_{(k)}} \geq C'\}$ over Ω_Δ given in (17.2). We note that $(\boldsymbol{\mu}, \boldsymbol{\sigma}) \in \Omega_\Delta$ implies that $\lambda_i \geq \Delta$ for $i = 1, \ldots, k$. By part (2) of Lemma 17.2.6, we get

$$\inf_{\Omega_\Delta} \Pr\{CD \mid R\} \approx \Pr\{F_{1,f^*_{(k)},\Delta} \geq C'\} = G(f^*_{(k)}), \tag{17.10}$$

where

$$G(f) = \left[1 + \exp\left[\frac{b\left\{\left(1 - \frac{2}{9f}\right)D - 1 + E\right\}}{\sqrt{E + \frac{2}{9f}D^2}}\right]\right]^{-1}, \tag{17.11}$$

$D = \left(\frac{C'}{1+\Delta}\right)^{1/3}$, and $E = \frac{2(1+2\Delta)}{9(1+\Delta)^2}$.

Now, the solution of $\frac{d}{df}G(f) = 0$ is:

$$f_0 = \left[9\left\{\frac{D(1-E) - 2E}{2D^2} - \frac{1}{2}\right\}\right]^{-1}. \tag{17.12}$$

It is easy to verify that:

1. If $f_0 < 0$, then for all $f > 0$, $\frac{d}{df}G(f) < 0$, and

2. If $f_0 > 0$, then $\frac{d}{df}G(f)$ is negative for $0 < f < f_0$ and positive for $f > f_0$.

Define

$$f_0^* = \begin{cases} f_{\text{tot}} & \text{if } f_0 \leq 0 \text{ or } f_0 \geq f_{\text{tot}}, \\ \max(f_0, f_{\text{min}}) & \text{if } 0 < f_0 < f_{\text{tot}}. \end{cases} \tag{17.13}$$

If $f_0 \leq 0$, then by (17.13), $f_0^* \geq f_i^*$ for $i = 1, \ldots, k$. Hence, $G((f_0^*) \geq \inf_{\Omega_\Delta} G(f_i^*)$. On the other hand, when $f_0 > 0$, three cases arise; namely, (i) $f_0 \leq f_{\text{min}}$, (ii) $f_{\text{min}} < f_0 < f_{\text{tot}}$, and (iii) $f_0 \geq f_{\text{tot}}$. When $0 < f_0 \leq f_{\text{min}}$, we have $f_0^* = f_{\text{min}} \leq f_i^*$ for all i. Hence, $G(f_0^*) \leq G(f_i^*)$ for all i. When $f_{\text{min}} < f_0 < f_{\text{tot}}$, we have $f_0^* = f_0$. This implies that $G(f_0^*) \leq G(f_i^*)$ for all i. Finally, when $f_0 \geq f_{\text{tot}}$, we have $f_0^* = f_{\text{tot}} \geq f_i^*$ for all i. Hence, $G(f_0^*) \leq G(f_i^*)$ for all i. Thus, in all cases, $G(f_0^*) \leq G(f_i^*)$ for $i = 1, \ldots, k$. Using this result in (17.10), we obtain

$$\inf_{\Omega_\Delta} \Pr\{CD|R\} \geq G(f_0^*). \tag{17.14}$$

Consequently, the P^*-condition is satisfied if

$$G(f_0^*) \geq P^*. \tag{17.15}$$

The results established in this section regarding the decision rule R are now stated as a theorem below.

Theorem 17.3.1

1. *Given $\frac{1}{k} < P^* < 1$ and $0 < \alpha < 1$, the critical value $C \approx 10\log_{10} C'$, where C' is given by (17.9) with A given in (17.8) and $b = 1.794148$.*

2. *For $\Delta > 0$ and Ω_Δ defined in (17.2),*

$$\inf_{(\boldsymbol{\mu}, \boldsymbol{\sigma}) \in \Omega_\Delta} Pr\{CD|R\} \geq P^*$$

   provided that, for C' given by (17.9), the sample sizes n_i satisfy: $G(f_0^*) \geq P^*$, where $G(f)$ is given by (17.11) and f_0^* is defined by (17.13).

Remark 17.3.1 We have determined C so that only the requirement (17.3) is approximately satisfied. The P^*-condition (17.4) will be satisfied if $P^* \leq G(f_0^*)$. We have not addressed here the problem of choosing C and the n_i so that (17.3) and (17.4) are simultaneously satisfied. However, we can give the value of $G(f_0^*)$ as the guaranteed probability of correct decision when $(\boldsymbol{\mu}, \boldsymbol{\sigma})$ belongs to the subset Ω_Δ of the alternative region. This is what is done in the example discussed in the next section.

17.4   Example

Consider the following data of Ott (1993, p. 840) on the nitrogen contents (X_{ij}) of red clover plants inoculated with $k = 3$ strains of Rhizobium (3dok1, 3dok5, 3dok7). Suppose we choose $\alpha = 0.05$. The computed values are given in Table 17.2. We get $C' = 2.754$ and so $C = 4.4$. Since $SN_1 > C$ and $SN_3 > C$, using the decision rule R, we reject $H_0 : \mu_1 = \mu_2 = \mu_3$ and select the subset containing treatments 1 and 3. We identify these two treatments as significantly contributing to the departure from homogeneity. If we choose $\alpha = 0.01$, then $C' = 5.3596$ which gives $C = 7.2913$. The conclusions will still be the same.

For a chosen value of Δ, the P^*-value that will be guaranteed is $G(f_0^*)$, which is equal to 0.86 for $\Delta = 2$ and 0.96 for $\Delta = 3$ in the case of $\alpha = 0.05$. The guaranteed P^* is equal t

Table 17.1: Data

3dok1	3dok5	3dok7
19.4	18.2	20.7
32.6	24.6	21.0
27.0	25.5	20.5
32.1	19.4	18.8
33.0	21.7	20.1
	20.8	21.3

Table 17.2: Computed values

	3dok1	3dok5	3dok7
n_i	5.00	6.00	7.00
f_i	4.00	5.00	6.00
$\overline{X}_i$	28.82	21.70	20.14
s_i^2	33.64	8.24	1.12
$\hat{\theta}_i$	5.27	−1.85	−3.41
$\widehat{\mathrm{Var}}(\hat{\theta}_i)$	3.16	1.38	0.97
$\hat{\theta}_i^2/\widehat{\mathrm{Var}}(\hat{\theta}_i)$	8.77	2.50	11.98
SN_i	9.43	3.98	10.79

References

1. Basu, D. (1995). On statistics independent of a complete sufficient statistic, *Sankhyā*, **15**, 377–380.

2. Graybill, F. A. (1976). *Theory and Application of the Linear Model*, Boston, MA: Duxbury Press.

3. Gupta, S. S., Huang, D. Y. and Panchapakesan, S. (1995). Multiple decision procedures in analysis of variance and regression analysis, *Technical Report No. 95-44C*, Center for Statistical Decision Sciences and Department of Statistics, Purdue University, West Lafayette, Indiana.

4. Huang, D. Y. (1996). Selection procedures in linear models, *Journal of Statistical Planning and Inference*, **54**, 271–277.

5. Lee, A. F. S. (1992). Optimal sample sizes determined by two-sample Welch's t test, *Communications in Statistics—Simulation and Computation*, **21**, 689–696.

6. Ott, R. L. (1993). *An Introduction to Statistical Methods and Data Analysis*, Belmont, CA: Wadsworth Publishing Company, Inc.

7. Scheffé, H. (1970). Practical solutions of the Behrens-Fisher problem, *Journal of the American Statistical Association*, **65**, 1501–1508.

8. Welch, B. L. (1937). The significance between the difference between two means when the population variances are unequal, *Biometrika*, **29**, 350–362.

9. Welch, B. L. (1947). The generalization of "Student's" problem when several different population variances are involved, *Biometrika*, **34**, 28–35.

PART V
RANKING AND SELECTION

18

A Sequential Multinomial Selection Procedure With Elimination

D. M. Goldsman, A. J. Hayter and T. M. Kastner

Georgia Institute of Technology, Atlanta, GA

Abstract: We discuss a multinomial procedure for selecting the best of a number of competing systems or alternatives. The new procedure is an augmentation of the Bechhofer and Kulkarni (1984) (BK) sequential indifference-zone procedure for selecting the most probable multinomial cell; the augmentation eliminates noncompetitive events from further consideration as sampling progresses. The advantages of this procedure over the BK procedure are highlighted with specific attention given to the reduction in the expected number of total samples required. We present Monte Carlo simulation results to illustrate the performance of the cell elimination augmentation.

Keywords and phrases: Multinomial selection, nonparametric selection, elimination procedure

18.1 Introduction

This article is concerned with nonparametric selection of the "best" of a number of alternatives. The general method we will use is to interpret in a nonparametric way the problem of selecting the most probable multinomial cell [see Bechhofer and Sobel (1958) and Bechhofer, Santner, and Goldsman (1995)]. By way of motivation, consider the following problems.

- (Surveys.) Which of t soft drink formulations is the most popular?

- (Manufacturing.) Which of t simulated factory layouts is most likely to yield the greatest production throughput?

- (Tactical military combat modeling.) Which of t simulated battlefield scenarios is most likely to produce the fewest friendly casualties?

Suppose the t categories (e.g., soft drink types) have associated (unknown) probabilities $p_1, p_2, \ldots, p_t$ of being selected as the "most desirable" on a given observation. In the soft drink example, the term "most desirable" corresponds to a respondent's favorite soda; in the manufacturing simulation example, "most desirable" corresponds to the simulated layout with the greatest throughput on a given vector-trial of the t simulated layouts; in the military example, "most desirable" corresponds to the scenario with the fewest casualties on a given vector-trial of the t simulations. In all our of examples, each trial (or vector-trial) corresponds to a multinomial observation with $\boldsymbol{p} = (p_1, p_2, \ldots, p_t)$ and $\sum_{i=1}^{t} p_i = 1$; and it is of interest to determine the most probable ("best") category, i.e., that cell having the highest probability of being the "most desirable" on a given trial (or vector-trial).

The remainder of this paper is organized as follows. §18.1.1 gives a single-stage procedure to find the most probable multinomial category; §18.1.2 presents a more efficient sequential version of the procedure. §18.2 is concerned with a new procedure that eliminates noncompetitive categories as sampling progresses. We compare the various procedures in §18.3, followed by a summary in §18.4.

18.1.1 Single-stage procedure

Various multinomial selection procedures seek to determine which of the competing categories is associated with the largest unknown multinomial cell probability. Bechhofer, Elmaghraby, and Morse (1959) (BEM) present a single-stage procedure for selecting the most probable event. The procedure tells the experimenter how many observations to take in order to satisfy the *indifference-zone* probability requirement,

$$P\{\text{CS} \,|\, \boldsymbol{p}\} \;\geq\; P^\star \;\; \text{whenever} \;\; p_{[t]}/p_{[t-1]} \;\geq\; \theta^\star, \tag{18.1}$$

where $P\{\text{CS}\}$ denotes the probability of *correct selection*, $p_{[1]} \leq p_{[2]} \leq \cdots \leq p_{[t]}$ are the ordered (but unknown) p_i's, and $P^\star$ and $\theta^\star$ are user-specified constants. The quantity $P^\star \in (1/t, 1)$ is the desired $P\{\text{CS}\}$; and $\theta^\star > 1$ is the smallest $p_{[t]}/p_{[t-1]}$ ratio that the experimenter deems as "worth detecting". The BEM procedure, $\mathcal{M}_{\text{BEM}}$, is outlined as follows.

Procedure $\mathcal{M}_{\text{BEM}}$

- Sampling Rule: Take a random sample of n multinomial observations, $\boldsymbol{x}_m = (x_{1m}, x_{2m}, \ldots, x_{tm})$ $(1 \leq m \leq n)$, in a single stage. Here $x_{im} = 1\,[0]$ if event i occurs [does not occur] on the mth observation; and $\sum_{i=1}^{t} x_{im} = 1$ $(1 \leq m \leq n)$.

- Terminal Decision Rule: Calculate the sample sums, $y_{in} = \sum_{j=1}^{n} x_{ij}$ $(1 \leq i \leq t)$, and select the category that yields the largest sample sum,

$\max_{1 \leq i \leq t} y_{in}$, as the one associated with $p_{[t]}$. If there is a tie between two or more cells, then randomize to declare the "winner".

Tabled values of n for combinations of $(t, \theta^\star, P^\star)$ for procedure $\mathcal{M}_{\mathrm{BEM}}$ are available in various sources, e.g., Bechhofer, Santner, and Goldsman (1995). The idea is to find the smallest n-value that satisfies the probability requirement (18.1). In particular, (18.1) must hold for $\boldsymbol{p}$'s *least favorable configuration* (LFC), i.e., the element of $\{\boldsymbol{p} \mid p_{[t]}/p_{[t-1]} \geq \theta^\star\}$ that minimizes $P\{\mathrm{CS} \mid \boldsymbol{p}\}$. The LFC can be regarded as a "worst-case" configuration for $\boldsymbol{p}$'s satisfying (18.1). For procedure $\mathcal{M}_{\mathrm{BEM}}$, Kesten and Morse (1959) proved that the LFC is given by a slippage configuration of the form $\boldsymbol{p} = (p, \ldots, p, \theta^\star p)$.

Consider again our soft drink survey example. After giving a taste test to n individuals, we will conclude the most popular product is that which has the largest cell total, $\max_i y_{in}$, after sampling. If, for instance, our survey is conducted among $t = 3$ competing colas, and we specify $P^\star = 0.95$ and $\theta^\star = 1.4$, then it turns out that $\mathcal{M}_{\mathrm{BEM}}$ requires the test to be administered to 186 people. In the context of the manufacturing and military simulation examples, procedure $\mathcal{M}_{\mathrm{BEM}}$ with $(t, \theta^\star, P^\star) = (3, 1.4, 0.95)$ would require 186 sets of 3 computer runs each to achieve the desired $P\{\mathrm{CS}\}$. However, notice that if after 150 observations the respective cell totals were $(83, 46, 21)$ for categories A, B, and C, say, then in the remaining 36 observations, it would not be possible for B or C to catch up to A—even if all remaining observations were in favor of B or C. This suggests that there are ways to improve upon $\mathcal{M}_{\mathrm{BEM}}$ by using a sequential rule to reduce the expected number of observations required to achieve the probability requirement (18.1).

18.1.2 Sequential procedure

Bechhofer and Kulkarni (1984) (BK) propose a closed sequential procedure, $\mathcal{M}_{\mathrm{BK}}$, that is more efficient than procedure $\mathcal{M}_{\mathrm{BEM}}$ in terms of sampling costs, yet maintains the same $P\{\mathrm{CS}\}$ as $\mathcal{M}_{\mathrm{BEM}}$. Sampling and collecting observations have an economic impact for experimenters interested in determining which category is best; procedure $\mathcal{M}_{\mathrm{BK}}$ provides a means of reducing these costs. The underlying improvement in this procedure is the use of *curtailment* during sampling if one category garners an insurmountable advantage over the others. Recall in the above example that cells B and C could not win after 150 observations. Procedure $\mathcal{M}_{\mathrm{BK}}$ identifies such situations and will terminate early and declare A the winner.

Procedure $\mathcal{M}_{\mathrm{BK}}$

- Specify n from procedure $\mathcal{M}_{\mathrm{BEM}}$ prior to the start of sampling.

- Sampling Rule: At the mth stage of sampling $(m \geq 1)$, take the random multinomial observation $\boldsymbol{x}_m = (x_{1m}, x_{2m}, \ldots, x_{tm})$.

- Stopping Rule: Calculate the sample sums, y_{im} $(1 \le i \le t)$, through stage m. Stop sampling at the first stage m where there exists a category i satisfying $y_{im} \ge y_{jm} + n - m$, for all $j \ne i$, i.e., terminate sampling when the cell in second place can at best *tie* the current leader.

- Terminal Decision Rule: Let N denote the value of m at the termination of sampling. If $N < n$, the procedure must terminate with a single category; select this category as that associated with $p_{[t]}$. If $N = n$ and several categories simultaneously satisfy the stopping rule criterion, randomize to break the tie.

BK show that, uniformly in $\boldsymbol{p}$,

1. $P\{\text{CS} \,|\, \mathcal{M}_{\text{BK}}, \boldsymbol{p}\} = P\{\text{CS} \,|\, \mathcal{M}_{\text{BEM}}, \boldsymbol{p}\}$.

2. $E(N \,|\, \mathcal{M}_{\text{BK}}, \boldsymbol{p}) \le E(N \,|\, \mathcal{M}_{\text{BEM}}, \boldsymbol{p}) = n$.

As an illustration, Table 18.1 shows all of the possible sample paths for a problem with $t = 3$ and $n = 5$ (independent of $P^\star$ and $\theta^\star$). The ordered triplets are the possible (y_{1m}, y_{2m}, y_{3m}) realizations. An asterisk in the table indicates a sample path configuration that would result in the termination of procedure $\mathcal{M}_{\text{BK}}$. The numbers shown in brackets depict the number of possible ways to arrive at the specific cell totals for each of the three categories. For instance, there is exactly one sample path that would lead to early termination at $(y_{1m}, y_{2m}, y_{3m}) = (0, 3, 0)$; and there is no way to get to $(4,0,0)$. Further, there are 12 possible sample paths that would have led to termination at $(2,1,1)$.

Using counting arguments based on the above sample path analysis, it is possible to calculate the exact $P\{\text{CS} \,|\, \mathcal{M}_{\text{BK}}, \boldsymbol{p}\}$ and $E(N \,|\, \mathcal{M}_{\text{BK}}, \boldsymbol{p})$. Table 18.2 summarizes these results for the case $\boldsymbol{p} = (0.6, 0.2, 0.2)$. The results show that, for this example, procedure $\mathcal{M}_{\text{BK}}$ yields a substantial savings in $E(N)$ over $\mathcal{M}_{\text{BEM}}$ at no loss in $P\{\text{CS}\}$.

Although not studied herein, other indifference-zone sequential procedures for the multinomial selection problem include Cacoullos and Sobel (1966), Ramey and Alam (1979), and Bechhofer and Goldsman (1986). Some fundamental multinomial procedures that employ the *subset* selection approach are given by Gupta and Nagel (1967), Panchapakesan (1971), Bechhofer and Chen (1991), and Chen and Hsu (1991).

Table 18.1: Sample paths for Bechhofer-Kulkarni procedure

$m = 1$	$m = 2$	$m = 3$	$m = 4$	$m = 5$
(1,0,0) [1]	(2,0,0) [1]	(3,0,0) [1]*	(4,0,0) [0]	(5,0,0) [0]
(0,1,0) [1]	(0,2,0) [1]	(0,3,0) [1]*	(0,4,0) [0]	(0,5,0) [0]
(0,0,1) [1]	(0,0,2) [1]	(0,0,3) [1]*	(0,0,4) [0]	(0,0,5) [0]
	(1,1,0) [2]	(2,1,0) [3]	(3,1,0) [3]*	(4,1,0) [0]
	(1,0,1) [2]	(2,0,1) [3]	(3,0,1) [3]*	(4,0,1) [0]
	(0,1,1) [2]	(1,2,0) [3]	(1,3,0) [3]*	(1,4,0) [0]
		(0,2,1) [3]	(0,3,1) [3]*	(0,4,1) [0]
		(1,0,2) [3]	(1,0,3) [3]*	(1,0,4) [0]
		(0,1,2) [3]	(0,1,3) [3]*	(0,1,4) [0]
		(1,1,1) [6]	(2,2,0) [6]	(3,2,0) [6]*
			(2,0,2) [6]	(2,3,0) [6]*
			(0,2,2) [6]	(3,0,2) [6]*
				(2,0,3) [6]*
				(0,3,2) [6]*
				(0,2,3) [6]*
			(2,1,1) [12]*	(2,2,1) [6]*
			(1,2,1) [12]*	(2,1,2) [6]*
			(1,1,2) [12]*	(1,2,2) [6]*

Table 18.2: $P\{CS\}$ and $E(N)$ results for $t = 3$, $n = 5$, and $\boldsymbol{p} = (0.6, 0.2, 0.2)$

	$\mathcal{M}_{BEM}$	$\mathcal{M}_{BK}$
$P\{CS\}$	0.7690	0.7690
$E(N)$	5.000	3.950

18.2   Cell Elimination Procedure for $t = 3$

In the current paper, we present a modification of the $\mathcal{M}_{BK}$ procedure that implements cell *elimination*—the removal of any of the t competing treatments from further consideration if the cumulative sum of the specific category is lagging too far behind the other treatment totals. Henceforth, we shall address the special case when the number of competing treatments (categories) is $t = 3$.

In the new procedure, the elimination of one competitor results in a two-way contest between the two remaining treatments. Sampling continues until a clear winner is determined or until the procedure would have naturally terminated.

In our soft drink example, suppose that cola C is not very popular after 135 observations, the cell totals being (52,47,36). Further suppose that we are taking up to a maximum of 150 observations. Since it is not possible for C to win, we can eliminate this competitor and continue sampling to determine whether A or B will win. A practical benefit of eliminating cola C is that in the upcoming samples, only two glasses of cola are required for each trial. Similarly, this type of elimination scheme would allow simulators to "turn off" one of three competing simulated systems if its performance lags behind the other two. Fewer colas or computer simulation runs equate to lower costs.

Define R as the total number of drinks served or simulation runs conducted during the experiment. Under procedure $\mathcal{M}_{\mathrm{BK}}$ (with no elimination), $E(R \mid \mathcal{M}_{\mathrm{BK}}) = 3E(N \mid \mathcal{M}_{\mathrm{BK}})$. The cell elimination procedure aims to reduce the expected total number of cola servings or simulation runs required, and under certain conditions increase the probability of correctly selecting the most probable category.

18.2.1   Description of procedure

The cell elimination procedure, $\mathcal{M}_{\mathrm{GHK}}$, is implemented as follows.

Procedure $\mathcal{M}_{\mathrm{GHK}}$

- Specify n prior to start of sampling.

- Sampling Rule: At the mth stage of sampling ($m \geq 1$), take a trinomial observation $\boldsymbol{x}_m = (x_{1m}, x_{2m}, x_{3m})$ or binomial observation $\boldsymbol{x}_m = (x_{i_1m}, x_{i_2m})$ depending on whether a cell j was eliminated for some $j \neq i_1$ or i_2.

- Stopping Rule: Calculate the sample sums, y_{im} ($1 \leq i \leq 3$), through stage m. Stop sampling at the first stage m where there exists a category i satisfying $y_{im} \geq y_{jm} + n - m$, for all $j \neq i$.

- Elimination Rule: If, at stage m, the stopping criterion has not yet been met, eliminate from further consideration any cell j satisfying $\max_i y_{im} \geq y_{jm} + n - m$.

- Terminal Decision Rule: Same as that for procedure $\mathcal{M}_{\mathrm{BK}}$.

Note that procedure $\mathcal{M}_{\mathrm{GHK}}$ is identical to $\mathcal{M}_{\mathrm{BK}}$ unless the elimination rule is implemented.

18.2.2   Operating characteristics

We are interested in evaluating key performance characteristics of procedure $\mathcal{M}_{\mathrm{GHK}}$ such as $P\{\mathrm{CS}\}$, $E(N)$, and $E(R)$. In order to calculate such quantities, one must determine how the probability vector $\boldsymbol{p}$ is updated to reflect the elimination of one of the competing treatments. We offer two possible methods to alter $\boldsymbol{p}$.

The first method preserves the relative probabilities of the cells that remain in contention. As an example, suppose we sample with replacement up to $n = 5$ balls from an urn containing 60 red balls, 20 blues, and 20 whites. The corresponding multinomial cell probabilities are $\boldsymbol{p} = (p_R, p_B, p_W) = (0.6, 0.2, 0.2)$. If, after the third stage, the cumulative cell sums were $\boldsymbol{y}_3 = (2, 1, 0)$, the elimination procedure would drop white from further consideration because in the remaining observations, the best that category can do is to tie. Define $p_{RB}^\star$ $[p_{BR}^\star]$ as the probability that a red [blue] ball is selected on a particular trial after the white balls have (somehow) been removed from the urn. Then the updated probability vector would be $\boldsymbol{p}^\star = (p_{RB}^\star, p_{BR}^\star) = (0.6/0.8, 0.2/0.8) = (0.75, 0.25)$.

Unfortunately, the first method may fail if there is "vote splitting" among the categories. Consider the problem of determining which of three colas, A, B, and C, is the most popular. Suppose that $\boldsymbol{p} = (0.4, 0.3, 0.3)$, indicating that cola A maintains a 40% market share and colas B and C each a 30% share. If, during the course of sampling, cola C happened to be eliminated from contention, then the updated probability vector based on the first approach would be $\boldsymbol{p}^\star = (p_{AB}^\star, p_{BA}^\star) = (0.571, 0.429)$. But what if B and C were very similar in taste and had been splitting the vote of the survey participants? Then it might very well turn out that $p_{AB}^\star < p_{BA}^\star$.

To address such vote-splitting problems, we propose (and actually use) a second method of updating the probability vector that may be more realistic than the first method — simply give a *complete ordering* of survey participant preferences. For example, the six possible orderings of categories A, B, and C that result from the cola experiment have associated (unknown) probabilities

- $q_1 = P\{\text{order } ABC\}$

- $q_2 = P\{\text{order } ACB\}$

- $q_3 = P\{\text{order } BAC\}$

- $q_4 = P\{\text{order } BCA\}$

- $q_5 = P\{\text{order } CAB\}$

- $q_6 = P\{\text{order } CBA\}$

In the context of procedure $\mathcal{M}_{\text{BK}}$, we have $p_A = q_1 + q_2$, $p_B = q_3 + q_4$, and $p_C = q_5 + q_6$. Furthermore, if category C, say, is eliminated, then it is clear that in the subsequent trials between categories A and B, the probability that a participant would select A over B is $p_{AB} = q_1 + q_2 + q_5$; the probability of taking B over A is $p_{BA} = q_3 + q_4 + q_6$. Similarly, using the obvious notation, trials between categories A, C result in participant selection probabilities $p_{AC} =$

$q_1+q_2+q_3$ and $p_{CA} = q_4+q_5+q_6$; and trials between B, C yield $p_{BC} = q_1+q_3+q_4$ and $p_{CB} = q_2 + q_5 + q_6$. Of course, $\boldsymbol{p}$ and $\boldsymbol{q}$ are *unknown*, so we must examine the performance of $\mathcal{M}_{\mathrm{GHK}}$ under various $\boldsymbol{p}$ and $\boldsymbol{q}$ configurations.

18.3   Comparison of $\mathcal{M}_{\mathrm{BK}}$ and $\mathcal{M}_{\mathrm{GHK}}$

18.3.1   Simulation results

We conducted Monte Carlo simulation experimentation to compare the performance of procedures $\mathcal{M}_{\mathrm{BK}}$ and $\mathcal{M}_{\mathrm{GHK}}$. We were particularly interested in studying the effects of cell elimination on the achieved probability of correct selection, $P\{\mathrm{CS}\}$, the expected number of stages, $E(N)$, and the expected number of total items on trial, $E(R)$. Written in FORTRAN, our code simulates procedures $\mathcal{M}_{\mathrm{BK}}$ and $\mathcal{M}_{\mathrm{GHK}}$ for the case $t = 3$; it requires the user to input n, the maximum number of multinomial observations to be taken, as well as the six-dimensional probability vector $\boldsymbol{q}$ (which immediately gives the corresponding $\boldsymbol{p}$-vector). Our experiments all used $n = 30$. For each choice of $\boldsymbol{q}$, we ran 100000 replications of both procedures. Without loss of generality, the program requires cell A to be the most probable.

Table 18.3 provides a comparison of $P\{\mathrm{CS}\}$, $E(N)$, and $E(R)$ for both procedures. All results are accurate to at least the last decimal place given in the table. In each case, the elimination procedure $\mathcal{M}_{\mathrm{GHK}}$ indeed (modestly) reduces $E(R)$ relative to $\mathcal{M}_{\mathrm{BK}}$. However, for a given $\boldsymbol{p}$, the $P\{\mathrm{CS}\,|\,\mathcal{M}_{\mathrm{GHK}}, \boldsymbol{q}\}$ *changes* according to the configuration of $\boldsymbol{q}$. For instance, in Table 18.3, ZWC refers to the *zero-weight* configuration of $\boldsymbol{q}$ in which category A receives no additional probability weight if B or C are eliminated during the course of sampling. For fixed $\boldsymbol{p}$, we believe that the ZWC minimizes $P\{\mathrm{CS}\,|\,\mathcal{M}_{\mathrm{GHK}}, \boldsymbol{q}\}$; this conclusion appears to be borne out by the simulation results of Table 18.3. The term EWC is the *equal-weight* configuration in which $q_3 = q_4 = p_B/2$ and $q_5 = q_6 = p_C/2$. The *full-weight* configuration, FWC, refers to the $\boldsymbol{q}$ in which category A receives *all* of the additional probability weight if B or C are eliminated. For fixed $\boldsymbol{p}$, we believe that the FWC maximizes $P\{\mathrm{CS}\,|\,\mathcal{M}_{\mathrm{GHK}}, \boldsymbol{q}\}$; this conclusion also seems to be backed up by Table 18.3.

Table 18.3: Simulation comparison of $\mathcal{M}_{\text{GHK}}$ and $\mathcal{M}_{\text{BK}}$

	p or q Configuration	$P\{CS\}$	$E(N)$	$E(R)$
BK	$p = (0.4, 0.3, 0.3)$	0.615	27.0	81.1
ZWC	$q = (0.4, 0.0, 0.0, 0.3, 0.0, 0.3)$	0.514	27.2	79.3
EWC	$q = (0.4, 0.0, 0.15, 0.15, 0.15, 0.15)$	0.609	27.1	79.0
FWC	$q = (0.4, 0.0, 0.3, 0.0, 0.3, 0.0)$	0.693	26.9	78.7
BK	$p = (0.5, 0.4, 0.1)$	0.718	25.9	77.7
ZWC	$q = (0.5, 0.0, 0.0, 0.4, 0.0, 0.1)$	0.664	26.0	73.1
EWC	$q = (0.5, 0.0, 0.2, 0.2, 0.05, 0.05)$	0.716	25.9	72.8
FWC	$q = (0.5, 0.0, 0.4, 0.0, 0.1, 0.0)$	0.767	25.8	72.6
BK	$p = (0.5, 0.3, 0.2)$	0.883	25.2	75.6
ZWC	$q = (0.5, 0.0, 0.0, 0.3, 0.0, 0.2)$	0.823	25.6	73.8
EWC	$q = (0.5, 0.0, 0.15, 0.15, 0.1, 0.1)$	0.876	25.3	73.2
FWC	$q = (0.5, 0.0, 0.3, 0.0, 0.2, 0.0)$	0.916	25.0	72.8
BK	$p = (0.5, 0.25, 0.25)$	0.909	25.1	75.2
ZWC	$q = (0.5, 0.0, 0.0, 0.25, 0.0, 0.25)$	0.852	25.4	73.9
EWC	$q = (0.5, 0.0, 0.125, 0.125, 0.125, 0.125)$	0.904	25.2	73.3
FWC	$q = (0.5, 0.0, 0.25, 0.0, 0.25, 0.0)$	0.935	24.9	72.8
BK	$p = (0.6, 0.2, 0.2)$	0.992	22.5	67.6
ZWC	$q = (0.6, 0.0, 0.0, 0.2, 0.0, 0.2)$	0.981	22.8	66.5
EWC	$q = (0.6, 0.0, 0.1, 0.1, 0.1, 0.1)$	0.990	22.6	66.1
FWC	$q = (0.6, 0.0, 0.2, 0.0, 0.2, 0.0)$	0.995	22.4	65.6

18.3.2 Discussion

For procedures $\mathcal{M}_{\text{BEM}}$ and $\mathcal{M}_{\text{BK}}$, a correct selection occurs when we choose the category associated with $p_{[3]}$ after the procedure terminates. For the cell elimination procedure $\mathcal{M}_{\text{GHK}}$, a correct selection requires that the best category outperform competitors in three-way and (if necessary) two-way contests. Notice in Table 18.3 that for $p = (0.5, 0.4, 0.1)$, we obtain $P\{CS \mid \mathcal{M}_{\text{BK}}\} = 0.718$. In the ZWC, $q = (0.5, 0.0, 0.0, 0.4, 0.0, 0.1)$, we see that $P\{CS \mid \mathcal{M}_{\text{GHK}}\}$ falls to 0.664. In this configuration, $p_{AB} = p_{AC} = 0.5$, so if either category B or C is eliminated, then cell A has a 50% chance of being beaten in subsequent binomial observations. In the EWC, $q = (0.5, 0.0, 0.2, 0.2, 0.05, 0.05)$, and we see that $P\{CS \mid \mathcal{M}_{\text{GHK}}\} = 0.716$ is practically the same as (within two standard errors of) $P\{CS \mid \mathcal{M}_{\text{BK}}\}$. Now p_{AB} and p_{AC} have increased to 0.55 and 0.70, respectively. The increase in the probability of selecting A in our procedure is due to the improved chances of A winning two-way contests. In the FWC, $q = (0.5, 0.0, 0.4, 0.0, 0.1, 0.0)$, and the $P\{CS\}$ is maximized. The higher values of p_{AB} and p_{AC} (0.60 and 0.90) account for the increase in $P\{CS\}$ to 0.767.

18.4 Summary

The closed sequential multinomial event selection procedure with cell elimination proposed herein reduces the expected total number of items on trial when compared to like values under procedure $\mathcal{M}_{\mathrm{BK}}$. This leads to a reduction in the cost of sampling. An important distinction regarding the $\mathcal{M}_{\mathrm{GHK}}$ procedure is that the probability vector of three-way orderings, $\boldsymbol{q}$, directly affects the $P\{\mathrm{CS}\}$. An ongoing research problem is to give conditions on $\boldsymbol{q}$ so that the $P\{\mathrm{CS}\}$ remains at least as high as that from procedure $\mathcal{M}_{\mathrm{BK}}$. We are also investigating obvious (and not-so-obvious) ways to extend procedure $\mathcal{M}_{\mathrm{GHK}}$ to the comparison of more than $t = 3$ categories.

Acknowledgements. We would like to thank Prof. Shanti S. Gupta for the inspiration he has given us over the years. Our work was supported by the National Science Foundation under grant no. DMI-9622269.

References

1. Bechhofer, R. E. and Chen, P. (1991). A note on a curtailed sequential procedure for subset selection of multinomial cells, *American Journal of Mathematical and Management Sciences*, **11**, 309–324.

2. Bechhofer, R. E., Elmaghraby, S. and Morse, N. (1959). A single-sample multiple decision procedure for selecting the multinomial event which has the highest probability, *Annals of Mathematical Statistics*, **30**, 102–119.

3. Bechhofer, R. E. and Goldsman, D. M. (1986). Truncation of the Bechhofer-Kiefer-Sobel sequential procedure for selecting the multinomial event which has the largest probability (II): extended tables and an improved procedure, *Communications in Statistics—Simulation and Computation*, **15**, 829–851.

4. Bechhofer, R. E. and Kulkarni, R. V. (1984). Closed sequential procedures for selecting the multinomial events which have the largest probabilities, *Communications in Statistics—Theory and Methods*, **13**, 2997–3031.

5. Bechhofer, R. E., Santner, T. J. and Goldsman, D. M. (1995). *Design and Analysis of Experiments for Statistical Selection, Screening and Multiple Comparisons*, New York: John Wiley & Sons.

6. Bechhofer, R. E. and Sobel, M. (1958). Non-parametric multiple-decision procedures for selecting that one of k populations which has the highest probability of yielding the largest observation (preliminary report), abstract, *Annals of Mathematical Statistics*, **29**, 325.

7. Cacoullos, T. and Sobel, M. (1966). An inverse-sampling procedure for selecting the most probable event in a multinomial distribution, In *Multivariate Analysis* (Ed. P. R. Krishnaiah), pp. 423–455, New York: Academic Press.

8. Chen, P. and Hsu, L. (1991). A composite stopping rule for multinomial subset selection, *British Journal of Math. and Statist. Psychology*, **44**, 403–411.

9. Gupta, S. S. and Nagel, K. (1967). On selection and ranking procedures and order statistics from the multinomial distribution, *Sankhyā, Series B*, **29**, 1–17.

10. Kesten, H. and Morse, N. (1959). A property of the multinomial distribution, *Annals of Mathematical Statistics*, **30**, 120–127.

11. Panchapakesan, S. (1971). On a subset selection procedure for the most probable event in a multinomial distribution, In *Statistical Decision Theory and Related Topics* (Eds. S. S. Gupta and J. Yackel), pp. 275–298, New York: Academic Press.

12. Ramey, J. T., Jr. and Alam, K. (1979). A sequential procedure for selecting the most probable multinomial event, *Biometrika*, **66**, 171–173.

19

An Integrated Formulation for Selecting the Best From Several Normal Populations in Terms of the Absolute Values of Their Means: Common Known Variance Case

S. Jeyaratnam and S. Panchapakesan

Southern Illinois University, Carbondale, IL

Abstract: Consider k (≥ 2) normal populations with unknown means μ_i ($i = 1, \ldots, k$) and known common variance σ^2. Let $\theta_i = |\mu_i|$. Rizvi (1971) investigated the problem of selecting the population associated with the largest θ_i under the indifference zone approach of Bechhofer (1954) and the subset selection approach of Gupta (1956, 1965). We consider here an integrated formulation of the problem combining features of the two classical approaches. For the proposed procedure we establish the form of the least favorable configurations over the indifference zone and the preference zone, and obtain results regarding determination of the sample size and constants associated with the procedure in order to guarantee minimum probability of correct decision for each zone. Properties of some operating characteristics are also studied.

Keywords and phrases: Indifference zone approach, subset selection approach, integrated formulation, normal populations, absolute values of means

19.1 Introduction

Let $\Pi_1, \ldots, \Pi_k$, be k (≥ 2) independent normal populations with unknown means $\mu_1, \ldots, \mu_k$, respectively, and common *known* variance σ^2. Let $\theta_i = |\mu_i|$, $i = 1, \ldots, k$. In communications theory, the $\frac{\theta_i}{\sigma}$ will represent the signal-to-noise ratios of k different electronic devices to be compared. A device is considered superior if it has a larger $\frac{\theta}{\sigma}$-value. Let $\theta_{[1]} \leq \theta_{[2]} \leq \cdots \leq \theta_{[k]}$ denote the ordered θ_i. It is assumed that there is no prior knowledge about the pairing of the ordered and unordered θ_i. Our goal is to select the t ($1 \leq t \leq k - 1$) best populations, namely, those associated with $\theta_{[k-t+1]}, \ldots, \theta_{[k]}$. The two usual approaches to this problem are the indifference zone (IZ) approach of Bechhofer (1954) and the subset selection (SS) approach of Gupta (1956, 1965). Let $\Omega = \{\boldsymbol{\theta} : \boldsymbol{\theta} = (\theta_1, \ldots, \theta_k), 0 < \theta_i < \infty, i = 1, \ldots, k\}$ and $\Omega(\delta^*) = \{\boldsymbol{\theta} : \theta_{[k-t+1]} - \theta_{[k-t]} \geq \delta^* > 0; \boldsymbol{\theta} \in \Omega\}$ for any specified δ^*. In the IZ approach, one wants to select exactly t populations with a requirement that the probability of a correct selection (PCS) is at least P^* whenever $\boldsymbol{\theta} \in \Omega(\delta^*)$. Here δ^* and P^* [with $\binom{k}{t}^{-1} < P^* < 1$] are specified in advance by the experimenter. The region $\Omega(\delta^*)$ is called the *preference zone* (PZ) and its complement w.r.t. Ω is the *indifference zone* (IZ). In the SS approach, the goal is to select a non-empty subset of the k populations so that the subset selected contains the t best ones (in which case a correct selection is said to occur) with a guaranteed probability P^* for all $\boldsymbol{\theta} \in \Omega$. The above problem was studied by Rizvi (1971) for $1 \leq t \leq k-1$ using the IZ approach and for $t = 1$ using the SS approach. While the IZ approach does not control the PCS when $\delta \equiv \theta_{[k-t+1]} - \theta_{[k-t]} < \delta^*$, the SS formulation does not (but it can) modify profitably the definition of a correct selection when $\delta \geq \delta^*$. This set the stage for an integrated formulation studied in the case of ranking normal populations according to their means by Chen and Sobel (1987a) and in the case of ranking the cells of a multinomial distribution by Chen and Sobel (1987b). A few other papers with slightly modified goal and/or procedure are by Chen (1988) and Chen and Panchapakesan (1994). In the present paper, the formulation is closer to Chen and Panchapakesan (1994). Section 19.2 presents the integrated formulation of the problem and the proposed rule R. Probabilities of correct decision are derived when $\boldsymbol{\theta} \in PZ$ and $\boldsymbol{\theta} \in IZ$ and their infima are obtained in Section 19.4. In the case of PZ, the results are for $1 \leq t < k$ where as, in the case of IZ, only $t = 1$ is considered. The determination of the necessary constants for the implementation of the procedure R is discussed in Section 19.5. The next section investigates some properties of the procedure. These relate to the behavior of $E(S)$, $P(S = 1)$, and $P(S = k)$, where S denotes the size of the selected subset.

19.2   Formulation of the Problem and Definition of Procedure R

When $\delta \geq \delta^*$ (i.e. $\boldsymbol{\theta} \in PZ$), the populations associated with $\theta_{[k-t+1]}, \ldots, \theta_{[k]}$ are defined to be δ^*-*best* populations. When $\delta < \delta^*$ (i.e. $\boldsymbol{\theta} \in IZ$), the populations associated with the t largest θ_i (which means that anyone of these θ_i is greater than or equal to any of the remaining $k - t$) are defined to be just t *best*. In case of more than one set of t populations qualifying as the t best, we assume that one such set is tagged as the t best.

Our goal is to select *exactly* the t populations which are δ^*-best and none other when $\boldsymbol{\theta} \in PZ$, and select a subset of at least t population so that the t best populations are included in the selected subset when $\boldsymbol{\theta} \in IZ$. A *correct decision* (CD) occurs if the claim regarding the selected subset is consistent with the goal. Let CD_1 and CD_2 denote the events of a correct decision when $\boldsymbol{\theta} \in PZ$ and $\boldsymbol{\theta} \in IZ$, respectively. It is required of any valid procedure that

$$P(CD_1) \geq P_1^* \text{ whenever } \boldsymbol{\theta} \in PZ \tag{19.1}$$

and

$$P(CD_2) \geq P_2^* \text{ whenever } \boldsymbol{\theta} \in IZ \tag{19.2}$$

where $\binom{k}{t}^{-1} < P_i^* < 1$, $i = 1, 2$. Here δ^*, P_1^*, and P_2^* are specified in advance.

We are only considering vector-at-a-time sampling, so that there is a common number of observations n from each of the k populations. Let $W_i = |\bar{X}_i|$, where $\bar{X}_i$ is the mean of the sample observations from Π_i ($i = 1, 2, \ldots, k$). Let the ordered W_i be denoted by $W_{[1]} \leq \cdots \leq W_{[k]}$. Let c and d be constants, with $0 < d < c$, to be determined (along with n) as a function of δ^*, P_1^* and P_2^* for any given values of k and t. Assume that c, d and n are already determined.

Procedure R: "If $W_{[k-t+1]} - W_{[k-t]} > c$, then select the t populations that yielded the t largest W_i's as δ^*-best populations. If $W_{[k-t+1]} - W_{[k-t]} \leq c$, then select all populations Π_i for which $W_i > W_{[k-t+1]} - d$ with the claim that the selected subset includes the t best populations."

19.3 Some Preliminary Results

The cdf $H(w; \theta_i)$ and the density function $h(w; \theta_i)$ of W_i are given by

$$H(w; \theta_i) = \Phi\left\{\frac{\sqrt{n}}{\sigma}(w - \theta_i)\right\} - \Phi\left\{\frac{\sqrt{n}}{\sigma}(-w - \theta_i)\right\}, \ w > 0 \tag{19.3}$$

and

$$h(w; \theta_i) = \frac{\sqrt{n}}{\sigma}\left[\varphi\left\{\frac{\sqrt{n}}{\sigma}(w - \theta_i)\right\} + \varphi\left\{\frac{\sqrt{n}}{\sigma}(w + \theta_i)\right\}\right], \ w > 0 \tag{19.4}$$

where Φ and φ are standard normal cdf and density function. It can be easily verified that $\{H(w; \theta)\}$, $\theta \geq 0$, is a stochastically increasing (SI) family in θ.

We now state two useful results as lemmas. Lemma 19.3.1 is an immediate consequence of a result in Lehmann (1986, p116, Problem 15). Lemma 19.3.2 is due to Gupta and Panchapakesan (1972).

Lemma 19.3.1 *Let $X_1, X_2, \ldots, X_k$ be independent having cdf's $G(\cdot|\theta_i)$, $i = 1, 2, \ldots, k$, where the family $\{G(\cdot|\theta)\}$ is stochastically increasing in θ. Let $\psi(x_1, \ldots, x_k)$ be non-decreasing (non-increasing) in x_i when the x_j, $j \neq i$, are kept fixed. Then $E[\psi(X_1, \ldots, X_k)]$ is non-decreasing (non-increasing) in θ_i when the θ_j, $j \neq i$, are kept fixed.*

Lemma 19.3.2 *Let $\{F(\cdot, \theta)\}$, $\theta \in \Theta$ (an interval on the real line), be a family of absolutely continuous distributions on the real line with continuous densities $f(\cdot, \theta)$. Let $\psi(x, \theta)$ be a bounded real-valued function possessing first partial derivatives ψ_x and ψ_θ with respect to x and θ respectively and satisfying certain regularity conditions. Then $E_\theta[\psi(X, \theta)]$ is nondecreasing in θ provided for all $\theta \in \Theta$,*

$$f(x, \theta)\frac{\partial}{\partial \theta}\psi(x, \theta) - \frac{\partial}{\partial \theta}F(x, \theta)\frac{\partial}{\partial x}\psi(x, \theta) \geq 0 \ for \ a\dot{e} \ x. \tag{19.5}$$

Remark. By reversing the inequality in (19.5), $E_\theta[\psi(x, \theta)]$ is nonincreasing in θ. The monotonicity of $E_\theta[\psi(x, \theta)]$ is strict, if the inequality in (19.5) is strict.

19.4 $P(CD_1|PZ)$ and $P(CD_2|IZ)$ and Their Infima

To derive $P(CD_1|PZ)$, which is $P(CD_1)$ for $\theta \in PZ$, we first note that CD_1 in the PZ occurs only if we select the t best populations as δ^*-best ones. Let $W_{(i)}$ denote the W-statistic associated with $\theta_{[i]}$ $(i = 1, 2, \ldots, k)$. Then

$$
\begin{aligned}
&P(CD_1|PZ) \\
&= \sum_{i=1}^{k-t} \sum_{j=k-t+1}^{k} \Pr\left[W_{(j)} = \min_{k-t+1 \leq \beta \leq k} W_{(\beta)} > c + \max_{1 \leq \alpha \leq k-t} W_{(\alpha)} = c + W_{(i)}\right] \\
&= \sum_{i=1}^{k-t} \sum_{j=k-t+1}^{k} \Pr[E_{ij}], \ \text{say.}
\end{aligned}
$$

Let

$$\Psi_{ij}(W_{(1)}, \ldots, W_{(k)}) = \begin{cases} 1 & \text{if the event } E_{ij} \text{ occurs,} \\ 0 & \text{otherwise.} \end{cases}$$

Then it is easy to see that, by Lemma 19.3.1, $P(CD_1|PZ)$ is decreasing in $\theta_{[\alpha]}(\alpha = 1, \ldots, k - t)$ and increasing in $\theta_{[\beta]}(\beta = k - t + 1, \ldots, k)$, each monotonicity holding when all the remaining θ's are kept fixed. Thus $P(CD_1|PZ)$ is minimized when $\boldsymbol{\theta}$ has a configuration given by

$$\theta_{[1]} = \cdots = \theta_{[k-t]} = \theta = \theta_{[k-t+1]} - \delta^* = \cdots = \theta_{[k]} - \delta^*. \tag{19.6}$$

Consequently we have

$$\inf P(CD_1|PZ) \tag{19.7}$$

$$= \inf_{\theta \geq 0} (k-t) \int_0^\infty [1 - H(u+c, \theta+\delta^*)]^t H^{k-t-1}(u, \theta)\, h(u, \theta)\, du$$

$$= \inf_{\theta \geq 0} \int_0^\infty [1 - H(u+c, \theta+\delta^*)]^t\, dH^{k-t}(u, \theta). \tag{19.8}$$

Noting that $F(u, \theta) = H^{k-t}(u, \theta)$ is the cdf of $V = \max\{W_{(1)}, \ldots, W_{(k-t)}\}$ under the configuration (19.6), and setting $\psi(u, \theta) = [1 - H(u+c, \theta+\delta^*)]^t$, it can be seen that the sufficient condition (19.5) of Lemma 19.3.2 reduces to

$$h(u+c, \theta+\delta^*)\frac{\partial}{\partial \theta}H(u, \theta) - h(u, \theta)\frac{\partial}{\partial \theta}H(u+c, \theta+\delta^*) \geq 0, \tag{19.9}$$

which can be verified to be true by straightforward algebra. Thus, the integral in (19.7) is increasing in θ and the least favorable configuration (LFC) for $P(CD_1|PZ)$ is $\boldsymbol{\theta}_{\delta^*} = (0, \ldots, 0, \delta^*, \ldots, \delta^*)$ with zero as the first $k-t$ components. It follows that the requirement (19.1) for $P(CD_1|PZ)$ is met if

$$(k-t) \int_0^\infty [1 - H(u+c, \delta^*)]^t H^{k-t-1}(u, 0)\, h(u, 0)\, du \geq P_1^*. \tag{19.10}$$

Now, the left-hand side of (19.10) can be written as

$$I(a, c) \equiv \int_0^\infty [2 - \Phi(y+ac-a\delta^*) - \Phi(y+ac+a\delta^*)]^t\, dF(y, 0),$$

where $F(y, 0) = [2\Phi(y) - 1]^{k-t}$ and $a = \sqrt{n}/\sigma$. It is easy to see that, as a increases from 0 to ∞, $I(a, c)$ decreases from $\binom{k}{t}^{-1}$ to 0 if $c > \delta^*$ and to $[2^t\binom{k}{t}]^{-1}$ if $c = \delta^*$. Thus, for $c \geq \delta^*$, the probability requirement (19.10) cannot be met with $P_1^* > \binom{k}{t}^{-1}$ as stipulated in our formulation. It should be noted that $a = 0$ corresponds to the no-data situation and, in this case, (19.10) can be satisfied with $P_1^* = \binom{k}{t}^{-1}$ by using a random device to choose t populations. When $c < \delta^*$, $I(a, c) \to 1$ as $a \to \infty$. However, the exact behavior of $I(a, c)$ as a increases is not established.

We note that

$$I(a, c) \geq \int_0^\infty [1 - \Phi\{y+ac-a\delta^*\}]^t\, dF(y, 0) = I_L(a, c), \text{ say.}$$

Since $I_L(a, c)$ is monotonically increasing in a and tending to 1 as $a \to \infty$, it is clear that, for a given $c < \delta^*$, there exists a positive integer $n_0(c)$ such that (19.10) is satisfied for all $n \geq n_0(c)$.

Regarding $P(CD_2|IZ)$, we consider here only the special case of $t = 1$ for which we will establish the LFC. First we note that, since $d < c$, our procedure R can be written in the form:

$$\text{"Select all } \Pi_i\text{'s for which } W_i > W_{[k]} - d\text{"} \tag{19.11}$$

where $W_{[k]}$ is the largest of W_i's.

Now,

$$
\begin{aligned}
&P(CD_2|IZ) \\
&\quad = \ \Pr[W_{(k)} > W_{(j)}] - d, \ j = 1, \ldots, k-1] \\
&\quad = \ \int_0^\infty \prod_{j=1}^{k-1} H(w+d, \theta_{[j]}) h(w, \theta_{[k]}) \ dw \\
&\quad \geq \ \int_0^\infty H^{k-1}(w+d, \theta_{[k]}) h(w, \theta_{[k]}) \ dw, \ \text{using the SI property}
\end{aligned}
$$

of the family $\{H(w, \theta)\}$, $\theta \geq 0$. Thus, we have

$$
\inf P(CD_2|IZ) = \inf_{\theta \geq 0} \int_0^\infty H^{k-1}(w+d, \theta) \, h(w, d) \ dw. \tag{19.12}
$$

We note here that the rule in (19.11) is of the same form as that of the subset selection rule of Rizvi (1971). His Lemma 2 establishes that the integral on the right-hand side of (19.12) decreases in θ and that its limit as $\theta \to \infty$ is given by

$$
\int_{-\infty}^\infty \Phi^{k-1}\left(u + \frac{d\sqrt{n}}{\sigma}\right) \varphi(u) \ du. \tag{19.13}
$$

The probability requirement (19.2) is satisfied if the integral in (19.13) is greater than or equal to P_2^*. Thus, we need to determine n, c and d with $d < c$ satisfying

$$
\int_0^\infty H^{k-1}(u, 0) \, h(u+c, \delta^*) \ du \geq P_1^* \tag{19.14}
$$

and

$$
\int_{-\infty}^\infty \Phi^{k-1}\left(u + \frac{d\sqrt{n}}{\sigma}\right) \varphi(u) \ du \geq P_2^*. \tag{19.15}
$$

19.5   Determination of n, c and d

In order to implement the rule R, we need to find n, c and d satisfying (19.14) and (19.15) for given k, P_1^* and P_2^* with the restriction that $d < c < \delta^*$. Since there are three constants, any one of them can be arbitrarily chosen first. We first note that values of $D > 0$ satisfying

$$
\int_{-\infty}^\infty \Phi^{k-1}(u + D)\varphi(u) \ du = P_2^* \tag{19.16}
$$

can be obtained for selected values of k and P_2^* from the tables of Bechhofer (1954), Gupta (1963), and Gupta, Nagel and Panchapakesan (1973). Table 19.1 provides values of $C = \frac{\sqrt{n}c}{\sigma}$ satisfying (19.14) with equality for $k = 2(1)10$, $P_1^* = 0.90, 0.95$, and $\Delta^* = \frac{\sqrt{n}\delta^*}{\sigma} = a(0.2)7$, where the starting value a depends on k and P^*, and a is the smallest value for which we get a positive value of c.

Letting $D = \frac{\sqrt{n}d}{\sigma}$, one can arbitrarily choose n (or d) and determine the smallest constant d (or n) satisfying (19.15) by solving for D from (19.16). However, this choice of n and d may not lead to a feasible solution for c because we need $\delta^* > c > d$. It should be noted that, if we have (n, c, d) satisfying (19.14) and (19.15) with $c < d$, it may not be possible to increase c to exceed d for the same n and c while being consistent with (19.14) and (19.15) because the left-hand side of (19.14) will then steadily decrease and might go below P_1^*. This necessitates a process of trial and error by changing the choice of n (or d) to start with. Alternatively, one can obtain a matching n and c from Table 19.1 and then determine the constant d from (19.16). However, the solution is not valid unless $d < c$. So we go by trial and error again.

As pointed out in Section 19.4, there exists, for a given $c < \delta^*$, a positive integer $n_0(c)$ such that (19.14) is satisfied for all $n \geq n_0(c)$. Since increasing n decreases d in (19.16), it follows that there exists an $n \geq n_0(c)$ yielding a feasible solution satisfying (19.14) and (19.15). Consequently, among all feasible solutions there is one with smallest n [not necessarily $\geq n_0(c)$] which can be taken as an optimal solution. If there are more than one solution with smallest n [but different c and d], we will choose the one with the smallest d as this will decrease S, the size of the selected subset. One can also evolve other minimax type criteria, for example, minimizing the sample size over the set of solutions which maximize $\Pr[S = 1]$ over a specified set of parametric configurations. This aspect is not addressed at this time.

19.6   Some Properties of Procedure R

We continue to consider the case of $t = 1$. As we have noted earlier, Procedure R in this case can be written in the form (19.11). Let S denote the size of the selected subset. Then the expected subset size is given by

$$E(S) = \sum_{i=1}^{k} \Pr[W_{(i)} > W_{(j)} - d, \ j = 1, \ldots, k; \ j \neq i]$$

$$= \sum_{i=1}^{k} \int_0^{\infty} \prod_{j \neq i} H(w + d, \theta_{[j]}) dH(w, \theta_{[i]}). \tag{19.17}$$

We can again use the results of Rizvi (1971) to claim that the supremum of $E(S)$ over the entire parameter space Ω is attained when $\theta_1 = \ldots = \theta_k = \theta (say)$

and that $E(S)$ is nonincreasing in θ. Thus

$$\sup_{\boldsymbol{\theta}\in\Omega} E(S) = 2k \int_0^\infty \left[2\Phi(u + \frac{\sqrt{n}d}{\sigma}) - 1\right]^{k-1} \varphi(u)\, du. \tag{19.18}$$

It is also interesting to consider $\Pr[S = 1]$ and $\Pr[S = k]$. Now

$$\begin{aligned}
\Pr[S = 1] &= \Pr[W_{[k]} - W_{[k-1]} > d] \\
&= \sum_{i=1}^k \Pr\{W_{(j)} < W_{(i)} - d \text{ for all } j \neq i\} \\
&= \sum_{i=1}^k \int_d^\infty \left\{\prod_{j\neq i} H(w - d, \theta_{[j]})\right\} h(w, \theta_{[i]})\, dw \\
&= \sum_{i=1}^k \int_0^\infty \left\{\prod_{j\neq i} H(w, \theta_{[j]})\right\} h(w + d, \theta_{[i]})\, dw \\
&= \sum_{i=1}^k A_i(\theta_{[1]}, \ldots, \theta_{[k]}), \text{ say.}
\end{aligned} \tag{19.19}$$

Integrating $A_k(\theta_{[1]}, \ldots, \theta_{[k]})$ by parts and regrouping terms, we get

$$\begin{aligned}
\Pr[S = 1] = 1 &+ \sum_{i=1}^{k-1} \int_0^\infty \left\{\prod_{\substack{j=1\\j\neq i}}^{k-1} H(w, \theta_{[j]})\right\} \\
&\times \left\{h(w + d, \theta_{[i]})H(w, \theta_{[k]}) - h(w, \theta_{[i]})H(w + d, \theta_{[k]})\right\} dw.
\end{aligned}$$

This gives

$$\begin{aligned}
\frac{\partial}{\partial\theta_{[k]}} &\Pr[S = 1] \\
&= \sum_{i=1}^{k-1} \int_0^\infty \left\{\prod_{\substack{j=1\\j\neq i}}^{k-1} H(w, \theta_{[j]})\right\} \\
&\times \left\{h(w + d, \theta_{[i]})\frac{\partial}{\partial\theta_{[k]}}H(w, \theta_{[k]}) - h(w, \theta_{[i]})\frac{\partial}{\partial\theta_{[k]}}H(w + d, \theta_{[k]})\right\} dw.
\end{aligned} \tag{19.20}$$

It follows from (19.20) that $\frac{\partial}{\partial\theta_{[k]}} \Pr[S = 1] \geq 0$ provided the following condition is satisfied: For $\theta_1 \leq \theta_2$,

$$h(w + d, \theta_1)\frac{\partial}{\partial\theta_2}H(w, \theta_2) - h(w, \theta_1)\frac{\partial}{\partial\theta_2}H(w + d, \theta_2) \geq 0 \text{ for all } w \geq 0. \tag{19.21}$$

Condition (19.21) is equivalent to $\psi(d) \geq \psi(0)$, where

$$
\begin{aligned}
\psi(d) &= \frac{\varphi(w + d + \theta_1) + \varphi(w + d - \theta_1)}{\varphi(w + d + \theta_2) - \varphi(w + d - \theta_2)} \\
&= -e^{-\frac{1}{2}[w(\theta_2 - \theta_1) + \theta_1^2 - \theta_2^2]} e^{-d(\theta_2 - \theta_1)} B(d)
\end{aligned}
$$

and

$$
B(d) = \frac{1 + e^{-2(w+d)\theta_1}}{1 - e^{-2(w+d)\theta_2}} .
$$

It is easy to show that $B(d)$, which is positive, is decreasing in d. Consequently, $\psi(d)$ is increasing in $d \geq 0$, which implies that $\Pr[S = 1]$ is increasing in $\theta_{[k]}$ when all other $\theta_{[i]}$'s are kept fixed.

Now, we consider the configuration $\theta_{[1]} \leq \theta_{[2]} \leq \cdots \leq \theta_{[m]} = \cdots = \theta_{[k]} = \theta$, where $1 \leq m \leq k - 1$. Then $\frac{\partial}{\partial \theta} \Pr[S = 1] = \sum_{i=m}^{k} \frac{\partial}{\partial \theta_{[i]}} \Pr[S = 1] \Big|_{\theta_{[m]} = \ldots = \theta_{[k]} = \theta}$.
For evaluating $\frac{\partial}{\partial \theta_{[i]}} \Pr[S = 1]$, $i = m, \ldots, k-1$, we first integrate $A_i(\theta_{[1]}, \ldots, \theta_{[k]})$ by parts and regroup terms as we did earlier for $\frac{\partial}{\partial \theta_{[k]}} \Pr[S = 1]$. Thus

$$
\frac{\partial}{\partial \theta} \Pr[S = 1]
$$
$$
= \sum_{\substack{\alpha = m \\ }}^{k} \sum_{\substack{i=1 \\ i \neq \alpha}}^{k} \int_0^{\infty} \left\{ \prod_{\substack{j=1 \\ j \neq \alpha, i}}^{k} H(w, \theta_{[j]}) \right\}
$$
$$
\times \left\{ h(w + d, \theta_{[i]}) \frac{\partial}{\partial \theta_{[\alpha]}} H(w, \theta_{[\alpha]}) - h(w, \theta_{[i]}) \frac{\partial}{\partial \theta_{[\alpha]}} H(w + d, \theta_{[\alpha]}) \right\} dw,
$$

which is ≥ 0, because condition (19.21) is satisfied. By successive applications of the above result, we see that $\Pr[S = 1]$ is minimized when $\theta_{[1]} = \ldots = \theta_{[k]} = \theta$ and is decreasing in the common value θ. Therefore,

$$
\inf_{\Omega} \Pr[S = 1] = k \int_0^{\infty} H^{k-1}(w, 0) h(w + d, 0) \, dw. \tag{19.22}
$$

The behavior of $\Pr[S = k]$ can be investigated in an analogous manner. We have

$$
\begin{aligned}
\Pr[S = k] &= \Pr[W_{[k]} - W_{[1]} \leq d] \\
&= \sum_{i=1}^{k} \Pr\left[W_{[k]} < W_{(i)} + d, W_{(i)} = W_{[1]} \right] \\
&= \sum_{i=1}^{k} \Pr\left[W_{(j)} < W_{(i)} + d \text{ and } W_{(j)} > W_{(i)} \text{ for all } j \neq i \right] \\
&= \sum_{i=1}^{k} \int_0^{\infty} \left[\prod_{j \neq i} \left\{ H(w + d, \theta_{[j]}) - H(w, \theta_{[j]}) \right\} \right] h(w, \theta_{[i]}) \, dw
\end{aligned}
$$

$$= \sum_{i=1}^{k} B_i(\theta_{[1]}, \ldots, \theta_{[k]}), \text{ say.}$$

Integrating $B_k(\theta_{[1]}, \ldots, \theta_{[k]})$ by parts and regrouping terms, we get

$$\Pr[S = k] = \sum_{i=1}^{k} \int_0^{\infty} \left[\prod_{j=i} \left\{ H(w + d, \theta_{[j]}) - H(w, \theta_{[j]}) \right\} \right]$$

$$\times \left[H(w + d, \theta_{[k]}) h(w; \theta_{[i]}) - H(w, \theta_{[k]}) h(w + d, \theta_{[i]}) \right] dw.$$

Now, it is easy to see that $\frac{\partial}{\partial \theta_{[k]}} \Pr[S = k] \leq 0$ because of (19.21) being satisfied. Thus $\Pr[S = k]$ is decreasing in $\theta_{[k]}$ when all other θ's are kept fixed. Similar to the case of $\Pr[S = 1]$, we can show (proof omitted) that, when $\theta_{[1]} \leq \theta_{[2]} \leq \cdots \leq \theta_{[m]} = \cdots = \theta_{[k]} = \theta$, $\Pr[S = k]$ is decreasing in θ when $\theta_{[1]}, \ldots, \theta_{[m-1]}$ are kept fixed. Consequently, $\Pr[S = k]$ is maximized when $\theta_1 = \cdots = \theta_k = \theta$ and

$$\sup_{\Omega} \Pr[S = k] = k \int_0^{\infty} [H(w + d, 0) - H(w, 0)]^{k-1} h(w, 0) \, dw. \qquad (19.23)$$

References

1. Bechhofer, R. E. (1954). A single-sample multiple decision procedure for ranking means of normal populations with known variances, *Annals of Mathematical Statistics*, **25**, 16–39.

2. Chen, P. (1988). An integrated formulation for selecting the most probable multinomial cell, *Annals of the Institute of Statistical Mathematics*, **40**, 615–625.

3. Chen, P. and Panchapakesan, S. (1994). An integrated formulation for selecting the best normal population and eliminating the bad ones, In *Proceedings in Computational Statistics: Short Communications* (Eds., R. Dutter and W. Grossmann), 11th Symposium on Computational Statistics, Vienna, Austria, pp. 18–19, Vienna, Austria: Vienna Institutes for Statistics et al.

4. Chen P. and Sobel, M. (1987a). An integrated formulation for selecting the t best of k normal populations, *Communications in Statistics—Theory and Methods*, **16**, 121–146.

5. Chen, P. and Sobel, M. (1987b). A new formulation for the multinomial selection problem, *Communications in Statistics-Theory and Methods*, **16**, 147–180.

6. Gupta, S. S. (1956). On a decision rule for a problem in ranking means, Mimeograph Series No. 150, Institute of Statistics, University of North Carolina, Chapel Hill, North Carolina.

7. Gupta, S. S. (1963). Probability integrals of the multivariate normal and multivariate t, *Annals of Mathematical Statistics*, **34**, 792–828.

8. Gupta, S. S. (1965). On some multiple decision (selection and ranking) rules, *Technometrics*, **7**, 225–245.

9. Gupta, S. S., Nagel, K. and Panchapakesan, S. (1973). On the order statistics from equally correlated normal random variables, *Biometrika*, **60**, 403–413.

10. Gupta, S. S. and Panchapakesan, S. (1972). On a class of subset selection procedures, *Annals of Mathematical Statistics*, **43**, 814–822.

11. Lehmann, E. (1986). *Testing Statistical Hypotheses*, Second Edition, New York: John Wiley & Sons.

12. Rizvi, M. H. (1971). Some selection problems involving folded normal distribution, *Technometrics*, **13**, 355–369.

Table 19.1: Values of $C = \frac{\sqrt{n}c}{\sigma}$ satisfying (19.14) with equality
$$P^* = 0.90$$

TABLE 1. VALUES OF $C = \frac{\sqrt{n}c}{\sigma}$ SATISFYING (4.8) WITH EQUALITY
$$P^* = 0.90$$

k	Δ*	C	k	Δ*	C	k	Δ*	C	k	Δ*	C
2	2.4	0.0960	4	3.0	0.1791	6	4.2	1.1558	8	5.6	2.4240
	2.6	0.2897		3.2	0.3789		4.4	1.3558		5.8	2.6241
	2.8	0.4879		3.4	0.5788		4.6	1.5553		6.0	2.8240
	3.0	0.6871		3.6	0.7789		4.8	1.7556		6.2	3.0241
	3.2	0.8874		3.8	0.9788		5.0	1.9558		6.4	3.2240
	3.4	1.0868		4.0	1.1789		5.2	2.1558		6.6	3.4240
	3.6	1.2871		4.2	1.3789		5.4	2.3558		6.8	3.6240
	3.8	1.4874		4.4	1.5783		5.6	2.5561		7.0	3.8240
	4.0	1.6873		4.6	1.7786		5.8	2.7558	9	3.4	0.1572
	4.2	1.8874		4.8	1.9789		6.0	2.9556		3.6	0.3572
	4.4	2.0874		5.0	2.1789		6.2	3.1562		3.8	0.5572
	4.6	2.2873		5.2	2.3789		6.4	3.3558		4.0	0.7572
	4.8	2.4880		5.4	2.5791		6.6	3.5558		4.2	0.9571
	5.0	2.6874		5.6	2.7783		6.8	3.7538		4.4	1.1572
	5.2	2.8868		5.8	2.9787		7.0	3.9559		4.6	1.3572
	5.4	3.0868		6.0	3.1789	7	3.2	0.0923		4.8	1.5572
	5.6	3.2874		6.2	3.3789		3.4	0.2923		5.0	1.7569
	5.8	3.4874		6.4	3.5789		3.6	0.4918		5.2	1.9572
	6.0	3.6872		6.6	3.7789		3.8	0.6920		5.4	2.1572
	6.2	3.8875		6.8	3.9789		4.0	0.8923		5.6	2.3572
	6.4	4.0872		7.0	4.1789		4.2	1.0917		5.8	2.5575
	6.6	4.2873	5	3.0	0.0266		4.4	1.2920		6.0	2.7572
	6.8	4.4873		3.2	0.2259		4.6	1.4923		6.2	2.9573
	7.0	4.6871		3.4	0.4261		4.8	1.6922		6.4	3.1578
3	2.6	0.1365		3.6	0.6263		5.0	1.8923		6.6	3.3572
	2.8	0.3349		3.8	0.8265		5.2	2.0923		6.8	3.5572
	3.0	0.5347		4.0	1.0261		5.4	2.2922		7.0	3.7572
	3.2	0.7346		4.2	1.2263		5.6	2.4923	10	3.4	0.0986
	3.4	0.9346		4.4	1.4265		5.8	2.6923		3.6	0.2987
	3.6	1.1346		4.6	1.6264		6.0	2.8922		3.8	0.4986
	3.8	1.3346		4.8	1.8264		6.2	3.0924		4.0	0.6982
	4.0	1.5342		5.0	2.0265		6.4	3.2922		4.2	0.8986
	4.2	1.7347		5.2	2.2263		6.6	3.4923		4.4	1.0986
	4.4	1.9346		5.4	2.4264		6.8	3.6921		4.6	1.2986
	4.6	2.1344		5.6	2.6265		7.0	3.8925		4.8	1.4988
	4.8	2.3345		5.8	2.8262	8	3.2	0.0239		5.0	1.6984
	5.0	2.5346		6.0	3.0266		3.4	0.2240		5.2	1.8986
	5.2	2.7348		6.2	3.2264		3.6	0.4239		5.4	2.0986
	5.4	2.9346		6.4	3.4265		3.8	0.6239		5.6	2.2985
	5.6	3.1346		6.6	3.6265		4.0	0.8240		5.8	2.4992
	5.8	3.3344		6.8	3.8271		4.2	1.0238		6.0	2.6985
	6.0	3.5341		7.0	4.0264		4.4	1.2239		6.2	2.8985
	6.2	3.7342	6	3.2	0.1558		4.6	1.4240		6.4	3.0990
	6.4	3.9348		3.4	0.3558		4.8	1.6240		6.6	3.2986
	6.6	4.1346		3.6	0.5558		5.0	1.8234		6.8	3.4986
	6.8	4.3348		3.8	0.7558		5.2	2.0240		7.0	3.6984
	7.0	4.5341		4.0	0.9557		5.4	2.2240			

Table 19.1: Values of $C = \frac{\sqrt{n}c}{\sigma}$ satisfying (19.14) with equality
$P^* = 0.95$

TABLE 1. VALUES OF $C = \frac{\sqrt{n}c}{\sigma}$ SATISFYING (4.8) WITH EQUALITY
$P^* = 0.95$

k	Δ*	C	k	Δ*	C	k	Δ*	C	k	Δ*	C
2	2.8	0.0371	3	7.0	4.1876	6	4.4	0.9245	8	5.6	2.0031
	3.0	0.2331	4	3.4	0.1392		4.6	1.1244		5.8	2.2036
	3.2	0.4315		3.6	0.3391		4.8	1.3246		6.0	2.4036
	3.4	0.6320		3.8	0.5393		5.0	1.5235		6.2	2.6045
	3.6	0.8319		4.0	0.7393		5.2	1.7236		6.4	2.8032
	3.8	1.0320		4.2	0.9393		5.4	1.9234		6.6	3.0032
	4.0	1.2314		4.4	1.1392		5.6	2.1244		6.8	3.2036
	4.2	1.4321		4.6	1.3394		5.8	2.3243		7.0	3.4034
	4.4	1.6315		4.8	1.5386		6.0	2.5246	9	3.8	0.1296
	4.6	1.8317		5.0	1.7385		6.2	2.7242		4.0	0.3296
	4.8	2.0317		5.2	1.9385		6.4	2.9240		4.2	0.5298
	5.0	2.2319		5.4	2.1392		6.6	3.1245		4.4	0.7298
	5.2	2.4316		5.6	2.3392		6.8	3.3244		4.6	0.9298
	5.4	2.6326		5.8	2.5394		7.0	3.5244		4.8	1.1296
	5.6	2.8316		6.0	2.7391	7	3.6	0.0728		5.0	1.3293
	5.8	3.0317		6.2	2.9389		3.8	0.2728		5.2	1.5298
	6.0	3.2320		6.4	3.1393		4.0	0.4728		5.4	1.7292
	6.2	3.4317		6.6	3.3382		4.2	0.6725		5.6	1.9286
	6.4	3.6318		6.8	3.5393		4.4	0.8725		5.8	2.1297
	6.6	3.8320		7.0	3.7397		4.6	1.0723		6.0	2.3297
	6.8	4.0322	5	3.6	0.1720		4.8	1.2729		6.2	2.5301
	7.0	4.2319		3.8	0.3722		5.0	1.4722		6.4	2.7294
3	3.0	0.1881		4.0	0.5726		5.2	1.6718		6.6	2.9293
	3.2	0.3877		4.2	0.7725		5.4	1.8728		6.8	3.1299
	3.4	0.5868		4.4	0.9726		5.6	2.0726		7.0	3.3289
	3.6	0.7877		4.6	1.1716		5.8	2.2720	10	3.8	0.0655
	3.8	0.9876		4.8	1.3719		6.0	2.4728		4.0	0.2643
	4.0	1.1876		5.0	1.5726		6.2	2.6728		4.2	0.4644
	4.2	1.3876		5.2	1.7724		6.4	2.8724		4.4	0.6638
	4.4	1.5876		5.4	1.9720		6.6	3.0730		4.6	0.8636
	4.6	1.7876		5.6	2.1726		6.8	3.2729		4.8	1.0644
	4.8	1.9876		5.8	2.3725		7.0	3.4728		5.0	1.2645
	5.0	2.1875		6.0	2.5736	8	3.6	0.0025		5.2	1.4644
	5.2	2.3880		6.2	2.7721		3.8	0.2029		5.4	1.6644
	5.4	2.5871		6.4	2.9721		4.0	0.4031		5.6	1.8654
	5.6	2.7877		6.6	3.1726		4.2	0.6036		5.8	2.0642
	5.8	2.9876		6.8	3.3723		4.4	0.8035		6.0	2.2639
	6.0	3.1875		7.0	3.5718		4.6	1.0036		6.2	2.4645
	6.2	3.3876	6	3.6	0.1243		4.8	1.2027		6.4	2.6642
	6.4	3.5876		3.8	0.3243		5.0	1.4031		6.6	2.8638
	6.6	3.7881		4.0	0.5245		5.2	1.6036		6.8	3.0645
	6.8	3.9873		4.2	0.7233		5.4	1.8034		7.0	3.2644

20

Applications of Two Majorization Inequalities to Ranking and Selection Problems

Y. L. Tong

Georgia Institute of Technology, Atlanta, GA

Abstract: Multivariate probability inequalities via majorization ordering have been found useful in a variety of statistical applications. This paper deals with applications of two such inequalities in ranking and selection problems. They provide upper and lower bounds for the true probability function of a correct selection, and the results apply to selection procedures under Bechhofer's indifference-zone formulation and Gupta's maximum-type subset formula.

Keywords and phrases: Probability bounds, indifference-zone formulation, subset formulation, majorization and Schur-concavity

20.1 Introduction and Summary

This paper concerns applications of two probability inequalities to ranking and selection problems under Bechhofer's indifference zone formulation and Gupta's maximum-type subset selection procedures. Under these formulations, the probability of a correct selection (PCS) involves the parameters and the number of the populations. Two majorization-related probability inequalities that depend on heterogeneity of the parameters and dimension reduction are shown to provide bounds under such ranking and selection procedures. In Section 2, the basic idea of majorization is reviewed and the inequalities are stated. Section 3 concerns bounds for the PCS functions under the indifference zone formulation, and results under the subset formulation are given in Section 4. In both cases the inequality for the distribution function yields upper bounds for the true PCS function for location parameter families, and the dimension-reduction inequality provides lower bounds under the least favorable configuration. A practical application of the lower bounds yields a conservative solution for determining the constant associated with the procedure based on probability tables that are already available for lower dimensions.

20.2   Majorization and Two Useful Probability Inequalities

20.2.1   A review of majorization

The notion of majorization deals with heterogeneity of the components of real vectors. Let

$$\mathbf{a} = (a_1, \ldots, a_m), \qquad \mathbf{b} = (b_1, \ldots, b_m)$$

be two m-dimensional real vectors, and let

$$a_{[1]} \geq \cdots \geq a_{[m]}, \qquad b_{[1]} \geq \cdots \geq b_{[m]}$$

denote the corresponding ordered values of the a_i's and the b_i's. $\mathbf{a}$ is said to majorize $\mathbf{b}$, in symbols $\mathbf{a} \succ \mathbf{b}$, if

$$\sum_{i=1}^{m_1} a_{[i]} \geq \sum_{i=1}^{m_1} b_{[i]}$$

holds for all $m_1 < m$ and equality holds for $m_1 = m$. This notion defines a partial ordering in the sense that if $\mathbf{a} \succ \mathbf{b}$, then the a_i's are more heterogeneous than the b_i's. The following fact provides a simple motivation for the idea involved:

Fact 20.2.1 *(i)* $(a_1, \ldots, a_m) \succ (\bar{a}, \ldots, \bar{a})$ *always holds where* $\bar{a} = \frac{1}{m}\sum_{i=1}^{m} a_i$ *is the arithmetic mean.*
(ii) If $b_i \geq 0$ $(i = 1, \ldots, m)$, *then* $(\sum_{i=1}^{m} b_i, 0, \ldots, 0) \succ (b_1, \ldots, b_m)$.
(iii) If $\mathbf{a} \succ \mathbf{b}$, *then* $\frac{1}{m-1}\sum_{i=1}^{m}(a_i - \bar{a})^2 \geq \frac{1}{m-1}\sum_{i=1}^{m}(b_i - \bar{b})^2$ *holds; thus the majorization ordering is stronger than the variance ordering.*

A real-valued function $h : \mathbb{R}^m \to \mathbb{R}$ is called a Schur-concave (Schur-convex) function if $\mathbf{a} \succ \mathbf{b}$ implies $h(\mathbf{a}) \leq (\geq)h(\mathbf{b})$. That is, subject to the condition that the sums of the a_i's and of the b_i's are equal, the value of a Schur-concave function is larger when the components of the vector are less heterogeneous. Schur-concave functions make their appearance in various applications in statistics, and a comprehensive reference on this topic is the Marshall-Olkin (1979) book. Some recent results can be found in Tong (1988), and special results for the multivariate normal distribution are given in Tong (1990, Section 4.4 and Chapter 7).

In the rest of this section we state two majorization inequalities which have been found useful for studying ranking and selection problems.

20.2.2 A Schur-Concavity property of the joint distribution function

In certain statistical applications, it can be easily verified that the joint density function $f(\mathbf{x}) \equiv f(x_1, \ldots, x_m)$ of an m-dimensional random vector $\mathbf{X} = (X_1, \ldots, X_m)$ is a Schur-concave function. For example, the following fact is known to be true.

Fact 20.2.2 *(i) If $\mathbf{X}$ is distributed according to a multivariate normal distribution with a common mean μ, a common variance $\sigma^2 > 0$, and a common correlation coefficient $\rho \in (-\frac{1}{(m-1)}, 1)$, then $f(\mathbf{x})$ is a Schur-concave function of $\mathbf{x}$.*
(ii) If $X_1, \ldots, X_m$ are i.i.d. random variables with a marginal density function $f^*(x)$ such that $\log f^*(x)$ is a concave function of x for x in the support of f^*, then $f(\mathbf{x}) = \prod_{i=1}^m f^*(x_i)$ is a Schur-concave function of $\mathbf{x} = (x_1, \ldots, x_m)$. [Note that this log-concavity condition of f^* is satisfied in most statistical applications.]

An important question is then whether this Schur-concavity property can be passed on from the joint density function to the joint distribution function. If the answer is in the affirmative, then probability inequalities follow immediately. The answer to this question was provided by Marshall-Olkin (1974, 1979, p. 100).

Theorem 20.2.1 *If the density function $f(\mathbf{x})$ is a Schur-concave function of $\mathbf{x}$, then the corresponding distribution function $F(\mathbf{x})$ is also Schur-concave. As a special case, the Schur-concavity of $f(\mathbf{x})$ implies*

$$F(x_1, \ldots, x_m) \equiv P[X_i \leq x_i, i = 1, \ldots, m] \leq F(\bar{x}, \ldots, \bar{x}), \quad \bar{x} = \frac{1}{m} \sum_{i=1}^m x_i.$$

$$(20.1)$$

20.2.3 A Schur-Convexity property for exchangeable random variables

The next result uses majorization in a different fashion, and it involves exchangeable random variables. An infinite sequence of random variables $\{X_i^*\}_{i=1}^\infty$ is said to be exchangeable if, for all finite m and for all permutations $(\pi_1, \ldots, \pi_m)$ of $\{1, \ldots, m\}$, the random vectors $(X_{\pi_1}^*, \ldots, X_{\pi_m}^*)$ and $(X_1^*, \ldots, X_m^*)$ have the same distribution. Let $\mathbf{X} = (X_1, \ldots, X_m)$ be an m-dimensional random vector. Its components, $X_1, \ldots, X_m$, are said to be exchangeable if there exists an exchangeable sequence $\{X_i^*\}_{i=1}^\infty$ such that $(X_1^*, \ldots, X_m^*)$ and $\mathbf{X}$ have the same distribution. By de Finetti's Theorem, exchangeability is equivalent to a mixture of conditionally i.i.d. random variables [see, e.g., Loève (1963, p. 365)]. In other words, $X_1, \ldots, X_m$ are exchangeable if and only if there exist random variables $U_1, \ldots, U_m, U_0$ such that (1) for every given $U_0 = u_0$ in the support of the density function of U_0 the random variables $U_1, \ldots, U_m$ are conditionally i.i.d., and (2) the unconditional distribution of $(U_1, \ldots, U_m)$ and the distribution of $(X_1, \ldots, X_m)$ are identical.

This mixing property provides an analytical tool for proving useful results via exchangeability. In a recent paper Tong (1997) proved a general theorem via this property by applying a majorization-related moment inequality. A special application of the general theorem yields the following result:

Theorem 20.2.2 *Let $X_1, \ldots, X_m$ be exchangeable random variables. Then*

$$P[X_i \in A, i = 1, \ldots, m] \geq$$
$$P[X_i \in A, i = 1, \ldots, m_1]P[X_i \in A, i = 1, \ldots, m - m_1] \quad (20.2)$$

and

$$P[X_i \in A, i = 1, \ldots, m] \geq \{P[X_i \in A, i = 1, \ldots, m_1]\}^{m/m_1} \quad (20.3)$$

hold for all Borel-measurable sets A and all $m_1 < m$.

The inequalities in (20.2) and (20.3) are special cases of the general result, and a direct proof is possible without applying the majorization-related moment inequality. The proof given below depends on the log-convexity property of the moment function of a non-negative random variable.

PROOF OF THEOREM 20.2.2. Let W be any nonnegative random variable such that $EW^m \equiv \mu_m$ exists for given $m > 0$ ($\mu_0 \equiv 1$). Then, as a function of t, μ_t is log-convex for $t \in [0, m]$, hence $\mu_m \geq (\mu_{m_1})^{m/m_1}$ holds for all $0 < m_1 < m$ [see, e.g., Loève (1963, p. 156)]. By de Finetti's theorem, $X_1, \ldots, X_n$ are exchangeable iff $(X_1, \ldots, X_m)$ and $(q(U_1, U_0), \ldots, q(U_m, U_0))$ are identically distributed for some Borel function q. Thus for every Borel set A we have

$$\begin{aligned} P\left[\bigcap_{i=1}^m (X_i \in A)\right] &= \left[P\bigcap_{i=1}^m (q(U_1, U_0) \in A)\right] \\ &= E^{U_0} P\left[\bigcap_{i=1}^m (q(U_i, U_0) \in A) \mid U_0\right] \\ &= E^{U_0} (P[q(U_1, U_0) \in A \mid U_0])^m \\ &\geq \left\{E^{U_0} (P[q(U_1, U_0) \in A \mid U_0])^{m_1}\right\}^{m/m_1} \\ &= \{P[X_i \in A, i = 1, \ldots, m_1]\}^{m/m_1} \end{aligned}$$

for all $m_1 < m$, where the inequality follows by letting $P[q(U_1, U_0) \in A \mid U_0]$ play the role of W. This establishes (20.3). The inequality in (20.2) follows immediately from

$$\{P[X_i \in A, i = 1, \ldots, m]\}^{m_1} \geq \{P[X_i \in A, i = 1, \ldots, m_1]\}^m$$

and

$$\{P[X_i \in A, i = 1, \ldots, m]\}^{m-m_1} \geq \{P[X_i \in A, i = 1, \ldots, m - m_1]\}^m.$$

∎

We note in passing that the inequalities in (20.2) and (20.3) can be treated as majorization-related inequalities because, in (20.2) the majorization ordering $(m, 0) \succ (m_1, m - m_1)$ leads to the corresponding probability inequality. A similar statement also holds for (20.3).

20.3 Probability Bounds Under the Indifference-Zone Formulation

Under Bechhofer's (1954) indifference-zone formulation, the problem concerns the selection of the population associated with the largest parameter such that the probability of a correct selection (PCS) is at least P^* (preassigned) under the least favorable configuration (LFC).

20.3.1 The PCS function and the LFC

For fixed n let $\mathbf{Z}_i = \{Z_{ij}\}_{j=1}^n$ be such that $Z_{i1}, \ldots, Z_{in}$ are i.i.d. random variables from a population whose distribution depends on a parameter θ_i $(i = 1, \ldots, k)$. Without loss of generality, we may assume that $\theta_k = \max_{1 \leq i \leq k} \theta_i$. Let $Y = \psi(X_1, \ldots, X_n)$ be a suitably-chosen statistic such that $Y_i = \psi(Z_{i1}, \ldots, Z_{in})$ has a p.d.f. $g_{n,\theta_i}(y)$ for $i = 1, \ldots, k$. It is generally assumed that the family of p.d.f.'s

$$\mathcal{G}_n = \{g_{n,\theta}(y) : \theta \in \Omega\}$$

is stochastically increasing (SI). Then, under Bechhofer's "natural" selection rule, the PCS function is given by

$$P_{m,n,\boldsymbol{\theta}}(CS) = P_{m,n,\boldsymbol{\theta}}[Y_i \leq Y_k, i = 1, \ldots, m] \equiv \beta_1(m, \boldsymbol{\theta}), \qquad (20.4)$$

where $\boldsymbol{\theta} = (\theta_1, \ldots, \theta_k)$ and $m = k - 1$. Let $d(\cdot, \cdot)$ denote a distance function such that $d(\theta_i, \theta_k) \downarrow \theta_i$ for all fixed $\theta_k > \theta_i$ and $d(\theta_i, \theta_k) \uparrow \theta_k$ for all fixed $\theta_i < \theta_k$. For a given θ_k let $d_0 > d(\theta_k, \theta_k)$ be preassigned and let $\Lambda(\theta_k)$ be a subset of $\Omega \times \cdots \times \Omega$ such that

$$\Lambda(\theta_k) = \{\boldsymbol{\theta} : d(\theta_i, \theta_k) \geq d_0\}.$$

If $\mathcal{G}_n$ is a SI family, then by the SI property of $\mathcal{G}_n$, we have

$$\inf_{\boldsymbol{\theta} \in \Lambda(\theta_k)} P_{m,n,\boldsymbol{\theta}}(CS) = P_{m,n,\boldsymbol{\theta}^*(\theta_k)}(CS), \qquad (20.5)$$

where $\boldsymbol{\theta}^*(\theta_k)$ represents the least-favorable configuration (LFC) for given $\theta_k \in \Omega$, and it is of the form

$$\theta_1 = \cdots = \theta_m, \qquad d(\theta_1, \theta_k) = d_0. \qquad (20.6)$$

20.3.2 A lower bound for the PCS function under LFC

With the above observation, we immediately obtain the following fact:

Corollary 20.3.1 *Assume that the $\mathbf{Z}_i$'s (and hence the Y_i's) are independent and that $\mathcal{G}_n$ is a SI family. For an arbitrary but fixed $\theta_k \in \Omega$ let $\boldsymbol{\theta}^*(\theta_k)$ be of the form in (20.6). Then when $\boldsymbol{\theta} = \boldsymbol{\theta}^*(\theta_k)$ holds, the random variables $Y_1 - Y_k, \ldots, Y_m - Y_k$ are exchangeable. Thus an application of* Theorem 20.2.2 *yields the following dimension-reduction inequalities:*

$$\beta_1(m, \boldsymbol{\theta}^*(\theta_k)) \geq \beta_1(m_1, \boldsymbol{\theta}_1^*(\theta_k))\beta_1(m - m_1, \boldsymbol{\theta}_2^*(\theta_k)), \qquad (20.7)$$

$$\beta_1(m, \boldsymbol{\theta}^*(\theta_k)) \geq [\beta_1(m_1, \boldsymbol{\theta}_1^*(\theta_k))]^{m/m_1}; \qquad (20.8)$$

where m_1 is any positive integer that satisfies $m_1 < m = k - 1$ and

$$\boldsymbol{\theta}_1^* = (\theta_1, \ldots, \theta_1, \theta_k), \qquad \boldsymbol{\theta}_2^* = (\theta_1, \ldots, \theta_1, \theta_k)$$

are (m_1+1)- and $(m-m_1+1)$-dimensional parameter vectors such that $d(\theta_1, \theta_k) = d_0$.

20.3.3 An upper bound for the true PCS function for location parameter families

For the location parameter families in which $d(\theta_i, \theta_k) = \theta_k - \theta_i$ and $g_{n,\theta}(y) = g_n(y - \theta)$ for $y, \theta \in \mathbb{R}$, the PCS function given in (20.4) depends on $\boldsymbol{\theta}$ only through

$$\boldsymbol{\delta} = (\delta_1, \ldots, \delta_m) \equiv (\theta_k - \theta_1, \ldots, \theta_k - \theta_m), \qquad (20.9)$$

where $\delta_i \geq 0$ ($i = 1, \ldots, m$). For any given parameter vector $\boldsymbol{\theta}$ in $\Omega \times \cdots \times \Omega$, the *true* probability of a correct section can be expressed as (from (20.4))

$$\begin{aligned} P_{m,n,\boldsymbol{\theta}}(CS) &= P[Y_i' \leq Y_k' + \delta_i, \ i = 1, \ldots, m] \\ &= \int_{-\infty}^{\infty} \left[\prod_{i=1}^{m} G_n(y + \delta_i)\right] g_n(y)dy \\ &\equiv \beta_2(\boldsymbol{\delta}) \qquad (20.10) \end{aligned}$$

where $Y_1', \ldots, Y_k'$ are i.i.d. random variables with p.d.f. $g_{n,\theta=0}(y) \equiv g_n(y)$ and a corresponding distribution function $G_n(y)$. An application of Theorem 20.2.1 then yields the following result:

Corollary 20.3.2 *Assume that $g_{n,\theta}(y) = g_n(y - \theta)$ for $y, \theta \in \mathbb{R}$. If (i) either the joint density function of $(Y_1' - Y_k', \ldots, Y_m' - Y_k')$ is a Schur-concave function or (ii) the (marginal) distribution function $G_n(y)$ is log-concave, then the true PCS function $\beta_2(\boldsymbol{\delta})$ given in (20.10) is a Schur-concave function of $\boldsymbol{\delta}$. As a special consequence,*

$$\beta_2(\boldsymbol{\delta}) \leq \beta_2(\bar{\boldsymbol{\delta}}) \tag{20.11}$$

holds for all m, where

$$\bar{\boldsymbol{\delta}} = \left(\theta_k - \sum_{i=1}^{m} \theta_i/m, \ldots, \theta_k - \sum_{i=1}^{m} \theta_i/m \right). \tag{20.12}$$

20.3.4 The normal family

In his pioneering work on ranking and selection problems, Bechhofer (1954) considered the case in which $Z_{i1}, \ldots, Z_{in}$ are i.i.d. $\mathcal{N}(\theta_i, \sigma^2)$ random variables and σ^2 is assumed to be known. Let $Y_i = \frac{1}{n} \sum_{i=1}^{n} Z_{ij}$ $(i = 1, \ldots, k)$ be the sample means, then $g_{n,\theta}$ is just the density function of an $\mathcal{N}(\theta, \sigma^2/n)$ distribution. It is well-known that in this special case the true PCS function is given by

$$
\begin{aligned}
P_{m,n,\boldsymbol{\theta}}(CS) &= P[Y_i' - Y_k' \leq \delta_i, i = 1, \ldots, m] \\
&= \int_{-\infty}^{\infty} \left[\prod_{i=1}^{m} \Phi\left(\sqrt{n}(y + \delta_i)/\sigma \right) \right] (\sqrt{n}/\sigma) \, \phi\left(\sqrt{n}y/\sigma \right) dy,
\end{aligned}
$$

where $(Y_1' - Y_k', \ldots, Y_m' - Y_k')$ has an $\mathcal{N}_m(\mathbf{0}, (2\sigma^2/n)\boldsymbol{\Sigma})$ distribution such that the diagonal elements of $\boldsymbol{\Sigma}$ are 1 and the off-diagonal elements are $\frac{1}{2}$, and ϕ and Φ are the $\mathcal{N}(0, 1)$ p.d.f. and distribution function, respectively. Consequently, its joint density function is Schur-concave [Tong (1990, p. 83)]. On the other hand, it is also known that $\log \Phi(\sqrt{n}y/\sigma)$ is a concave function of y [see Tong (1978)]. Thus conditions (i) and (ii) in Corollary 20.3.2 are both satisfied, and an upper bound on the true PCS function can be obtained by applying the result given in Corollary 20.3.2. A special result is that, for all population means $\theta_1, \ldots, \theta_m, \theta_k$, the true PCS function is maximized when the means of the inferior populations are all equal to their (arithmetic) average.

On the other hand, an application of Corollary 20.3.1 yields a lower bound on the PCS function, hence an upper bound on the sample size required to guarantee a probability requirement under the LFC. Since tables for the equi-coordinate percentage points for exchangeable normal random variables with lower dimensions are already available [e.g., Bechhofer (1954), Gupta (1963b), Gupta, Nagel and Panchapakesan (1973) and Tong (1990)], Corollary 20.3.1 implies that a conservative solution can be obtained based on these existing tables when k is large. The details are left to the reader.

20.4   Bounds for PCS Functions Under the Subset Selection Formulation

20.4.1   Gupta's maximum-type selection rules

Under Gupta's (1956, 1963a, 1965) subset selection formulation, a subset of the k populations (with the size of the subset being random) is selected, and a selection is said to be correct if the best population with parameter θ_k is included in the selected subset. In particular, he proposed the following selection rules for location and scale parameter families:

R_1 (location parameter families): Include the i-th population in the selected subset iff $Y_i \geq \max_{1 \leq i' \leq k} Y_{i'} - c_1$,

R_2 (scale parameter families): Include the i-th population in the selected subset iff $Y_i \geq c_2 \max_{1 \leq i' \leq k} Y_{i'}$

where $c_1 > 0$ and $c_2 \in (0, 1)$. This type of selection rules are known as Gupta's maximum-type rules, and they possess certain optimal properties. For example, they are known to be approximately Bayes-optimal under reasonable loss functions [see, e.g., Gupta and Panchapakesan (1979, 1985)]. It is also known that when $\mathcal{G}_n = \{g_{n,\theta}(y) : \theta \in \Omega\}$ is a SI family, then the LFC (under which the PCS function is minimized over all $\boldsymbol{\theta} = (\theta_1, \dots, \theta_k) \in \Omega \times \cdots \times \Omega$) is of the form

$$\boldsymbol{\theta} = \boldsymbol{\theta}^* \equiv (\theta, \dots, \theta)$$

for some $\theta \in \Omega$. Furthermore, without loss of generality we may assume that $\theta = 0$ ($\theta = 1$) for the location (the scale) parameter family for the purpose of evaluating the PCS.

In the rest of this section we illustrate how Theorems 20.2.1 and 20.2.2 yield bounds for the PCS functions for the maximum-type selection rules for location and scale parameter families.

20.4.2   Location parameter families

It is easy to see that if $\mathcal{G}_n$ is a family of p.d.f.'s with a location parameter, then under selection rule R_1 the true PCS function is given by

$$
\begin{aligned}
P_{m,n,\boldsymbol{\theta}}(CS) &= P_{\boldsymbol{\theta}}\left[Y_k \geq \max_{1 \leq i \leq m} Y_i - c_1\right] \\
&= \int_{-\infty}^{\infty} \prod_{i=1}^{m} G_n(y + c_1 + \delta_i) g_n(y) dy \\
&\equiv \gamma_1(m, \boldsymbol{\delta}),
\end{aligned}
\tag{20.13}
$$

where $m = k - 1$, $\boldsymbol{\theta} = (\theta_1, \ldots, \theta_k) \in \Omega \times \cdots \times \Omega$, and $\delta_i = \theta_k - \theta_i$ $(i = 1, \ldots, m)$. An application of Theorem 20.2.1 then yields the following result:

Corollary 20.4.1 *Assume that the conditions in* Corollary 20.3.2 *are satisfied. Then for every* m, $\gamma_1(m, \boldsymbol{\delta})$ *is a Schur-concave function of* $\boldsymbol{\delta}$. *In particular, the r.h.s. of* (20.13) *is maximized when the* δ_i's *are substituted with* $\bar{\delta} = \frac{1}{m} \sum_{i=1}^m \delta_i$.

On the other hand, when the δ_i values are the same, the PCS function on the r.h.s. of (20.13) involves exchangeable random variables. Thus an application of Theorem 20.2.2 also yields:

Corollary 20.4.2 *Assume that* $\delta_1 = \cdots = \delta_m = \delta^* \geq 0$ *holds (i.e., the* δ_i's *are in a slippage configuration). Then under Gupta's maximum-type selection rules of the form* R_1 *the following dimension-reduction inequalities hold:*

$$\gamma_1(m, (\delta^*, \ldots, \delta^*)) \geq \gamma_1(m_1, (\delta^*, \ldots, \delta^*))\gamma_1(m - m_1, (\delta^*, \ldots, \delta^*)) \quad (20.14)$$

and

$$\gamma_1(m, (\delta^*, \ldots, \delta^*)) \geq [\gamma_1(m_1, (\delta^*, \ldots, \delta^*))]^{m/m_1}; \quad (20.15)$$

where the vectors $(\delta^*, \ldots, \delta^*)$ *are of appropriate dimensions.*

20.4.3 Scale parameter families

Similarly, consider the situation in which $\mathcal{G}_n$ is a scale parameter family such that

$$\mathcal{G}_n = \{g_n(y, \theta) = \frac{1}{\theta} g_n(y/\theta) : y \geq 0, \ \theta > 0\}$$

for some p.d.f. g_n. By letting $\lambda_i = \theta_k/\theta_i$ $(i = 1, \ldots, m)$, it follows that the PCS function under Gupta's maximum-type selection rules R_2 is of the form

$$
\begin{aligned}
P_{m,n,\boldsymbol{\theta}}(CS) &= \int_0^\infty \left[\prod_{i=1}^m G_n(y/c_2\theta_i) \right] \theta_k^{-1} g_n(y/\theta_k) dy \\
&= \int_0^\infty \left[\prod_{i=1}^m G_n(\lambda_i u/c_2) \right] g_n(u) du \\
&\equiv \gamma_2(m, \boldsymbol{\lambda}),
\end{aligned}
\quad (20.16)
$$

where $\boldsymbol{\lambda} = (\lambda_1, \ldots, \lambda_m)$. Thus an application of Theorem 20.2.2 yields the following result:

Corollary 20.4.3 *Assume that* $\lambda_1 = \cdots = \lambda_m = \lambda^* \geq 1$ *(i.e., the* λ_i's *are in a slippage configuration). Then under Gupta's maximum-type selection rules we have*

$$\gamma_2(m, (\lambda^*, \ldots, \lambda^*)) \geq \gamma_2(m_1, (\lambda^*, \ldots, \lambda^*))\gamma_2(m - m_1, (\lambda^*, \ldots, \lambda^*)) \quad (20.17)$$

and

$$\gamma_2(m, (\lambda^*, \ldots, \lambda^*)) \geq [\gamma_2(m_1, (\lambda^*, \ldots, \lambda^*))]^{m/m_1}, \quad (20.18)$$

where the vectors $(\lambda^*, \ldots, \lambda^*)$ *are also of appropriate dimensions.*

In practical applications, the result in Corollary 20.4.3 can be applied to obtain a conservative solution for determining the constant associated with the procedure based on probability tables that are already available for lower dimensions. The details are left to the reader.

Acknowledgements. This paper is dedicated to Professors Shanti S. Gupta and Milton Sobel, and to the memory of Professor Robert E. Bechhofer; the pioneers and leaders in the area of ranking and selection problems. The author thanks Mercer University for providing library facilities while, during the course of this research, the Georgia Tech campus was transformed into the Centennial Olympic Village and was not easily accessible to its faculty members.

References

1. Bechhofer, R. E. (1954). A single-sample multiple decision procedure for ranking means of normal populations with known variances, *Annals of Mathematical Statistics*, **25**, 16–39.

2. Gupta, S. S. (1956). On a decision rule for a problem in ranking means, Mimeograph Series No. 150, Institute of Statistics, University of North Carolina, Chapel Hill, NC.

3. Gupta, S. S. (1963a). On a selection and ranking procedure for gamma populations, *Annals of the Institute of Statistical Mathematics*, **14**, 199–216.

4. Gupta, S. S. (1963b). Probability integrals of multivariate normal and multivariate t, *Annals of Mathematical Statistics*, **34**, 792–828.

5. Gupta, S. S. (1965). On some multiple decision (selection and ranking) rules, *Technometrics*, **7**, 225–245.

6. Gupta, S. S., Nagel, K. and Panchapakesan, S. (1973). On the order statistics from equally correlated normal random variables, *Biometrika*, **60**, 403–413.

7. Gupta, S. S. and Panchapakesan, S. (1979). *Multiple Decision Procedures: Theory and Methodology of Selecting and Ranking Populations*, New York: John Wiley & Sons.

8. Gupta, S. S. and Panchapakesan, S. (1985). Subset selection procedures: Review and assessment. *American Journal of Mathematical Management and Sciences*, **5**, 235–311.

9. Loève, M. (1963). *Probability Theory*, Third edition, Princeton: NJ: D. Van Nostrand.

10. Marshall, A. W. and Olkin, I. (1974). Majorization in multivariate distributions, *Annals of Statistics*, **2**, 1189–1200.

11. Marshall, A. and Olkin, I. (1979). *Inequalities: Theory of Majorization and Its Applications*, New York: Academic Press.

12. Tong, Y. L. (1978). An adaptive solution to ranking and selection problems, *Annals of Statistics*, **6**, 658–672.

13. Tong, Y. L. (1988). Some majorization inequalities in multivariate statistical analysis, *SIAM Review*, **30**, 602–622.

14. Tong, Y. L. (1990). *The Multivariate Normal Distribution*, New York and Berlin: Springer-Verlag.

15. Tong, Y. L. (1997). Dimension-reduction inequalities for exchangeable random variables, with applications to non-normal inference (submitted for publication).

PART VI

Distributions and Applications

21

Correlation Analysis of Ordered Observations From a Block-Equicorrelated Multivariate Normal Distribution

Marlos Viana and Ingram Olkin

The University of Illinois at Chicago, Chicago, IL
Stanford University, Stanford, CA

Abstract: Inferential procedures are developed for correlations between the order statistics from two normally distributed random vectors that are permutation symmetric. These models are used in the analysis of intra-ocular pressure data.

Keywords and phrases: Covariance matrices, order statistics, doubly stochastic matrices, multivariate analysis, intra-ocular pressure data

21.1 Introduction

Multivariate extreme observations are often present in vision research. Two common examples are the determination of corneal astigmatism and the assessment of extreme visual acuities. Corneal astigmatism is obtained by keratometry, which is the measurement of corneal curvature of a small area using a sample of four reflected points of light along an annulus 3 to 4 mm in diameter, centered about the line of sight. Computer-analyzed topography is similar to keratometry in that the relative separation of reflected points of light are used to calculate the curvature (Y) of the measured surface. Using a pattern of concentric rings and sampling at imaginary ring-semimeridiam intersections, a numerical model of the measured surface may be obtained. At any specified concentric ring, the extreme curvatures represent the steepest and flattest curvatures and their positive difference α is related to the amount of astigmatism at that ring or refractive aperture. Given p curvatures $(\mathbf{Y})$ corresponding to a subregion $\mathcal{G}$, let $\mathcal{Y} = (Y_{(1)}, Y_{(2)}, \ldots, Y_{(p)})$ denote the ordered curvatures ranked from the smallest, flattest curvature $Y_{(1)}$, to the largest, steepest curvature $Y_{(p)}$. Then

$$\alpha = Y_{(p)} - Y_{(1)}$$

is the curvature range [for further details, see Viana, Olkin and McMahon (1993)].

Extreme observations between fellow eyes are often used to describe visual acuity. Normally, a single joint measure Y_1, Y_2 of visual acuity is made in each eye, together with one or more covariates $\mathbf{X}$, such as the subject's age or physical condition. Because visual acuities generally are unequal, of interest are not the measures Y_1, Y_2 but rather the extreme visual acuities, the "best" acuity $Y_{(1)}$ and the "worst" acuity $Y_{(2)}$. Consequently, there is interest in making inferences on the covariance structure of $\mathcal{Y}, \mathbf{X}$. Such models have been considered recently by Olkin and Viana (1995). However, when $\mathcal{Y}_i$ represents a vector of ordered visual acuities

$$Y_{i(1)} \leq Y_{i(2)} \leq \cdots \leq Y_{i(p)}$$

at different time-points, $i = 1, 2$, e.g., pre-treatment and post-treatment, then the structure of interest becomes

$$\Psi = \begin{bmatrix} \mathrm{Cov}(\mathcal{Y}_1) & \mathrm{Cov}(\mathcal{Y}_1, \mathcal{Y}_2) \\ \mathrm{Cov}(\mathcal{Y}_2, \mathcal{Y}_1) & \mathrm{Cov}(\mathcal{Y}_2) \end{bmatrix} = \begin{bmatrix} \Psi_{11} & \Psi_{12} \\ \Psi_{21} & \Psi_{22} \end{bmatrix}. \qquad (21.1)$$

In general, the correlations of interest are

$$\eta_{ii,jk} = \mathrm{Cor}(Y_{i(j)}, Y_{i(k)}), \quad \eta_{12,jk} = \mathrm{Cor}(Y_{1(j)}, Y_{2(k)}).$$

Note that the same structure is required to express, for example, the correlation between astigmatism of fellow eyes or at two different time-points. In the present work, we consider the covariance structure described by Ψ when

$$\Sigma = \begin{bmatrix} \Sigma_{11} & \Sigma_{12} \\ \Sigma_{21} & \Sigma_{22} \end{bmatrix} = \begin{bmatrix} \mathrm{Cov}(\mathbf{Y}_1) & \mathrm{Cov}(\mathbf{Y}_1, \mathbf{Y}_2) \\ \mathrm{Cov}(\mathbf{Y}_2, \mathbf{Y}_1) & \mathrm{Cov}(\mathbf{Y}_2) \end{bmatrix}$$

has a block-equicorrelated structure. In the bivariate case, for example, Σ is determined by

$$\Sigma_{11} = \sigma_1^2 \begin{bmatrix} 1 & \rho_1 \\ \rho_1 & 1 \end{bmatrix}, \quad \Sigma_{22} = \sigma_2^2 \begin{bmatrix} 1 & \rho_2 \\ \rho_2 & 1 \end{bmatrix}, \quad \Sigma_{12} = \sigma_1\sigma_2 \begin{bmatrix} \gamma & \phi \\ \phi & \gamma \end{bmatrix},$$

which assumes that (pre-treatment) measurements $\mathbf{Y}_1$ on fellow eyes have a common variance σ_1^2 and a common correlation ρ_1, that corresponding (post-treatment) measurements $\mathbf{Y}_2$ on fellow eyes have a common variance σ_2^2 and a common correlation ρ_2, that corresponding (same-eye) measurements have a common correlation γ between time-points, whereas adjacent (contralateral-eye) measurements have a common correlation ϕ between time-points.

The following matrix notation will be used. Denote by $\boldsymbol{v}_{ij}$ the $p_i \times p_j$ matrix with all entries equal to 1, by $\boldsymbol{I}_{ij}$ the $p_i \times p_j$ ($p_i = p_j$) matrix with ones along the main diagonal entries and zeros in the remaining entries, and let

$$J_{ij}(g,h) = \sigma_i \sigma_j [h\boldsymbol{v}_{ij} + (g-h)\boldsymbol{I}_{ij}].$$

Then $J_{ii}(g,h)$ indicates the matrix of dimension $p_i \times p_i$ with common main diagonal entries equal to $\sigma_i^2 g$ and common off diagonal entries equal to $\sigma_i^2 h$, whereas $J_{ij}(g,g)$ indicates the $p_i \times p_j$ matrix with all entries equal to $\sigma_i \sigma_j g$.

In this paper, we consider permutation-symmetric random vectors $\boldsymbol{Y}_1$ of dimension p_1 and $\boldsymbol{Y}_2$ of dimension p_2 jointly normally distributed with covariance structure Σ

$$\Sigma = \begin{bmatrix} J_{11}(1,\rho_1) & J_{12}(\gamma,\phi) \\ J_{21}(\gamma,\phi) & J_{22}(1,\rho_2) \end{bmatrix}, \tag{21.2}$$

subject to the restriction $\phi = \gamma$. Under this model, we show that

$$\Psi = \begin{bmatrix} \Sigma_{11}\mathcal{C}_1 & \Sigma_{12} \\ \Sigma_{21} & \Sigma_{22}\mathcal{C}_2 \end{bmatrix},$$

where $\mathcal{C}_i$ is the covariance matrix among p_i ordered independent standard normal variates.

The covariance matrix Ψ under model (21.2) restricted by $\gamma = \phi$ is derived for multivariate distributions commonly generated through certain affine transformations $\boldsymbol{Y} = \boldsymbol{m} + \boldsymbol{TU}$ [see Guttman (1954) and Tong (1990, p. 183)]. In Section 21.4, we review the inferential results for the covariance matrix Σ. The corresponding large-sample distributions for the maximum likelihood estimators associated with Σ and Ψ are derived in Section 21.5. A numerical application is discussed in Section 21.6 and selected derivations are outlined in Section 21.8.

21.2 The Covariance Structure of Ordered Affine Observations

Given a $p \times m$ real non-negative constant matrix $\boldsymbol{T}$ and a random vector $\boldsymbol{U}$ with $m \geq p$ components, denote by

$$\boldsymbol{Y} = \boldsymbol{m} + \boldsymbol{TU}, \quad \boldsymbol{m} \in \mathbf{R}^p,$$

the multivariate (affine) transformation with generating random vector $\boldsymbol{U}$. It is assumed that the components of $\boldsymbol{U}$ are jointly independent with $E(\boldsymbol{U}) = \boldsymbol{0}$ and variances 1.

Example 21.2.1 If the distribution of Y is multivariate normal of dimension p with vector of means m and common non-negative correlation ρ, then $Y = m + TU$, with generating vector $U \sim N(0, I)$ of dimension $p+1$ and transformation matrix $T = [\sqrt{\rho}e, \quad \sqrt{1-\rho}\,I]$, where e has dimension p and I is the identity matrix also of dimension p.

Example 21.2.2 If the distribution of $Y = (Y_1, Y_2)$ of dimension $q = p_1 + p_2$ is multivariate normal with vector of means (m_1, m_2) and block-equicorrelated covariance structure (21.2) restricted by $\phi = \gamma$, then $Y = m + TU$ with generating vector $U \sim N_{q+1}(0, I)$ and transformation matrix

$$T = \begin{bmatrix} a_0 e_1 & a I_1 & \alpha v_{12} \\ b_0 e_2 & \beta v'_{12} & b I_2 \end{bmatrix}, \tag{21.3}$$

where e_i is the $p_i \times 1$ vector of ones, for appropriate choices of $a_0, a, \alpha, b_0, b, \beta \in \mathbf{R}$. In fact, to see this, note that if

$$Y = \begin{bmatrix} Y_1 \\ Y_2 \end{bmatrix} = \begin{bmatrix} m_1 \\ m_2 \end{bmatrix} + T \begin{bmatrix} U_0 \\ U_1 \\ U_2 \end{bmatrix}, \tag{21.4}$$

then $Y \sim N(0, \Sigma)$, with

$$\begin{aligned} \Sigma_{11} &= J_{11}(a_0^2 + p_2\alpha^2 + a^2, a_0^2 + p_2\alpha^2), \\ \Sigma_{12} &= J_{12}(a_0 b_0 + a\beta + \alpha b, a_0 b_0 + a\beta + \alpha b), \\ \Sigma_{22} &= J_{22}(b_0^2 + p_1\beta^2 + b^2, b_0^2 + p_1\beta^2). \end{aligned}$$

Given $\sigma_1, \sigma_2, \rho_1, \rho_2, \gamma$, set

$$a^2 = \sigma_1^2(1 - \rho_1), \quad b^2 = \sigma_2^2(1 - \rho_2),$$

and solve

$$\sqrt{\sigma_1^2\rho_1 - p_2\alpha^2}\sqrt{\sigma_2^2\rho_2 - p_1\beta^2} + \beta\sigma_1\sqrt{1 - rho_1} + \alpha\sigma_2\sqrt{1 - \rho_2} = \sigma_1\sigma_2\gamma$$

for α and β, to yield

$$a_0 = \sqrt{\sigma_1^2\rho_1 - p_2\alpha^2}, \quad b_0 = \sqrt{\sigma_2^2\rho_2 - p_1\beta^2}.$$

Proposition 21.2.1 *If $Y = TU$ with transformation matrix $T = [a_0 e, \quad a I]$, $a \neq 0$, and arbitrary generating vector $U = (U_0, U_1)$, then*

$$Cov(\mathcal{Y}) = a_0^2 v_{11} + a^2 Cov(\mathcal{U}_1),$$

where $Cov(\mathcal{U}_1)$ is the covariance matrix of the ordered components

$$\mathcal{U}_1 = (U_{(1)}, \ldots, U_{(p)})$$

of U_1, when $a > 0$, and of $\mathcal{U}_1 = (U_{(p)}, \ldots, U_{(1)})$ when $a < 0$.

PROOF. Because

$$\mathcal{Y} = T \begin{bmatrix} U_0 \\ \mathcal{U}_1 \end{bmatrix}, \tag{21.5}$$

where $\mathcal{U}_1$ is defined as above according to the sign of a, we have

$$\text{Cov}(\mathcal{Y}) = T \begin{bmatrix} 1 & \text{Cov}(U_0, \mathcal{U}_1)' \\ \text{Cov}(U_0, \mathcal{U}_1) & \text{Cov}(\mathcal{U}_1) \end{bmatrix} T'.$$

Moreover, since the components of $(U_0, \boldsymbol{U}_1)$ are independent, then U_0 and $\mathcal{U}_1$ are independent and hence

$$\text{Cov}(\mathcal{Y}) = T \begin{bmatrix} 1 & \boldsymbol{0}' \\ \boldsymbol{0} & \text{Cov}(\mathcal{U}_1) \end{bmatrix} T' = a_0^2 \boldsymbol{v}_{11} + a^2 \text{Cov}(\mathcal{U}_1).$$

∎

Note that if, in addition to the conditions of Proposition 21.2.1, $\text{Cov}(\mathcal{U}_1)$ is proportional to a doubly stochastic matrix, then so is $\text{Cov}(\mathcal{Y})$. Also note that Proposition 21.2.1 holds when the mean of $\boldsymbol{Y}$ is a permutation-symmetric vector $\boldsymbol{m}$.

Proposition 21.2.2 *If*

$$Y = \begin{bmatrix} \boldsymbol{Y}_1 \\ \boldsymbol{Y}_2 \end{bmatrix} = T \begin{bmatrix} U_0 \\ \boldsymbol{U}_1 \\ \boldsymbol{U}_2 \end{bmatrix} \equiv TU,$$

with generating vector $\boldsymbol{U}$ and transformation matrix (21.3) then

$$\begin{aligned} Cov(\mathcal{Y}_1) &= a_0^2 \boldsymbol{v}_{12} + a^2 Cov(\mathcal{U}_1) + \alpha^2 \boldsymbol{v}_{12} Cov(\mathcal{U}_2) \boldsymbol{v}'_{12}, \\ Cov(\mathcal{Y}_1, \mathcal{Y}_2) &= a_0 b_0 \boldsymbol{v}_{12} + a\beta Cov(\mathcal{U}_1) \boldsymbol{v}_{12} + \alpha b \boldsymbol{v}_{12} Cov(\mathcal{U}_2), \\ Cov(\mathcal{Y}_2) &= b_0^2 \boldsymbol{v}_{22} + \beta^2 \boldsymbol{v}'_{12} Cov(\mathcal{U}_1) \boldsymbol{v}_{12} + b^2 Cov(\mathcal{U}_2), \end{aligned}$$

where $Cov(\mathcal{U}_i)$ is the covariance matrix of the ordered components

$$\mathcal{U}_i = (U_{(1)}, \ldots, U_{(p_i)})$$

of $\boldsymbol{U}_i$, $i = 1, 2$, and the ordered vectors $\mathcal{U}_1, \mathcal{U}_2$ of dimension p_1 and p_2, respectively, are defined according to the sign of a and b, respectively.

PROOF. Because $e_i' U_i = e_i' \mathcal{U}_i$, note that

$$\begin{bmatrix} \mathcal{Y}_1 \\ \mathcal{Y}_2 \end{bmatrix} = T \begin{bmatrix} U_0 \\ \mathcal{U}_1 \\ \mathcal{U}_2 \end{bmatrix}, \tag{21.6}$$

where $\mathcal{U}_1$ and $\mathcal{U}_2$ are defined according to the sign of a and b, respectively. Moreover, since the components of $(U_0, \boldsymbol{U}_1, \boldsymbol{U}_2)$ are independent, then U_0, $\mathcal{U}_1$ and $\mathcal{U}_2$ are also independent, and hence

$$\mathrm{Cov}\left(\begin{bmatrix} \mathcal{Y}_1 \\ \mathcal{Y}_2 \end{bmatrix}\right) = \boldsymbol{T} \begin{bmatrix} 1 & \boldsymbol{0}' & \boldsymbol{0}' \\ \boldsymbol{0} & \mathrm{Cov}(\mathcal{U}_1) & \boldsymbol{0}' \\ \boldsymbol{0} & \boldsymbol{0} & \mathrm{Cov}(\mathcal{U}_2) \end{bmatrix} \boldsymbol{T}'. \tag{21.7}$$

Direct evaluation of (21.7) leads to the proposed covariance matrices.   ∎

Corollary 21.2.1 *Under the conditions of Proposition 21.2.2, if $\mathrm{Cov}(\mathcal{U}_i)$ is stochastic (all row and column sums equal to 1), then*

$$\begin{aligned} Cov(\mathcal{Y}_1) &= \Sigma_{11} \, Cov(\mathcal{U}_1), \\ Cov(\mathcal{Y}_1, \mathcal{Y}_2) &= \Sigma_{12}, \\ Cov(\mathcal{Y}_2) &= \Sigma_{22} \, Cov(\mathcal{U}_2). \end{aligned}$$

Because $\mathrm{Cov}(\mathcal{U}_i)$ is symmetric and stochastic (and thereby is symmetric and doubly stochastic) and Σ_{ii} is a linear combination of $\boldsymbol{v}_{ii}$ and $\boldsymbol{I}$, the matrices $\mathrm{Cov}(\mathcal{U}_i)$ and Σ_{ii} commute and consequently their product is also symmetric. Also note that Proposition 21.2.2 holds when the means of $\boldsymbol{Y}_i$ are permutation-symmetric vectors $\boldsymbol{m}_i$, $i = 1, 2$.

Corollary 21.2.2 *If the distribution of $\boldsymbol{Y} = (\boldsymbol{Y}_1, \boldsymbol{Y}_2)$ of dimension $q = p_1 + p_2$ is multivariate normal with permutation-symmetric means $\boldsymbol{m}_1$ and $\boldsymbol{m}_2$, and block-equicorrelated covariance structure (21.2) with $\phi = \gamma$, then*

$$\begin{aligned} Cov(\mathcal{Y}_i) &= \Sigma_{ii} \mathcal{C}_i, \quad i = 1, 2 \\ Cov(\mathcal{Y}_1, \mathcal{Y}_2) &= \Sigma_{12}, \end{aligned}$$

where $\mathcal{C}_i$ is the covariance matrix among p_i ordered independent standard normal variates.

PROOF. If $\boldsymbol{Y}$ has the block-equicorrelated structure of $(\boldsymbol{Y}_1, \boldsymbol{Y}_2)$, then $\boldsymbol{Y} = \boldsymbol{m} + \boldsymbol{TU}$ for a suitable choice of $\boldsymbol{T}$ and $\boldsymbol{U} \sim N(0, \boldsymbol{I})$, as shown in Example 21.2.2. In addition, the fact that the generating distribution is normal implies that $\mathrm{Cov}(\mathcal{U}_i)$ is stochastic [e.g., David (1981, p. 39) and Arnold, Balakrishnan and Nagaraja (1992)]. Applying Corollary 21.2.1, the result follows.   ∎

To obtain a partial converse to Corollary 21.2.2, we need the following result:

Proposition 21.2.3 *Let $Z_1, \ldots, Z_p$ be independent, identically distributed random variables with zero means and variance σ^2, $Z_{(1)} \leq \cdots \leq Z_{(p)}$ its ordered values and $\mathcal{Z} = (Z_{(1)}, \ldots, Z_{(p)})$. Then $\mathrm{Cov}(\mathcal{Z})$ is doubly stochastic for all $p = 2, 3, \ldots$ if and only if the distribution of Z is normal.*

PROOF. The proof of doubly stochasticity for independent normal random variables is provided in David (1981, p. 39) and Arnold, Balakrishnan and Nagaraja (1992, p. 91). Suppose, conversely, that $\text{Cov}(\mathcal{Z})$ is doubly stochastic for all $p = 2, 3, \ldots$. Then

$$1 = \sum_{j=1}^{p} \text{Cov}(Z_{(i)}, Z_{(j)}), \quad i = 1, \ldots, p, \quad p = 2, \ldots.$$

But this is a characterization of the normal distribution [Govindarajulu (1966, Corollary 3.3.1)] [see also Johnson, Kotz and Balakrishnan (1994, Chapter 13)]. ∎

Proposition 21.2.4 *If $\boldsymbol{Y} = \boldsymbol{m} + \boldsymbol{TU}$ as in Proposition 21.2.2 with $U_0 \sim N(0, 1)$, $p_1 = p_2 = p$, $a\beta + \alpha b \neq 0$ and $\Sigma_{12} = \Psi_{12}$ for all $p \geq 2$, then $\boldsymbol{Y} \sim N(\boldsymbol{m}, \boldsymbol{TT}')$.*

PROOF. The conditions $p_1 = p_2 = p$, $a\beta + \alpha b \neq 0$ and $\Sigma_{12} = \Psi_{12}$ for all $p \geq 2$ imply that $\text{Cov}(\mathcal{U}_i)$ is stochastic for all $p \geq 2$. From Proposition 21.2.3 and using the fact that the components of $\boldsymbol{U}$ are jointly independent, we obtain that $\boldsymbol{U}_1$ and $\boldsymbol{U}_2$ are jointly normal and independent. Because, in addition, it is assumed that $U_0 \sim N(0, 1)$, the generating vector $\boldsymbol{U}$ is normal $N(\boldsymbol{0}, \boldsymbol{I})$ (of dimension $2p + 1$). Therefore, $\boldsymbol{Y} \sim N(\boldsymbol{m}, \boldsymbol{TT}')$, concluding the proof. ∎

21.3 Correlation Parameters of Ψ When $\phi = \gamma$

From Corollary 21.2.2, the correlation $\eta_{ii,jk}$ between $Y_{i(j)}$ and $Y_{i(k)}$ is

$$\eta_{ii,jk} = \frac{\rho_i + (1 - \rho_i)c_{i,jk}}{\sqrt{\rho_i + (1 - \rho_i)c_{i,jj}}\sqrt{\rho_i + (1 - \rho_i)c_{i,kk}}}, \tag{21.8}$$

where $c_{i,jk}$ is the jk-th entry of the covariance matrix $\mathcal{C}_i$ among order statistics from p_i independent standard normal variates, $i = 1, 2$, $j = 1, \ldots, p_1$, $k = 1, \ldots, p_2$.

The correlation $\eta_{12,jk}$ between $Y_{1(j)}$ and $Y_{2(k)}$ is

$$\eta_{12,jk} = \frac{\gamma}{\sqrt{\rho_1 + (1 - \rho_1)c_{1,jj}}\sqrt{\rho_2 + (1 - \rho_2)c_{2,kk}}}. \tag{21.9}$$

The conditional covariance matrix $\Psi_{22.1} = \Psi_{22} - \Psi_{21}\Psi_{11}^{-1}\Psi_{12}$ is

$$\Psi_{22.1} = \Sigma_{22}\mathcal{C}_2 - \frac{p_1\gamma^2}{\sigma_1^2(1 + (p_1 - 1)\rho_1)}\boldsymbol{v}_{22}, \tag{21.10}$$

whereas the matrix $\Psi_{21}\Psi_{11}^{-1}$ of regression coefficients is

$$\Psi_{21}\Psi_{11}^{-1} = \frac{\gamma}{\sigma_1^2(1+(p_1-1)\rho_1)}\boldsymbol{v}_{21}.$$

21.4 Inferences on Σ

We assume that $p_1 = p_2 = p$. The matrix of variances and covariances among the $2p$ components of $(\mathbf{X},\mathbf{Y})$ is determined by (21.2), where the range of the parameters is

$$
\begin{aligned}
[\gamma+(p-1)\phi]^2 &< [1+(p-1)\rho_1][1+(p-1)\rho_2], \\
(\gamma-\phi)^2 &< (1-\rho_1)(1-\rho_2), \\
\frac{-1}{p-1} &< \rho_1 < 1,\ \frac{-1}{p-1} < \rho_2 < 1,
\end{aligned}
$$

so as to guarantee that the covariance matrix Σ is positive definite. If the joint distribution of $(\mathbf{Y},\mathbf{X})$ is multivariate normal with covariance matrix (21.2), then the maximum likelihood estimate $\hat\Delta$ of $\Delta = (\sigma_1^2,\sigma_2^2)$ is given by

$$\hat\Delta = (\frac{\mathrm{tr}S_{11}}{p},\frac{\mathrm{tr}S_{22}}{p}),$$

where $\mathrm{tr}(S)$ indicates the trace of the corresponding sample covariance matrix S, whereas the maximum likelihood estimate $\hat\Theta$ of $\Theta = (\rho_1,\rho_2,\gamma,\phi)$ has components

$$
\begin{aligned}
\hat\rho_i &= \frac{\bar{S}_{ii}-\mathrm{tr}S_i}{(p-1)\mathrm{tr}S_i},\ i=1,2, \\
\hat\gamma &= \frac{\mathrm{tr}(S_{12})}{\sqrt{\mathrm{tr}S_{22}}\sqrt{\mathrm{tr}S_{11}}}, \\
\hat\phi &= \frac{\bar{S}_{12}-\mathrm{tr}(S_{12})}{(p-1)\sqrt{\mathrm{tr}S_{22}}\sqrt{\mathrm{tr}S_{11}}}.
\end{aligned}
\tag{21.11}
$$

Here, and throughout the paper, $\bar{S}$ indicates the sum of all entries of the corresponding sample covariance matrix S.

When $\gamma = \phi$, the MLE $\hat\gamma$ of γ becomes

$$\hat\gamma = \frac{\bar{S}_{12}}{p\sqrt{\mathrm{tr}S_{22}}\sqrt{\mathrm{tr}S_{11}}},$$

a weighted combination of the unrestricted estimates $\hat\gamma$ and $\hat\phi$ in (21.11) with corresponding weights $p(p-1)$ and p.

21.5   Large-Sample Distributions

Proposition 21.5.1 *If the joint distribution of* $(\mathbf{Y}, \mathbf{X})$ *is multivariate normal with covariance matrix (21.2), then*

i. *the asymptotic joint distribution of* $\sqrt{N}(\hat{\Delta} - \Delta)$ *is bivariate normal with means zero, variances*

$$var_\infty(\hat{\sigma_i}^2) = \frac{2\sigma_i^4\left(1 + (p-1)\rho_i^2\right)}{p}, \quad i = 1, 2,$$

and covariance

$$Cov_\infty(\hat{\sigma}_1^2, \hat{\sigma}_2^2) = \frac{2\,\sigma_2^2\sigma_1^2\left(\gamma^2 + (p-1)\phi^2\right)}{p} \;;$$

ii. *the asymptotic joint distribution of* $\sqrt{N}(\hat{\Theta} - \Theta)$ *is 4-variate normal, with means zero, variances*

$$var_\infty(\hat{\rho}_i) = \frac{2\left(1 + (p-1)\rho_i\right)^2\left(1 - \rho_i\right)^2}{p(p-1)}, \quad i = 1, 2,$$

$$
\begin{aligned}
var_\infty(\hat{\gamma}) \;=\; & [2 + \gamma^2\rho_1^2 p - \gamma^2\rho_1^2 + 2\,\gamma^2\phi^2 p - 2\,\gamma^2\phi^2 + \gamma^2\rho_2^2 p \\
& - \gamma^2\rho_2^2 - 4\,\gamma\,\rho_1\,p\phi + 4\,\gamma\,\rho_1\,\phi - 4\,\gamma^2 + 4\,\gamma\,\phi\,\rho_2 - 4\,\gamma\,\phi\,p\rho_2 \\
& + 2\,\gamma^4 + 2\,\phi^2 p - 2\,\phi^2 + 2\,\rho_1\,p\rho_2 - 2\,\rho_1\,\rho_2]/(2\,p),
\end{aligned}
$$

$$
\begin{aligned}
var_\infty(\hat{\phi}) \;=\; & [2 + 2\,\gamma^2 + 12\,\phi^2 + 2\,\rho_2\,p + 2\,\rho_1\,p - 8\,\gamma\,\phi - 4\,\rho_2 - 4\,\rho_1 \\
& - 12\,\phi^2 p + 2\,\phi^4 p^2 - 2\,\phi^2\rho_2^2 p + \phi^2\rho_1^2 p^2 + \phi^2\rho_2^2 p^2 - 4\,\phi^4 p \\
& + 2\,\rho_1\,p^2\rho_2 - 2\,\phi^2\rho_1^2 p + \phi^2\rho_2^2 + \phi^2\rho_1^2 + 2\,\phi^4 - 4\,\gamma\,\phi\,p\rho_2 \\
& + 12\,\rho_1\,p\phi^2 + 12\,\phi^2 p\rho_2 - 4\,\phi^2 p^2\rho_2 - 4\,\rho_1\,p^2\phi^2 - 4\,\gamma\,\rho_1\,p\phi \\
& - 2\,\gamma^2\phi^2 + 6\,\rho_1\,\rho_2 - 8\,\rho_1\,\phi^2 - 8\,\phi^2\rho_2 + 2\,\phi^2 p^2 + 2\,\gamma^2\phi^2 p \\
& + 4\,\gamma\,\rho_1\,\phi + 4\,\gamma\,\phi\,\rho_2 - 6\,\rho_1\,p\rho_2 + 4\,\gamma\,\phi\,p]/[2\,p(p-1)],
\end{aligned}
$$

and covariances

$$
\begin{aligned}
Cov_\infty(\hat{\rho}_1, \hat{\rho}_2) \;=\; & [-8\,\gamma\,\phi - 4\,\gamma\,\rho_1\,p\phi + 4\,\gamma\,\rho_1\,\phi + 2\,\gamma^2 + 2\,\gamma^2\rho_1\,p\rho_2 \\
& - 4\,\rho_1\,p\phi^2\rho_2 + 4\,\gamma\,\phi\,\rho_2 - 2\,\gamma^2\rho_1\,\rho_2 - 4\,\gamma\,\phi\,p\rho_2 - 6\,\phi^2 p \\
& + 6\,\phi^2 + 6\,\rho_1\,p\phi^2 - 4\,\rho_1\,\phi^2 + 6\,\phi^2 p\rho_2 - 4\,\phi^2\rho_2 + 2\,\phi^2 p^2 \\
& + 2\,\rho_1\,\phi^2\rho_2 + 2\,\rho_1\,p^2\phi^2\rho_2 - 2\,\phi^2 p^2\rho_2 - 2\,\rho_1\,p^2\phi^2 \\
& + 4\,\gamma\,\phi\,p]/[p(p-1)],
\end{aligned}
$$

$$
\begin{aligned}
Cov_\infty(\hat{\rho}_i, \hat{\gamma}) &= [-4\rho_i\phi - 2\gamma^2\phi + \gamma^3\rho_i - \gamma\rho_i\phi^2 + \gamma\rho_i p\phi^2 \\
&\quad - \gamma\phi^2 p - \gamma\rho_i^3 - \gamma\rho_i - \gamma\rho_i^2 p + 2\gamma\rho_i^2 + \gamma\rho_i^3 p + 2\phi \\
&\quad + 2\rho_i p\phi + 2\gamma\phi^2 - 2\phi\rho_i^2 p + 2\phi\rho_i^2]/p, \quad i = 1,2,
\end{aligned}
$$

$$
\begin{aligned}
Cov_\infty(\hat{\rho}_i, \hat{\phi}) &= [2\gamma + 9\rho_i\phi + 2\gamma\rho_i^2 - 6\phi\rho_i^2 - 4\phi - 2\phi^3 + \gamma^2\rho_i p\phi \\
&\quad + \rho_i\phi^3 + \phi\rho_i^3 + 3\phi^3 p + 2\phi p - \phi^3 p^2 - 2\gamma\phi^2 p \\
&\quad + 2\gamma\phi^2 - 9\rho_i p\phi + \phi\rho_i^3 p^2 - 2\phi\rho_i^3 p - 2\gamma\rho_i^2 p \\
&\quad + 2\rho_i p^2\phi + \rho_i p^2\phi^3 - 2\rho_i p\phi^3 - \gamma^2\rho_i\phi + 9\phi\rho_i^2 p \\
&\quad - 3\phi\rho_i^2 p^2 + 2\gamma\rho_i p - 4\gamma\rho_i]/[p(p-1)], \quad i = 1,2,
\end{aligned}
$$

$$
\begin{aligned}
Cov_\infty(\hat{\gamma}, \hat{\phi}) &= [-2\gamma^2\rho_2 - 2\gamma\phi + 2\gamma^3\phi - 2\gamma\rho_1 p\phi + 4\gamma\rho_1\phi + 4\gamma\phi\rho_2 \\
&\quad - 2\gamma\phi p\rho_2 + 2\rho_1 + \phi p\gamma\rho_2^2 + \phi p\gamma\rho_1^2 - \phi\gamma\rho_1^2 + 2\phi^3 p\gamma \\
&\quad - \phi\gamma\rho_2^2 - 2\gamma\phi^3 + 2\rho_2 + 2\phi^2 p - 4\phi^2 + 2\rho_1 p\rho_2 \\
&\quad - 4\rho_1\rho_2 - 2\rho_1 p\phi^2 + 2\rho_1\phi^2 - 2\phi^2 p\rho_2 \\
&\quad + 2\phi^2\rho_2 - 2\gamma^2\rho_1]/(2p).
\end{aligned}
$$

The proof is presented in Section 21.8.

21.5.1 Special cases

Under the conditions of Proposition 21.5.1 we obtain, for example, for the bivariate case ($p = 2$),

$$
\mathrm{var}_\infty(\hat{\Delta}) = \begin{bmatrix} \sigma_1^4(\rho_1^2+1) & \sigma_2^2\sigma_1^2(\phi^2+\gamma^2) \\ \sigma_2^2\sigma_1^2(\phi^2+\gamma^2) & \sigma_2^4(\rho_2^2+1) \end{bmatrix},
$$

$$
\mathrm{var}_\infty(\hat{\rho}_1 \mid \rho_1 = 0) = \mathrm{var}_\infty(\hat{\rho}_2 \mid \rho_2 = 0) = \frac{2}{p(p-1)},
$$

$$
\mathrm{var}_\infty(\hat{\gamma} \mid \gamma = 0) = \frac{1+(p-1)(\phi^2+\rho_1\rho_2)}{p},
$$

$$
\mathrm{var}_\infty(\hat{\phi} \mid \phi = 0) = \frac{p^2\rho_1\rho_2 - 3p\rho_1\rho_2 + 3\rho_1\rho_2 - 2\rho_2 - 2\rho_1 + \gamma^2 + p\rho_2 + p\rho_1 + 1}{p(p-1)},
$$

$$
Cov_\infty(\hat{\Theta} \mid \Theta = 0) = \mathrm{diag}(\frac{2}{p(p-1)}, \frac{2}{p(p-1)}, \frac{1}{p}, \frac{1}{p(p-1)}).
$$

The case $\rho_1 = \rho_2$ is important. However, Proposition 21.5.1 cannot be adapted easily. To see this, take the simplest case $p = 2$, $\gamma = \phi = 0$; then the MLE of the common correlation ρ is the solution of a quadratic equation [see Viana (1994), Olkin and Pratt (1958), Olkin and Siotani (1964) and Olkin (1967)].

From (21.8) and (21.9), note that when $p_1 = p_2 = 2$, the three distinct correlations of interest are

$$\begin{aligned}
\eta_{11,12} &= \text{Cor}(Y_{1(1)}, Y_{1(2)}), \\
\eta_{22,12} &= \text{Cor}(Y_{2(1)}, Y_{2(2)}), \\
\eta_{12,jk} &= \text{Cor}(Y_{1(j)}, Y_{2(k)}), \quad j, k = 1, 2.
\end{aligned}$$

Let $\Lambda = (\eta_{11,12}^2, \eta_{22,12}^2, \eta_{12,jk}^2)$.

Proposition 21.5.2 *If the distribution of $Y = (Y_1, Y_2)$ is multivariate normal with permutation-symmetric bivariate means m_1 and m_2, and block-equicorrelated covariance structure (21.2) with $\phi = \gamma$, then the asymptotic joint distribution of $\sqrt{N}(\hat{\Lambda} - \Lambda)$ is normal with means zero, variances*

$$Var_\infty(\hat{\eta}_{ii,12}^2) = \frac{(0.6817\rho_i + 0.3183)^2 (1 + \rho_i)^2 (0.2312\rho_i + 0.4954)^2 (1 - \rho_i)^2}{(0.3183\rho_i + 0.6817)^8},$$
$$i = 1, 2,$$

$$\begin{aligned}
Var_\infty(\hat{\eta}_{12,jk}^2) = &[(0.2028\gamma^4\rho_1\,\rho_2 - 0.899\gamma^2\rho_1\,\rho_2{}^2 + 3.230\rho_2 - 2.289\gamma^2 \\
&+ 3.225\rho_1 - 3.173\gamma^2\rho_2 + 0.750\rho_2{}^2 + 0.4338\gamma^4\rho_2 \\
&- 0.716\gamma^2\rho_2{}^2 + 0.7575\rho_1{}^2 - 0.7171\gamma^2\rho_1{}^2 + 3.188\rho_1{}^2\rho_2{}^2 \\
&+ 0.4342\gamma^4\rho_1 + 0.9292\gamma^4 - 3.933\gamma^2\rho_1\,\rho_2 + 6.472\rho_1\,\rho_2 \\
&+ 3.928\rho_1{}^2\rho_2 - 3.177\gamma^2\rho_1 + 3.923\rho_1\,\rho_2{}^2 + 0.7575\rho_1{}^3\rho_2 \\
&- 0.8968\gamma^2\rho_1^2\,\rho_2 - 0.198\gamma^2\,\rho_1^2\rho_2^2 + 3.455 + 0.750\rho_1\,\rho_2{}^3 \\
&+ 0.703\rho_1{}^3\rho_2{}^2 + 0.696\rho_2{}^3\rho_1{}^2 + 0.1688\rho_1{}^3\rho_2{}^3)\gamma^2] \\
&/[32\,(0.3183\rho_1 + 0.6817)^4\,(0.3183\rho_2 + 0.6817)^4],
\end{aligned}$$

and covariances

$$\begin{aligned}
Cov_\infty(\hat{\eta}_{11,12}^2, \hat{\eta}_{22,12}^2) = &[(0.492 + 0.237\rho_2)\,(1 - \rho_2) \\
&(0.6817\rho_2 + 0.3183)\,(0.6817\rho_1 + 0.3183)\,(1 - \rho_1) \\
&(0.492 + 0.237\rho_1)\,\gamma^2] \\
&/[2\,(0.3183\rho_2 + 0.6817)^4\,(0.3183\rho_1 + 0.6817)^4],
\end{aligned}$$

$$\begin{aligned}
Cov_\infty(\hat{\eta}_{ii,12}^2, \hat{\eta}_{12,jk}^2) = &[(-0.03\rho_1\,\rho_2{}^2 - 0.06\rho_2{}^2 + 1.43 - 0.318\gamma^2\rho_2 - 0.6817\gamma^2 \\
&+ 0.66\rho_1 + 1.36\rho_2 + 0.64\rho_1\,\rho_2)\,(0.6817\rho_2 + 0.3183) \\
&(1 - \rho_2)\,(0.492 + 0.237\rho_2)\,\gamma^2] \\
&/[8\,(0.3183\rho_1 + 0.6817)^2\,(0.3183\rho_2 + 0.6817)^6].
\end{aligned}$$

In particular, under the conditions of Proposition 21.5.2, when $\rho_1 = \rho_2 = 0$ we obtain

$$\begin{aligned}
\mathrm{Var}_\infty(\hat{\eta}^2_{ii,12}) &= 0.5330, \quad i = 1,2, \\
\mathrm{Var}_\infty(\hat{\eta}^2_{12,jk}) &= 0.6701\left(0.9288\gamma^4 + 3.455 - 2.291\gamma^2\right)\gamma^2, \\
\mathrm{Cov}_\infty(\hat{\eta}^2_{11,12}, \hat{\eta}^2_{22,12}) &= 0.2666\gamma^2, \\
\mathrm{Cov}_\infty(\hat{\eta}^2_{ii,12}, \hat{\eta}^2_{12,jk}) &= 0.4227\left(1.423 - 0.6813\gamma^2\right)\gamma^2, \quad i = 1,2.
\end{aligned}$$

21.6   Intra-ocular Pressure Data

In the study described by Sonty, Sonty and Viana (1996), intra-ocular pressure (IOP) measurements at pre-treatment (Y_1) and post-treatment (Y_2) conditions were obtained from fellow glaucomatous eyes of $N = 15$ subjects on topical beta blocker therapy. We want to estimate the covariance structure between pre- and post-treatment ordered IOP. This is important to assess and linearly predict the best (worst) post-treatment IOP from pre-treatment ocular pressure measurements. The observed covariance matrices for IOP between fellow eyes are:

$$S_{11} = \begin{bmatrix} 12.410 & 7.019 \\ 7.019 & 12.924 \end{bmatrix}, \quad S_{22} = \begin{bmatrix} 17.029 & 15.371 \\ 15.371 & 17.352 \end{bmatrix},$$

whereas the sample cross-covariance matrix is

$$S_{12} = \begin{bmatrix} 11.671 & 9.348 \\ 8.200 & 10.076 \end{bmatrix}.$$

From Section 21.4, the estimated variances are

$$\hat{\sigma}_1^2 = \frac{\mathrm{tr}S_{11}}{p} = 12.667, \quad \hat{\sigma}_2^2 = \frac{\mathrm{tr}S_{22}}{p} = 17.190,$$

whereas the estimated correlations are

$$\hat{\rho}_1 = \frac{\bar{S}_{11} - \mathrm{tr}S_{11}}{(p-1)\mathrm{tr}S_{11}} = 0.553, \quad \hat{\rho}_2 = \frac{\bar{S}_{22} - \mathrm{tr}S_{22}}{(p-1)\mathrm{tr}S_{22}} = 0.894,$$

$$\hat{\gamma} = \frac{\mathrm{tr}(S_{12})}{\sqrt{\mathrm{tr}S_{22}}\sqrt{\mathrm{tr}S_{11}}} = 0.736, \quad \hat{\phi} = \frac{\bar{S}_{12} - \mathrm{tr}(S_{21})}{(p-1)\sqrt{\mathrm{tr}S_{22}}\sqrt{\mathrm{tr}S_{11}}} = 0.594.$$

The corresponding large-sample estimated covariance matrices are

$$\text{Cov}_\infty(\hat{\Delta}) = \begin{bmatrix} 209.503 & 194.764 \\ 194.764 & 531.624 \end{bmatrix},$$

$$\text{Cov}_\infty(\hat{\Theta}) = \begin{bmatrix} 0.481 & 0.071 & 0.131 & 0.284 \\ 0.071 & 0.040 & 0.025 & 0.072 \\ 0.131 & 0.025 & 0.141 & 0.174 \\ 0.284 & 0.072 & 0.174 & 0.287 \end{bmatrix}.$$

The large-sample approximate 95% confidence intervals for γ and ϕ are, respectively, $(0.543, 0.929)$ and $(0.324, 0.864)$, thus suggesting a common cross-correlation γ. Under the restricted model ($\gamma = \phi$), we have

$$\hat{\gamma} = \frac{\bar{S}_{12}}{p\sqrt{\text{tr}S_{22}}\sqrt{\text{tr}S_{11}}} = 0.665.$$

21.6.1 Inferences on Ψ under $\phi = \gamma$

The values of $\mathcal{C}_i$ have been tabulated; see, for example, Sarhan and Greenberg (1962), Owen (1962) and Harter and Balakrishnan (1996). When $p_1 = p_2 = 2$, note that $c_{i,jj} = 0.6817$, $c_{i,12} = 1 - c_{i,jj} = 0.3183$, $i, j = 1, 2$. From equation (21.8), the estimated correlations $\eta_{ii,jk}$ between $Y_{i(j)}$ and $Y_{i(k)}$ are

$$\hat{\eta}_{11,12} = \frac{\hat{\rho}_1 + (1 - \hat{\rho}_1)0.3183}{\hat{\rho}_1 + (1 - \hat{\rho}_1)0.6817} = 0.8106,$$

$$\hat{\eta}_{22,12} = \frac{\hat{\rho}_2 + (1 - \hat{\rho}_2)0.3183}{\hat{\rho}_2 + (1 - \hat{\rho}_2)0.6817} = 0.9463;$$

From equation (21.9), the estimated correlation $\hat{\eta}_{12,jk}$ between $Y_{1(j)}$ and $Y_{2(k)}$ is

$$\hat{\eta}_{12,jk} = \frac{\hat{\gamma}}{\sqrt{\hat{\rho}_1 + (1 - \hat{\rho}_1)0.6817}\sqrt{\hat{\rho}_2 + (1 - \hat{\rho}_2)0.6817}} = 0.7347.$$

From (21.10), the estimated conditional covariance matrix $\hat{\Psi}_{22.1}$ is

$$\hat{\Psi}_{22.1} = \begin{pmatrix} 16.562 & 15.858 \\ 15.858 & 16.562 \end{pmatrix}.$$

From Proposition 21.5.2, the asymptotic joint distribution of the squared-correlations $\hat{\Lambda} = (\hat{\eta}_{11,12}^2, \hat{\eta}_{22,12}^2, \hat{\eta}_{12,jk}^2)$ is normal with mean Λ and covariance matrix

$$\frac{1}{15} \begin{bmatrix} 0.3090 & 0.0062 & 0.0771 \\ 0.0062 & 0.0225 & 0.0180 \\ 0.0771 & 0.0180 & 0.3571 \end{bmatrix}.$$

The resulting large-sample approximate 95% confidence intervals for $\eta_{11,12}$, $\eta_{22,12}$ and $\eta_{12,jk}$ are $(0.608, 0.971)$, $(0.904, 0.986)$ and $(0.480, 0.920)$, respectively. The results suggest a positive association between pre- and post-ordered IOP measurements, estimated as 0.7374. The estimated correlation between pre-treatment extreme IOP measurements is 0.81, whereas the estimated correlation between post-treatment extreme IOPs is 0.94. Note that

$$\eta_{ii,12} = 0.467, \quad \text{when } \rho_i = 0, \quad i = 1, 2,$$

which is the well-known positive dependence between extreme fellow observations [see Tong (1990, p. 130)].

21.7   Conclusions

We have described the covariance structure between the order statistics of p_1 permutation-symmetric normal variates $\boldsymbol{Y}_1$ and the order statistics of p_2 permutation-symmetric normal variates $\boldsymbol{Y}_2$, when $\boldsymbol{Y}_1$ and $\boldsymbol{Y}_2$ are jointly normally distributed with constant cross-covariance Σ_{12} structure. The central result shows that the covariance structure of ordered fellow observations $\boldsymbol{Y}_i$ (e.g., fellow eyes) has the form $\Sigma_{ii}\mathcal{C}$ where Σ_{ii} is the permutation symmetric covariance structure of fellow observations and $\mathcal{C}$ is the doubly symmetric stochastic covariance structure of ordered independent permutation symmetric normal random variates. In addition, the cross-covariance structure between ordered $\boldsymbol{Y}_1$ and ordered $\boldsymbol{Y}_2$ remains equal to the original cross-covariance structure Σ_{12}.

21.8   Derivations

To prove Proposition 21.5.1, we need to express the distribution of the sample covariance matrix S in a suitable canonical form. We start with the assumption that

$$nS = n \begin{bmatrix} S_{11} & S_{12} \\ S_{21} & S_{22} \end{bmatrix}$$

has a Wishart distribution $W_{2p}(\Sigma, n)$, where

$$\Sigma = D \begin{bmatrix} \rho_1 \boldsymbol{e}\boldsymbol{e}' + (1 - \rho_1)\boldsymbol{I} & \phi\boldsymbol{e}\boldsymbol{e}' + (\gamma - \phi)\boldsymbol{I} \\ \phi\boldsymbol{e}\boldsymbol{e}' + (\gamma - \phi)\boldsymbol{I} & \rho_2 \boldsymbol{e}\boldsymbol{e}' + (1 - \rho_2)\boldsymbol{I} \end{bmatrix} D,$$

and $D = \mathrm{diag}(\sigma_1, \ldots, \sigma_1, \sigma_2, \ldots, \sigma_2)$. Let

$$\Gamma = \begin{bmatrix} Q & 0 \\ 0 & Q \end{bmatrix},$$

where Q is any $p \times p$ orthonormal real matrix with first row constant and equal to $1/\sqrt{p}$. Then

$$\Gamma\Sigma\Gamma' = D \begin{bmatrix} M_{11} & M_{12} \\ M_{21} & M_{22} \end{bmatrix} D,$$

where

$$
\begin{aligned}
M_{ii} &= \mathrm{diag}(1 + (p-1)\rho_i, 1 - \rho_i, \ldots, 1 - \rho_i), \\
M_{12} &= \mathrm{diag}(\gamma + (p-1)\phi, \gamma - \phi, \ldots, \gamma - \phi).
\end{aligned}
$$

After transposition, we obtain

$$\Sigma_0 = \Gamma_0 \Sigma \Gamma_0' = D_0 \, \mathrm{diag}(E, F, \ldots, F) \, D_0,$$

where

$$
E = \begin{bmatrix} 1 + (p-1)\rho_1 & \gamma + (p-1)\phi \\ \gamma + (p-1)\phi & 1 + (p-1)\rho_2 \end{bmatrix}, \quad
F = \begin{bmatrix} 1 - \rho_1 & \gamma - \phi \\ \gamma - \phi & 1 - \rho_2 \end{bmatrix},
$$

and $D_0 = \mathrm{diag}(\sigma_1, \sigma_2, \ldots, \sigma_1, \sigma_2)$. Accordingly, $S^* = \Gamma_0 S \Gamma_0'$ has a Wishart distribution $W_{2p}(\Sigma_0/n, n)$, which can be decomposed into p independent Wishart components of dimension 2, corresponding to the non-zero block diagonals of Σ_0. Specifically, we obtain

$$
E^* = \begin{bmatrix} S_{11}^* & S_{12}^* \\ S_{21}^* & S_{22}^* \end{bmatrix} \sim W_2(E/n, n),
$$

$$
F_j^* = \begin{bmatrix} S_{2j-1,2j-1}^* & S_{2j-1,2j}^* \\ S_{2j,2j-1}^* & S_{2j,2j}^* \end{bmatrix} \sim W_2(F/n, n), \quad j = 2, \ldots, p,
$$

with $M_1, M_2, \ldots, M_p$ jointly independent. The canonical form of interest is then $nE^* \sim W_2(E, n)$, independent of

$$
nF^* = n \sum_{j=2}^{p} F_j^* \sim W_2(F, (p-1)n).
$$

Let

$$
\beta = \begin{bmatrix} E \\ \cdots\cdots \\ (p-1)F \end{bmatrix} = \begin{bmatrix} \sigma_1^2 \left(1 + (p-1)\rho_1\right) \\ \sigma_1 \sigma_1 \left(\gamma + (p-1)\phi\right) \\ \sigma_2^2 \left(1 + (p-1)\rho_2\right) \\ \cdots\cdots\cdots \\ (p-1)\sigma_1^2 \left(1 - \rho_1\right) \\ (p-1)\sigma_1 \sigma_2 \left(\gamma - \phi\right) \\ (p-1)\sigma_2^2 \left(1 - \rho_2\right) \end{bmatrix}.
$$

It then follows that $\rho_2 = \mathbf{f}(\beta)$ is determined by

$$\sigma_1^2 = \frac{\beta_1 + \beta_4}{p}, \quad \sigma_2^2 = \frac{\beta_3 + \beta_6}{p},$$

whereas $\Theta = \mathbf{g}(\beta)$ is determined by

$$\rho_1 = \frac{(p-1)\beta_1 - \beta_4}{(p-1)(\beta_1 + \beta_4)}, \quad \rho_2 = \frac{(p-1)\beta_3 - \beta_6}{(p-1)(\beta_3 + \beta_6)},$$

$$\gamma = \frac{\beta_2 + \beta_5}{\sqrt{\beta_1 + \beta_4}\sqrt{\beta_3 + \beta_6}}, \quad \phi = \frac{(p-1)\beta_2 - \beta_5}{(p-1)\sqrt{\beta_1 + \beta_4}\sqrt{\beta_3 + \beta_6}}.$$

In particular, when $\gamma = \phi$,

$$\gamma = \frac{\beta_2}{\sqrt{\beta_1 + \beta_4}\sqrt{\beta_3 + \beta_6}}.$$

Maximum Likelihood Estimates. Direct computation shows that

$$\operatorname{tr}(S_{11}) = \sum_{j=1,3,2p-1} S_{j,j}^*,$$

$$\operatorname{tr}(S_{22}) = \sum_{j=2,4,2p} S_{j,j}^*,$$

$$\operatorname{tr}(S_{12}) = \sum_{j=1,3,2p-1} S_{j,j+1}^*,$$

$$\bar{S}_{11} - \operatorname{tr}(S_1) = (p-1)S_{11}^* - \sum_{j=3,5,2p-1} S_{j,j}^*,$$

$$\bar{S}_{22} - \operatorname{tr}(S_2) = (p-1)S_{22}^* - \sum_{j=4,6,2p} S_{j,j}^*,$$

$$\bar{S}_{12} - \operatorname{tr}(S_{12}) = (p-1)S_{12}^* - \sum_{j=3,5,2p-1} S_{j,j+1}^*,$$

so that

$$\hat{E} = \frac{1}{p}\begin{bmatrix} \bar{S}_{11} & \bar{S}_{12} \\ \bar{S}_{21} & \bar{S}_{22} \end{bmatrix},$$

$$(p-1)\hat{F} = \begin{bmatrix} p\operatorname{tr}(S_{11}) - \bar{S}_{11} & p\operatorname{tr}(S_{12}) - \bar{S}_{12} \\ p\operatorname{tr}(S_{21}) - \bar{S}_{21} & p\operatorname{tr}(S_{22}) - \bar{S}_{22} \end{bmatrix},$$

thus showing that $\hat{\beta} = (\hat{E}, (p-1)\hat{F})$ of β does not depend on the particular choice of Γ_0. The estimates indicated on Proposition 21.5.1 follow by expressing the corresponding MLEs $\hat{\rho}_2 = \mathbf{f}(\hat{\beta}), \quad \hat{\Theta} = \mathbf{g}(\hat{\beta})$ accordingly.

Large-sample distributions. Because

$$\sqrt{n}(E^* - E) \overset{\mathcal{L}}{\to} N_2(0, T_1),$$

independent of

$$\sqrt{n}(F^* - (p-1)F) \overset{\mathcal{L}}{\to} N_2(0, (p-1)T_2),$$

where T_1 is the covariance matrix of E^* and T_2 is the common covariance matrix of F_j^*, $j = 2 \ldots, p$, it follows that

$$\sqrt{n}\left(\begin{bmatrix} E^* \\ F^* \end{bmatrix} - \beta\right) \overset{\mathcal{L}}{\to} N_4(0, T),$$

where T is the block matrix with diagonal $T_1, (p-1)T_2$. The limiting joint distribution of $\hat{\rho}_2 = \mathbf{f}(M)$ and $\hat{\Theta} = \mathbf{g}(M)$ follows from the standard delta method.

References

1. Arnold, B. C., Balakrishnan, N. and Nagaraja, H. N. (1992). *A First Course in Order Statistics*, New York: John Wiley & Sons.

2. David, H. A. (1981). *Order Statistics*, Second edition, New York: John Wiley & Sons.

3. Govindarajulu, Z. (1966). Characterization of normal and generalized truncated normal distributions using order statistics, *Annals of Mathematical Statistics*, **37**, 1011–1015.

4. Guttman, L. (1954). A new approach to factor analysis: The radex, In *Mathematical Thinking in the Social Sciences* (Ed., P. F. Lazarsfeld), pp. 258–348, Glencoe, IL: The Free Press.

5. Harter, H. L. and Balakrishnan, N. (1996). *CRC Handbook of Tables for the Use of Order Statistics in Estimation*, Boca Raton, FL: CRC Press.

6. Johnson, N. L., Kotz, S. and Balakrishnan, N. (1995). *Continuous Univariate Distributions–Vol. 1*, Second edition, New York: John Wiley & Sons.

7. Olkin, I. (1967). Correlations revisted, In *Improving Experiments: Design and Statistical Analysis* (Ed., J. Stanley), pp. 102–128, Seventh Annual Phi Delta Kappa Symposium on Educational Research, Chicago: Rand and McNally.

8. Olkin, I. and Pratt, J. (1958). Unbiased estimation of certain correlation coefficients, *Annals of Mathematical Statistics*, **29**, 201–211.

9. Olkin, I. and Siotani, M. (1964). Testing for the equality of correlation coefficients for various multivariate models, *Technical Report*, Laboratory for Quantitative Research in Education, Stanford University, Stanford, CA.

10. Olkin, I. and Viana, M. A. G. (1995). Correlation analysis of extreme observations from a multivariate normal distribution, *Journal of the American Statistical Association*, **90**, 1373–1379.

11. Owen, D. B. (1962). *Handbook of Statistical Tables*, Reading, MA: Addison-Wesley.

12. Sarhan, A. E. and Greenberg, B. G. (Eds.) (1962). *Contributions to Order Statistics*, New York: John Wiley & Sons.

13. Sonty, S. P., Sonty, S. and Viana, M. A. G. (1996). The additive ocular hypotensive effect of topical 2% dorzolamide on glaucomatous eyes on topical beta blocker therapy, *Investigative Ophthalmology and Visual Sciences*, **37**, S1100.

14. Tong, Y. L. (1990). *The Multivariate Normal Distribution*, Second edition, New York: Springer-Verlag.

15. Viana, M. A. G. (1994). Combined maximum likelihood estimates for the equicorrelation coefficient, *Biometrics*, **50**, 813–820.

16. Viana, M. A. G., Olkin, I. and McMahon, T. (1993). Multivariate assessment of computer analyzed corneal topographers, *Journal of the Optical Society of America—A*, **10**, 1826–1834.

22

On Distributions With Periodic Failure Rate and Related Inference Problems

B. L. S. Prakasa Rao

Indian Statistical Institute, New Delhi, India

Abstract: The connections between the Integrated Cauchy Functional Equation, "almost lack of memory" property of a distribution and periodic failure rate of a distribution are explained. Some properties of distributions with periodic failure rates are discussed. The one to one correspondence between the distributions with periodic failure rates and the distributions with the "almost lack of memory" property is studied following the work of Chukova, Dimitrov and others. Applications of these distributions to modeling of minimal repair systems and related inference problems are given.

Keywords and phrases: Periodic failure rate, "Almost lack of memory" property, reliability, integrated Cauchy Functional equation, minimal repair systems, nonhomogeneous Poisson process, estimation

22.1 Introduction

Exponential distribution is widely used for modeling and statistical analysis of reliability data in view of its "lack of memory" property or equivalently its constant failure rate. However there are situations where the assumption about the constancy of failure rate is not appropriate and the failure rate $\lambda(t)$ may be a nonconstant function of t, the time to failure. The function may be increasing or decreasing or might be decreasing initially but eventually increasing. Here we consider the situation when the failure rate is periodic. This can be incorporated for instance where a failed component is replaced by another component of the same age as in minimal repair system [cf., Baxter (1982), Lau and Prakasa Rao (1990)] with the time to failure following a distribution with periodic failure rate. It is also of interest in modelling processes which are periodic in nature such as periodic (say) monthly, yearly maintenance of the industrial equipment or in formulating insurance policies for risk assessment such as fixing time periods for payment of insurance premia etc.

In Section 22.2, we relate the work on the Integrated Cauchy functional equation [cf., Ramachandran and Lau (1991)] with distributions with) "almost lack of memory" property [cf., Chukova and Dimitrov (1992)]. Some properties of distributions with periodic failure rates are derived in Section 22.3. The connection between a distribution with periodic failure rate and almost lack of memory is explored in Section 22.4. A representation for distributions with periodic failure rates is given in Section 22.5 following Dimitrov, Chukova and Khalil (1995) with detailed proofs given by us. Non homogeneous Poisson process (NHPP) with periodic intensities for modeling minimal repair systems is discussed in Section 22.6 following Chukova, Dimitrov and Garriado (1993). Estimation of parameters for distributions with periodic failure rates is investigated in Section 22.7. Nonparametric and parametric estimation problems of intensity for a NHPP are discussed in Section 22.8.

A preliminary version of this paper was presented as an invited lecture at the conference on "Statistical Inference in Life Testing and Reliability using A priori Information held at Varanasi, India [cf., Prakasa Rao (1995)].

22.2 Integrated Cauchy Functional Equation and Distributions With Periodic Failure Rates

It is well known that if a random variable $X \geq 0$ has the property that

$$P(X > x + y | X > y) = P(X > x), \quad x > 0, \ y > 0 \qquad (22.1)$$

then X has the exponential distribution [cf., Galambos and Kotz (1978)]. This property is generally known as the "memorylessness" property or the "lack of memory" property. Ramachandran (1979) proved that the exponential distribution also enjoys the "strong memorylessness property" or the "strong lack of memory" property. He showed that if a random variable X is exponentially distributed and $Y \geq 0$ is a random variable independent of X, then

$$P(X > Y + x | X > Y) = P(X > x) \qquad (22.2)$$

for every $x \geq 0$. It is known that the equation (22.1) characterizes the exponential distribution and this is a consequence of the fact that if a function $g(\cdot)$ is continuous, then the only solution of the Cauchy Functional Equation (CFE)

$$g(x + y) = g(x)g(y), \quad x > 0, \ y > 0 \qquad (22.3)$$

is $g(x) = e^{\alpha x}$ for some $\alpha \in R$. Motivated by this result, investigations were made to find conditions under which the equation (22.2) characterizes the exponential distribution. It can be shown that the equation (22.2) leads to what is known as the Integrated Cauchy Functional Equation (ICFE)

$$\int_0^\infty g(x+y)d\sigma(y) = g(x), \quad x \geq 0 \tag{22.4}$$

where σ is a positive σ-finite Borel measure on R_+ nondegenerate at $\{0\}$ and $g(\cdot) : R_+ \to R$ is locally integrable. The following result is due to Lau and Rao (1982).

Theorem 22.2.1 *[Lau and Rao (1982).] Suppose $\alpha(\{0\}) < 1$ and g is a non-trivial locally integrable solution of the ICFE given by (22.4). Then*

$$g(x) = p(x)e^{\alpha x} \ a.e. \tag{22.5}$$

with respect to the Lebesgue measure on R where $\alpha \in R$ which is uniquely determined by the equation

$$\int_0^\infty e^{\alpha y}d\sigma(y) = 1 \tag{22.6}$$

and $p(\cdot)$ satisfies the condition $p(x+y) = p(x)$ for all $y \in supp\,(\sigma)$.

For a proof of Theorem 22.2.1, see Ramachandran and Lau (1991). For related results, see Prakasa Rao and Ramachandran (1983) and Ramachandran and Prakasa Rao (1984).

Suppose the equation (22.2) holds and that $P(X > Y) > 0$, where Y is a non-negative random variable with distribution function G. Let $T(x) = 1 - F(x)$ where F is the distribution of X. The relation (22.2) reduces to

$$c\,T(x) = \int_0^\infty T(x+y)dG(y) \tag{22.7}$$

where $c = P(X > Y)$. The problem is to determine the class of distributions F for which (22.7) holds given $G(\cdot)$ and to check the conditions under with (22.7) characterizes the exponential distribution. The following results are useful in the sequel.

Lemma 22.2.1 *[Marsagalia and Tubilla (1975).] Let $g : R_+ \to R$ be a continuous function with $g(0) = 1$. Suppose the set*

$$A = \{t \geq 0 : g(x+t) = g(x)g(t) \text{ for all } x \geq 0\} \tag{22.8}$$

contains a point other than zero. Then one of the following holds:

(i) $A = \{0\} \cup [\gamma, \infty)$ for some $\gamma > 0$ and $g(x) = 0$ for $x \geq \gamma$;

(ii) $A = [0, \infty)$ and $g(x) = e^{\alpha x}$ some $\alpha \in R$;

(iii) $A = \{n\rho : n \geq 0\}$ for some $\rho > 0$ and $g(x) = p(x)e^{\alpha x}$ for some $\alpha \in R$ and the function $p(\cdot)$ has period ρ.

Theorem 22.2.2 *[Marsagalia and Tubilla (1975).] Let A be a nonempty subset of $[0, \infty)$ and $g : R_+ \to R$ be a continuous function satisfying*

$$g(x + y) = g(x)g(y), \ x \geq 0, \ y \in A$$

with $g(0) = 1$. Then one of the following holds:

(i) *there exists $\gamma > 0$ such that $g(x) = 0$ for $x \geq \gamma$;*

(ii) *$g(x) = e^{\alpha x}$ for some $\alpha \in R$.*

(iii) *$g(x) = p(x)e^{\alpha x}$ where $p(0) = 1$ and $p(\cdot)$ has period ρ, the greatest common divisor of A.*

The following result can be obtained by arguments similar to those used to prove Lemma 22.2.1 and Theorem 22.2.2.

Theorem 22.2.3 *Let A be a nonempty subset of $[0, \infty)$ and $h : [0, \infty) \to R$ be continuous satisfying*

$$h(x + y) = h(x), \ x \geq 0, \ y \in A.$$

Then either $h(\cdot)$ is a constant or $h(\cdot)$ has period d where d is the greatest common divisor of A.

As a consequence of Theorems 22.2.1 and 22.2.3, we have the following corollary.

Corollary 22.2.1 *Let σ and $g(x) = p(x)e^{\alpha x}$ be as in Theorem 22.2.1. Then either $p(\cdot)$ is a constant or $p(\cdot)$ has period d where $d > 0$ is the greater common divisor of support (σ).*

This corollary follows from Theorem 22.2.3 in view of the observation that $p(x + y) = p(x)$ for $x \geq 0$ and $y \in$ support (σ).

For proofs of Lemma 22.2.1, Theorems 22.2.2 and 22.2.3, and Corollary 22.2.1, see Ramachandran and Lau (1991). An application of Theorem 22.2.1 and Corollary 22.2.1 yield the following theorem giving the class of distributions satisfying the equation (22.7).

Theorem 22.2.4 *Suppose that the equation (22.7) holds and $G(0) < c < 1$ where $c = P(X > Y)$. Define α by the relation*

$$\int_0^\infty e^{-\alpha y}\, dG(y) = c$$

and $A_d = \{nd : n \geq 0\}$ for $d > 0$. Then either $F(x) = 1 - e^{-\alpha x}, x \geq 0$ if support (G) is not contained in A_d for any $d > 0$, or $F(x) = 1 - p(x)e^{-\alpha x}, x \geq 0$ where $p(\cdot)$ is right continuous and has period d if support $(G) \subset A_d$ for some $d > 0$.

As a consequence of Theorem 22.2.4, it follows that, if the equation (22.2) holds, then the distribution of X is either exponential for some $\alpha > 0$ or

$$1 - F(x) = p(x)e^{-\alpha x}, \ x \geq 0$$

for some $\alpha > 0$ where $p(\cdot)$ is a periodic function with some period d and the distribution of Y has the support contained in the set $\{nd : n \geq 0\}$.

Remarks. Apparently, unaware of the work done by Rao, Shanbag, Ramachandran, Lau and others in the area of ICFE, Chukova and Dimitrov (1992) introduced the following notion of a distribution having the "almost lack of memory" property and studied properties of the distributions described above. This was also pointed out recently by Rao, Sapatinas and Shanbag (1994).

Definition 22.2.1 A random variable X is said to have the "*almost lack of memory*" property (ALM property) if there exists a sequence of constants $\{a_n\}$ such that the equation (22.1) holds for $y = a_n$, $n \geq 1$ and $x \geq 0$.

In view of the Lemma 22.2.1, the equation (22.1) holds for the sequence $y = a_n, n \geq 1$ and $x \geq 0$ only if $a_n = nd$ for some $d > 0$ excluding the trivial cases $P(X \geq d) = 0$ and $P(X \geq d) = 1$. Dimitrov, Chukova and Khalil (1995) give a survey of the results obtained by them during the last few years.

22.3 Distributions With Periodic Failure Rates

We shall now study the properties of a nonnegative random variable X with distribution function $F(\cdot)$ defined by

$$1 - F(x) = p(x)e^{-\alpha x}, \ x \geq 0 \tag{22.9}$$

where $\alpha > 0$ and $p(\cdot)$ is a nonnegative periodic function with period d and $p(0) = 1$. Suppose $p(\cdot)$ is differentiable. Clearly the probability density function f of X is given by

$$\begin{aligned} f(x) &= e^{-\alpha x}(\alpha p(x) - p'(x)), \ x \geq 0 \\ &= 0 \quad \text{otherwise.} \end{aligned} \tag{22.10}$$

It follows from the properties of a probability density function that the periodic function $p(x)$ should satisfy the condition

$$p'(x) \leq \alpha p(x) \text{ a.e.}$$

with respect to the Lebesgue measure on R a priori. The failure rate or hazard rate λ is given by

$$\lambda(x) = \frac{f(x)}{1 - F(x)} = \frac{\alpha p(x) - p'(x)}{p(x)}, \qquad (22.11)$$

for $x \geq 0$ with $F(x) < 1$. Since $p(\cdot)$ is periodic with period d, it is easy to check that $p'(\cdot)$ is also periodic with period d and hence the failure rate $\lambda(\cdot)$ is periodic with period d. Observe that the exponential distribution has the constant failure rate. As a consequence of (22.10), it follows that

$$\int_0^\infty e^{-\alpha x}(\alpha p(x) - p'(x))dx = 1 \qquad (22.12)$$

or equivalently

$$\sum_{j=0}^\infty \int_{jd}^{(j+1)d} e^{-\alpha x}(\alpha p(x) - p'(x))dx = 1. \qquad (22.13)$$

Note that

$$\sum_{j=0}^\infty \int_{jd}^{(j+1)d} e^{-\alpha x}(\alpha p(x) - p'(x))dx$$

$$= \sum_{j=0}^\infty \int_0^d e^{-\alpha(y+jd)}(\alpha p(y+jd) - p'(y+jd))dy$$

$$= \sum_{j=0}^\infty e^{-\alpha jd} \int_0^d (\alpha p(y) - p'(y))e^{-\alpha y}dy$$

(since p and p' are periodic with period d)

$$= \left(\sum_{j=0}^\infty e^{-\alpha jd}\right)R(\alpha, d)$$

$$= \frac{R(\alpha, d)}{1 - e^{\alpha d}} \qquad (22.14)$$

where

$$R(\alpha, d) = \int_0^d (\alpha p(y) - p'(y))e^{-\alpha y}dy. \qquad (22.15)$$

Relations (22.13) and (22.14) imply that

$$R(\alpha, d) = 1 - e^{-\alpha d}. \tag{22.16}$$

Hence it follows that the function $p(\cdot)$ is nonnegative, $p(0) = 1$, periodic with period d and satisfies the relation

$$\int_0^d (\alpha p(y) - p'(y))e^{-\alpha y}dy = 1 - e^{-\alpha d}. \tag{22.17}$$

It was noted earlier that

$$\alpha p(x) - p'(x) \geq 0, \quad x \geq 0, \text{ a.e.} \tag{22.18}$$

with respect to the Lebesgue measure on R. It can be checked that

$$E(X) = \int_0^\infty p(z)e^{-\alpha z}dz \tag{22.19}$$

and

$$\text{Var}(X) = 2\int_0^\infty zp(z)e^{-\alpha z}dz - \{\int_0^\infty p(z)e^{-\alpha z}dz\}^2. \tag{22.20}$$

Define

$$r(\alpha, d) = \int_0^d p(z)e^{-\alpha z}dz \tag{22.21}$$

and

$$s(\alpha, d) = \int_0^d zp(z)e^{-\alpha z}dz. \tag{22.22}$$

Then, it is easy to verify that

$$E(X) = \frac{r(\alpha, d)}{1 - e^{\alpha d}} \tag{22.23}$$

and

$$\text{Var}(X) = 2\left\{\frac{s(\alpha, d)}{1 - e^{\alpha d}} + \frac{de^{\alpha d}r(\alpha, d)}{(1 - e^{-\alpha d})^2}\right\} - \left(\frac{r(\alpha, d)}{1 - e^{-\alpha d}}\right)^2 \tag{22.24}$$

using the fact that $p(\cdot)$ is periodic with period d. In general

$$E(X^r) = r\int_0^\infty z^{r-1}p(z)e^{-\alpha z}dz. \tag{22.25}$$

22.4   Distributions With Periodic Failure Rates and "Almost Lack of Memory" Property

Let us now suppose that X is a nonnegative random variable with periodic failure rate $\lambda(\cdot)$ with period d. Then

$$\lambda(x) = \frac{f(x)}{1 - F(x)} = \frac{f(x + d)}{1 - F(x + d)} = \lambda(x + d), \ \ 0 \le x < \infty. \qquad (22.26)$$

Integrating both sides of (22.26) with respect to x on $[0, z]$, we have

$$\int_0^z \frac{f(x)}{1 - F(x)} dx = \int_0^z \frac{f(x + d)}{1 - F(x + d)} dx, \ \ 0 \le z < \infty$$

or equivalently

$$-\log(1 - F(z)) = -\log(1 - F(z + d)) + \log(1 - F(d)), \ 0 \le z < \infty$$

using the fact that $F(0) = 0$. Hence

$$\frac{1}{1 - F(z)} = \frac{1 - F(d)}{1 - F(z + d)}, \ \ 0 \le z < \infty. \qquad (22.27)$$

Let $S(z) = 1 - F(z)$. Relation (22.27) implies that

$$S(z + d) = S(d)S(z), \ \ 0 \le z < \infty. \qquad (22.28)$$

Repeated application of (22.28) shows that

$$S(z + nd) = S(nd)S(z), \ \ 0 \le z < \infty \qquad (22.29)$$

which proves that the random variable X has the "Almost Lack of Memory" property. Further more

$$S(nd) = [S(d)]^n, \ \ n \ge 1. \qquad (22.30)$$

An application of Lemma 22.2.1 implies that

$$S(x) = p(x)e^{-\alpha x}, \ \ x \ge 0 \qquad (22.31)$$

for some $\alpha \in R$ and some nonnegative function $p(x)$ periodic with period d. Since $S(x) \to 0$ as $x \to \infty$, it follows that $\alpha > 0$. Further more $p(0) = 1$.

We have seen above that if a nonnegative random variable has a periodic failure rate, then it has the almost lack of memory property. The converse is also true from the following analysis. Suppose that

$$P(X > x + nd | X > nd) = P(X > x), \ \ x \ge 0, \ n \ge 1. \qquad (22.32)$$

Then

$$(1 - F(x + nd)) = (1 - F(x))(1 - F(nd)), \quad x \geq 0, \ n \geq 1. \tag{22.33}$$

Note that

$$1 - F(x) = e^{-\int_0^x \lambda(y)dy} \tag{22.34}$$

where $\lambda(\cdot)$ is the failure rate of X. Hence it follows that

$$\{\int_0^x \lambda(y)dy\} + \{\int_0^{nd} \lambda(y)dy\} = \int_0^{x+nd} \lambda(y)dy, \quad x \geq 0, \ n \geq 1$$

from (22.33). Therefore

$$\int_0^x \lambda(y)dy \ = \ \int_{nd}^{x+nd} \lambda(y)dy, \quad x \geq 0, \ n \geq 1$$

$$= \ \int_0^x \lambda(y + nd)dy, \quad x \geq 0, \ n \geq 1. \tag{22.35}$$

If f is continuous, then $\lambda(\cdot)$ is continuous and it follows that

$$\lambda(x) = \lambda(x + nd), \quad x \geq 0, \ n \geq 1. \tag{22.36}$$

Otherwise, we can conclude that

$$\lambda(x + nd) = \lambda(x) \ \text{a.e.} \tag{22.37}$$

with respect to the Lebesgue measure on R for every $n \geq 1$.

Hence we have the following theorem.

Theorem 22.4.1 *A nonnegative random variable X with a continuous probability density function has a periodic failure rate iff it has the "almost lack of memory" property.*

22.5 A Representation for Distributions With Periodic Failure Rates

Let X be a nonnegative random variable with distribution function F, density function f and with periodic failure rate $\lambda(\cdot)$ with period d. Let $\gamma = 1 - F(d) = S(d)$. Define a new random variable Y with density function

$$f_Y(y) \ = \ \frac{f(y)}{F(d)}, \ 0 \leq y \leq d$$

$$= \ 0 \quad \text{otherwise.} \tag{22.38}$$

Note that $f_Y(\cdot)$ is the density of the conditional distribution of the random variable X restricted to $[0, d]$. Further more the equation (22.38) implies that

$$
\begin{aligned}
f(y) &= f_Y(y)F(d) \\
&= f_Y(y)(1 - \gamma), \quad 0 \le y \le d. \tag{22.39}
\end{aligned}
$$

Suppose $kd \le y < (k+1)d$ for some integer $k \ge 1$. Since $\lambda(y - kd) = \lambda(y)$ by the periodicity of the failure rate, it follows that

$$
\frac{f(y)}{S(y)} = \frac{f(y)}{1 - F(y)} = \frac{f(y - kd)}{1 - F(y - kd)} = \frac{f(y - kd)}{S(y - kd)}
$$

or equivalently

$$
\begin{aligned}
f(y) &= \frac{f(y - kd)}{F(d)} \cdot \frac{S(y)}{S(y - kd)} \cdot F(d) \\
&= f_Y(y - kd)S(kd)F(d) \\
&= f_Y(y - kd)[S(d)]^k F(d) \\
&= f_Y(y - kd)\gamma^k(1 - \gamma) \tag{22.40}
\end{aligned}
$$

from the relations (22.29) and (22.30). Hence the density $f(\cdot)$ of X can be represented in the form

$$
f(y) = f_Y(y - [\tfrac{y}{d}]d)(1 - \gamma)\gamma^{[\frac{y}{d}]}, \quad 0 \le y < \infty \tag{22.41}
$$

where $[d]$ denotes the greatest integer less than or equal to x and $\gamma = p(X > d)$. Note that, for any $0 \le x < \infty$,

$$
\begin{aligned}
F(x) &= \int_0^x f(y)dy \\
&= \int_0^{[\frac{x}{d}]d} f(y)dy + \int_{[\frac{x}{d}]d}^x f(y)dy \\
&= F\left(\left[\frac{x}{d}\right]d\right) + \int_{[\frac{x}{d}]d}^x f(y)dy \\
&= 1 - \{1 - F([\tfrac{x}{d}]d)\} + (1 - \gamma)\gamma^{[\frac{x}{d}]}F_Y(x - [\tfrac{x}{d}]d) \\
&= 1 - \{1 - F(d)\}^{[\frac{x}{d}]} + (1 - \gamma)\gamma^{[\frac{x}{d}]}F_Y(x - [\tfrac{x}{d}]d) \\
&= 1 - \gamma^{[\frac{x}{d}]} + (1 - \gamma)\gamma^{[\frac{x}{d}]}F_Y(x - [\tfrac{x}{d}]d) \tag{22.42}
\end{aligned}
$$

from the relations (22.30) and (22.41) where $F_Y(\cdot)$ denotes the distribution function of Y. Relations (22.41) and (22.42) give representations for the density and the distribution function of a nonnegative random variable with a periodic failure rate with period d and $\gamma = p(X > d)$.

Let Y be a random variable as defined by (22.38) and let Z be a random variable *independent* of Y with

$$P(Z = k) = (1 - \gamma)\gamma^k, \quad k \geq 0.$$

It is easy to check that X can be represented as $Y + dZ$. In fact the probability density function of $Y + dZ$ is given by

$$\sum_{k=0}^{\infty} f_Y(x - dk)P(Z = k)$$

$$= \sum_{k=0}^{\infty} f_Y(x - dk)(1 - \gamma)\gamma^k$$

$$= f_Y(x - [\frac{x}{d}]d)(1 - \gamma)\gamma^{[\frac{x}{d}]}$$

$$= f(x) \tag{22.43}$$

from (22.41). In other words, starting with any probability density with support on $[0, d]$ and $0 < \gamma < 1$, one can construct a random variable X with periodic failure rate d using the representation $X = Y + dZ$. Relations (22.41), (22.42) and the above observations are due to Dimitrov, Chukova and Khalil (1995). However they have not given proofs.

22.6 Nonhomogeneous Poisson Process and Periodic Intensity

A system may be considered as a collection of components which is required to perform one or more functions. A repairable system is a system which, after it has failed to perform properly, can be restored to satisfactory performance by methods other than replacement of the entire system [cf., Crowder *et al.* (1991)].

Consider the following process of minimal repair actions: if an item fails at time t, then the failure is counted and an operating item of the same age t is immediately put in operation in replacement of the failed component. This process of replacement can be modeled by a nonhomogeneous Poisson process (NHPP) with cumulative intensity

$$\Lambda_X(t) = \int_0^t \lambda_X(u)du = -\log(1 - F_X(t))$$

where X is a nonnegative random variable with distribution function $F_X(\cdot)$ and hazard or failure rate $\lambda_X(\cdot)$. Assume that the distribution of X has periodic failure rate with period d or equivalently the "almost lack of memory" property.

Let $N_{[\tau,\tau+t)}$ denote the number of failures in the interval $[\tau,\tau+t)$. If the initial item is a new one at time $t_0 = 0$, then we interpret τ as the age of the item at the beginning of the observation period $[\tau,\tau+t)$ and $N_{[\tau,\tau+t)}$ gives the total number of failures after t time units. Note that the random variables $N_{[t_1,t_2)}$ and $N_{[t_3,t_4)}$ need not be independent even if $t_1 < t_2 \leq t_3 < t_4$. The following result is due to Chukova, Dimitrov and Garriado (1993).

Theorem 22.6.1 *Under the conditions stated above,*

(i) $P[N_{[d,d+t)} = k] = P[N_{[0,t)} = k]$, $k \geq 0$, $t > 0$;

(ii) $P[N_{[d,d+t)} = k]|N_{[0,d)} = \ell] = P[N_{[d,d+t)} = k]$, $k \geq 0$, $\ell \geq 0$, $t > 0$;

(iii) *the random variable $N_{[0,t)}$ can be represented in the form*

$$\sum_{n=1}^{[t/d]} N_{[0,d)}^{(n)} + N_{[0,t-[\frac{t}{d}]d)}$$

where $N_{[0,d)}^{(n)}$, $1 \leq n \leq [\frac{t}{d}]$ are independent and identically distributed (i.i.d.) Poisson random variables with parameter

$$\Lambda(d) = \int_0^d \lambda(u)du;$$

and

(iv) $N_{[0,t-[\frac{t}{d}]d)}$ *is a Poisson random variable with parameter $\Lambda(t-[\frac{t}{d}]d)$ independent of $N_{[0,d)}^{(n)}$, $1 \leq n \leq [\frac{t}{d}]$.*

22.7 Estimation of Parameters for Distributions With Periodic Failure Rates

Suppose $X_1, X_2, \ldots, X_n$ are i.i.d. nonnegative random variables with density function given by (22.10). It would be of interest to estimate the parameter α assuming that the period of $p(\cdot)$ is known.

22.7.1 Method of moments

Equating the sample mean $\overline{X}_n$ to the mean of X_1, we have the equation

$$\overline{X}_n = \frac{r(\alpha, d)}{1 - e^{-\alpha d}} \tag{22.44}$$

where

$$r(\alpha, d) = \int_0^d p(z)e^{-\alpha z}dz. \tag{22.45}$$

This may need to be solved numerically to get an estimator of α.

22.7.2 Method of maximum likelihood

The likelihood function is given by

$$L(\alpha) = \prod_{i=1}^n e^{-\alpha X_i}(\alpha p(X_i) - p'(X_i)) \tag{22.46}$$

and hence

$$\log\, L(\alpha) = -\alpha\sum_{i=1}^n X_i + \sum_{i=1}^n \log(\alpha p(X_i) - p'(X_i)).$$

Therefore

$$\frac{d\,\log L(\alpha)}{d\alpha} = -\sum_{i=1}^n X_i + \sum_{i=1}^n \frac{1}{\alpha p(X_i) - p'(X_i)}\,\frac{d}{d\alpha}\,(\alpha p(X_i) - p'(X_i)).$$

If the functional form of $p(\cdot)$ is known, then solving the equation $\frac{d\,\log L(\alpha)}{d\alpha} = 0$ and checking whether the solution leads to a maximum of the likelihood, a maximum likelihood estimator of α can be obtained.

An alternate way of looking at the inference problem is via the representation given by (22.41).

Suppose $X_1, X_2, \ldots, X_n$ are i.i.d. nonnegative random variable with periodic failure rate $\lambda(\cdot)$ with known period d. Let $f_X(\cdot)$ be the probability density function of X_1. Let $\gamma = P(X > d)$. It is known that

$$f_X(x) = f_Y(x - [\frac{x}{d}]d)(1 - \gamma)\gamma^{[\frac{x}{d}]}, \ x \geq 0 \tag{22.47}$$

from (22.40) where $f_Y(\cdot)$ denotes the conditional density of X given that $X \leq d$. The joint likelihood function is given by

$$L(\gamma) = \prod_{i=1}^n f_Y(X_i - [\frac{X_i}{d}]d)(1 - \gamma)^n\gamma^{\sum_{i=1}^n[\frac{X_i}{d}]} \tag{22.48}$$

and hence

$$\log L(\gamma) = \sum_{i=1}^{n} \log f_Y(X_i - [\frac{X_i}{d}]d) + n \log(1 - \gamma) + (\sum_{i=1}^{n} [\frac{X_i}{d}]) \log \gamma.$$

Observe that

$$\frac{d \log L(\gamma)}{d\gamma} = -\frac{n}{1-\gamma} + \frac{\sum_{i=1}^{n} [\frac{X_i}{d}]}{\gamma}$$

which shows that

$$\hat{\gamma} = \frac{\sum_{i=1}^{n} [\frac{X_i}{d}]}{n + \sum_{i=1}^{n} [\frac{X_i}{d}]} \tag{22.49}$$

is a maximum likelihood estimator of γ since

$$\left. \frac{d^2 \log L(\gamma)}{d\gamma^2} \right|_{\gamma=\hat{\gamma}} < 0.$$

Example 22.7.1 Suppose

$$
\begin{aligned}
f_Y(y) &= \frac{\lambda e^{-\lambda y}}{1 - e^{-\lambda d}} & 0 \le y \le d \\
&= 0 & \text{otherwise}
\end{aligned}
\tag{22.50}
$$

and the random variable X has the density

$$
\begin{aligned}
f_X(\cdot) &= \frac{\lambda e^{-\lambda(x - [\frac{x}{d}]d)}}{1 - e^{-\lambda d}} (1 - \gamma) \gamma^{[\frac{x}{d}]}, & x \ge 0 \\
&= 0 & \text{otherwise}
\end{aligned}
\tag{22.51}
$$

where $0 < \gamma < 1$. It follows from the earlier discussion that $\gamma = P(X > d)$ and that X has a periodic failure rate with period d. Given an i.i.d. sample $X_i, 1 \le i \le n$ distributed as X, we have seen above that the MLE of γ is given by

$$\hat{\gamma} = \frac{\sum_{i=1}^{n} [\frac{X_i}{d}]}{n + \sum_{i=1}^{n} [\frac{X_i}{d}]}.$$

Let L be the likelihood function based on $X_i, \ 1 \le i \le n$. It is easy to see that

$$\log L = n \log \lambda - n \log(1 - e^{-\lambda d}) - \lambda \sum_{i=1}^{n}(X_i - [\frac{X_i}{d}]d) + n \log(1 - \gamma) + (\sum_{i=1}^{n} [\frac{X_i}{d}]) \log \gamma$$

and hence

$$\frac{d \log L}{d\lambda} = \frac{n}{\lambda} - \frac{nde^{-\lambda d}}{1 - e^{-\lambda d}} - \sum_{i=1}^{n}(X_i - [\frac{X_i}{d}]d). \tag{22.52}$$

A solution of the equation

$$\frac{d \log L}{d\lambda} = 0 \tag{22.53}$$

leads to an estimator for λ. It needs to be checked whether it maximizes the likelihood. One can get an approximate solution by the following approach.

Note that the equation

$$\frac{d \log L}{d\lambda} = 0$$

leads to the equation

$$\frac{n}{\lambda} - \frac{nde^{-\lambda d}}{1 - e^{-\lambda d}} = \sum_{i=1}^{n}(X_i - [\frac{X_i}{d}]d) \tag{22.54}$$

or equivalently

$$\frac{n}{\lambda} - nde^{-\lambda d}(1 + e^{-\lambda d} + e^{-2\lambda d} + \cdots) = \sum_{i=1}^{n}(X_i - [\frac{X_i}{d}]d). \tag{22.55}$$

As a first approximation for $\hat{\lambda}$, we have $\hat{\lambda}_0$ given by

$$\frac{n}{\hat{\lambda}_0} \simeq \sum_{i=1}^{n}(X_i - [\frac{X_i}{d}]d) \text{ or } \hat{\lambda}_0 \simeq \frac{n}{\sum_{i=1}^{n}(X_i - [\frac{X_i}{d}]d)}$$

and as a second approximation for $\hat{\lambda}$, we have $\hat{\lambda}_1$ given by

$$\frac{n}{\hat{\lambda}_1} \simeq \frac{nde^{-\hat{\lambda}_0 d}}{1 - e^{-\hat{\lambda}_0 d}} + \sum_{i=1}^{n}(X_i - [\frac{X_i}{d}]d)$$

or

$$\hat{\lambda}_1 \simeq \frac{n}{\frac{nde^{-\hat{\lambda}_0 d}}{1-e^{-\hat{\lambda}_0 d}} + \sum_{i=1}^{n}(X_i - [\frac{X_i}{d}]d)}. \tag{22.56}$$

It needs to be checked how good these approximations are for $\hat{\lambda}$.

22.8 Estimation for NHPP With Periodic Intensity

Suppose $\{N_t, t \geq 0\}$ is a nonhomogeneous Poisson process with intensity function $\lambda(\cdot)$ which is periodic with period d as discussed in Section 22.6. Let the process be observed over a time period $[0, nd)$ and suppose that the number of failures in the interval $[(i-1)d, id)$ is k_i, for $1 \leq i \leq n$. It follows from Theorem 22.6.1 that the random variables

$$N_{[(i-1)d, id)}, \ 1 \leq i \leq n$$

are i.i.d. Poisson random variables with parameter

$$\Lambda(d) = \int_{0}^{d} \lambda(u)du. \tag{22.57}$$

The likelihood function L, given $(k_1 k_2, \ldots, k_n)$, is

$$L = \prod_{i=1}^{n} \frac{[\Lambda(d)]^{k_i} e^{-\Lambda(d)}}{k_i!} \tag{22.58}$$

or equivalently

$$\log \, L = \left(\sum_{i=1}^{n} k_i \right) \log \, \Lambda(d) - n \, \Lambda(d) - \log \left(\prod_{i=1}^{n} k_i! \right). \tag{22.59}$$

22.8.1 Parametric estimation

Maximizing the function (22.59), one can get the maximum likelihood estimators of the parameters involved in the function $\Lambda(\cdot)$.

Example 22.8.1 Suppose

$$\lambda(t) = e^{\beta_0 + \beta_1 t}, \ \ 0 \le t \le d \tag{22.60}$$

and that $\lambda(\cdot)$ is periodic with period d. This gives a simple model for a happy system $(\beta_1 < 0)$ or a sad system $(\beta_1 > 0)$. It β_1 is near zero, then $\lambda(t)$ is approximately linear over a short time period. A happy system is one in which the times between failures tend to increase and a sad system is one where they tend to decrease [cf., Crowder *et al.* (1991)].

Estimation of parameters β_0 and β_1 is of interest in this model. It is easy to check that

$$\log \, L = \left(\sum_{i=1}^{n} k_i \right) \log \left(\int_0^d e^{\beta_0 + \beta_1 t} dt \right) - n \int_0^d e^{\beta_0 + \beta_1 t} dt - \log \left(\prod_{i=1}^{n} k_i! \right). \tag{22.61}$$

Differentiating with respect to β_0 and β_1, both the likelihood equations reduce to

$$\sum_{i=1}^{n} k_i = n \, \frac{e^{\beta_0} (e^{\beta_1 d} - 1)}{\beta_1} = n \, \Lambda(d). \tag{22.62}$$

It is clear that there exists no unique solution for (β_0, β_1). However, by using the method of moments, one can get the relations

$$\frac{\sum_{i=1}^{n} k_i}{n} = \Lambda(d) = \frac{e^{\beta_0} (e^{\beta_1 d} - 1)}{\beta_1} \tag{22.63}$$

and

$$\frac{1}{n} \sum_{i=1}^{n} k_i^2 = \Lambda(d) + \Lambda^2(d) = \frac{e^{\beta_0} (e^{\beta_1 d} - 1)}{\beta_1} + \left[\frac{e^{\beta_0} (e^{\beta_1 d} - 1)}{\beta_1} \right]^2. \tag{22.64}$$

Using the fact that $k_i, 1 \le i \le n$ are i.i.d. Poisson with parameter $\Lambda(d)$ and solving the equations (22.63) and (22.64), we can estimate β_0 and β_1.

Example 22.8.2 Suppose

$$\lambda(t) = \gamma \delta t^{\delta-1}, \ \ 0 \le t \le d \tag{22.65}$$

where $\gamma > 0$ and $\delta > 0$. This model represents a sad system if $\delta > 1$ and a happy system if $0 < \delta < 1$. If $\delta = 2$, then $\lambda(\cdot)$ is linear. Here

$$\log L = \left(\sum_{i=1}^{n} k_i\right) \log \left(\int_0^d \gamma \delta t^{\delta-1} dt\right) - n \int_0^d \gamma \delta t^{\delta-1} dt - \log \left(\prod_{i=1}^{n} k_i!\right). \tag{22.66}$$

Differentiating with respect to γ and δ, both the likelihood equations reduce to

$$\sum_{i=1}^{n} k_i = n\gamma d^{\delta} = n\Lambda(d). \tag{22.67}$$

It is clear that there exists no unique solution for (γ, δ). However, by using the method of moments, one can get the relations

$$\frac{\sum_{i=1}^{n} k_i}{n} = \Lambda(d) = \gamma d^{\delta} \tag{22.68}$$

and

$$\frac{1}{n}\sum_{i=1}^{n} k_i^2 = \Lambda(d) + \Lambda^2(d) = \gamma d^{\delta} + (\gamma d^{\delta})^2. \tag{22.69}$$

Using again the fact that $k_i, 1 \le i \le n$ are i.i.d. Poisson with parameter $\Lambda(d)$, and solving these equations, one can estimate γ and δ.

22.8.2 Nonparametric estimation of $\Lambda(d) = \int_0^d \lambda(t)dt$

Since the path of the NHPP process on $[0, nd]$ can be considered as a combination of n independent realizations of identical NHPP processes on $[(j-1)d, jd), 1 \le j \le n$, an estimate of $\Lambda(d)$ is

$$\tilde{\Lambda}(d) = \frac{\sum_{i=1}^{n} k_i}{n}. \tag{22.70}$$

Note that $E(\hat{\Lambda}(d)) = \Lambda(d)$ since each $k_i, 1 \le i \le n$ is Poisson with parameter $\Lambda(d)$. Let $(t_{(1)}, t_{(2)}, \ldots, t_{(k)})$ be the order statistics of the superpositions of n realizations with $t_{(0)} = 0$ and $t_{(k+1)} = d$ where $k = \sum_{i=1}^{n} k_i$. Then

$$\hat{\Lambda}(t) = \frac{ik}{(k+1)n} + \left[\frac{k(t - t_{(i)})}{(k+1)n(t_{(i+1)} - t_{(i)})}\right], t_{(i)} < t \le t_{(i+1)}, \ 0 \le i \le k \tag{22.71}$$

gives a piecewise linear estimator of the function $\Lambda(t)$, $0 \leq t \leq d$. This estimator passes through the points $(t_{(i)}, \frac{ik}{(k+1)n})$, $1 \leq i \leq k+1$. A modification needs to be made in case of ties. It is easy to see that

$$\lim_{k \to \infty} \hat{\Lambda}(t) = \Lambda(t) \text{ a.s., } 0 \leq t \leq d \qquad (22.72)$$

and an approximate $100(1-\alpha)\%$ confidence interval for $\Lambda(t)$ is

$$\hat{\Lambda}(t) - Z_{\alpha/2}\sqrt{\frac{\hat{\Lambda}(t)}{n}} < \Lambda(t) < \hat{\Lambda}(t) + Z_{\alpha/2}\sqrt{\frac{\hat{\Lambda}(t)}{n}}$$

where $Z_{\alpha/2}$ is the $1 - \frac{\alpha}{2}$ fractile of the standard normal distribution. This approach is similar to the one given in Leemis and Shih (1993).

References

1. Baxter, L. A. (1982). Reliability applications of the relevation transform, *Naval Research Logistics Quarterly*, **29**, 323–330.

2. Chukova, S. and Dimitrov, B. (1992). On distribution having the almost lack of memory property, *Journal of Applied Probability*, **29**, 691–698.

3. Chukova, S., Dimitrov, B. and Garriado, J. (1993). Renewal and non-homogeneous Poisson processes generated by distributions with periodic failure rate, *Statistics and Probability Letters*, **17**, 19–25.

4. Crowder, M. J., Kimber, A. C., Smith, R. L. and Sweeting, T. J. (1991). *Statistical Analysis of Reliability Data*, London: Chapman & Hall.

5. Dimitrov, B., Chukova, S. and Khalil, Z. (1995). Definitions, characterizations and structural properties of probability distributions similar to the exponential, *Journal of Statistical Planning and Inference*, **43**, 271–287.

6. Galambos, J. and Kotz, S. (1978). *Characterization of Probability Distributions*, Lecture Notes in Mathematics # 675, Berlin: Springer-Verlag.

7. Lau, K. S. and Prakasa Rao, B. L. S. (1990). Characterization of exponential distribution by the relevation transform, *Journal of Applied Probability*, **27**, 726–729. [Addendum ibid. **29** (1992), 1003–1004.]

8. Lau, K. S. and Rao, C. R. (1982). Integrated Cauchy functional equation and characterizations of the exponential law, *Sankhyā, Series A*, **44**, 72–90.

9. Leemis, L. M. and Shih, L. H. (1993). Variate generation for Monte Carlo analysis of reliability and life time models, In *Advances in Reliability* (Ed., A. P. Basu), pp. 247–256, Amsterdam: North-Holland.

10. Marsagalia, G. and Tubilla, A. (1975). A note on the lack of memory property of the exponential distribution, *Annals of Probability*, **3**, 352–354.

11. Prakasa Rao, B. L. S. (1995). On distributions with periodic failure rate and related inference problems, Key note address at the Conference "Statistical Inference in Life Testing and Reliability", September 1995, Banaras Hindu University, Varanasi, India.

12. Prakasa Rao, B. L. S. and Ramachandran, B. (1983). On a characterization of symmetric stable processes, *Aequationes Mathematicae*, **26**, 113–119.

13. Ramachandran, B. (1979). On the "strong memorylessness property" of the exponential and geometric probability laws, *Sankhyā, Series A*, **49**, 244–251.

14. Ramachandran, B. and Lau, K. S. (1991). *Functional Equations in Probability Theory*, Boston: Academic Press.

15. Ramachandran, B. and Prakasa Rao, B. L. S. (1984). On the equation $f(x) = \int_{-\infty}^{\infty} f(x+y)d\mu(y)$, *Sankhyā, Series A*, **46**, 326–338.

16. Rao, C. R., Sapatinas, T. and Shanbag, D. (1994). The integrated Cauchy functional equation: Some comments on recent papers, *Advances in Applied Probability*, **26**, 825–829.

23

Venn Diagrams, Coupon Collections, Bingo Games and Dirichlet Distributions

Milton Sobel and Krzysztof Frankowski

University of California, Santa Barbara, CA
University of Minnesota, Minneapolis, MN

Abstract: This chapter is an extension of the coupon collectors problem which arose from the free baseball cards that used to be inserted in a penny package of chewing (bubble) gum. The setting is that of a bingo game with a number caller. Each player has a list of numbers and waits for all his numbers to be called. These lists may be overlapping to any degree; hence, the need for Venn diagrams. The contest usually ends with the first winner, but need not. If not, we have to specify whether or not we use curtailment, i.e., stop as soon as the results are known. The calculations include many concepts including:

1. Expected waiting time for the first winner (in terms of numbers called) and for the entire contest (if it is different),
2. Probability of a single winner, i.e., without ties,
3. Expected number of integers not yet observed when the contest ends.

In addition, for "with replacement", we also find:

1. Expected maximum frequency among the numbers called,
2. Expected number of singletons among the numbers called,
3. Expected number of integers repeated among the numbers called.

In summary, it is generally not realized how useful the Dirichlet integrals and their hypergeometric analogues can be to solve challenging, non-trivial problems; this chapter illustrates one more important use of these integrals.

Keywords and phrases: Coupon-collecting, Dirichlet integrals, Bingo games, Venn diagrams

23.1 Introduction

We consider a collection of problems associated with a coupon-collecting contest. The original *coupon collection problem*, may be formulated as follows: A child buys bubble gum and inside the wrapper is a free baseball card selected at random with equal probabilities from an entire set of (say) 24 cards. The basic problem is to find the expectation (and perhaps the distribution) of the number of bubble gums needed to collect the entire set of 24 different cards.

In our Venn-diagram extension of this problem, we can have any number of competing players and each player is waiting for the numbers on his list to be called. They need not be waiting for the same numbers. There may be numbers that neither of say $j = 2$ players wants and if there are some numbers that both players want then we have to specify whether one calling is sufficient for both (which we assume below) or how it is determined which player gets that number and whether we are waiting for that number to be called again.

Assume that there are 3 players denoted by X, Y and Z. The list of numbers wanted by X alone is of size n_1; the list wanted by X and Y (and not Z) is of size n_{12} and the list wanted by all three is of size n_{123}, etc. N will be used to denote the size of the union of all these wanted numbers; the total list wanted by X will be noted by n_x so that $n_x = n_1 + n_{12} + n_{13} + n_{123}$, etc.

The two main sampling disciplines of interest in this paper are sampling with replacement (W/R) and sampling without replacement (W/O R), corresponding to multinomial and hypergeometric sampling, respectively.

If sampling is W/O R, the content resembles a bingo game where the contestants (say) X, Y and Z have lists (bingo cards) with overlapping numbers on them and the first one to complete his card is the winner.

We regard this as the first stage of the contest and may want to continue for more "winners." If sampling is W/R, the contest resembles a coupon-collecting context where the contestants (say) X, Y and Z may have lists of different sizes with considerable overlap. In our formulation, any 2 or more contestants can all be satisfied with a single calling of a number common to all their lists.

In most instances, the solution of a problem for $k = 2$ and/or $k = 3$ is sufficient to indicate the more general solution for $k \geq 3$ players; in such instances, we will append the solution for the general case of k players without any detailed derivation.

For each of the sampling disciplines, we are interested in several problems as follows; for some problems, only one discipline applies as in Problems 4 and 5 below. For simplicity, we usually assume that there are at most 3 players (or contestants) denoted by X, Y and Z.

1. The probability that X (alone) wins the contest, i.e., that all his numbers are called before all those of Y and before all of those of Z are called.

Adding the results for X, Y and Z gives the probability of a single winner (without any ties) for the first stage. We may also then obtain the probability of a paired tie and of a triple tie in the first stage. Since this entire set of probabilities is disjoint and exhaustive, their sum is unity and this makes for a useful check on our results.

2. The expected waiting time $E\{WT\}$ in terms of observations needed to complete (i) the first stage and (ii) the entire contest, in case (i) and (ii) do not coincide. Here, we have to indicate whether we are going to stop (or not) as soon as the result becomes known even though the pertinent sampling is not yet completed; we refer to these as sampling with or without curtailment.

3. The expected number of the integers in N (and T) not yet observed (i) at the end of the first stage and (ii) at the end of the entire contest. Here, we assume that there are U numbers not on anyone's list and that the total $T = N + U$, where N is the size of the union of the lists of all the players.

4. For sampling W/R, we are interested in the expected maximum frequency $E\{MF\}$ (i) at the end of the first stage and (ii) at the end of the entire contest, whether or not curtailment is used.

5. When sampling W/R, the expected number of singletons $E\{S\}$ and the expected number of integers repeated $E\{R\}$; this is considered only for the first stage without curtailment.

To illustrate the use of many future formulas, we consider 3 simple numerical examples:

Example 23.1.1 2 players: $n_1 = n_2 = 2$, $n_{12} = 1 \Rightarrow N = 5$, $n_x = 3$.

Example 23.1.2 3 players: $n_1 = n_2 = n_3 = 2$, $n_{12} = n_{13} = n_{23} = 1$, $n_{123} = 3 \Rightarrow N = 12$, $n_x = 7$.

Example 23.1.3 3 players: $n_1 = n_2 = n_3 = 2$, $n_{12} = n_{13} = n_{23} = 1$, $n_{123} = 1 \Rightarrow N = 10$, $n_x = 5$.

23.2   Mathematical Preliminaries and Tools

As a background for this paper, we define the two types of Dirichlet integrals and some of their usages in a multinomial setting. Type 1 includes the I, the J and the IJ and is used for fixed sample size problems [Sobel, Uppuluri and

Frankowski (1977)]. Type 2 includes the C, the D and the CD; they are the main tools of this paper. They are defined, studied and tabulated in Sobel, Uppuluri and Frankowski (1985) with many examples of their usage and we repeat the properties used in this paper briefly.

For any positive integers b, m and b-vectors $\boldsymbol{a} = (a_1, a_2, \ldots, a_b)$ (real) and $\boldsymbol{r} = (r_1, r_2, \ldots, r_b)$ (positive integers) we define

$$C_{\boldsymbol{a}}^{(b)}(\boldsymbol{r}; m) = \frac{\Gamma(m+R)}{\Gamma(m) \prod_{i=1}^{b} \Gamma(r_i)} \int_0^{a_1} \cdots \int_0^{a_b} \frac{\prod_{i=1}^{b} x_i^{r_i-1} \, dx_i}{(1 + \sum_{i=1}^{b} x_i)^{m+R}}, \qquad (23.1)$$

where $R = \sum_{i=1}^{b} r_i$. In terms of the multinomial cell probabilities p_i, we have $a_i = p_i/p_0$ $(i = 1, 2, \ldots, b)$ where $0 < p_0 = 1 - \sum_{i=1}^{b} p_i < 1$; we refer to the b cells as *blue cells*. The i-th blue cell has cell probability p_i and one additional cell, called the *counting cell*, has cell probability p_0. We sample from this multinomial and stop as soon as the counting cell reaches frequency $m \geq 1$; this is referred to as *at stopping time* or *ast*. Then, (23.1) is the probability that the i-th blue cell reaches or exceeds its quota r_i ast.

The dual probability (or D-integral) is the probability that the i-th blue cell has frequency *less than* r_i ast $(i = 1, 2, \ldots, b)$, i.e., none of the b cells reaches its quota, and is given by the same integral as in (23.1) except that the limits are all from a_i to ∞. For $b = 0$, we define $C^{(0)} = D^{(0)} = 1$. In most of our present applications, we have all cell probabilities equal to p_0 $(p_i = p_0 = p$; we call it a *homogeneous case* when this occurs) and therefore all $a_i = 1$. In general whenever we have a vector having all equal components, we use a scalar notation for vectors $\boldsymbol{a}$ or $\boldsymbol{r}$ and write $C_a^{(b)}(r, m)$ or $D_a^{(b)}(r, m)$ using a simple " , " instead of " ; " as a separator. In the hypergeometric case, we use similar notation $HC_{M,N}^{(b)}(r, m)$ and $HD_{M,N}^{(b)}$ retaining analogous probability interpretation, where N denotes the total population and M the population in a single "batch", so that in the homogeneous case $N = bM$.

Next, we list some specially useful formulae for these Dirichlet integrals.

In the homogeneous case $(a = 1)$ with $\boldsymbol{r} = \boldsymbol{r}_0 = (1, \ldots, 1, r, \ldots, r)$ in which there are c one's with $0 \leq c \leq b$ and the $b - c$ remaining components $r_j = r$, a straightforward integration of the c variables (each of which has r-value 1) gives for the D-integral

$$D_1^{(b)}(\boldsymbol{r}_0; m) = \frac{1}{(1+c)^m} D_{1/(1+c)}^{(b-c)}(r, m), \qquad (23.2)$$

and for the C-integral

$$C_1^{(b)}(\boldsymbol{r}_0; m) = \sum_{\alpha=0}^{c} \frac{(-1)^\alpha \binom{c}{\alpha}}{(1+\alpha)^m} C_{1/(1+\alpha)}^{(b-c)}(r, m). \qquad (23.3)$$

The special case with $b = c$ of these formulae is used in many applications, which are given by

$$D_1^{(b)}(1, m) = \frac{1}{(1+b)^m} \tag{23.4}$$

and for the C-integral

$$C_1^{(b)}(1, m) = \sum_{\alpha=0}^{b} \frac{(-1)^\alpha \binom{b}{\alpha}}{(1+\alpha)^m} . \tag{23.5}$$

In particular, for $m = 1$ and $m = 2$,

$$D_1^{(b)}(1, 1) = C_1^{(b)}(1, 1) = \frac{1}{(1+b)} \quad \text{and} \quad C_1^{(b-1)}(1, 2) = \frac{1}{b} \sum_{k=1}^{b} \frac{1}{k} . \tag{23.6}$$

The hypergeometric analogue is

$$HC_{1,K}^{(b-1)}(1, 2) = \frac{(K-b+1)! \, b!}{K!} , \tag{23.7}$$

which for $K = b + 1$ reduces to the obvious result

$$HC_{1,b+1}^{(b-1)}(1, 2) = \frac{2}{b+1} , \tag{23.8}$$

since there are exactly 2 items we can terminate with to satisfy the condition that we have seen $b - 1$ items before sampling stops. The following equalities, called the *inclusion-exclusion relation*, allow change from D to C and vice versa and have the form

$$C^{(b)} = \sum_{\alpha=0}^{b} (-1)^\alpha \binom{b}{\alpha} D^{(\alpha)} \quad \text{and} \quad D^{(b)} = \sum_{\alpha=0}^{b} (-1)^\alpha \binom{b}{\alpha} C^{(\alpha)} \tag{23.9}$$

provided the arguments and subscript are the same on both sides of the equality. Symbolically, we can always replace C by $1 - D$ or D by $1 - C$ in any analytic expression and we can treat the superscript like any power of a binomial expansion. The resulting algebra sometimes involves the product of two or more Cs or Ds. Usually the r-values in the two different C's (or in a C and D pair) refer to different cells in the same experiment and for this we define a *star product* denoted by $\star$; we replace this product by a single multiple integral with the variables in each C (or D) having its own original r values. This star product is consistent with our definition and has nice algebraic properties such as commutativity and associativity and is used consistently throughout the paper whenever we have a sum of products of C or D operators. We notice here that the star product defines together with the addition operation a commutative ring.

In Sections 23.4 and 23.5, we need expectation formula for the waiting time to see each of the b blue cells reach its quota $r_i = r$ $(i = 1, 2, \ldots, b)$. We have shown in [Sobel, Uppuluri and Frankowski (1985, Eq. (5.8))] that for the multinomial case this expected waiting time $E\{WT\}$ is given by

$$E\{WT\} = \frac{br}{p}\, C_a^{(b-1)}(r, r+1), \qquad (23.10)$$

where p is the common cell probability, $a = p/p_0$ and p_0 is the counting cell probability. In most of our applications, a will be 1 and hence $p = 1/b$.

The corresponding hypergeometric result shown in Sobel and Frankowski (1995) is

$$E\{WT\} = br \left(\frac{N+1}{M+1}\right) HC_{M,1+bM}^{(b-1)}(r, r+1), \qquad (23.11)$$

where M is the common batch size for each of the b blue cells and N is the total population size.

23.3   Probability of a Single Winner or No Tie

As before, let X, Y and Z denote the 3 players and let n_i denote the set of numbers assigned only to player i $(i = 1, 2, 3)$, let n_{12} denote the common set of numbers to both players 1 and 2, etc. Let $n_x = n_1 + n_{12} + n_{13} + n_{123}$ with similar definitions for n_y and n_z. Using the Dirichlet integral C^b for $C_1^{(b)}(1, 1)$, for sampling with replacement, the probability of no tie when the first stage is completed is

$$\Pr[\text{No Tie}] = n_x C^{n_x-1} \star [1 - C^{n_{23}} + C^{n_{23}} \star (1 - C^{n_2}) \star (1 - C^{n_3})] + 2ST, \qquad (23.12)$$

where $2ST$ indicates two similar terms, one for Y and one for Z.

The operations leading to and from Eq. (23.12) form our first application of the star product operator, operating this time on the C-functions. We use these Dirichlet operators like indicators of random variables for events. The first factor C^{n_x-1} indicates that just before the first stage ended we saw all but the terminal number each at least once. The multiplier n_x gives the number of possibilities for the terminal number each at least once. The multiplier n_x gives the number of possibilities for the terminal number for X. The bracket after C^{n_x-1} indicates that we are intersecting the above event with the event that either (*cf.* + sign) the n_{23} set is incomplete (note the complement of $C^{n_{23}}$) or we have the conjunction that the n_{23} set is complete but the n_2 and n_3 sets are both incomplete. One of these has to happen for X to be a solo winner.

One word of caution is in order for the star operation. All star operations have to be carried out before the corresponding integrals are evaluated, because only then the terms have clear meaning for the final result. Naturally, the same remark applies to powers of sums of C's and D's. They should all be regarded as star powers.

From (23.12), after evaluation of C integrals using (23.6), we have

$$\Pr[\text{No Tie}] = 1 - \frac{n_x}{N - n_3} - \frac{n_x}{N - n_2} + \frac{n_x}{N} + 2ST, \qquad (23.13)$$

where N is the totality of numbers assigned to all 3 players. Summing the 3 similar terms,

$$\begin{aligned}
\Pr[\text{No Tie}] \quad = \quad & 1 + \frac{n_{12} + n_{13} + n_{23} + 2n_{123}}{N} - \frac{n_{23} + n_{123}}{N - n_1} \\
& - \frac{n_{13} + n_{123}}{N - n_2} - \frac{n_{12} + n_{123}}{N - n_3} .
\end{aligned} \qquad (23.14)$$

A three-way tie in the first stage occurs only when the last observation of the first stage comes from n_{123} and this is easily seen to have probability

$$\Pr[\text{Triple Tie}] = \frac{n_{123}}{N} . \qquad (23.15)$$

In order to have exactly one 2-way tie in the first stage (say, between players X and Y), the last observation in the first stage must come from n_{12} or n_{123}. Hence, we obtain for the first stage

$$\begin{aligned}
\Pr[\text{Tie for } X \text{ and } Y] \quad = \quad & (n_{12} + n_{123}) C^{n_{12} + n_{123} - 1} \\
& \star [C^{n_1 + n_2 + n_{13} + n_{23}} - C^{N - n_{12} - n_{123}}] \\
= \quad & \frac{n_{12} + n_{123}}{N - n_3} - \frac{n_{12} + n_{123}}{N} = \frac{(n_{12} + n_{123})n_3}{N(N - n_3)} .
\end{aligned} \qquad (23.16)$$

Hence, summing over the three possibilities, we get for the first stage

$$\begin{aligned}
& \Pr[\text{Exactly One 2-way Tie}] \\
= \quad & \frac{(n_{12} + n_{123})n_3}{N(N - n_3)} + \frac{(n_{13} + n_{123})n_2}{N(N - n_2)} + \frac{(n_{12} + n_{123})n_1}{N(N - n_1)} .
\end{aligned} \qquad (23.17)$$

To illustrate these computations, we consider Example 23.1.2 ($n_1 = n_2 = n_3 = 2$, $n_{12} = n_{13} = n_{23} = 1$ and $n_{123} = 3$); then from (23.14), (23.15) and (23.17), we obtain

$$\Pr[\text{No Tie}] = 3 \left(1 - \frac{7}{10} - \frac{7}{10} + \frac{7}{12} \right) = \frac{11}{20} , \qquad (23.18)$$

$$\Pr[\text{Triple Tie}] \quad = \quad \frac{5}{20} , \qquad (23.19)$$

$$\Pr[\text{Exactly One 2-way Tie}] \quad = \quad \frac{3(4)2}{12(10)} = \frac{4}{20}, \qquad (23.20)$$

and the sum of these 3 probabilities is unity, as a partial check on our results.

The above results were obtained by assuming (i) sampling with replacement and (ii) that one calling (or observation) of a particular number will satisfy all players waiting for that number. Under this assumption, it clearly does not matter whether we sample with or without replacement. Hence, the 3 results in (23.14), (23.15) and (23.17) remain valid (i.e., exactly the same) for sampling W/O R. Furthermore, even if we decided to replace one preassigned set and not another, these probability results dealing with the order of the winners and the ties among them all remain exactly the same.

A corollary of (23.12) and (23.13) gives the probability that a particular player finishes first without ties, viz. for the first player X (with a total of 3 players), we have

$$\Pr[X \text{ wins}] = 1 - \frac{n_x}{N - n_3} - \frac{n_x}{N - n_2} + \frac{n_x}{N}. \qquad (23.21)$$

It is not difficult to generalize this result to any number of players. Thus, for 4 players W, X, Y and Z, we obtain (letting W denote player 1)

$$\begin{aligned}\Pr[W \text{ wins}] \quad = \quad & 1 - \frac{n_w}{N - n_3 - n_4 - n_{34}} - \frac{n_w}{N - n_2 - n_4 - n_{24}} \\ & - \frac{n_w}{N - n_2 - n_3 - n_{23}} + \frac{n_w}{N - n_2} + \frac{n_w}{N - n_3} \\ & + \frac{n_w}{N - n_4} - \frac{n_w}{N}. \end{aligned} \qquad (23.22)$$

Even in (23.21) with 3 players and no pairwise overlap, this result is not intuitively clear. Thus, if Player 1 has $2d$ numbers and the two other players each have d numbers without any pairwise overlap, the numerical result for the probability that Player 1 finishes first is

$$\Pr[\text{Player 1 wins}] = 1 - \frac{2}{3} - \frac{2}{3} + \frac{1}{2} = \frac{1}{6}. \qquad (23.23)$$

Denote the above model with $(2d, d, \ldots, d)$ numbers assigned and no pairwise overlap as Model 1 and the corresponding model with $(\frac{d}{2}, d, \ldots, d)$ as Model 2. For these two models, we give numerical results for a total of j players ($j = 2, 3, 4, 5, 10$) in the following table. (The reader might try to guess these answers before looking at the table.)

j $\rightarrow$	2	3	4	5	10
Model 1 $\rightarrow$	0.333333	0.166667	0.100000	0.066667	0.018182
Model 2 $\rightarrow$	0.666667	0.533333	0.457143	0.406349	0.283773

The probabilities for Model 1 were computed from $\Pr[\text{Model 1}] = \frac{2}{j(j+1)}$ and for Model 2 from $\Pr[\text{Model 2}] = 2^{2j-1}/(j\binom{2j}{j})$. Both of the rows in above table are approaching zero as j increases, but for Model 1 as $2/j^2$ and for the Model 2 as $\frac{1}{2}\sqrt{\pi/j}$.

23.4 The Expected Waiting Time $\{WT\}$ to Complete the First Stage

For the expected waiting time to complete the first stage, we consider the following disjoint events:

Events 1,2,3: X is the solo winner in the First Stage $+ 2ST$ (one for Y and one for Z).

Events 4,5,6: X and Y (but not Z) are tied in the First Stage $+ 2ST$ (one for X, Y and one for Y, Z).

Event 7: X, Y and Z are all tied in the First Stage.

We again use the star product operator acting on the C-function. The major difference here (from the operation in Section 23.3) is that we are calculating expected waiting time (not a probability) and the result in (23.10) with $r = 1$ tells us that the resulting C-function will now have arguments (1,2) not (1,1) and again because of the equal probabilities the subscript a will be 1.

For Event 1, just before the terminal observation we must have seen all but one of the numbers on the X's list *and* either n_{23} is incomplete or else the set of size n_{23} is complete but the sets corresponding to n_2 and n_3 are both incomplete. Hence, for the first 3 of the 7 events, we have

$$N\{n_x C_1^{n_x-1}(1,2) \star [1 - C_1^{n_{23}}(1,2) + C_1^{n_{23}}(1,2) \star (1 - C_1^{n_2}(1,2))$$
$$\star (1 - C_1^{n_3}(1,2))]\} + 2ST. \qquad (23.24)$$

For Event 4, we must have seen (just before the terminal number) all the numbers in the union of X's and Y's lists but not all the numbers in the list of size n_3. Hence, for the next 3 events (4, 5, 6), we have

$$N\left\{(n_{12} + n_{123})C_1^{(N-n_3-1)}(1,2) \star \left[1 - C_1^{(n_3)}(1,2)\right]\right\} + 2ST. \qquad (23.25)$$

Note that the terminal number must be common to both X and Y, which explains the coefficient.

For the final event of the 7, we must have seen all but one of the N numbers just before the terminal observation and the terminal number comes from the set of size n_{123}. Hence, the contribution for this event is

$$N\left\{n_{123}C_1^{(N-1)}(1,2)\right\}. \tag{23.26}$$

Putting all of these together, we obtain the expected waiting time for the first stage as

$$
\begin{aligned}
E\{WT\} &= N\{n_x[C_1^{(n_x-1)}(1,2) - C_1^{(n_x+n_2+n_{23}-1)}(1,2) \\
&\quad - C_1^{(n_x+n_3+n_{23}-1)}(1,2) + C_1^{(N-1)}(1,2)] \\
&\quad + 2ST + (n_{12}+n_{123})[C_1^{(N-n_3-1)}(1,2) - C_1^{(N-1)}(1,2)] \\
&\quad + 2ST + n_{123}\,C_1^{(N-1)}(1,2)\}.
\end{aligned}
\tag{23.27}
$$

It should be noted that (23.27) assumes no curtailment, i.e., we play to the end even if we know the winner (or the result) sooner.

To illustrate the use of this formula (and future formulas), we consider our 3 examples from Section 23.1 above.

For Example 23.1.1, we obtain for the first stage when sampling W/R using (23.6)

$$
\begin{aligned}
E_M\{WT\} &= 5\{3[C_1^{(2)}(1,2) - C_1^{(4)}(1,2) \times 2 + C_1^{(4)}(1,2)\} \\
&= 30C_1^{(2)}(1,2) - 25C_1^{(4)}(1,2) \\
&\approx 6.916667.
\end{aligned}
\tag{23.28}
$$

For Example 23.1.2 and Example 23.1.3, respectively, we obtain

$$
\begin{aligned}
E_M\{WT\} &= 12\{3[7C_1^{(6)}(1,2) - 14C_1^{(9)}(1,2) + 7C_1^{(11)}(1,2)] \\
&\quad + 12[C_1^{(9)}(1,2) - C_1^{(11)}(1,2)] + 3C_1^{(11)}(1,2)\} \\
&= 252C_1^{(6)}(1,2) - 360C_1^{(9)}(1,2) + 144C_1^{(11)}(1,2) \\
&\approx 25.138528,
\end{aligned}
\tag{23.29}
$$

$$
\begin{aligned}
E_M\{WT\} &= 150C_1^{(4)}(1,2) - 240C_1^{(7)}(1,2) + 100C_1^{(9)}(1,2) \\
&\approx 16.253968.
\end{aligned}
\tag{23.30}
$$

In the result (23.27) and in the above three examples, the sampling is W/R and there are no unassigned numbers ($u = 0$), so that $T = N$. If $u \geq 1$, then the only effect it has on the formula (23.27) is to replace the initial multiplier N by the larger T. It does not affect the probability that anyone wins the first stage or the probability of one or more ties. The only effect that u has on Problem 2, ($E\{WT\}$), with replacement is to replace the initial constant N by T. Thus, if

$u = 5$ in the three examples above, then the new answers are, respectively,

$$E_M\{WT\} = \begin{cases} \frac{10}{5} \ (6.916667) & \approx \ 13.833333, \\[2mm] \frac{17}{12} \ (25.138528) & \approx \ 35.612915, \\[2mm] \frac{15}{10} \ (16.253968) & \approx \ 24.380952. \end{cases} \tag{23.31}$$

The general case for sample W/O R is obtained from (23.27) using (23.11) with $M = 1$ to (i) replace every $C_1^{(b-1)}(1,2)$ by $HC_{1,b+1}^{(b-1)}(1,2)$ and (ii) to replace the initial constant N by $(N+1)/(M+1) = (N+1)/2$. If there are unassigned numbers present, then we use $(T+1)/2$ in (ii) above. We illustrate this with the same three examples (Examples 23.1.1, 23.1.2 and 23.1.3) as above with $u = 0$ and $u = 5$. For $u = 0$, we obtain for sampling W/O R in Examples 23.1.1, 23.1.2 and 23.1.3, respectively,

$$\begin{aligned} &E_H\{WT\} \\ &= \begin{cases} 3\left[6HC_{1,4}^{(2)}(1,2) - 5HC_{1,6}^{(4)}(1,2)\right] = 4, \\[2mm] \frac{13}{2}\left[21HC_{1,8}^{(6)}(1,2) - 30HC_{1,11}^{(9)}(1,2) + 12HC_{1,13}^{(11)}(1,2)\right] \approx 10.670455, \\[2mm] \frac{11}{2}\left[15HC_{1,6}^{(4)}(1,2) - 24HC_{1,9}^{(7)}(1,2) + 10HC_{1,11}^{(9)}(1,2)\right] \approx 8.166667. \end{cases} \end{aligned}$$
$$\tag{23.32}$$

For $u = 5$, we merely change the multipliers and the corresponding numerical results are $\frac{11}{2}\,(4)/3 \approx 7.333333$, $\frac{18}{13}\,(10.670455) \approx 14.774476$ and $\frac{16}{11}\,(8.166667) \approx 11.878788$, respectively.

Consider the last term in (23.27). Under curtailment, we do not wish to wait for all the numbers in n_{123} if it is already clear that we are getting a triple tie; no curtailment is possible for the first six terms of (23.27). So under curtailment, we replace the last term of (23.27) by

$$(N - n_{123})\left[C_1^{(N-n_{123}-1)}(1,2) - C_1^{(N-1)}(1,2)\right], \tag{23.33}$$

so that the amount of reduction (ΛR) due to curtailment in the first stage is

$$AR = \begin{cases} NC_1^{(N-1)}(1,2) - (N - n_{123})C_1^{(N-n_{123}-1)}(1,2) & \text{for 3 players} \\[2mm] NC_1^{(N-1)}(1,2) - (N - n_{12})C_1^{(N-n_{12}-1)}(1,2) & \text{for 2 players.} \end{cases} \tag{23.34}$$

The numerical results for the amount of reduction under curtailment in our three examples are

$$AR = \begin{cases} 5C_1^{(4)}(1,2) - 3C_1^{(2)}(1,2) = 0.450000 & \text{(Example 23.1.1)} \\[2mm] 12C_1^{(11)}(1,2) - 9C_1^{(8)}(1,2) \approx 0.274242 & \text{(Example 23.1.2)} \\[2mm] 10C_1^{(9)}(1,2) - 9C_1^{(8)}(1,2) = 0.100000 & \text{(Example 23.1.3).} \end{cases} \tag{23.35}$$

23.5   The Expected Waiting Time $\{WT\}$ to Complete the Entire Contest

There are several types of curtailment in a 3-person complete contest. Before enumerating these, we may wish to see the total $E\{WT\}$ for all N numbers without curtailment and W/R in our three examples above for $u = 0$ and $u = 5$. For $u = 0$,

$$
\begin{aligned}
E_M\{WT\} &= N^2 C_1^{(N-1)}(1,2) \\
&= N \sum_{i=1}^{N} \frac{1}{i} \approx \begin{cases} 11.416667 & \text{(Example 23.1.1)} \\ 37.238528 & \text{(Example 23.1.2)} \\ 29.289683 & \text{(Example 23.1.3)} \end{cases} \quad (23.36)
\end{aligned}
$$

where (23.10) and (23.6) was used with $r = 1$, $b = N$ and $p = 1/N$.

For $u = 5$, these answers become

$$
E_M\{WT\} = T N C_1^{(N-1)}(1,2) \approx \begin{cases} 22.833333 & \text{(Example 23.1.1)} \\ 52.754582 & \text{(Example 23.1.2)} \\ 43.934524 & \text{(Example 23.1.3)}. \end{cases} \quad (23.37)
$$

Of course, if we sample W/O R the answers are simply 5, 12, 10 for $u = 0$ and 10, 17, 15 for $u = 5$.

For 3 players, there are three types of curtailment: 3 cases for each of Types 1 and 2, and 1 case for Type 3. Type 1 is to stop after 2 winners whether they make a tie or not. Type 2 is to stop if it is clear that 2 players are going to tie for second place. Type 3 is to stop if it is clear that there will be a 3-way tie in the very first stage. These seven cases are all disjoint and one of them must occur, i.e., they are also exhaustive. Hence, the result for $E\{WT\}$ for the entire contest with all possible curtailments is

$$
\begin{aligned}
E_M\{WT\} &= N\{(N - n_3)[C_1^{(N-n_3-1)}(1,2) - C_1^{(N-1)}(1,2)] + 2ST \\
&+ (N - n_{23})[C_1^{(N-n_{23}-1)}(1,2) - C_1^{(N-1)}(1,2)] + 2ST \\
&+ (N - n_{123})[C_1^{(N-n_{123}-1)}(1,2) - C_1^{(N-1)}(1,2)]\}. \quad (23.38)
\end{aligned}
$$

For Example 23.1.2, (23.38) evaluates to

$$
\begin{aligned}
E_M\{WT\} &= 360[C_1^{(9)}(1,2) - C_1^{(11)}(1,2)] + 396[C_1^{(10)}(1,2) - C_1^{(11)}(1,2)] \\
&+ 108[C_1^{(8)}(1,2) - C_1^{(11)}(1,2)] \approx 24.675896, \quad (23.39)
\end{aligned}
$$

which is to be compared with 37.238528 in (23.36). For $u = 5$, this increases to 34.957519 as against 52.754582 in (23.37). The corresponding results for Example 23.1.3 with $u = 0$ are 18.956349 (with curtailment) as against 29.289683 (without curtailment) in (23.36), and with $u = 5$ are 28.28434524 as against 43.934524 in (23.37).

To obtain the corresponding results W/O R, we change the formulas obtained W/R and use (23.11). Then, the (23.38) $E\{WT\}$ result for the entire contest sampling with curtailment and W/O R takes the form

$$E_H\{WT\} = \frac{n_3(N - n_3)}{N - n_3 + 1} + 2ST + \frac{n_{23}(N - n_{23})}{N - n_{23} + 1}$$
$$+ 2ST + \frac{n_{123}(N - n_{123})}{N - n_{123} + 1} . \qquad (23.40)$$

For our two examples, Example 23.1.2 and Example 23.1.3 of 3-person games, we obtain using (23.40)

$$E_H\{WT \mid \text{Example 23.1.2}\} = \frac{60}{11} + \frac{11}{4} + \frac{27}{10} \approx 10.904545 \quad (23.41)$$

$$E_H\{WT \mid \text{Example 23.1.3}\} = \frac{16}{3} + \frac{27}{10} + \frac{9}{10} \approx 8.933333. \quad (23.42)$$

We describe here another, more general method, which can also be used for evaluation of higher moments. It has the advantage that the C-functions remain with their original $(1,1)$ arguments with subscript 1 and never change to the $(1,2)$ form. A separate factor t indicates the last item sampled. $(Ct)^{b-1}$ indicates that we saw $b-1$ items before the last item sampled at stopping time. D indicates that we did not observe an item and $(D+Ct)^b - (Ct)^b$ indicates that we may have seen some of the b items but not all the b items. Our generating function $G(t)$ takes the form

$$G(t) = (N - n_3)t(Ct)^{N-n_3-1} \star [(D + Ct)^{n_3} - (Ct)^{n_3}] + 2ST$$
$$+ (N - n_{23})t(Ct)^{N-n_{23}-1} \star [(D + Ct)^{n_{23}} - (Ct)^{n_{23}}] + 2ST$$
$$+ (N - n_{123})t(Ct)^{N-n_{123}-1} \star [(D + Ct)^{n_{123}} - (Ct)^{n_{123}}]. \quad (23.43)$$

After differentiation with respect to t, we set $t = 1$ so that $D+Ct = D+C = 1$. Any C functions can be evaluated since $C^b = 1/(b+1)$. The result of this is exactly the same as the result in (23.40).

23.6 The Expected Number of Integers Not Seen in the First Stage

Since we are dealing only with events in the first stage, the question of curtailment does not enter. We carry out the computation for the case of sampling W/O R and claim that the answer is exactly the same W/R. For the 2-player case with $u = 0$, we can omit the possibility of a tie since in that case we see all the numbers by the end of the first stage. We use a generating function $G(t)$ approach and consider the 2-player case first. At stopping time, if X wins, then we can stop with any one of $n_1 + n_{12}$ different numbers and we have seen exactly $n_1 + n_{12} - 1$ before that point. Then, we have for our generating function

$$G(t) = (n_1 + n_{12})C^{n_1 + n_{12} - 1} \star (C + Dt)^{n_2 + u} + 1ST, \qquad (23.44)$$

where u is the number of unassigned integers in the initial set from which numbers are randomly drawn. Differentiating with respect to t and setting $t = 1$, gives for the expected number of integers not seen in the first stage,

$$E\{\# \text{ Unseen Integers}\} = \frac{n_2 + u}{T - n_2 + 1} + \frac{n_1 + u}{T - n_1 + 1}. \qquad (23.45)$$

Here, we used the fact that symbolically $D = 1 - C$ and $C^b = 1/(b + 1)$; $T = N + u$ as defined previously.

For our Example 23.1.1 where $u = 0$, the answer is 1 and this is easily checked since we stop after $j = 3$, 4 and 5 observations with probabilities $\frac{1}{5}$, $\frac{3}{5}$ and $\frac{1}{5}$, respectively. As mentioned above, the same result holds for sampling W/R.

Using the 7-term expression (23.27) for the stopping situation when the first stage is completed for 3 players, we form a generating function $G(t)$ by putting $Dt + C$ for numbers that may or may not have been called. The result for sampling W/O R, without curtailment and with u unassigned integers, is

$$\begin{aligned}
G(t) = \ & n_x[C^{n_x - 1} \star (Dt + C)^{T - n_x} - C^{N - n_3 - 1} \star (Dt + C)^{n_3 + u} \\
& - C^{N - n_2 - 1} \star (Dt + c)^{n_2 + u} + C^{N - 1} \star (Dt + c)^u] + 2ST \\
& + (n_{12} + n_{123})[C^{N - n_3 - 1} \star (Dt + C)^{n_3 + u} - C^{N - 1} \star (Dt + C)^u] \\
& + 2ST + n_{13}C^{N - 1} \star (Dt + C)^u.
\end{aligned} \qquad (23.46)$$

Of these 16 terms, 4 terms do not contain t and disappear when we differentiate with respect to t. All the remaining 12 terms have the form $C^{a-1} \star (Dt + C)^b$, where the common arguments of C and D are (1,1) and the common subscript is 1. Differentiating this form with respect to t and setting $t = 1$ gives the result

$$bC^{a-1} \star D = b(C^{a-1} - C^a) = b\left(\frac{1}{a} - \frac{1}{a+1}\right) = \frac{b}{a(a+1)}. \qquad (23.47)$$

Substituting this in the twelve terms of (23.46) that contain t with appropriate a and b values, we obtain the desired result

$$\begin{aligned}
E\{\# \text{ Unseen Integers}\} = \ & \frac{T - n_x}{n_x + 1} - \frac{(n_1 + n_{13})(n_3 + u)}{(N - n_3)(N - n_3 + 1)} \\
& - \frac{n_x(n_2 + u)}{(N - n_2)(N - n_2 + 1)} - \frac{n_{12}}{N(N + 1)} \\
& + 2ST - \frac{2n_{123}u}{N(N + 1)}.
\end{aligned} \qquad (23.48)$$

For our two 3-players examples, Example 23.1.2 and Example 23.1.3, with $u = 0$ in both, the numerical results are $\frac{117}{88} \approx 1.329545$ and $\frac{11}{6} \approx 1.833333$, respectively.

Obviously, for the entire contest without curtailment, with 3 players and with $u = 0$, we wait for all the numbers to be seen and hence the expected number unseen is zero. If u is positive, then we wait for all N assigned numbers and the generating function for the unseen numbers is

$$N\, C^{N-1} \star (Dt + C)^u \tag{23.49}$$

and by the method used above, the expectation answer is

$$E\{\# \text{ Unseen}\} = \frac{u}{N+1} . \tag{23.50}$$

For the entire contest with curtailment, the generating function is

$$\begin{aligned}
(n_x + n_y - n_{12})C^{N-n_3-1} &\star [(Dt + C)^{n_3+u} - C^{n_3+u}] \\
+ n_{23}C^{N-n_{23}-1} &\star [(Dt + C)^{n_{23}+u} - C^{n_{23}+u}] + 2ST \\
+ n_{123}C^{N-1} &\star (Dt + C)^u.
\end{aligned} \tag{23.51}$$

Proceeding as before, the desired result for the $E\{\# \text{ Unseen}\}$ for the entire contest with curtailment is from (23.51), using (23.47),

$$\begin{aligned}
E\{\# \text{ Unseen}\} =\ & \frac{(n_x + n_y - n_{12})(n_3 + u)}{(N - n_3)(N - n_3 + 1)} + \frac{n_{23}(n_{23} + u)}{(N - n_{23})(N - n_{23} + 1)} \\
& + 2ST + \frac{n_{123}u}{N(N+1)} .
\end{aligned} \tag{23.52}$$

For our two 3-players examples, Example 23.1.2 and Example 23.1.3, we obtain $\frac{161}{220} \approx 0.731818$ and $\frac{47}{60} \approx 0.783333$, respectively. The former is about 6.10% of 12 and the latter is about 7.83% of 10; these n-values are too small to say anything about the asymptotic behaviour as n (and N) approach infinity.

In all of the results that contain u, such as (23.52), we may be interested in the expected $\#$ unseen from the u unassigned. Then, we subtract the answer for $u = 0$ from the total including u to get the portion coming from the unassigned. Thus, if we add 5 to the 3-player Example 23.1.2 so that $u = 5$, $N = 12$, $T = 17$, we obtain from (23.51)

$$E\{\# \text{ Unseen}\} = \left(\frac{13(7)}{110} + \frac{6}{132} \right) 3 + \frac{15}{156} \approx 2.714336, \tag{23.53}$$

and, using the result obtained above, the part coming from the five unassigned numbers is $\frac{7763 - 2093}{2860} \approx 1.982517$.

23.7 Maximum Frequency for 2 Players in the First Stage

When sampling with replacement, the question of maximum frequency of observation among all N numbers at the end of the first stage is perhaps the most interesting and the most challenging of our problems. We start with 2 players and first derive an expression for $\Pr[\text{Max Freq} \leq r]$ using Dirichlet operators. Omitting the subscript 1 on D, the "star power" expression $[D(r+1,1) - D(1,1)]^{N-1}$ indicates that N specified items all showed up, one of the N appear exactly once at stopping time and that all remaining frequencies were less than $r+1$. Then, $D^N(r+1,1) - [D(r+1,1) - D(1,1)]^N$ indicates that all frequencies were less than $r+1$, but *not* all of the N items showed up by stopping time. For 2 players, we can stop with player 1 winning or player 2 winning or with a tie, and hence we write

$$
\begin{aligned}
\Pr[\text{Max Freq} \leq r] \;=\; & (N - n_2)[D(r+1,1) - D(1,1)]^{N-n_2-1} \\
& \star \{D^{n_2}(r+1,1) - [D(r+1,1) - D(1,1)]^{n_2}\} \\
& + 1ST + n_{12}[D(r+1,1) - D(1,1)]^{N-1}. \quad (23.54)
\end{aligned}
$$

In the first expression above, we expand both bracketed quantities. We also replace $D[r+1,1]$ by $[D(r+1,1) - D(r,1)]$ to obtain $\Pr[\text{Max Freq} = r]$. Then, multiply the r-th term by r to obtain the expectation $E\{\text{Max Freq}\}$. Any power of $D(1,1)$, say $D(1,1)^\alpha$, is the probability that a specified item will appear before α distinct items and hence is $1/(\alpha+1)$. However, this operation changes the "counting cell" probability to $(1+\alpha)/N$ and the ratio of ordinary cell probabilities to that of the counting cell becomes $1/(1+\alpha)$. After some manipulations, the result for expectation becomes

$$
\begin{aligned}
E\{\text{Max Freq}\} \;=\; & (N - n_2) \sum_{\alpha=0}^{N-n_2-1} \sum_{\beta=1}^{n_2} \frac{(-1)^{\alpha+\beta-1}}{1+\alpha+\beta} \binom{N-n_2-1}{\alpha} \\
& \times \binom{n_2}{\beta} F_{\alpha+\beta}(N) + 1ST \\
& + n_{12} \sum_{i=0}^{N-1} \frac{(-1)^i}{1+i} \binom{N-1}{i} F_i(N), \quad (23.55)
\end{aligned}
$$

where

$$
F_i(N) \;=\; \sum_{r=1}^{\infty} r \left[D_{\frac{1}{1+i}}^{(N-i-1)}(r+1,1) - D_{\frac{1}{1+i}}^{(N-i-1)}(r,1) \right]
$$
$$
(i = 0, 1, \ldots, N-2). \quad (23.56)
$$

The value of $F_i(N)$ is zero for $i = N - 1$ and we eliminate these terms below. Setting $\alpha + \beta = i$ in (23.55), we can eliminate one summation in (23.55) above and obtain the simpler form

$$E\{\text{Max Freq}\} = (N - n_2) \sum_{i=1}^{N-2} \frac{(-1)^i}{1+i} \left[\binom{N-1}{i} - \binom{N-n_2-1}{i} \right] F_i(N)$$

$$+ 1ST + n_{12} \sum_{i=0}^{N-1} \frac{(-1)^i}{1+i} \binom{N-1}{i} F_i(N), \qquad (23.57)$$

where $F_i(N)$ is given by (23.56). After some simplifications, we can also write (23.57) as

$$E\{\text{Max Freq}\} = \sum_{i=2}^{N-1} (-1)^i \left[\binom{N}{i} - \binom{N-n_1}{i} - \binom{N-n_2}{i} \right]$$

$$\times F_{i-1}(N) + n_{12} F_0(N). \qquad (23.58)$$

The formula (23.56) is not suitable for accurate computations. We can transform it as follows: first, we "telescope" the sum to obtain

$$F_i(N) = \sum_{r=1}^{\infty} \left[1 - D_{\frac{1}{1+i}}^{(N-i-1)}(r, 1) \right]. \qquad (23.59)$$

Then, using result (2.29b) and (2.30) of Sobel, Uppuluri and Frankowski (1985), we obtain

$$F_i(N) = (N - i - 1) \sum_{r=1}^{\infty} \left(\frac{1}{2+i} \right)^r D_{\frac{1}{2+i}}^{(N-i-2)}(r, r). \qquad (23.60)$$

It is interesting to point out, as a corollary, that (23.57) gives an answer to one-player problem which is not at all trivial. If a single die is tossed (only) until all six faces appear (each at least once), what is the expected maximum frequency among all the six faces observed? By setting $n_1 = n_2 = 0$ and $n_{12} = 6$ in (23.57), we obtain from the last term of (23.57)

$$E\{\text{Max Freq}\} = 6 \sum_{i=0}^{4} \frac{(-1)^i}{1+i} \binom{5}{i} F_i(6) \approx 4.46349. \qquad (23.61)$$

This should be compared with the expected frequency in any non-terminating cell. Since $E\{WT\} = 14.7$ exactly [*cf.* Sobel, Uppuluri and Frankowski (1985, p. 65)], the expected frequency for a non-terminating cell is $(14.7 - 1)/5 = 2.74$. One might surmise that the expected smallest frequency and the expected largest frequency *before termination* should lie equally distanced about 2.74;

Values of $F_i(N)$

i	$N = 3$	i	$N = 4$	i	$N = 5$	i	$N = 6$	i	$N = 7$	i	$N = 8$
0	1.447214	0	1.724464	0	1.922036	0	2.074153	0	2.197166	0	2.300067
		1	0.788675	1	0.984975	1	1.131177	1	1.246521	1	1.341187
				2	0.551551	2	0.709394	2	0.831074	2	0.929089
						3	0.426777	3	0.560481	3	0.666455
								4	0.349071	4	0.465702
										5	0.295766

i	$N = 9$	i	$N = 10$	i	$N = 11$	i	$N = 12$	i	$N = 13$	i	$N = 14$
0	2.388296	0	2.465383	0	2.533735	0	2.595068	0	2.650645	0	2.701420
1	1.421139	1	1.490144	1	1.550717	1	1.604612	1	1.653100	1	1.697128
2	1.010606	2	1.080064	2	1.140374	2	1.193541	2	1.240991	2	1.283777
3	0.753382	3	0.826571	3	0.889467	3	0.944412	3	0.993060	3	1.036617
4	0.560272	4	0.639083	4	0.706191	4	0.764335	4	0.815437	4	0.860885
5	0.399489	5	0.485201	5	0.557623	5	0.619927	5	0.674329	5	0.722423
6	0.256818	6	0.350358	6	0.428896	6	0.496064	6	0.554388	6	0.605683
		7	0.227062	7	0.312324	7	0.384889	7	0.447611	7	0.502533
				8	0.203561	8	0.281939	8	0.349430	8	0.408314
						9	0.184515	9	0.257069	9	0.320181
								10	0.168759	10	0.236313
										11	0.155502

(For $i = N - 2$, $F_{N-2}(N) = 1/(N - 1)$ and for $i = N - 1$, $F_{N-1}(N) = 0$; both of these values are omitted).

using (23.61), this would make the expected smallest frequency among non-terminating cells about 1.0165. The reader may wish to check on this surmise; it may depend in the general case on the number of players and on whether or not $n_{12} = 0$.

For our 2-player example (Example 23.1.1), the result (23.58) gives

$$E\{\text{Max Freq}\} = F_0(5) + 4F_1(5) - 8F_2(5) + 5F_3(5) \approx 2.69953. \qquad (23.62)$$

In the Dirichlet analysis of this problem for 3 players, we use the same method as for the 2-player analogue. Suppose X is the single winner. Then at the end of the first stage, either the set of the size n_{23} is not completed or if it is, then the sets of size n_2 and n_3 are not completed. Thus, the probability that the Max Freq $\leq r$ at stopping time becomes

$$\begin{aligned}
&\Pr[\text{Max Freq} \leq r] \\
&= \; n_x[D(r + 1, 1) - D(1, 1)]^{n_x - 1} \\
&\quad \star \{(D^{n_{23}}(r + 1, 1) - [D(r + 1, 1) - D(1, 1)]^{n_{23}})D^{n_2 + n_3}(r + 1, 1) \\
&\quad + [D(r + 1, 1) - D(1, 1)]^{n_{23}}(D^{n_2}(r + 1, 1) - [D(r + 1, 1) - D(1, 1)]^{n_2}) \\
&\quad \star (D^{n_3}(r + 1, 1) - [D(r + 1, 1) - D(1, 1)]^{n_3})\} + 2ST \\
&\quad + (n_{12} + n_{123})[D(r + 1, 1) - D(1, 1)]^{N - n_3 - 1} \\
&\quad \star (D^{n_3}(r + 1, 1) - [D(r + 1, 1) - D(1, 1)]^{n_3}) + 2ST \\
&\quad + n_{123}[D(r + 1, 1) - D(1, 1)]^{N - 1} \qquad (23.63)
\end{aligned}$$

and the powers in (23.63) are all star powers.

The last term in (23.63) above represents a triple tie for the first stage and we denote it by T_3. The previous term represents a 2-way tie and we denote it by T_2. The part of the first term in (23.63) with the set of size n_{23} not completed is denoted by T_{11} and the part with the set of size n_{23} completed is denoted by T_{12}; then, (23.63) can be written as

$$
\Pr[\text{Max Freq} \leq r] = n_x(T_{11} + T_{12}) + 2ST + (n_{12} + n_{123})T_2 \\
+ 2ST + n_{123}T_3. \tag{23.64}
$$

We proceed exactly as in the case of 2 players: first, we compute $\Pr[\text{Max Freq} = r]$, then multiply by r and sum from $r = 1$ to $r = \infty$. A common expression enters every term, namely, $F_i(N)$ defined above. We replace each T-expression by the corresponding expectation or E-expression. After some algebraic manipulation, the result is

$$
E\{\text{Max Freq}\} = n_x(E_{11} + E_{12}) + 2ST + (n_{12} + n_{123})E_2 \\
+ 2ST + n_{123}E_3, \tag{23.65}
$$

where

$$
E_{11} = \sum_{i=1}^{n_x+n_{23}-1} \frac{(-1)^{i-1}}{1+i} \left[\binom{n_x + n_{23} - 1}{i} - \binom{n_x - 1}{i} \right] F_i(N),
\tag{23.66}
$$

$$
E_{12} = \sum_{i=2}^{N-2} \frac{(-1)^{i-1}}{1+i} \left[\binom{N-1}{i} - \binom{N - n_2 - 1}{i} \right. \\
\left. - \binom{N - n_3 - 1}{i} + \binom{N - n_2 - n_3 - 1}{i} \right] F_i(N),
\tag{23.67}
$$

$$
E_2 = \sum_{i=1}^{N-2} \frac{(-1)^{i-1}}{1+i} \left[\binom{N-1}{i} - \binom{N - n_3 - 1}{i} \right] F_i(N),
\tag{23.68}
$$

$$
E_3 = \sum_{i=0}^{N-1} \frac{(-1)^{i-1}}{1+i} \binom{N-1}{i} F_i(N).
\tag{23.69}
$$

For our two examples of 3 players, we now give the results of the above using our table of values of $F_i(N)$. For Example 23.1.2 (where $N = 12$), we get $E_{11} \approx 0.072160$, $E_{12} \approx 0.030826$, $E_2 \approx 0.076650$, $E_3 \approx 0.514534$. Combining these gives one of the desired answers as

$$
E\{\text{Max Freq}\} = 21(E_{11} + E_{12}) + 12E_2 + 3E_3 \\
\approx 4.626110 \text{ for Example 23.1.2.}
\tag{23.70}
$$

For Example 23.1.3 (where $N = 10$), we get $E_{11} \approx 0.114445$, $E_{12} \approx 0.053463$, $E_2 \approx 0.103383$ and $E_3 \approx 0.571792$, and the other desired answer is

$$
\begin{aligned}
E\{\text{Max Freq}\} \quad &= \quad 15(E_{11} + E_{12}) + 6E_2 + 1E_3 \\
&\approx \quad 3.71071 \text{ for Example 23.1.3.} \tag{23.71}
\end{aligned}
$$

These numerical results of $E\{\text{Max Freq}\}$ for Examples 23.1.2 and 23.1.3 are "reasonable" and "consistent" with the fact that under sampling W/R for a first stage winner, $E\{WT\} \approx 25$ in (23.29) for Example 23.1.2 and $E\{WT\} \approx 16$ in (23.30) for Example 23.1.3. A Monte Carlo check on Example 23.1.3 for (23.71) gave 3.721 based on 2000 experiments.

23.8   The Singleton Problem in the First Stage

Since the problem of singletons(S) is more involved for 3 players and also for the entire contest and since it is trivial for sampling W/O R, we restrict our attention here to the 2-player and 1-player formulas for $E\{S\}$ in the first stage, sampling W/R.

In analogy with the "star" product of D's used in Section 23.7, we use a "star" product of C's in this section. By straightforward integration, we have for $N \geq 2$

$$
\begin{aligned}
G_N(N-2) \quad &= \quad C_1^{(N-2)}(1,1) \star C_1^{(1)}(2,1) \\
&= \quad \sum_{\alpha=0}^{N-2} \frac{(-1)^\alpha \binom{N-2}{\alpha}}{1+\alpha} \; C_{\frac{1}{1+\alpha}}^{(1)}(2,1), \tag{23.72}
\end{aligned}
$$

and by a probability argument we have another expression for $G_N(N-2)$ as

$$
\begin{aligned}
G_N(N-2) \quad &= \quad \frac{1}{N} - \frac{1}{4} \; C_{1/2}^{(N-2)}(1,2) \\
&= \quad \frac{1}{N} - \frac{1}{N(N-1)} \sum_{\alpha=2}^{N} \frac{1}{\alpha} \tag{23.73}
\end{aligned}
$$

where the relation of C to the finite harmonic series is known from Sobel, Uppuluri and Frankowski (1985, p. 94).

Let $g_N = (1/N) - G_N(N-2)$ so that $1 + N(N-1)g_N = \sum_{\alpha=1}^{N} \frac{1}{\alpha} = H_N$. (We write H_N for the harmonic series up to N and h_N for the same series starting with 2.)

The contributions to $E(S)$ in the first stage come from four components:

1. The terminal observation is a singleton. It adds 1 and is not counted below.

2. $E_{WX}(S)$ is the NT (non-terminal) contribution from the winner when X is the winner, so that (see **4** below for the case of a tie) the result for $E_W(S)$ and 1 similar term(ST) where $n_1 \Rightarrow n_2$ is

$$E_W(S) = (N - n_2)(N - n_2 - 1)[g_{N-n_2} - g_N] + 1ST. \qquad (23.74)$$

To derive (23.74), we start with $(N - n_2)C^{N-1} \star [1 - C^{n_2}]$ which indicates that X is the winner. We replace C^{N-1} by its equivalent

$$\{[C(1,1) - C(2,1)] + C(2,1)\}^{N-1}, \qquad (23.75)$$

and expand the bracket in powers of α; the bracket represents NT singletons and α is the number of them obtained. Hence, if we multiply by α and sum on α, we obtain a result for $E_{WX}(S)$, which combined with $E_{WY}(S)$, takes the form

$$\begin{aligned} E_N(S) &= (N - n_2)(N - n_2 - 1)C^{N-n_2-2}(1,1) \star (1 - C^{n_2}) \\ &\quad \star [C(1,1) - C(2,1)] + 1ST, \end{aligned} \qquad (23.76)$$

where ST denotes a similar term in which Y is the winner and the contribution is from Y. Using (23.72) and (23.73), this can be written as

$$\begin{aligned} E_W(S) &= (N - n_2)(N - n_2 - 1)\left[\frac{1}{N - n_2} - \frac{1}{N} - G_{N-n_2}(N - n_2 - 2) \right. \\ &\quad \left. + G_N(N - 2)\right] + 1ST, \end{aligned} \qquad (23.77)$$

from which (23.74) follows. The derivations for case 3 and case 4 below are similar and we omit them.

3. $E_{LX}(S)$ is the contribution from the loser when X is the winner, so that $E_L(S) = E_{LX}(S) + E_{LY}(S)$. The result for this case is

$$E_L(S) = n_2(N - n_2)[g_{N-n_2+1} - g_N] + 1ST. \qquad (23.78)$$

4. $E_T(S)$ is the NT contribution from both players in the case of a tie. The result for this case is

$$E_T(S) = n_{12}(N - 1)g_N. \qquad (23.79)$$

The final result for $E(S)$ (for the first stage, sampling W/R) is the sum of 1 plus (23.74) plus (23.78) plus (23.79); it can be expressed as

$$\begin{aligned} E(S) &= H_{N-n_1} + H_{N-n_1} - H_N + \frac{n_2}{N - n_2 + 1} h_{N-n_2+1} \\ &\quad + \frac{n_1}{N - n_1 + 1} h_{N-n_1+1}. \end{aligned} \qquad (23.80)$$

As an interesting special case, we can put $n_1 = n_2 = 0$, and $n_{12} = N = 6$, obtaining the 1-player problem of tossing a fair die until all 6 sides are seen, each at least once. Then using only components 1 and 4, we have for the expected total number of singletons ast

$$
\begin{aligned}
E(S) &= 1 + 6(5)g_6 = 1 + \frac{15}{2} C_1^{(4)}(1,2) \\
&= H_6 = 2.45,
\end{aligned}
\tag{23.81}
$$

and more generally with a regular die of N sides, using (23.73),

$$
\begin{aligned}
E(S) &= 1 + N(N+1)\left[\frac{1}{N(N-1)} \sum_{\alpha=1}^{N} \frac{1}{\alpha}\right] \\
&= H_N = \frac{1}{N} E\{WT\}.
\end{aligned}
\tag{23.82}
$$

Thus for any regular polyhedron, $E(S)$ and $E\{WT\}$ are simply related by the factor N. Hence we see easily that for large N, $E(S) \approx \log N$ and $E\{WT\} \approx \frac{1}{N}(\log N)$. Although the numerical results for $E\{\text{Max Freq}\}$ behave in a similar manner, we have no asymptotic results for this case. A Monte Carlo check for the fair die with 150,000 experiments gave an average of 2.4505 singletons per experiment.

For 3 players, the value of $E(S)$ may be computed as the sum of 8 terms coming from the winner [which we call $E_W(S)$ below] and 18 terms from the loser [which we call $E_L(S)$]. Since the method of derivation is similar to that of 2 players, we only give the final results and then apply them to our Examples 23.1.2 and 23.1.3.

For the winner contribution we have, including the terminal singleton,

$$
\begin{aligned}
E_W(S) = {}& 1 + n_{123}(N-1)g_N \\
& + \{(n_{12} + n_{123})(N - n_3 - 1)[g_{N-n_3} - g_N] + 2ST\} \\
& + \{n_x(n_x - 1)[g_{n_x} - g_{N-n_3} - g_{N-n_2} + g_N] + 2ST\}.
\end{aligned}
\tag{23.83}
$$

For the loser contribution (when, say, X is the winner), we separate it into four disjoint parts according as at the end of the first stage

1. n_2 and n_3 are both completed but n_{23} is not,
2. only one of n_2, n_3 and n_{23} is completed,
3. none of n_2, n_3 and n_{23} is completed,
4. there are two tied winners, say, X and Y.

Denote them by $E_L(Si)$, $i = 1, 2, 3, 4$, respectively. Then

$$
E_L(S1) = n_x(n_2 + n_3)[g_{N-n_{23}} - g_N] + n_x n_{23}[g_{N-n_{23}+1} - g_N] + 2ST,
\tag{23.84}
$$

$$E_L(S2) = n_x n_2[g_{n_x+n_2} - g_{N-n_{23}} - g_{N-n_3} + g_N] + 2ST$$
$$+ n_x n_3[g_{n_x+n_3} - g_{N-n_{23}} - g_{N-n_2} + g_N] + 2ST$$
$$+ n_x n_{23}[g_{n_x+n_{23}} - g_{N-n_2} - g_{N-n_3} + g_N] + 2ST, \qquad (23.85)$$

$$E_L(S3) = n_x n_2[g_{n_x+1} - g_{n_x+n_2} - g_{n_x+n_3+1} - g_{n_x+n_{23}+1}$$
$$+ g_{N-n_{23}} + g_{N-n_2} + g_{N-n_3+1} - g_N]$$
$$+ n_x n_3[g_{n_x+1} - g_{n_x+n_3} - g_{n_x+n_2+1} - g_{n_x+n_{23}+1}$$
$$+ g_{N-n_{23}} + g_{N-n_2} + g_{N-n_3+1} - g_N]$$
$$+ n_x n_{23}[g_{n_x+1} - g_{n_x+n_{23}} - g_{n_x+n_2+1} - g_{n_x+n_3+1}$$
$$+ g_{N-n_2} + g_{N-n_3} + g_{N-n_{23}+1} - g_N], \qquad (23.86)$$

$$E_L(S4) = n_{12} n_3[g_{N-n_3+1} - g_N] + 2ST. \qquad (23.87)$$

Summing all of these four parts to form $E_L(S)$ and adding them to $E_W(S)$, we obtain the final formula for $E(S) = E_W(S) + E_L(S)$. Applying this result to our two examples with 3 players gives us the following table:

$E(S)$ results for 3 players

	Example 23.1.2	Example 23.1.3
Winner Contribution	4.32718	3.03544
Loser Contribution	0.52162	0.99409
Total $E(S)$	4.84880	4.02953
N	12	10
$E(S)$ as % of N	40.4%	40.3%

A Monte Carlo check for Example 23.1.3 with 30,000 experiments gave an average of 4.05 singletons per experiment.

23.9 Concluding Remarks and Summary

In the previous sections, we tried to show the usefulness of Dirichlet operators in many different coupon-collecting contest applications. The methodology of using a C-expression like (23.12) to obtain exact expectation formulas is a new idea in this paper. We have not exploited this method to its full since it can also be used to obtain higher moments and even generating functions. We can also obtain a joint generating function for singletons (S) and doubletons (T), which gives us covariances and correlations. A rough illustration of this is obtained by considering the one-player problem of waiting to see all N sides of a fair die. Then, using a star power expression,

$$G(s,t) = N\left\{[C(1,1) - C(2,1)]s + [C(2,1) - C(3,1)]t + C(3,1)\right\}^{N-1}$$

(23.88)

gives us a "star" joint generating function for NT (non-terminal) singletons [the first bracket in (23.88)] and doubletons [the second bracket in (23.88)]. Differentiating with respect to s and t and setting them equal to 1 gives us $E(S)$ for NT-singletons, $E(T)$, $E(ST)$, $E\{S(S-1)\}$ and $E\{T(T-1)\}$ from which we can obtain $\sigma^2(S)$, $\sigma^2(T)$ and also covariances and correlations. The $E(S)$ result for $N = 6$ is 1.45 as expected.

If we consider an N-sided fair die and stop as soon as we see each face at least r times, we can show that the expected number of faces X_r that appear exactly r times can be written in the simple form

$$E\{X_r\} = 1 + \frac{N(N-1)}{4^r}\binom{2r-1}{r-1}C_{1/2}^{(N-2)}(r, 2r),$$

(23.89)

where the C-values are tabulated in Sobel, Uppuluri and Frankowski (1985). This result is consistent with our result in (23.81) for $N = 6$ and $r = 1$, where the result is 2.45 exactly. For $N = 6$ and $r = 2, 3, 4, 5$ and 6, the values from (23.89) are 2.011156, 1.815397, 1.699531, 1.621170 and 1.563752, respectively. The reader is invited to derive (23.89) using star products or by any other method and also to find an asymptotic expression for (23.89) as $r \to \infty$.

In a recent paper [Sobel and Frankowski (1994)], the authors gave a partly expository and partly historical report on the *Sharing problem* which has been around for 500 years. In contrast to that, the present paper deals with off-shoots of the sharing problem that are to the best of our knowledge mostly new and have not been proposed or solved. Although we use small sample sizes to illustrate the problems, they are all sequential, challenging problems with no standard methodology available to solve them. One important distinction between the new offshoots and the old problem is that the lists of numbers for competing players need no longer be disjoint and is best described in the form of a Venn diagram.

Some further variations and extensions of coupon-collecting can be found in the recent delightful book of Blom, Holst and Sandell (1994) which includes a list of references on coupon-collecting and other interesting probability problems.

References

1. Blom, G., Holst, L. and Sandell, D. (1994). *Problems and Snapshots From the World of Probability*, New York: Springer-Verlag.

2. Sobel, M., Uppuluri, V. R. R. and Frankowski, K. (1977). Dirichlet Distribution Type-1, *Selected Tables in Mathematical Statistics*, Volume 4, Providence, RI: American Mathematical Society.

3. Sobel, M., Uppuluri, V. R. R. and Frankowski, K. (1985). Dirichlet Integrals of Type-2 and Their Applications, *Selected tables in Mathematical Statistics*, Volume 9, Providence, RI: American Mathematical Society.

4. Sobel, M. and Frankowski, K. (1994). The 500th anniversary of the sharing problem (The oldest problem in the theory of probability), *American Mathematical Monthly*, 833–847 (see also pp. 896, 910).

5. Sobel, M. and Frankowski, K. (1994). Hypergeometric analogues of multinomial Type-1 Dirichlet problems, *Congressus Numerantium*, **101**, 65–82.

6. Sobel, M. and Frankowski, K. (1995). Hypergeometric analogues of multinomial Type-2 problems via Dirichlet methodolgy, *Congressus Numerantium*, **106**, 171–191.

PART VII

Industrial Applications

24

Control Charts for Autocorrelated Process Data

Ajit C. Tamhane and Edward C. Malthouse

Northwestern University, Evanston, IL

Abstract: We show that the conventional X-chart based on the independence assumption is not robust because of the bias present in the moving range estimate of the standard deviation when the data are autocorrelated. We give a simple way to correct for this bias which results in a modified X-chart. Next we show that the frequently made proposal of control charting the residuals obtained by fitting a time-series model to autocorrelated data suffers from loss of power under certain situations commonly met in practice. This chart also requires a sufficiently large initial sample to obtain a good fit when used for on-line monitoring. A simulation study is performed to compare the performances of alternative control charts for data following the ARMA(1,1) model.

Keywords and phrases: Statistical process control, serial correlation, ARMA models, ARIMA models, average run length, simulation

24.1 Introduction

The standard control charts are based on the so-called Shewhart model

$$x_t = \mu + e_t \quad (t = 1, 2, \ldots) \tag{24.1}$$

where x_t is a process measurement at time t, μ is a fixed process mean and the e_t are uncorrelated measurement errors. In practice, however, the assumption of uncorrelated errors is often violated, for example, in discrete parts manufacturing involving automated measurements made at very high (often 100%) sampling rates. In this case, autocorrelation is the result of inertial or residual effects in the measuring instrument. In continuous processes, the underlying process mean itself may wander randomly thus causing autocorrelation. The mathematical and simulation analyses in this paper are restricted to the case where the process mean is stable except when a disturbance causes a shift in the process mean making the process go *out of control*; measurement errors are assumed to be the sole source of autocorrelation.

It is well-known that the conventional control chart for individual measurements (the so-called X-chart) based on the Shewhart model is not robust in the presence of autocorrelation, often exhibiting excessive false alarms [see Montgomery and Mastrangelo (1991) and the discussion following it]. A common solution proposed to overcome this problem is to fit a time-series model to the data and plot the residuals from this fit on a control chart [Alwan (1991), Alwan (1992), Alwan and Roberts (1989), Berthouex, Hunter and Pallesen (1978), and Montgomery and Friedman (1989)]. Because the residuals are approximately uncorrelated, standard control chart techniques, including the popular Western Electric runs rules, can be applied to this *residuals chart* to detect *special causes* and take corrective *local actions*. In addition, it is recommended that the fitted values be plotted on a separate chart to gain a better understanding of the systematic time-series behavior of the process. This chart can be used to detect the *common causes* that result in this behavior and take suitable corrective *system actions* to reduce the process variance.

The purpose of this paper is two-fold. First, we point out that the main reason behind the excessive false alarms experienced by a *conventional X-chart* is the underestimation of the process standard deviation by the moving range estimate when the data are positively autocorrelated (the more common case in practice). We propose a simple solution to fix this problem which entails approximate unbiasing of the moving range estimate. The resulting *modified X-chart* controls the false alarm rate even in the presence of autocorrelation in the data.

Second, we show that the idea of charting residuals, although attractive in principle, can result in loss of power for detecting mean shifts in some types of time-series models (which also correspond to positively autocorrelated data). When these models are fitted to the data, computation of residuals involves approximate differencing of successive observations causing suppression of the mean shift effect apparent in the residuals, in the extreme case its cancellation altogether.

The alternative control charts are illustrated on two practical data sets. The analytical results are supported by an extensive simulation study done under the ARMA(1,1) model.

24.2   Modified X-chart

Let $x_1, x_2, \ldots, x_T$ be successive measurements from a process under control, i.e., the x_t follow the model (24.1), but the e_t follow a stationary stochastic process. In the remainder of this paper, the marginal distribution of the e_t will be assumed to be $N(0, \sigma_e^2)$. Thus, the x_t are marginally $N(\mu, \sigma_e^2)$ and have a stable autocorrelation structure when the process is under control.

The control limits for the X-chart are usually based on the average of moving ranges given by

$$\overline{MR}_x = \frac{1}{T-1} \sum_{t=2}^{T} |x_t - x_{t-1}|.$$

Now it is well-known that

$$E |x_t - x_{t-1}| = \frac{2}{\sqrt{\pi}} \sigma_e \sqrt{1-\rho} = d_2 \sigma_e \sqrt{1-\rho}$$

where $d_2 = 2/\sqrt{\pi} = 1.128$ and $\rho = \operatorname{corr}(x_t, x_{t-1})$ is the lag one autocorrelation coefficient. Therefore, the usual moving range estimate of σ_e, $\overline{MR}_x/d_2$, gives an underestimate (overestimate) if $\rho > 0$ ($\rho < 0$). The consequence is that for $\rho > 0$, the control limits are too narrow resulting in too many false alarms, while for $\rho < 0$, the control limits are too wide resulting in an insensitive control chart [Maragah and Woodall (1992)]. In practice, $\rho > 0$ is more prevalent and hence the commonly observed incidence of excessive false alarms, by which we mean that the false alarm rate is $\gg 0.27\%$, the nominal false alarm rate for a three-sigma control chart.

An approximately unbiased estimate of σ_e is given by

$$\hat{\sigma}_e = \frac{\overline{MR}_x}{d_2 \sqrt{1-\hat{\rho}}}$$

where

$$\hat{\rho} = \frac{\sum_{t=2}^{T} (x_t - \bar{x})(x_{t-1} - \bar{x})}{\sum_{t=1}^{T} (x_t - \bar{x})^2}$$

is the lag one sample autocorrelation coefficient and $\bar{x} = (1/T) \sum_{t=1}^{T} x_t$. Note that this is a *model-independent* estimate of ρ, i.e., it does not depend on a particular time-series model fitted to the data. One should then use the following three-sigma control limits for the modified X-chart:

$$\bar{x} \pm 3\hat{\sigma}_e = \bar{x} \pm 3 \frac{\overline{MR}_x}{d_2 \sqrt{1-\hat{\rho}}}.$$

The process will be deemed out of control when a data point x_t falls outside these limits. These modified control limits will maintain a nominal false alarm rate without sacrificing the sensitivity to shifts in the mean. This assumes, of course, that T is sufficiently large so that the sampling errors in $\bar{x}$ and especially in $\hat{\sigma}_e$ are negligible. We will comment on the accuracy of this approximation in Section 24.5 dealing with the simulation study.

Now consider additional data, $x_{T+1}, x_{T+2}, \ldots$, after the control limits have been set. We will assume the following model for a shift in the process mean occurring at some unknown change point $\tau > T$:

$$E(x_t) = \begin{cases} \mu & \text{for } t < \tau \\ \mu + \delta & \text{for } t \geq \tau \end{cases} \tag{24.2}$$

For $t > T$, let

$$z_t^x = \frac{x_t - \bar{x}}{\hat{\sigma}_e} = \frac{\hat{e}_t}{\hat{\sigma}_e}$$

be the standardized statistic for the modified X-chart. For large T, the *power*, defined as the probability of detecting this shift on a single observation, can be approximated by

$$1 - P\{-3 \leq z_t^x \leq +3\} \approx \Phi[-3 + E(z_t^x)] + \Phi[-3 - E(z_t^x)],$$

where $\Phi(\cdot)$ is the standard normal c.d.f. and

$$E(z_t^x) \approx \frac{\delta}{\sigma_e} \text{ for } t > \tau.$$

As before, this approximation uses the fact that for large T, the sampling errors in $\bar{x}$ and $\hat{\sigma}_e$ are negligible. Note that the above power expression is monotonically increasing in $|E(z_t^x)|$.

In control charting application, the *average run length (ARL)*, which is the expected number of samples required to declare an out of control process signal, is a more appropriate performance measure than the power. For independent samples, of course, the ARL equals $(1 - \text{power})^{-1}$. In the simulation study presented in Section 24.5, we compare the alternative control charts based on their ARL's.

24.3   Residuals Control Chart

We now turn to the proposal of control charting the approximately independent residuals from fitting a time-series model to the autocorrelated data. Assuming the ARMA(1,1) model for the measurement errors:

$$e_t = \phi e_{t-1} + a_t - \theta a_{t-1},$$

we investigate the power of the residuals control chart; here, ϕ and θ are the autoregressive and moving average parameters ($|\phi|, |\theta| < 1$), respectively, and the a_t are i.i.d. $N(0, \sigma_a^2)$. For later use, it is convenient to write the ARMA(1,1) model in the form

$$(1 - \phi B)e_t = (1 - \theta B)a_t,$$

where B is the backshift operator, i.e., $B^k(x_t) = x_{t-k}$.

Let $\hat{\mu} = \bar{x}$ and $\hat{e}_t = x_t - \bar{x}$. Further, let $\hat{\phi}$ and $\hat{\theta}$ be the usual estimates of ϕ and θ, respectively, obtained by following the steps of identification, estimation and diagnostic checking [see, for example, Box and Jenkins (1976)]. Then the residuals r_t for the fitted ARMA(1,1) model are given by

$$
\begin{aligned}
r_t &= \left(\frac{1 - \hat{\phi}B}{1 - \hat{\theta}B}\right)\hat{e}_t \\
&= \hat{a}_t \\
&= \hat{e}_t - \hat{\phi}\hat{e}_{t-1} + \hat{\theta}\hat{a}_{t-1} \\
&= \hat{e}_t - \hat{\phi}\hat{e}_{t-1} + \hat{\theta}r_{t-1}.
\end{aligned} \tag{24.3}
$$

Let $r_1, r_2, \ldots, r_T$ be the residuals obtained from the first T observations, $x_1, x_2, \ldots, x_T$, from a process under control. It is easy to see that, for large T, the r_t are approximately i.i.d. $N(0, \sigma_a^2)$. Let $\bar{r} = (1/T)\sum_{t=1}^{T} r_t$ and

$$
\overline{\mathrm{MR}}_r = \frac{1}{T-1}\sum_{t=2}^{T} |r_t - r_{t-1}|.
$$

The three-sigma control limits for the residuals chart are given by

$$
\bar{r} \pm 3\hat{\sigma}_a = \bar{r} \pm 3\frac{\overline{\mathrm{MR}}_r}{d_2}.
$$

Note that since the r_t are approximately uncorrelated, no correction for autocorrelation is necessary as was the case with the X-chart. Also, one may use zero as the center line of the chart instead of $\bar{r}$.

To find the power expression for the residuals chart, we need to evaluate the expected value of the corresponding standardized statistic,

$$
z_t^r = \frac{r_t}{\hat{\sigma}_a}.
$$

For the shift model (24.2), when T is large, we have

$$
E(\hat{e}_t) \approx \begin{cases} 0 & \text{for } t < \tau \\ \delta & \text{for } t \geq \tau. \end{cases}
$$

Now, from (24.3) it can be shown that

$$
r_t = \hat{e}_t + (\hat{\theta} - \hat{\phi})\{\hat{e}_{t-1} + \hat{\theta}\hat{e}_{t-2} + \hat{\theta}^2\hat{e}_{t-3} + \cdots\}.
$$

Hence,

$$
\begin{aligned}
E(r_{\tau+k}) &\approx \delta[1 + (\theta - \phi)\{1 + \theta + \theta^2 + \cdots + \theta^{k-1}\}] \\
&= \delta\left[1 + \frac{(\theta - \phi)(1 - \theta^k)}{(1 - \theta)}\right].
\end{aligned}
$$

Furthermore, for the ARMA(1,1) model we have

$$\sigma_a^2 = \left(\frac{1 - \phi^2}{1 + \theta^2 - 2\phi\theta} \right) \sigma_e^2. \tag{24.4}$$

Therefore, putting $t = \tau + k$, we obtain

$$E(z_t^r) \approx \frac{E(r_t)}{\sigma_a} \approx \frac{\delta}{\sigma_e} \left[\left\{ \frac{1 + \theta^2 - 2\phi\theta}{1 - \phi^2} \right\}^{1/2} \left\{ \frac{1 - \theta + (\theta - \phi)(1 - \theta^k)}{1 - \theta} \right\} \right]. \tag{24.5}$$

Then, for large T, the power of the residuals chart can be approximated by

$$1 - P\{-3 \le z_t^r \le +3\} \approx \Phi[-3 + E(z_t^r)] + \Phi[-3 - E(z_t^r)]. \tag{24.6}$$

Comparison between the powers of the modified X-chart and the residuals chart reduces to the comparison between $|E(z_t^x)| = |\delta/\sigma_e|$ and $|E(z_t^r)|$. Straightforward algebra shows that for large k,

$$|E(z_t^x)| \ge |E(z_t^r)| \Longleftrightarrow \phi \ge \theta.$$

For the ARMA(1,1) model, the condition $\phi \ge \theta$ is equivalent to $\rho \ge 0$, since

$$\rho = \frac{(\phi - \theta)(1 - \phi\theta)}{1 + \theta^2 - 2\phi\theta}. \tag{24.7}$$

Thus for an AR(1) model with $\phi > 0$, the modified X-chart will be more powerful than the residuals chart, while for an MA(1) model with $\theta > 0$, the residuals chart will be more powerful. In fact, by substituting $\theta = 0$ in (24.5) we obtain the following expression for an AR(1) model for any k:

$$E(z_t^r) \approx \frac{\delta}{\sigma_e} \sqrt{\frac{1 - \phi}{1 + \phi}}.$$

From this expression it is seen that for a highly positively correlated AR(1) process ($\phi \approx 1$), a situation not uncommon in practice, z_t^r has expectation close to zero, and hence has very little power for detecting a shift in the process mean. Ryan (1991) and Harris and Ross (1991) have noted this result previously.

The above analysis applies to an ARMA model, which is suitable for modeling a stationary process. Box and Jenkins (1976) have argued that many nonstationary processes can be modeled using an ARIMA model. Suppose that $e_t = x_t - \mu$ follows the ARIMA(p, d, q) model. This means that the d-th order difference of e_t, denoted symbolically by $\nabla^d e_t = (1 - B)^d e_t$, follows an ARMA$(p, q)$ model:

$$(1 - \Phi_p(B))\nabla^d e_t = (1 - \Theta_q(B))a_t$$

where $\Phi_p(B) = \phi_1 B + \phi_2 B^2 + \cdots + \phi_p B^p$ and $\Theta_q(B) = \theta_1 B + \theta_2 B^2 + \cdots + \theta_q B^q$. A general expression for the residual r_t is given by

$$r_t = \hat{a}_t = (1 - \hat{\Theta}_q(B))^{-1}(1 - \hat{\Phi}_p(B))\nabla^d \hat{e}_t$$

where $\hat{e}_t = x_t - \hat{\mu} = x_t - \bar{x}$ and the carets on Φ_p and Θ_q denote that the unknown parameters in the corresponding expressions are replaced by their sample estimates. From the above expression, it is readily apparent that any shift δ in the mean of x_t is cancelled out by the differencing operation. For example, for the special case $p = 1, q = 0, d = 1$, we have

$$\begin{aligned} r_t &= (1 - \hat{\phi}B)(\hat{e}_t - \hat{e}_{t-1}) \\ &= \hat{e}_t - (1 + \hat{\phi})\hat{e}_{t-1} + \hat{\phi}\hat{e}_{t-2}. \end{aligned}$$

For the shift model (24.2), it is easy to see that

$$E(r_{\tau+k}) \approx \begin{cases} \delta & \text{for } k = 0 \\ -\phi\delta & \text{for } k = 1 \\ 0 & \text{for } k \geq 2. \end{cases}$$

Thus, if a shift is not detected within one period after it has occurred then the chance of detecting it later is very small (equal to the false alarm rate).

24.4 Examples

We give two examples of application of the three control charts discussed in this paper. Both these examples involve a *retrospective surveillance* as opposed to *on-line monitoring*. Specifically, the data used to create a control chart are tested for lack of control using the same chart. The simulation study reported in the following section uses the control charts in an on-line monitoring mode.

Example 24.4.1 Montgomery and Mastrangelo (1991) present a plot of 100 autocorrelated observations in their Figure 1. The control limits, 84.5 ± 8.3, shown in that figure are based on the usual moving range formula, and are too narrow resulting in many points falling outside the control limits. The lag one sample autocorrelation coefficient for these data is $\hat{\rho} = 0.860$. (Montgomery and Mastrangelo also fit an AR(1) model to the data with estimated autoregressive parameter $\hat{\phi} = 0.847$. This provides a *model-dependent* estimate $\hat{\rho} = \hat{\phi} = 0.847$.) Therefore, the adjusted control limits are $84.5 \pm 8.3/\sqrt{1 - 0.860} = 84.5 \pm 22.2$, and no points are observed to fall outside these control limits. This result agrees with the residuals control chart obtained by fitting an AR(1) model to the data which also shows that the process is under control.

Example 24.4.2 As another example, we calculate modified control limits for the chemical concentration data [given as Series A in Box and Jenkins (1976)] consisting of 197 readings taken every two hours. These data were also used by Alwan and Roberts (1989) to illustrate their residuals chart method. For these data, we obtain $\bar{x} = 17.0574, \overline{MR}_x = 0.2847$ and $\hat{\rho} = 0.5451$. The control limits for the conventional X-chart are $[16.30, 17.81]$, and there are 10 data points that breach these limits. On the other hand, the modified control limits are $[15.93, 18.18]$, and there is only one data point (no. 192) that breaches these limits. The time order plot of these data with both sets of control limits is shown in Figure 24.1. Alwan and Roberts (1989) fitted an ARIMA(0,1,1) model and found using the residuals chart that two data points (nos. 43 and 64) are out of control. An ARMA(1,1) model gives equally good fit and the same results. The corresponding residuals chart is shown in Figure 27.2.

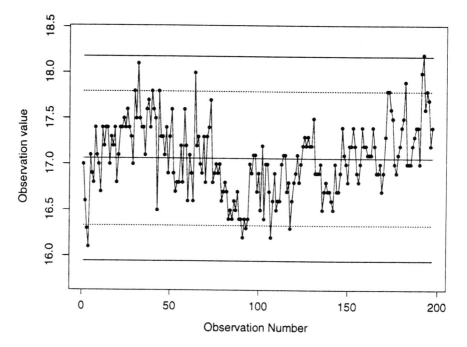

Figure 24.1: Conventional and modified X-charts for the Box-Jenkins Series
A data (dotted line: conventional X-chart limits; solid line: modified
X-chart limits)

24.5 Simulation Study

We conducted a simulation study to compare the performances of the three control charts under the ARMA(1,1) model with ϕ and θ values chosen from the set $\{-0.8, -0.4, 0, 0.4, 0.8\}$. Thus, a total of twenty-five (ϕ, θ) combinations were studied. The ρ value for each (ϕ, θ) combination was calculated using (24.7), and is also tabulated. For each (ϕ, θ) combination, we made 10,000 independent simulations, each simulation consisting of a time-series of 2100 normally distributed observations. The first 1100 observations were generated with no mean shift $(\mu = 0)$, while the last 1000 observations were generated with a mean shift of $\mu = \delta = 0.5\sigma_e$. Throughout, σ_e was fixed at 1, and σ_a was calculated using (24.4). The same time-series of observations was tested using each of the control charts as described below.

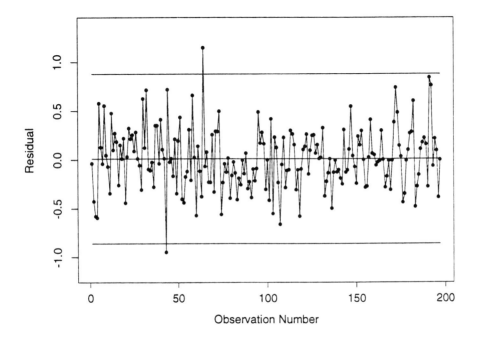

Figure 24.2: Residuals control chart for the Box-Jenkins Series A data

The first 100 observations (the initial sample) were used to set up the control limits for each chart (i.e., $T = 100$). The next 1000 observations were used to find the run length (i.e., the number of observations until an observation falls outside the control limits) for each chart under no mean shift. Finally, the last 1000 observations were used to find the run length for each chart under mean shift. If no breaching of the control limits was observed even after 1000 observations, the corresponding run length was set equal to 1000. This truncation slightly underestimates the ARL's, but not seriously enough to qualitatively affect the comparisons between the three control charts.

For the residuals chart, estimation of ϕ and θ based on an initial sample of 100 observations is required for each simulation. This was done by calling SAS PROC ARIMA routine (with the default option of maximum likelihood estimation) during each simulation. Convergence was not achieved in this estimation using the default parameters in SAS (in particular, the maximum number of iterations $= 15$) in 821 out of 250,000 simulations. Thus, the residuals chart could not be constructed in these 0.33% of the simulations. Even when convergence was achieved, $\hat{\phi}$ and $\hat{\theta}$ were found to be corrupted by large sampling errors in many cases. Note that the model identification step was omitted in this estimation procedure, it being assumed that the correct model to be fitted is known to be ARMA(1,1). If this step were included, then fitting the time-series model would be even more difficult.

To reduce extraneous variation and obtain more precise comparisons, the same pseudo-random numbers were used to generate the time-series data for all (ϕ, θ) combinations. The ARL estimates under no mean shift and under mean shift are given in Table 24.1 and Table 24.2, respectively. The standard errors of the ARL estimates are quite small, never exceeding 4.0, and are therefore not reported.

First, look at Table 24.1. The nominal ARL for a three-sigma control chart under no mean shift in the independence case is $1/.0027 = 370.4$. We see that the conventional X-chart is highly sensitive to autocorrelation in the data, with extremely low ARL values (excessive false alarms) when $\rho > 0$. The residuals chart and the modified X-chart both maintain the ARL around 300. Although this is less than 370.4, it should be remembered that this nominal value is based on the assumption that the true values of the parameters μ, σ_e, ϕ and θ are known. Since these parameters are actually estimated, it is not surprising that the ARL falls below the nominal value. If the initial sample size is increased to about 500, then the ARL values are much closer to the nominal value. Small amount of underbias still remains, however, because of truncation of the run lengths at 1000. For the modified X-chart, the ARL is generally < 370.4 when $\rho \geq 0$; for $\rho < 0$, the ARL is always > 370.4. For the residuals chart, the ARL is < 370.4 for all (ϕ, θ) combinations except $(\phi, \theta) = (0.8, -0.8)$.

Next, turn to Table 24.2. The extremely low ARL values for the conventional X-chart when $\rho > 0$ are quite meaningless since this chart does not control the ARL under no mean shift. Comparing the ARL values for the modified X-chart with those for the residuals chart, we see that for $\rho > 0$, the modified X-chart has smaller ARL's, while for $\rho \leq 0$, the residuals chart has smaller ARL's. This result is, of course, anticipated in view of the analytical comparison of the powers of the two charts. Although the residuals chart has extremely low ARL's for large negative ρ values, such ρ values are almost never encountered in practice. A far more common case is positive autocorrelation, for which the modified X-chart has an upper hand.

Table 24.1: Average run lengths[†] under no shift in the process mean for ARMA(1,1) data

θ	ϕ				
	-0.8	-0.4	0	0.4	0.8
	$\rho = 0$	$\rho = 0.272$	$\rho = 0.488$	$\rho = 0.695$	$\rho = 0.899$
	367.08	118.08	38.10	13.45	4.62
-0.8	349.79	335.27	332.29	327.73	303.92
	299.40	340.85	335.12	349.36	468.74
	$\rho = -0.523$	$\rho = 0$	$\rho = 0.345$	$\rho = 0.627$	$\rho = 0.880$
	803.60	367.08	76.96	18.79	5.34
-0.4	423.63	349.79	331.73	321.91	300.00
	309.68	283.98	302.15	302.11	322.80
	$\rho = -0.800$	$\rho = -0.400$	$\rho = 0$	$\rho = 0.400$	$\rho = 0.800$
	889.86	751.82	367.08	59.00	8.84
0	479.88	377.45	349.79	321.86	295.64
	311.38	300.84	283.23	304.71	308.23
	$\rho = -0.880$	$\rho = -0.627$	$\rho = -0.345$	$\rho = 0$	$\rho = 0.523$
	908.23	869.39	716.09	367.08	34.78
0.4	508.19	411.81	370.85	349.79	294.90
	313.43	301.26	289.25	285.33	316.11
	$\rho = -0.899$	$\rho = -0.695$	$\rho = -0.488$	$\rho = -0.272$	$\rho = 0$
	913.26	892.42	814.17	660.59	367.08
0.8	519.73	425.75	383.82	376.09	349.79
	334.38	297.24	279.08	264.69	316.93

[†] The top entry in each cell is the ARL for the conventional X-chart, the middle entry is the ARL for the modified X-chart, and the bottom entry is the ARL for the residuals chart.

Table 24.2: Average run length[†] under shift, $\delta = 0.5\sigma_e$, in the process mean for ARMA(1,1) data

θ	ϕ				
	-0.8	-0.4	0	0.4	0.8
	$\rho = 0$	$\rho = 0.272$	$\rho = 0.488$	$\rho = 0.695$	$\rho = 0.899$
	199.41	61.12	23.28	10.03	3.74
-0.8	180.03	177.33	183.72	201.90	231.49
	145.95	212.94	235.31	274.61	413.07
	$\rho = -0.523$	$\rho = 0$	$\rho = 0.345$	$\rho = 0.627$	$\rho = 0.880$
	653.39	199.41	42.48	13.43	4.38
-0.4	250.97	180.03	177.29	193.78	228.07
	79.67	140.12	200.81	236.66	287.28
	$\rho = -0.800$	$\rho = -0.400$	$\rho = 0$	$\rho = 0.400$	$\rho = 0.800$
	775.49	563.33	199.41	34.66	7.34
0	286.53	198.73	180.03	180.91	217.79
	14.98	75.37	140.38	220.93	273.15
	$\rho = -0.880$	$\rho = -0.627$	$\rho = -0.345$	$\rho = 0$	$\rho = 0.523$
	802.21	725.39	514.15	199.41	24.80
0.4	301.34	211.85	193.60	180.03	199.09
	2.24	14.25	50.80	142.76	269.86
	$\rho = -0.899$	$\rho = -0.695$	$\rho = -0.488$	$\rho = -0.272$	$\rho = 0$
	805.91	763.30	634.68	444.47	199.41
0.8	304.37	218.28	195.87	194.72	180.03
	1.19	2.60	6.24	19.05	178.18

† The top entry in each cell is the ARL for the conventional X-chart, the middle entry is the ARL for the modified X-chart, and the bottom entry is the ARL for the residuals chart.

24.6 Conclusions and Recommendations

It is obvious that the conventional X-chart must not be used with autocorrelated data because of its lack of robustness. The choice is between the modified X-chart and the residuals chart [coupled with the fitted values chart for diagnosing common causes as recommended by Alwan and Roberts (1989)]. Based on the power consideration alone, the use of the modified X-chart would be preferred since it is more powerful for detecting mean shifts than the residuals chart for positively autocorrelated data. However, there are other practical considerations which also support the use of the modified X-chart as discussed below.

First, the residuals chart requires a rather large initial sample in order to fit a good time-series model. The usual prescription of the initial sample size of 20–25 is clearly inadequate; a sample size of at least 100 is needed for the modified X-chart and an even larger sample size is needed for the residuals chart. To get such a large initial sample is not a problem in high volume manufacturing, but in batch processing where each batch constitutes an observation, this is clearly impractical.

Second, considerable time-series modeling skill is needed to use the residuals chart. Although such modeling is useful for gaining process undertanding, not many engineers, leave alone shop personnel, possess this skill. It has been suggested that in such cases a statistical package having capabilities for automated fitting of time-series models may be used. But blind use of a package is unlikely to aid in the understanding of common causes underlying the process. If a wrong model is fitted, then the residuals chart may indicate special causes when the true cause is the model misspecification. Based on all these considerations, the use of the modified X-chart is recommended for detecting special causes.

However, this recommendation must be tempered by the fact that only a special kind of lack of control, namely, a mean shift, is considered here. There are other ways a process may go out of control, such as a sudden increase in the process variance, which may be more readily detectable with the residuals chart. Since the residuals are generally so useful for diagnostic purposes in all areas of data analyses, it would be foolhardy to dismiss their use outright in the present context. A further detailed study is warranted.

As a final remark, we note that we have assumed the usual definition of an in-control process, namely, if the same system of common cause variation (i.e., the same probability law) governs the process then it is in-control. For autocorrelated data, this definition permits excessive variation around the target which may be unacceptable in practice. MacGregor (1991) says "... (the wandering of the process) may be due to common cause variation, but surely the extent of the drifts away from the target would indicate that something better can be done than simply declaring them to be in-control on the basis of residuals charts." He goes on to suggest that a better course of action is the adaptive minimum mean square error (MMSE) control [MacGregor (1990)]. But this requires the availability of a relatively easily manipulable control variable. Perhaps, we need to reformulate the definition of statistical control when the process data are not i.i.d. This basic question has not been addressed in the literature thus far and awaits solution.

Acknowledgements. Computations in this paper were partially supported by the NSF Grant DMS-9505799.

References

1. Alwan, L. C. (1991). Autocorrelation: Fixed versus variable control limits, *Quality Engineering*, **4**, 167–188.

2. Alwan, L. C. (1992). Effects of autocorrelation on control chart performance, *Communications in Statistics—Theory and Methods*, **21**, 1025–1049.

3. Alwan, L. C. and Roberts, H. V. (1989). Time series modeling for statistical process control. In *Statistical Process Control in Automated Manufacturing* (Eds., J. B. Keats and N. F. Hubele), pp. 87–95, New York: Marcel Dekker.

4. Berthouex, P. M., Hunter, W. G. and Pallesen, L. (1978). Monitoring sewage treatment plants: Some quality control aspects, *Journal of Quality Technology*, **10**, 139–149.

5. Box, G. E. P. and Jenkins, G. M. (1976). *Time Series Analysis, Forecasting and Control*, Oakland, CA: Holden Day.

6. Box, G. E. P., Jenkins, G. M. and MacGregor, J. F. (1974). Some recent advances in forecasting and control. *Journal of the Royal Statistical Society, Series C*, **23**, 158–179.

7. Harris, T. J. and Ross, W. H. (1991). Statistical process control procedures for correlated observations, *Canadian Journal of Chemical Engineering*, **69**, 48–57.

8. MacGregor, J. F. (1990). A different view of the funnel experiment, *Journal of Quality Technology*, **22**, 255–259.

9. MacGregor, J. F. (1991). Discussion of Montgomery and Mastrangelo (1991), *Journal of Quality Technology*, **23**, 198–199.

10. Maragah, H. D. and Woodall, W. H. (1992). The effect of autocorrelation on the retrospective X-chart, *Journal of Statistical Computation and Simulation*, **40**, 29–42.

11. Montgomery, D. C. and Friedman, D. J. (1989). Statistical process control in a computer-integrated manufacturing environment, In *Statistical Process Control in Automated Manufacturing* (Eds., J. B. Keats and N. F. Hubele), pp. 67–86, New York: Marcel Dekker.

12. Montgomery, D. C. and Mastrangelo, C. M. (1991). Some statistical process control methods for autocorrelated data (with Discussion), *Journal of Quality Technology*, **23**, 179–202.

13. Ryan, T. P. (1991). Discussion of Montgomery and Mastrangelo (1991), *Journal of Quality Technology*, **23**, 200–202.

25

Reconstructive Estimation in a Parametric Random Censorship Model With Incomplete Data

Tommaso Gastaldi

Università degli Studi di Roma "La Sapienza", Rome, Italy

Abstract: We explore theoretically an approach to estimation, in a multivariate random censorship model with incomplete data, based on the reconstruction of the missing information. Simulation results are also presented.

Keywords and phrases: Competing risk model, multi-component system, r-out-of-r system, incomplete data, masked data, order statistic, failure rate, survival, ML estimator, censoring, exponential distribution

25.1 Introduction and Summary

The setup we consider is a special case of the model with *partially classified* data studied by Gastaldi and Gupta (1994), and consists of a useful generalization of the well-known *competing risk model*. [Early references on competing risks are Mendenhall and Hader (1958) and Cox (1959); for surveys or other references see, for instance, David (1974), David and Moeschberger (1978), and Basu and Klein (1982).] The generalization consists of the following two points:

1. several independent concurrent censoring variables, instead of only one, are considered which are subject to mutual censoring, in the sense that any variable can be censored by any other variable,

2. the information about the variable which actually generated the observed data, and hence censored the observation of the remaining variables, is possibly incomplete, in a way which will be explained in Section 25.2.

Such a model is variously referred to in the literature as *competing risk model with incomplete information on the cause of failure*, or *multivariate random censorship model with incomplete data*, or model with *masked data*. A critical review can be found in Gastaldi (1994). Fundamental references are Miyakawa (1984), Usher and Hodgson (1988), Guess, Usher and Hodgson (1991), Usher (1993), and Gupta and Gastaldi (1996). This model finds application in many situations, especially in reliability and biological problems [see, Dinse (1982, 1986), Usher and Hodgson (1988), and Baxter (1995).] Without loss of generality, we prefer to use a terminology which is more suitable to the reliability applications. The model will be described in detail in Section 25.2, while here we make an informal introduction to our subject matter.

The common approach to the estimation of the unknown parameters in a multivariate random censorship model with incomplete data consists of deriving the maximum likelihood estimators (MLEs). Unfortunately, although several efforts have been made along this direction, there are very few results really usable in practice, and these too mainly for the negative exponential distribution. In fact, the likelihood equations, in general, yield mathematically untractable systems. Most of the complexity that arises in the solution of the ML equations is due to the incompleteness of the information on the components which caused the system failure. The above considerations lead us to explore possible estimation procedures which allow us to reconstruct the missing information and estimate the unknown parameters. The idea of treating the missing data as parameters and maximizing the likelihood over the missing data and parameters is not particularly strange and a critical discussion on it can be found in Little and Rubin (1987, pp. 92–95). The results which follow from these techniques find application in reliability and biological problems, where the model in consideration is often used. In fact, as we shall see, this approach allows the use of the estimators developed for the standard competing risk model, thus overcoming the practical impossibility of deriving the estimators in the model with incomplete data.

25.2   Model and Notation

Consider n systems $\Sigma_1, \ldots, \Sigma_n$ which are being tested. Each system is made of r (≥ 2) components $C_1, \ldots, C_r$. Denote by $I \equiv \{1, 2, \ldots, n\}$ the set of subscripts for the systems, and by $J \equiv \{1, 2, \ldots, r\}$ the set of subscripts for the components.

The random lifelength of component C_j, $j \in J$, in system Σ_i, $i \in I$, is denoted by T_{ij}. The nr random variables T_{ij}, $i = 1, \ldots, n$, $j = 1, \ldots, r$, are assumed to be independent. For each $j \in J$, the n random variables $T_{1j}, \ldots, T_{nj}$ (lifelengths of components C_j, in systems $\Sigma_1, \ldots, \Sigma_n$) are identically distributed, with common absolutely continuous distribution $F_j(t; \boldsymbol{\theta}_j)$ and probability density $f_j(t; \boldsymbol{\theta}_j)$ indexed by the (vector) parameter $\boldsymbol{\theta}_j$.

Given a distribution $F(t; \boldsymbol{\theta})$, we denote by $\bar{F}(t; \boldsymbol{\theta}) \equiv 1 - F(t; \boldsymbol{\theta})$ the *reliability* (or *survival*) function, and by $\lambda(t; \boldsymbol{\theta}) \equiv f(t; \boldsymbol{\theta})/F(t; \boldsymbol{\theta})$ the *failure rate* (or *hazard*) function.

In each system, the components are assumed to be *in series*, that is, for each $i \in I$, the random lifelength of system Σ_i, which is denoted by T_i, is given by:

$$T_i \equiv \min_{j \in J} T_{ij}. \tag{25.1}$$

It is easy to see that the T_i, $i = 1, \ldots, n$, are i.i.d. with common distribution:

$$F_T(t; \boldsymbol{\theta}_1, \ldots, \boldsymbol{\theta}_r) = 1 - \prod_{j \in J} \bar{F}_j(t; \boldsymbol{\theta}_j) \tag{25.2}$$

and probability density:

$$f_T(t; \boldsymbol{\theta}_1, \ldots, \boldsymbol{\theta}_r) = \sum_{j \in J} \lambda_j(t; \boldsymbol{\theta}_j) \bar{F}_T(t; \boldsymbol{\theta}_1, \ldots, \boldsymbol{\theta}_r). \tag{25.3}$$

For each $i \in I$, let K_i denote the random index of the component causing the failure of system Σ_i, and let κ_i denote the corresponding realization.

It is assumed that, for each system Σ_i, $i \in I$, there corresponds a random set of indices S_i whose realization s_i includes the index of the component which caused the system failure (at time t_i), i.e.,

$$\Pr\{K_i \in S_i\} = 1. \tag{25.4}$$

Only when $s_i = \{j\}$, $j \in J$, we will know that $\kappa_i = j$. In any other case where s_i contains more than one index, we can only state that $\kappa_i \in s_i$, while the true cause of failure is *masked* from our knowledge.

The observations (T_i, S_i), $i = 1, \ldots, n$, are assumed to be independent, and the observed data is denoted by

$$\mathcal{D} \equiv \{(t_1, s_1), (t_2, s_2), \ldots, (t_n, s_n)\}. \tag{25.5}$$

Following a consolidated convention, we denote by $n_1, n_2, \ldots, n_r, n_{12}, n_{13}, \ldots,$ $n_{r-1,r}, \ldots, n_{12\cdots r}$ the number of observations (t_i, s_i), $i \in \{1, \ldots, n\}$, such that $s_i = \{1\}$, $s_i = \{2\}, \ldots, s_i = \{r\}$, $s_i = \{1, 2\}$, $s_i = \{1, 3\}, \ldots, s_i = \{r - 1, r\}, \ldots, s_i = \{1, \ldots, r\}$, respectively, i.e., $n_{\pi_1 \cdots \pi_h} \equiv |\{i \in I \mid s_i = \{\pi_1, \ldots, \pi_h\}\}|$, $2 \le h \le r$. The order of the subscripts in the $n_{\pi_1 \cdots \pi_h}$'s (as well as the order of the elements in the sets s_i) is unimportant, so that $n_{\pi_1 \cdots \pi_h}$ is assumed to denote the same frequency for every permutation of $(\pi_1, \ldots, \pi_h)$. We also denote by $\mathcal{N}_r$ the space of all the possible realizations of the set S_i, $i \in I$.

The above model contains a large class of practical frameworks, obtainable by specifying the actual process generating the sets S_i, which we might refer to as *masking process*. There are, in fact, several situations which may cause incompleteness of the data. For instance, for each $i \in I$, after the failure of system Σ_i at time t_i, a (possibly incomplete) serial inspection of the components might be carried out. In such a case, S_i is formed by either the index of the failed component (if this has been identified) or the subscripts of the components which have not been inspected (when the failed component has not yet been identified).

Whatever is the nature of the masking process, we assume that it is *informative* about the unknown parameters and formally express this requirement by the following condition.

Condition for informative masking: for each $i \in I$, we have:

$$\left[\bigcup_{k \in s_i} (K_i = k) \right] \Rightarrow [S_i = s_i] \tag{25.6}$$

for any *possible* s_i.

A comprehensive discussion on the above condition is contained in Gastaldi (1994). Here, we will only recall that (25.6) (whose converse holds by definition) "dictates that the plan of the experiment be such that if $\mathcal{C}_k$ is a possible failed component of system Σ_i then the index k must be found in s_i. From this point of view, we can consider such a condition as a way to formalize the idea that the masking should not be a meaningless random process which just comes to cover the true cause of failure, but, instead, be a result of a meaningful plan of experiment which insures that in s_i there will be the indices of the components which could likely have failed and not other indices, which just come randomly to mask and confuse our data" [adjusted quotation from Gastaldi (1994, p. 15)].

Under informative masking, the likelihood function of the parameters given the observed data (25.5) is

$$L(\boldsymbol{\theta}_1, \ldots, \boldsymbol{\theta}_r \mid \mathcal{D}) = \prod_{i \in I} \sum_{j \in s_i} \lambda_j(t_i; \boldsymbol{\theta}_j) \bar{F}_T(t_i; \boldsymbol{\theta}_i, \ldots, \boldsymbol{\theta}_r) \tag{25.7}$$

[cf., Gastaldi (1994, p. 12)], and, for instance, in the negative exponential case (the only one where the likelihood equations have been derived), the MLEs are obtained by solving the following set of r simultaneous equations:

$$\frac{\partial \ln[L(\lambda_1, \ldots, \lambda_r \mid \mathcal{D})]}{\partial \lambda_j} = 0 = \sum_{\boldsymbol{\zeta} \in \mathcal{N}_r} \frac{n_{\boldsymbol{\zeta}}}{\sum_{w \in \boldsymbol{\zeta}} \lambda_w} 1_{[j \in \boldsymbol{\zeta}]} - \sum_{i \in I} t_i, \quad j = 1, \ldots, r,$$

$$\tag{25.8}$$

(the parameters $\boldsymbol{\theta}_1, \ldots, \boldsymbol{\theta}_r$ to be estimated are the failure rates $\lambda_1, \ldots, \lambda_r$; $1_{\mathcal{A}}$ is the indicator function of the event $\mathcal{A}$), which is a nonlinear system analytically untractable [see Gastaldi (1996)].

25.3 ML Simultaneous Estimation of Parameters and Causes of Failure

In our approach, the causes of failure of the systems, which in general are unknown, are treated as if they were parameters which have to be estimated. Denote by $N(t_i, \delta t_i) = [t_i, t_i + \delta t_i]$ a neighborhood of t_i. The likelihood function of the parameters $\boldsymbol{\theta}_1, \ldots, \boldsymbol{\theta}_r$ and of the causes of failure $\kappa_1, \ldots, \kappa_n$, with $\kappa_i \in s_i$, is given by:

$$
\begin{aligned}
&\mathcal{L}(\boldsymbol{\theta}_1, \ldots, \boldsymbol{\theta}_r; \kappa_1, \ldots, \kappa_n \mid \mathcal{D}) \\
&= \lim_{\delta t_1 \to 0, \ldots, \delta t_n \to 0} \frac{\Pr\left[\bigcap_{i \in I}\{(T_i \in N(t_i, \delta t_i)) \cap (S_i = s_i)\}\right]}{\prod_{i \in I} \delta t_i},
\end{aligned} \tag{25.9}
$$

where

$$
\Pr\left[\bigcap_{i \in I}\{(T_i \in N(t_i, \delta t_i)) \cap (S_i = s_i)\}\right] \tag{25.10}
$$

(by the condition for informative masking)

$$
\prod_{i \in I} \Pr\left[(T_i \in N(t_i, \delta t_i)) \cap \left(\bigcup_{j \in s_i}(K_i = j)\right)\right] \tag{25.11}
$$

(since $\mathcal{C}_{\kappa_i}$ is the failed component in system Σ_i)

$$
\begin{aligned}
&\prod_{i \in I} \Pr\left[(T_i \in N(t_i, \delta t_i)) \cap (K_i = \kappa_i)\right] \\
&= \prod_{i \in I}\left\{\Pr\left[(T_{i\kappa_i} \in N(t_i, \delta t_i)) \cap \left(\bigcap_{w \in J, w \neq \kappa_i}(T_{iw} > t_i + \delta t_i)\right)\right]\right. \\
&\quad + \Pr\left[(T_{i\kappa_i} \in N(t_i, \delta t_i)) \cap \left(\bigcap_{w \in J, w \neq \kappa_i}(T_{iw} \geq T_{i\kappa_i})\right)\right. \\
&\qquad \left.\left. \cap \left(\bigcup_{w \in J, w \neq \kappa_i}(T_{iw} \in N(t_i, \delta t_i))\right)\right]\right\} \\
&= \prod_{i \in I} \Pr\left[(T_{i\kappa_i} \in N(t_i, \delta t_i)) \cap \left(\bigcap_{w \in J, w \neq \kappa_i}(T_{iw} > t_i + \delta t_i)\right)\right] + o(\delta t_i)
\end{aligned} \tag{25.12}
$$

where $o(\delta t_i)$ denotes, for each system Σ_i, $i \in I$, the probability of some other component(s) $\mathcal{C}_w$, $w \in J - \{\kappa_i\}$, falling in the interval $N(t_i, \delta t_i)$ after or at the same time as C_{κ_i}, which vanishes as $\delta t_i \to 0$. Since, for each $w \in J$, the T_{iw}'s, $i \in I$, are i.i.d. with common distribution $F_w(t; \boldsymbol{\theta}_w)$, expression (25.12) is equal to:

$$\prod_{i\in I} \Pr\left[T_{i\kappa_i} \in N(t_i, \delta t_i)\right] \prod_{w\in J,\, w\neq \kappa_i} \bar{F}_w(t_i + \delta t_i; \boldsymbol{\theta}_w) + o(\delta t_i).$$

Hence, by (25.9) and (25.2), we have:

$$\mathcal{L}(\boldsymbol{\theta}_1, \dots, \boldsymbol{\theta}_r; \kappa_n \mid \mathcal{D}) = \prod_{i\in I} \lambda_{\kappa_i}(t_i; \boldsymbol{\theta}_{\kappa_i}) \bar{F}_T(t_i; \boldsymbol{\theta}_1, \dots, \boldsymbol{\theta}_r). \qquad (25.13)$$

Given the observed data $\mathcal{D}$, denote by $\mathcal{H}$, $\mathcal{H} \subset J^n$, the space of all the sets $\boldsymbol{\kappa} \equiv (\kappa_1, \dots, \kappa_n)$, such that $\kappa_i \in s_i$ for all $i \in I$.

The MLEs are obtained as the values $\boldsymbol{\theta}_1^*, \dots, \boldsymbol{\theta}_r^*; \kappa_1^*, \dots, \kappa_n^*$ for which the likelihood (25.13) is maximized:

$$\begin{aligned}
\mathcal{L}(\boldsymbol{\theta}_1^*, \dots, \boldsymbol{\theta}_r^*; \kappa_1^*, \dots, \kappa_n^* \mid \mathcal{D}) &= \sup_{\boldsymbol{\theta}_1, \dots, \boldsymbol{\theta}_r; \kappa_1, \dots, \kappa_n} \mathcal{L}(\boldsymbol{\theta}_1, \dots, \boldsymbol{\theta}_r; \kappa_1, \dots, \kappa_n \mid \mathcal{D}) \\
&= \max_{\boldsymbol{\kappa} \in \mathcal{H}} \mathcal{L}(\boldsymbol{\theta}_1^*(\boldsymbol{\kappa}), \dots, \boldsymbol{\theta}_r^*(); \boldsymbol{\kappa} \mid \mathcal{D}) \qquad (25.14)
\end{aligned}$$

where it is useful to notice that $\boldsymbol{\theta}_1^*(\boldsymbol{\kappa}), \dots, \boldsymbol{\theta}_r^*(\boldsymbol{\kappa})$ coincide with the ML (vector) estimates we obtain in the standard competing risk model, whenever the causes of failure of the systems $\Sigma_1, \dots, \Sigma_n$ are assumed to be the components labeled $\kappa_1, \dots, \kappa_n$, respectively. Therefore, the estimation can be reduced to a problem of maximization over a finite set, whenever we are able to compute the MLEs for the standard competing risk model, while parametric estimation based on the incomplete data is possible, in general, only in the negative exponential case, through numerical methods.

25.4   The Exponential Case

Since the case when the lifetimes are exponentially distributed is the most common in industrial and reliability applications, it is worthwhile to discuss it in detail. We shall also discuss an estimation procedure considerably less complex, from the computational point of view, than (25.14).

Notation and definitions

Assume

$$F_j(t; \boldsymbol{\theta}_j) \equiv F_j(t; \lambda_j) = (1 - e^{-\lambda_j t}) 1_{[0,\infty)}(t), \qquad (25.15)$$

where $1_A(\ )$ is the indicator function of the set A, and denote by $\lambda_\Sigma = \sum_{j\in J} \lambda_j$ the *system failure rate*. Consider the set $\mathcal{P}_r$ of the $r!$ permutations of the first

r integers $\{1, 2, \ldots, r\}$, and denote by $\boldsymbol{\pi} \equiv (\pi_1, \ldots, \pi_r)$ an element of $\mathcal{P}_r$. Let $\Lambda \equiv (0, \infty)^r$ be the parameter space, $\boldsymbol{\lambda} \in \Lambda$. For each $\boldsymbol{\pi} \in \mathcal{P}_r$, define $\Lambda(\boldsymbol{\pi})$ as the subset of the parameter space where $\pi_j = \text{rank of } \lambda_j$, for all $j \in J$. Finally, for each $j \in J$, denote by $n_j(\boldsymbol{\kappa})$ the number of systems failed due to failure of the j-th component, based on the reconstruction $\boldsymbol{\kappa}$.

Remark 25.4.1 (i) The $r!$ sets $\Lambda(\boldsymbol{\pi})$, $\boldsymbol{\pi} \in \mathcal{P}_r$, are not disjoint (common points are the $\boldsymbol{\lambda}$'s with ties) and their union is equal to Λ. (ii) Given any two permutations $\boldsymbol{\pi}$ and $\boldsymbol{\pi}'$, it is always possible to establish a one-to-one correspondence between the two sets $\Lambda(\boldsymbol{\pi})$ and $\Lambda(\boldsymbol{\pi}')$, since any element $\boldsymbol{\lambda} \in \Lambda(\boldsymbol{\pi})$ can be put into correspondence with its permutation $\boldsymbol{\lambda}' \in \Lambda(\boldsymbol{\pi}')$.

In the exponential case (25.15), the likelihood function (25.13) can be written as:

$$\mathcal{L}(\lambda_1, \ldots, \lambda_r; \ \kappa_1, \ldots, \kappa_n \mid \mathcal{D}) = \prod_{i \in I} \lambda_{\kappa_i} e^{-\lambda_\Sigma t_i} \tag{25.16}$$

and (25.14) becomes

$$\max_{\boldsymbol{\kappa} \in \mathcal{H}} \mathcal{L}\left(\frac{n_1(\boldsymbol{\kappa})}{\sum_{u \in I} t_u}, \ldots, \frac{n_r(\boldsymbol{\kappa})}{\sum_{u \in I} t_u}; \ \boldsymbol{\kappa} \mid \mathcal{D} \right)$$
$$= \max_{\boldsymbol{\kappa} \in \mathcal{H}} \prod_{i \in I} \frac{n_{\kappa_i}(\boldsymbol{\kappa})}{\sum_{u \in I} t_u} \ e^{-n t_i / \sum_{w \in I} t_w}. \tag{25.17}$$

As an alternative approach, the problem of finding the MLEs, i.e., the values $\lambda_1^*, \ldots, \lambda_r^*, \kappa_1^*, \ldots, \kappa_n^*$ such that

$$\mathcal{L}(\lambda_1^*, \ldots, \lambda_r^*; \ \kappa_1^*, \ldots, \kappa_n^* \mid \mathcal{D}) = \sup_{\boldsymbol{\lambda} \in \Lambda} \max_{\boldsymbol{\kappa} \in \mathcal{H}} \mathcal{L}(\lambda_1, \ldots, \lambda_r; \ \kappa_1, \ldots, \kappa_n \mid \mathcal{D})$$

can be solved through the following steps.

1. For any $\boldsymbol{\lambda} \in \Lambda$, find $\kappa_1^*(\boldsymbol{\lambda}), \ldots, \kappa_n^*(\boldsymbol{\lambda})$, such that

$$\mathcal{L}(\lambda_1, \ldots, \lambda_r; \ \kappa_1^*(\boldsymbol{\lambda}), \ldots, \kappa_n^*(\boldsymbol{\lambda}) \mid \mathcal{D})$$
$$= \max_{\boldsymbol{\kappa} \in \mathcal{H}} \mathcal{L}(\lambda_1, \ldots, \lambda_r; \ \kappa_1(\boldsymbol{\lambda}), \ldots, \kappa_n(\boldsymbol{\lambda}) \mid \mathcal{D})$$
$$\equiv \mathcal{M}(\lambda_1, \ldots, \lambda_r) \mid \mathcal{D}).$$

2. Compute $\lambda_1^*, \ldots, \lambda_r^*$, such that

$$\mathcal{M}(\lambda_1^*, \ldots, \lambda_r^* \mid \mathcal{D}) = \sup_{\boldsymbol{\lambda} \in \Lambda} \mathcal{M}(\lambda_1, \ldots, \lambda_r \mid \mathcal{D}).$$

3. Determine the ML reconstruction as $\kappa_i^* = \kappa_i^*(\boldsymbol{\lambda}^*)$, $i = 1, \ldots, n$.

The above procedure is different from that indicated by formula (25.14), where, conversely, one first finds the MLEs of the unknown parameters as functions of the reconstruction, $\boldsymbol{\theta}_1^*(\boldsymbol{\kappa}), \ldots, \boldsymbol{\theta}_r^*(\boldsymbol{\kappa})$, and then finds the ML reconstruction $\kappa_1^*, \ldots, \kappa_n^*$. In the exponential case, it is possible and convenient to apply a reduced version of the above procedure, due to a special property of the ML reconstruction, expressed by the following Theorem 25.4.1, which leads to a considerable simplification of the estimation process. This property characterizes the exponential distribution, and, as some reflection on it will disclose, it is essentially due to the fact that the failure rate functions are independent of the time.

Theorem 25.4.1 *For any given $\boldsymbol{\lambda} \in \Lambda$, the conditional ML reconstruction $\kappa_i^*(\boldsymbol{\lambda})$, $i = 1, \ldots, n$, is unique and constant over the sets $\Lambda(\boldsymbol{\pi})$, $\boldsymbol{\pi} \in \mathcal{P}_r$, provided that $\boldsymbol{\lambda}$ has no ties. That is,*

$$\boldsymbol{\kappa}^*(\boldsymbol{\lambda}) = \boldsymbol{\lambda}^*(\boldsymbol{\pi}) \qquad \textit{for all} \qquad \boldsymbol{\lambda} \in \bar{\Lambda}(\boldsymbol{\pi})$$

where

$$\bar{\Lambda}(\boldsymbol{\pi} \equiv \Lambda(\boldsymbol{\pi}) \cap \left(\bigcup_{\boldsymbol{\pi}' \in \mathcal{P}_r - \{\boldsymbol{\pi}\}} \Lambda(\boldsymbol{\pi}') \right)^c .$$

In other words, the reconstruction $\kappa_1^*(\boldsymbol{\lambda}), \ldots, \kappa_n^*(\boldsymbol{\lambda})$ depends only on the ranks of the failure rates and not on their magnitudes.

PROOF. For any given $\boldsymbol{\lambda} \in \Lambda$, the likelihood function (25.16) is maximized, with respect to $\boldsymbol{\kappa} \in \mathcal{H}$, by choosing, for each variable κ_i, $i = 1, \ldots, n$, the value, say $\kappa_i^*(\boldsymbol{\lambda})$, equal to the index in s_i to which corresponds the highest failure rate:

$$\kappa_i^*(\boldsymbol{\lambda}) \equiv \boldsymbol{\kappa}^*(\boldsymbol{\pi}) = \{\kappa_i \in s_i \mid \lambda_{\kappa_i} = \max_{k \in s_i} \lambda_k\}, \qquad i = 1, \ldots, n.$$

It follows that, for all the configurations $\boldsymbol{\lambda}$ with the same ranking $\boldsymbol{\pi}$ of failure rates, we have the same point of maximum $\boldsymbol{\kappa}^*(\boldsymbol{\pi})$ (the maximum likelihood, clearly, does depend on $\boldsymbol{\lambda}$).   ∎

Theorem 25.4.1 suggests that the following estimation procedure could be used in the negative exponential case.

1. For each $\boldsymbol{\pi} \in \mathcal{P}_r$, determine the (conditional) ML reconstruction as:

$$\kappa_i^*(\boldsymbol{\pi}) = \left\{\kappa_i \in s_i \mid \lambda_{\kappa_i}^0(\boldsymbol{\pi}) = \max_{\kappa \in s_i} \lambda_\kappa^0(\boldsymbol{\pi})\right\}, \qquad i = 1, \ldots, n,$$

where $(\lambda_1^0(\boldsymbol{\pi}), \ldots, \lambda_r^0(\boldsymbol{\pi})) \equiv \boldsymbol{\lambda}^0(\boldsymbol{\pi})$ is an arbitrary point in the set $\bar{\Lambda}(\boldsymbol{\pi})$; for instance, we might choose $\boldsymbol{\lambda}^0(\boldsymbol{\pi}) = \boldsymbol{\pi}$.

2. Find the ranking $\hat{\pi} \in \mathcal{P}_r$ to which corresponds the reconstruction $\hat{\kappa}$ such that:

$$\mathcal{L}(\boldsymbol{\lambda}^0(\hat{\pi}); \hat{\kappa} \mid \mathcal{D}) = \max_{\pi \in \mathcal{P}_r} \mathcal{L}(\lambda_1^0(\pi), \ldots, \lambda_r^0(\pi); \kappa_1^*(\pi), \ldots, \kappa_n^*(\pi) \mid \mathcal{D}).$$

Here, one may notice that $\hat{\pi}$ provides an estimate of the failure rates ranking:

$$\hat{\pi}_j = \text{ estimated rank of } \lambda_j, \qquad j = 1, \ldots, r.$$

3. Find the MLEs $\hat{\boldsymbol{\lambda}}$ as in the standard competing risk model, by assuming the reconstruction $\hat{\kappa}$ as the set of subscripts of the true causes of failure:

$$\mathcal{L}(\hat{\boldsymbol{\lambda}}; \hat{\kappa} \mid \mathcal{D}) = \sup_{\boldsymbol{\lambda} \in \Lambda} \mathcal{L}(\lambda_1, \ldots, \lambda_r; \hat{\kappa}_1, \ldots, \hat{\kappa}_n \mid \mathcal{D}).$$

As anticipated, the above estimation procedure is computationally more convenient than (25.14), where the number of configurations to be examined is equal to the cardinality of the set $\mathcal{H}$, i.e., $\prod_{i \in I} |s_i|$. In fact, the number of comparisons necessary to find the reconstruction are reduced to $r!$, which is independent of n, and, in practical applications, the number of system components r is usually small, as compared with the sample size n.

25.5 An Example

Concepts preliminary to the simulation

Before presenting an example of reconstruction for simulated masked data, we caution the reader that generating correctly the incomplete data (25.5) requires some attention. For instance, some authors have used the following procedure:

1. for each system Σ_i, $i \in I$, generate the components lifelengths and then take the smallest among them as the lifelength t_i,

2. add randomly other possible causes of failure to the true cause of failure of each system, in order to obtain the sets s_i, $i \in I$.

The above procedure would generate data whose distribution violates the condition for an informative masking, and, in general, the data generated as shown above has a likelihood different from that we have obtained [in particular, expression (25.10) is not equal to (25.11).] For the masking to be "informative," it should arise from the experiment, *and not be generated randomly after the experiment*. In order to provide a meaningful mechanism of generation of the sets s_i where the condition of informative masking is satisfied, we shall now consider, following Gupta and Gastaldi (1996), the incompleteness resulting from a censored search for the cause of failure.

Censored search for the cause of failure

Assume that, after the failure of each system Σ_i, $i = 1, \ldots, n$, a sequential search for the failed component within the system is carried out by checking, one by one, the system components, taken in the same order as their subscripts appear in the *ordered* set $\omega_i(t_i) \equiv \langle j_{i1}, \ldots, j_{ir} \rangle$. We will refer to such a set as the *(component) inspection strategy*. Assume that $\alpha_{ij}(t_i)$ represents the time needed to carry out the failure analysis on component $\mathcal{C}_j$ failed at time t_i in system Σ_i, referred to as the *component checking time*. Let $\beta_i(t_i)$ denote the maximum length of time available for the failure analysis on system Σ_i, referred to as the *system checking time limit*.

Denote by $C_i(t_i)$ the random set formed by all those and only those subscripts of the components that can be inspected, given the time constraint $\beta_i(t_i)$, i.e., $C_i(t_i)$ is the possibly empty random set of all the j_{ih} in $\omega_i(t_i)$ satisfying the conditions:

$$\sum_{s=1}^{h} \alpha_{ij_{is}}(t_i) \leq \beta_i(t_i) \qquad \text{and} \qquad h \leq r - 1 \tag{25.18}$$

taken *in the same order as they appear in* $\omega_i(t_i)$. Let $\nu_i(t_i)$ indicate the total number of subscripts in $\omega_i(t_i)$ satisfying (25.18). Clearly, $\nu_i(t_i)$, in general, is a function of $\omega_i(t_i)$, $\beta_i(t_i)$, and $\alpha_{i1}(t_i), \ldots, \alpha_{ir}(t_i)$; however, for simplicity, these dependencies will be disregarded in our notation. By assumption, the cardinality of $C_i(t_i)$ is at most equal to $r - 1$, because the inspection of $r - 1$ components ensures the determination of the cause of failure.

For convenience of notation, we use the generalized indicator function [Gupta and Gastaldi (1996)]:

$$\text{If}(\mathcal{E};\ r_1, r_2) \equiv \begin{cases} r_1 & \text{if event } \mathcal{E} \text{ is true} \\ r_2 & \text{if event } \mathcal{E} \text{ is false} \end{cases} \tag{25.19}$$

where r_1, r_2, can be objects of any type.

Consider the following decomposition of $\omega_i(t_i)$:

$$\omega_i(t_i) = C_i(t_i) + \bar{C}_i(t_i) \tag{25.20}$$

where

$$\begin{aligned} C_i(t_i) &\equiv \text{If}\left((\nu_i(t_i) = 0); \emptyset, \langle j_{i1}, \ldots, j_{i,\nu_i(t_i)} \rangle\right), \\ \bar{C}_i(t_i) &\equiv \langle j_{i,\nu_i(t_i)+1}, \cdots j_{ir} \rangle, \end{aligned}$$

and "+" is a symbol for concatenation.

The random set S_i is defined as:

$$S_i \equiv \text{If}\left((K_i \in C_i(t_i));\ \{K_i\}, \bar{C}_i(t_i)\right)$$

i.e., it is formed by all those and only those possible causes of failure isolated after a failure analysis was conducted, for a period not longer than $\beta_i(t_i)$, on $r-1$ components $C_{j_i 1}, \ldots, C_{j_i, r-1}$ of system Σ_i, each of which requires a time exactly equal to $\alpha_{ij}(t_i)$ to be inspected. Clearly, the cardinality of S_i is either 1 or $(r - \nu_i(t_i))$.

An important problem, relevant to the above framework, is that of the optimal choice of the inspection strategy in order to obtain the *best* (in some sense to be made precise) estimates of the unknown parameters. Based on the discussion contained in Gupta and Gastaldi (1996), we will consider a *random* order of inspection of the components, which varies (randomly) for each failed system.

Simulation results and concluding remarks

For our simulation study, we generated life-tests of size n, and applied the last procedure discussed in Section 25.4. Simulated data for each life-test have been obtained as follows. For each system, we have simulated exponential life-lengths for the r components. The number of components, sample size, mean lifelengths, failure rates, component checking times, and system checking time limits are reported in the tables along with the simulation results. For simplicity, we considered component checking times $\alpha_{ij}(t_i)$'s constant with respect to the failure time as well as to the system. Similarly, the system checking times $\beta_i(t_i)$'s have been set constant for all systems and failure times. In each life-test, we have generated lifelength and failure of all systems and simulated a censored search for the cause of failure according to a *random* inspection strategy.

Based on the simulations we have carried out, of which two are reported below, we observed that the applied procedure leads usually to a better estimation (than the MLEs based on the incomplete data) of the higher failure rates, since a larger amount of data is made available for their estimation. We also noticed that, when the values of the unknown parameters are close to each other, the estimated ranking $\hat{\pi}$, due to its discrete nature, is too unstable and should not be relied upon.

While the techniques based on the reconstructed data are the only way to carry out parametric estimation under masking and distribution other than the negative exponential; specifically, for the latter case, it is advisable using both techniques based on incomplete data and reconstructed data and make an appropriate use of the indication one gets, taking into account that, usually, shorter lifelengths are estimated better by the estimators based on the reconstructed data.

Table 25.1: First simulation results

Life Tests' Parameters and Statistics

Simulation id. number: 001
Life test size: 300

Systems' Features

Number of system components: 5
System checking time limit: 13.00

Components' Features

Component number	1	2	3	4	5
Mean lifelength	165	270	495	595	5
Failure rate	0.00606	0.00370	0.00202	0.00168	0.20000
Failure rate rank	4	3	2	1	5
Inspection time	1.00	4.50	6.50	3.50	12.00

Statistics on the Simulated Lifelengths

Component number	1	2	3	4	5
Sample means	164.66	269.83	494.57	595.26	4.61
Sample failure rate	0.00607	0.00371	0.00202	0.00168	0.21698
Sample failure rate rank	4	3	2	1	5
Sample std	165.17	270.29	495.13	594.90	4.80

Permutation (ranks)   Loglikelihood

1 2 3 5 4   −19405.3623
1 2 3 4 5   −19365.1965
1 2 4 5 3   −19439.5965

[here multiple lines (76) have been omitted]

4 3 2 1 5   −19359.5335
4 3 1 5 2   −19485.0201
4 3 1 2 5   −19358.8404 * <= maximum
4 3 5 2 1   −19510.7020
4 3 5 1 2   −19470.4995

[here multiple lines (33) have been omitted]

Permutation (ranks)	Loglikelihood
5 1 4 3 2	−19486.0487
5 1 2 3 4	−19401.9308
5 1 2 4 3	−19445.3708

t_i	s_i	Inspection strategy	Failed	Reconstruction & success indicator[†]
2.95501	{25}	{ 4 1 3 | 2 5 }	5	5 *
8.40062	{145}	{ 3 2 | 1 4 5 }	5	5 *
1.05305	{5}	{ 1 5 | 3 2 4 }	5	5 -
25.41086	{25}	{ 3 4 1 | 2 5 }	5	5 *
0.05093	{345}	{ 1 2 | 3 4 5 }	5	5 *
1.41593	{245}	{ 1 3 | 2 4 5 }	5	5 *
1.81063	{35}	{ 2 4 1 | 3 5 }	5	5 *
6.32284	{1345}	{ 2 | 1 3 4 5 }	5	5 *
2.96306	{245}	{ 3 1 | 2 4 5 }	5	5 *
6.92815	{25}	{ 1 4 3 | 2 5 }	5	5 *
4.09108	{135}	{ 4 2 | 1 3 5 }	5	5 *
2.76151	{25}	{ 4 3 1 | 2 5 }	5	5 *
8.76076	{35}	{ 2 4 1 | 3 5 }	5	5 *
1.84667	{5}	{ 5 | 2 4 1 3 }	5	5 -
8.70233	{125}	{ 4 3 | 1 2 5 }	5	5 *
12.83945	{35}	{ 2 1 4 | 3 5 }	5	5 *
1.42853	{1235}	{ 4 | 1 2 3 5 }	5	5 *
2.90415	{45}	{ 2 3 1 | 4 5 }	5	5 *
3.70079	{125}	{ 3 4 | 1 2 5 }	5	5 *
1.48431	{1245}	{ 3 | 1 2 4 5 }	5	5 *
2.09125	{25}	{ 4 1 3 | 2 5 }	5	5 *
0.65925	{35}	{ 4 2 1 | 3 5 }	5	5 *
3.12510	{1345}	{ 2 | 1 3 4 5 }	5	5 *
8.64039	{5}	{ 5 | 3 1 2 4 }	5	5 -
4.11304	{35}	{ 1 4 2 | 3 5 }	5	5 *
6.30476	{5}	{ 1 5 | 2 4 3 }	5	5 -
5.00904	{5}	{ 5 | 3 2 4 1 }	5	5 -
1.27659	{35}	{ 2 1 4 | 3 5 }	5	5 *
0.74472	{5}	{ 5 | 2 4 1 3 }	5	5 -
0.89881	{245}	{ 3 1 | 2 4 5 }	5	5 *
2.51949	{125}	{ 3 4 | 1 2 5 }	5	5 *
1.68318	{1235}	{ 4 | 1 2 3 5 }	5	5 *
8.22434	{1245}	{ 3 | 1 2 4 5 }	5	5 *

t_i	s_i	Inspection strategy	Failed	Reconstruction & success indicator[†]
0.18339	{5}	{ 5 1 | 4 3 2 }	5	5 -
9.20167	{245}	{ 1 3 | 2 4 5 }	4	5 w
5.16085	{35}	{ 2 4 1 | 3 5 }	5	5 *
1.06596	{1345}	{ 2 | 1 3 4 5 }	5	5 *

[here multiple lines (255) have been omitted]

t_i	s_i	Inspection strategy	Failed	Reconstruction & success indicator
1.06610	{1245}	{ 3 | 1 2 4 5 }	5	5 *
5.59530	{245}	{ 3 1 | 2 4 5 }	5	5 *
4.01618	{35}	{ 2 4 1 | 3 5 }	5	5 *
2.71450	{5}	{ 1 5 | 2 4 3 }	5	5 -
14.22565	{5}	{ 5 | 3 1 2 4 }	5	5 -
9.95210	{5}	{ 5 | 4 2 1 3 }	5	5 -
0.35014	{45}	{ 2 3 1 | 4 5 }	5	5 *

[†] * = exact, w = wrong, − = no choice

Number of found reconstructions: 1

Correct reconstruction:

Estimated ranking	Total	Among masked
4 3 1 2 5	293 out of 300 (98%)	222 out of 229 (97%)

Counts corresponding to the reconstruction:

Estimated ranking	n1	n2	n3	n4	n5	Total
4 3 1 2 5	8	3	1	2	286	300

Parameters:

	0.00606061	0.00370370	0.00202020	0.00168067	0.20000000

Estimators:

4 3 1 2 5	0.00604994	0.00226873	0.00075624	0.00151248	0.21628530
MD-MLE	0.00747189	0.00386451	0.00149267	0.00320705	0.21083658

Note: MD-MLE are the ML estimators based on masked data computed numerically through the Gauss-Seidel method. On the preceding line there are the estimators based on the reconstructed data corresponding to the estimated ranking.

Table 25.2: Second simulation results

Life Tests' Parameters and Statistics

Simulation id. number:	002
Life test size:	300

Systems' Features

Number of system components:	5
System checking time limit:	13.00

Components' Features

Component number	1	2	3	4	5
Mean lifelength	165	270	495	595	5
Failure rate	0.00606	0.00370	0.00202	0.00168	0.20000
Failure rate rank	4	3	2	1	5
Inspection time	1.00	4.50	6.50	3.50	12.00

Statistics on the Simulated Lifelengths

Component number	1	2	3	4	5
Sample means	164.97	269.51	494.38	595.00	5.15
Sample failure rate	0.00606	0.00371	0.00202	0.00168	0.19422
Sample failure rate rank	4	3	2	1	5
Sample std	164.58	269.92	494.52	595.09	4.54

Permutation (ranks)   Loglikelihood

Permutation	Loglikelihood
1 2 3 5 4	−21983.8926
1 2 3 4 5	−21943.7267
1 2 4 5 3	−22017.2637

[here multiple lines (74) have been omitted]

4 3 2 1 5	−22022.1170
4 3 1 5 2	−22110.1090
4 3 1 2 5	−21939.4501 * <= maximum
4 3 5 2 1	−22069.2133
4 3 5 1 2	−21939.4501 * <= maximum
4 3 5 2 1	−22088.1516
4 3 5 1 2	−22047.2559

Permutation (ranks)   Loglikelihood

[here multiple lines (33) have been omitted]

5 1 4 3 2	−22067.8794
5 1 2 3 4	−21984.2848
5 1 2 4 3	−22028.8756

t_i	s_i	Inspection strategy	Failed	Reconstruction & success indicator[†]
7.42170	{1345}	{ 2 | 1 3 4 5 }	5	5 *
8.79900	{1}	{ 2 3 1 | 5 4 }	1	1 -
7.51176	{125}	{ 3 4 | 1 2 5 }	5	5 *
4.74631	{1345}	{ 2 | 1 3 4 5 }	5	5 *
0.22940	{3}	{ 3 1 4 | 5 2 }	3	3 -
17.24333	{145}	{ 3 2 | 1 4 5 }	5	5 *
0.22598	{234}	{ 5 1 | 2 3 4 }	3	2 w
4.73101	{5}	{ 5 | 2 4 1 3 }	5	5 -
13.44542	{5}	{ 5 | 3 1 2 4 }	5	5 -
3.53096	{35}	{ 1 4 2 | 3 5 }	5	5 *

[here multiple lines (290) have been omitted]

[†] * = exact, w = wrong, − = no choice

Number of found reconstructions: 2

Correct reconstructions:

Estimated ranking	Total	Among masked
4 3 2 1 5	293 out of 300 (98%)	225 out of 232 (97%)
4 3 1 2 5	293 out of 300 (98%)	225 out of 232 (97%)

Counts corresponding to the reconstructions:

Estimated ranking	n1	n2	n3	n4	n5	Total
4 3 2 1 5	8	3	1	1	287	300
4 3 1 2 5	8	3	1	1	287	300

Parameters:

0.00606061	0.00370370	0.00202020	0.00168067	0.20000000

Estimators:

4 3 2 1 5	0.00535321	0.00200746	0.00066915	0.00066915	0.19204657
4 3 1 2 5	0.00535321	0.00200746	0.00066915	0.00066915	0.19204657
MD-MLE	0.00517423	0.00452820	0.00178859	0.00180782	0.18744669

References

1. Basu, A. P. and Klein, J. P. (1982). Some recent results in competing risks theory, In *Survival Analysis* (Eds., J. Crowley and R. A. Johnson), pp. 216–229, IMS Lecture Notes-Monograph Series 2, Hayward, CA: IMS.

2. Baxter, L. A. (1995). Estimation subject to block censoring, *IEEE Transactions on Reliability*, **44**, 489–495.

3. Cox, D. R. (1959). The analysis of exponentially distributed lifetimes with two types of failures, *Journal of the Royal Statistical Society, Series B*, **21**, 411–421.

4. David, H. A. (1974). Parametric approaches to the theory of competing risks, In *Reliability and Biometry: Statistical Analysis of Lifelengths* (Eds., F. Proschan and R. J. Serfling), pp. 275-290, Philadelphia: SIAM.

5. David, H. A. and Moeschberger, M. L. (1978). *The Theory of Competing Risks*, Griffin's Statistical Monographs & Courses No. 39, London: Charles W. Griffin.

6. Dinse, G. E. (1982). Nonparametric estimation for partially-incomplete times and types of failure data, *Biometrics*, **38**, 417–431.

7. Dinse, G. E. (1986). Nonparametric prevalence and mortality estimators for animal experiments with incomplete cause-of-death data, *Journal of the American Statistical Association*, **81**, 328–336.

8. Gastaldi, T. (1994). Improved maximum likelihood estimation for component reliabilities with Miyakawa-Usher-Hodgson-Guess' estimators under censored search for the cause of failure, *Statistics & Probability Letters*, **19**, 5–18.

9. Gastaldi, T. (1996). Note on closed-form MLEs of failure rates in a fully parametric random censorship model with incomplete data, *Statistics & Probability Letters*, **26**, 309–314.

10. Gastaldi, T. and Gupta, S. S. (1994). Minimax type procedures for nonparametric selection of the best population with partially classified data, *Communications in Statistics—Theory and Methods*, **23**, 2503–2531.

11. Guess, M. F., Usher, J. S. and Hodgson, T. J. (1991). Estimating system and component reliabilities under partial information on cause of failure, *Journal of Statistical Planning and Inference*, **29**, 75–85.

12. Gupta, S. S. and Gastaldi, T. (1996). Life testing for multi-component system with incomplete information on the cause of failure: A study on some inspection strategies, *Computational Statistics & Data Analysis*, **22**, 373–393.

13. Little, R. J. A. and Rubin D. B. (1987). *Statistical Analysis with Missing Data*, New York: John Wiley & Sons.

14. Mendenhall, W. and Hader, R. J. (1958). Estimation of parameters of mixed exponentially distributed failure time distributions from censored life test data, *Biometrika*, **45**, 504–520.

15. Miyakawa, M. (1984). Analysis of incomplete data in a competing risks model, *IEEE Transactions on Reliability*, **33**, 293–296.

16. Usher, J. S. (1987). Estimating component reliabilities from incomplete accelerated life test data, *Unpublished Ph.D. Dissertation*, Department of Industrial Engineering, North Carolina State University, Raleigh, NC.

17. Usher, J. S. (1993). On the problem of masked system life data, In *Advances in Reliability* (Ed., A. P. Basu), pp. 435–443, Amsterdam: Elsevier Science Publishers B.V.

18. Usher, J. S. and Guess, F. M. (1989). An iterative approach for estimating component reliability from masked system life data, *Quality & Reliability Engineering International*, **5**, 257–261.

19. Usher, J. S. and Hodgson, T. J. (1988). Maximum likelihood analysis of component reliability using masked system life data, *IEEE Transactions on Reliability*, **37**, 550–555.

26

A Review of the Gupta–Sobel Subset Selection Rule for Binomial Populations With Industrial Applications

Gary C. McDonald and W. Craig Palmer

General Motors Research and Development Center, Warren, MI

Abstract: In 1960, Gupta and Sobel proposed a subset selection rule for selecting the best of k binomial populations. In applying the Gupta–Sobel procedure a number of practical considerations arise. Operating characteristics for this rule, and for an analogous rule when data are subject to contamination, are reviewed. Some extensions and variations of the Gupta–Sobel selection procedures are discussed. In particular, rules for selecting populations when bounds on the probability parameters are applicable are noted. Computational and theoretical difficulties arising in the case of unequal sample sizes are identified.

Keywords and phrases: Probability of selection, operating characteristics, contaminated data, restricted inference, sample size determination

26.1 Introduction

In 1960, Gupta and Sobel proposed a subset selection procedure for binomial populations. The Gupta–Sobel procedure has many applications in industry. However, in applying the Gupta–Sobel procedure a number of practical considerations arise. This article reviews, and in some cases, extends what is known about subset selection procedures for binomial populations.

For practical applications, understanding the operating characteristics (OC's) of the procedure is critical. OC's for ranking and selection procedures are like power functions for hypothesis testing. When designing experiments, OC's play an important role in determining the necessary sample size.

In many reliability applications, the inference can be restricted (a priori)

to a certain portion of the parameter space. It will be shown that this can substantially improve the OC's of the Gupta–Sobel procedure.

Data contamination is a frequent practical concern; i.e., when the success and failure of the Bernoulli experiment is reported with probability of truthfulness less than one. For example, with consumer survey data, contamination might occur due to respondent error in filling out the questionnaire or when asking a societally sensitive question such as seat belt usage. The implications of data contamination for the Gupta–Sobel procedure are reviewed.

The last portion of the article addresses the challenging case of decision problems with unequal sample sizes. Unequal sample sizes are typical in a broad class of industrial studies, particularly in market research. Some surprising theoretical and empirical results will be noted for this commonplace situation requiring the application of the Gupta–Sobel procedure.

26.2   Gupta–Sobel Procedure

A practical application of the basic Gupta–Sobel procedure arises when selecting component suppliers. The decision of which supplier(s) to select is a function of many variables including: quality, cost, service, reliability. An approach to this selection problem is to use a statistical selection procedure based on one of these characteristics, say quality, to choose a subset of proposed suppliers using the quality performance of a random sample of their components. The selected subset of suppliers is then further screened by the remaining important characteristics.

The formulation and notation required for the Gupta–Sobel procedure are given in the supplier selection context. Of course, the problem could be stated in the context of simply selecting a binomial population. The i-th supplier (or population) will be denoted by π_i and the number of suppliers indicated by k. The sampling results from each supplier will be characterized by a binomial model with sample size n and conformance probability p_i. The ordered values of the p_i's will be delineated with bracketed subscripts, and the best supplier will be defined as that one which has the largest conformance probability, $p_{[k]}$. In case more than one supplier is characterized by $p_{[k]}$, one of these is considered to be tagged at random and designated the best so as to remove ambiguity. The random variable X_i will denote the number of conforming items from π_i based on testing a random sample of n items. In other words, X_i is the number of successes in the binomial experiment.

The statistical goal is to choose a nonempty subset of suppliers (or populations) which includes the best one with a probability no less than a preassigned value P^*. Note that, for a meaningful problem, $P^* > 1/k$. A correct selection (CS) occurs when the selected subset includes the best supplier, and the probability of a correct selection is denoted by P(CS). The probability of including a non-best supplier, the so-called probability of non-best selections will be denoted by P(NBS). In general this probability varies from supplier to supplier. In the special case where each of the non-best suppliers have identical binomial parameters, then this probability will be the same for each non-best supplier (or population) and the notation P(NBS) will be unambiguous.

The statistical goal for the selection procedure requires a "P^*-condition" to be met; i.e., however the selection is made the procedure should insure that

$$\inf_{\Omega} P(CS) \geq P^*,$$

where Ω denotes the unrestricted parameter space of the true (but unknown) configuration of p_i's.

The Gupta–Sobel procedure is both intuitively appealing and simple to implement. The ith supplier is selected for inclusion in the subset if X_i is greater than or equal to the maximum of the X's less a constant d. In other words,

$$\text{Select } \pi_i \Leftrightarrow X_i \geq \max_{1 \leq j \leq k} X_j - d.$$

The d-value is chosen as the smallest nonnegative integer assuring the P^*-condition. Values for d are extensively tabulated as a function of the number of suppliers (k), sample size (n), and the P^* value [Gupta and Sobel (1960), Gupta and McDonald (1986).] These tabulations involve exact computations as well as extensive computations based on normal approximations. The Gupta–Sobel procedure will always result in a nonempty subset being chosen since the supplier(s) associated with the maximum X_i will always be chosen.

Gupta and Sobel (1960) proved that the inf P(CS) is attained when all of the true (unobserved) p_i's are equal, called the least favorable configuration. This fact greatly simplifies the search for the value of d which satisfies the P^*-condition.

$$\inf_{\Omega} P(CS) = \inf_{0 \leq p \leq 1} \sum_{l=0}^{n} \binom{n}{l} p^l q^{n-l} \left[\sum_{r=0}^{d+l} \binom{n}{r} p^r q^{n-r} \right]^{k-1}$$

26.3 Operating Characteristics

Operating characteristics (OC's) for ranking and selection procedures are like power functions for hypothesis testing. OC's are functions of the underlying parameter values and are messy to calculate. Thus, OC's are usually calculated under parameter configuration assumptions that can be easily stated and characterized; e.g., a slippage configuration where the probability parameter of the best supplier is δ units greater than all others assumed equal, or an equi-spaced configuration where the differences between two adjacent ordered probability parameters is δ units. Under these assumptions the probability of choosing the i-th ordered supplier (called the individual selection probabilities) can be calculated and, hence, the expected size of the selected subset (which is the sum of the individual selection probabilities). The sample size n can now be chosen to make the expected size of the selected subset small. Formulae, tables, and graphs are extensively available for such assessments [see, e.g., Gupta and McDonald (1986).]

As an illustration, suppose that there are five suppliers for a certain product under consideration and that each submits a random sample of 30 for evaluation. Suppose further that the results of the testing are: the first supplier had 21 of the 30 items acceptable, the second 25, the third 24, the fourth 22, and the fifth 28. If we take (arbitrarily) the constant d to be 5, then the Gupta–Sobel selection procedure can be stated as follows: choose all suppliers such that $X_i \geq \max X_j - 5 = 23$ (since $\max X_j = 28$). With $d=5$ the P^*-value would be 0.75, so there would be at least a 75% chance of correctly including the best supplier in the selected subset no matter what the true underlying parameter configuration is. Here all suppliers but the first and fourth would be selected. How good is this selection procedure?

Table 4 from Gupta and McDonald (1986) provides an answer to the question under an assumption about the true parameter configuration. In this table, a slippage configuration is assumed where four of the population parameters are assumed to be equal to $p = 0.75$ and one is assumed to have "slipped to the right" by $\delta = 0.05$, yielding its p-value at $p = 0.80$. Entering the table for $k = 5$, $d = 5$, and $n = 30$ corresponding to our illustration, the selection procedure is found to yield a probability of including the best supplier P(CS), of 0.95. The probability of including a non-best supplier, P(NBS), is 0.83. Hence, the expected subset size is $4(0.83) + 0.95 = 4.27$.

26.4   Restricted Procedures

There are some important alternatives to the basic Gupta–Sobel selection procedure. These alternatives are developed primarily to improve the OC's of the basic selection procedure.

For some applications, particularly in reliability, the range of the true p_i's is known a priori. The restricted procedure, as described in Gupta and McDonald (1986), limits the applicability of the inference to a subspace of Ω (the unrestricted parameter space), to Ω' in which it is known that the underlying parameters reside. In particular, $\Omega' = \{\underline{p} \in \Omega : p_i \geq p_0, i = 1, \ldots, k\}$. Here all of the p_i's are known a priori to be greater than or equal to p_0.

By restricting the scope of the inference, the OC's of the selection procedure can be improved substantially for both equal and unequal sample sizes. The restricted Gupta–Sobel selection rule is similar in form to the basic Gupta–Sobel selection rule:

$$\text{Select } \pi_i \Leftrightarrow X_i \geq \max_{1 \leq j \leq k} X_j - d'.$$

Based on asymptotic calculations, the required constant d' to be used in this restricted alternative is a simple linear function of the constant d extensively tabulated for the unrestricted Gupta–Sobel procedure. It can be shown that asymptotically $d' = (2d+1)\sqrt{p_0 q_0} - 0.5 < d$. Thus the size of the subset chosen with the restricted alternative will be no greater than that chosen with the unrestricted Gupta–Sobel procedure.

Other alternatives for improving the OC's of the Gupta–Sobel procedure have been developed. Gupta and Liao (1993) developed a Bayesian formulation which incorporates a prior distribution over the largest population probability parameter, $p_{[k]}$, and a computational method for determining the required constant d.

26.5   Data Contamination

This section of the article will extend the Gupta–Sobel procedure to the situation where the reported data itself may be in error. Data contamination is a real concern in practical applications, particularly when dealing with consumer data. The material in this section is based primarily on McDonald (1994a,b).

The model on which this analysis is based is the standard binomial model with one slight modification. In the binomial context, consider a population of elements in which each has or has not a specific characteristic – call it A. Under data contamination, it may not be possible to ascertain the exact presence or absence of A in an individual element. For example, if A is a personally sensitive issue being ascertained by survey or interview, the respondent may be inclined to lie. This situation can be modeled by introducing a hypothetical auxiliary experiment where an event B occurs with probability ξ.

Now this auxiliary experiment is combined with the basic binomial model and the observed data is assumed to be filtered by a response mechanism which passes correctly the presence or absence of characteristic A if the event B occurs. If B does not occur, then the mechanism passes the information incorrectly. So now the observed variable, say Y_i, is zero if A is not indicated and one if A is indicated. The actual presence of A in the ith element, however, is not necessarily indicated by the variable Y_i due to the contamination introduced by the auxiliary experiment. Under contamination, $P(Y_i = 1) = p\xi + (1-p)(1-\xi)$.

Let $S = \sum Y_i$. Since the random variables Y_i's do not reflect the true state-of-nature, neither does S, the sum of those variables. However, there is information in these quantities that will permit estimation of the parameters of interest. An unbiased estimator for p, the true underlying binomial parameter, is given by:

$$\hat{p} = [E(S/n) - (1 - \xi)]/(2\xi - 1), \xi > 1/2.$$

Now that the basic model for contamination has been set, we return to the problem of subset selection. First note that it is necessary to impose the condition that $\xi > 1/2$. This is necessary to insure that there is consistency in the ordering of the populations; i.e., the observed data must be truthful at least most of the time.

Under data contamination, a subset selection rule analogous in form to the basic Gupta–Sobel procedure can be constructed using the S_i's defined earlier;

$$\text{Select } \pi_i \Leftrightarrow S_i \geq \max_{1 \leq j \leq k} S_j - d''.$$

Again, the d''-value is chosen as the smallest nonnegative integer assuring the P^*-condition.

McDonald (1994b) shows that for large samples, using the normal approximation given in Gupta and Sobel (1960), the equation for inf P(CS) involving d in the basic Gupta–Sobel procedure is identical to the equation for inf P(CS) for d'' when the data are contaminated. No new tables are required. The P^*-condition is preserved for the Gupta–Sobel procedure even when the data are contaminated.

However, the OC's of the procedure used with contaminated data do differ from those of the Gupta–Sobel procedure used with no contamination. This is best illustrated by computing individual selection probabilities (ISP's) under a specified assumption about the underlying population parameters. Consider a slippage configuration where k-1 of the populations have equal p-values and the best population, characterized by $p_{[k]}$, has a p-value δ units greater than the others. The formulae required to compute these ISP's, assuming a slippage configuration, involve rather complex integral expressions as given in McDonald(1994b).

The tables in McDonald(1994b) give the P(CS), P(NBS), and the expected size of the subset. In reviewing the entries of these tables it is possible to draw qualitative conclusions about the impact of data contamination. Its presence serves to reduce the P(CS), increase the P(NBS), and increase the expected subset size. Contamination adversely affects all of the OC's.

26.6 Unequal Sample Sizes

We now turn to some of the challenges that arise with unequal sample sizes. Our work on unequal sample sizes was motivated by a practical problem encountered when applying binomial subset selection procedures to consumer research data routinely collected by General Motors. This problem is particularly interesting in that it comprehends all of the components that have already been discussed - the need for a selection procedure, sample size determination to ensure adequate OCs, the potential for restricting the range on the parameter space which yields the inf P(CS), the possibility of data contamination, and the new challenge of unequal sample sizes. Here, we will focus on the unequal sample size aspect of the problem.

The Gupta–Sobel subset selection rule for unequal sample sizes depends upon the observed proportions ($\hat{p}_i = X_i/n_i$) instead of observed successes (X_i) used for equal sample sizes;

$$\text{Select } \pi_i \Leftrightarrow \frac{X_i}{n_i} \geq \max_{1 \leq j \leq k} \left(\frac{X_j}{n_j} \right) - d''',$$

where d''' satisfies the P^* condition: $\inf_\Omega P(CS) \geq P^*$.

With unequal sample sizes, the equation for inf P(CS) is considerably more complicated. The equation for inf P(CS) is:

$$\inf_\Omega P(CS) = \inf_{0 \leq p \leq 1} \sum_{l=0}^{n_{(k)}} \binom{n_{(k)}}{l} p^l q^{n_{(k)}-l} \prod_{j=1}^{k-1} \left[\sum_{r=0}^{\left[d''' n_{(j)} + \frac{n_{(j)}}{n_{(k)}} l \right]} \binom{n_{(j)}}{r} p^r q^{n_{(j)}-r} \right],$$

where $n_{(j)}$ denotes the n_i associated with the jth smallest among the true unknown success probabilities. This equation needs to be checked k times, assuming each of the populations as the best. That is, P(CS) needs to be checked assuming that each of the k populations appears in the first term of the equation and the other $k-1$ populations in the latter term.

The critical similarity between equal and unequal sample sizes is proven in Gupta and Sobel(1960). The inf P(CS) occurs when the true p_i's are equal. However, there are several notable differences. With equal n, and $k = 2$, the inf P(CS) occurs at $p = 1/2$. For unequal sample sizes this analytic result has not been proven. Gupta and McDonald(1986) showed empirically that the common p yielding the inf P(CS) always occurs at values for $p \leq 1/2$ for equal n. With unequal sample sizes, we have encountered cases where the inf P(CS) occured at $p > 1/2$. For both equal and unequal sample sizes, asymptotically (as sample sizes grow large) the inf P(CS) occurs at $p = 1/2$. Finally, as noted previously, extensive tables are available for equal sample sizes. For unequal sample sizes a computational procedure is required. The remainder of this section discusses several challenges we encountered while implementing the Gupta–Sobel procedure for unequal sample sizes.

26.6.1   Computational procedure

At the time these procedures were developed (1960's), carrying out the exact computations for unequal sample sizes was tedious for small sample sizes and practically impossible for larger samples. The improvements in computing speed now make exact computations practical even for large, unequal sample sizes.

The computational procedure differs between equal and unequal n. For equal n, one simply goes to a table to find the value of d''' that satisfies the P^*-condition. For unequal n, the computer program computes the P(CS) for each value of d''' equal to the relevant observed differences in the samples (max $\hat{p}_i - \hat{p}_j$). The computed P(CS)'s for each subset size is then checked against the prestated P^*-condition.

One interesting aside is worth noting. During testing, we were surprised at how slowly the P(CS) converged to $1/k$ for $d''' = 0$. When dealing with large sample approximations, the difference between $P(X_1 > X_2)$ versus $P(X_1 \geq X_2)$ is typically ignored. The probability $P(X_1 = X_2)$ is not negligible even with equal n as large as 500. For example, with $k = 2$ the P(CS)= 0.5126 with $P(X_1 = X_2) = 0.0252$. For a simple case ($k = 2, p = 0.5$), we derived the equation for $P(X_1 = X_2)$. Using a result given in Feller (1968, eqn 12.11) and Stirling's approximation, for this simple case $P(X_1 = X_2) \cong \sqrt{1/n\pi}$.

For general investigations (e.g., OC computations) a larger grid of values needs to be considered for unequal sample sizes than for equal sample sizes. For equal n, P(CS) "jumps" at increments of $1/n$. For unequal n, P(CS) "jumps" at overlaying grids of $1/(n_i n_j)$ with some overlap when the n_i's are not all prime.

When designing experiments, it will usually be easier to work with the equal n tables. The total sample size required could be estimated without knowing in advance the sample sizes that will be obtained for each population. The Gupta–Sobel equal n tables may provide a usable approximation for unequal n.

Gupta and Sobel (1960) explored an approach using the average of the unequal sample sizes as a surrogate equal sample size. We also explored a case where all of the sample sizes are assumed to be equal to the minimum sample size. This area of research has largely been unexplored. Other heuristics may lead to practical uses of the published equal sample size tables for unequal sample sizes.

26.6.2 Lack of stochastic order

Some of the characteristics of the Gupta–Sobel procedure for unequal sample sizes can be demonstrated using a simple example. The example, chosen for numerical convenience, is for three populations with sample sizes of 2, 2, and 4. The grid of potential d'''-values is multiples of $1/4$.

The relevant test statistics are denoted by $T_i = \max(X_j/n_j) - X_i/n_i$. The ith population would be chosen by the selection rule if $T_i \leq d'''$. The following four graphs display the probability of the first population being chosen (i.e., $P(T_1 \leq d''')$) and the probability of the third population being chosen (i.e., $P(T_3 \leq d''')$) for $d'''=0, 1/4, 1/2, 3/4$. The probability of the second population being chosen is the same as the first population being chosen since both populations have a sample size of 2. These calculations assume that the binomial populations all have a common equal value of the binomial probability parameter p, the configuration which yields the inf P(CS).

On each of these charts the two curves cross. Since these probability curves are polynomials and continuous functions, the implication is that the P(NBS) can be larger than the P(CS) for some values of the probability parameters!

The complexity introduced by unequal sample sizes creates several other disconcerting properties. First, local minima in P(CS) as a function of p can occur. When searching for the common value of p that gives the inf P(CS) so care must be taken that the global minimum is found.

With equal sample sizes, we empirically know that the inf P(CS) occurs at common p less than $1/2$. for unequal n, cases regularly occur where the inf P(CS) is obtained for common p values greater than $1/2$.

Intuitively, one might expect that either the population associated with the smallest or largest sample size would be "tagged" (chosen as the unobserved best when computing the inf P(CS)). We have encountered cases where an intermediate (in terms of sample size) population yields the inf P(CS) when tagged. Note that the computation of P(CS) only depends on the sample sizes. The observed number of successes does not enter into the equation.

Note again, for unequal n the equation for P(CS) is a polynomial of degree $\sum n_i$ and is quite complicated. None of these cases present a problem – one can still find the d'''-value producing the inf P(CS) that satisfies P^*-condition. However, greater caution is required when working with unequal sample sizes than with equal sample sizes.

Finally, we conclude with a particularly difficult case we encountered with actual General Motors data. In this case $k = 4$ with $n_1 = 110$, $n_2 = 131$, $n_3 = 194$, $n_4 = 224$. The observed $\hat{p}_i$'s were extremely low; ranging from 0.0089 to 0.0182 ($X_i = 2$ for each sample). Clearly, any statistical procedure will have problems drawing an inference from such data.

Sheet1

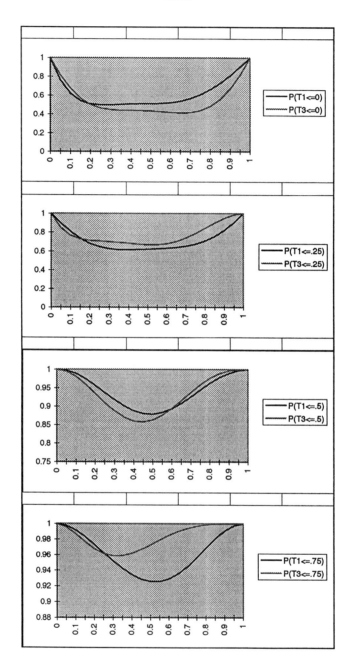

Figure 26.1: Operating characteristics (a) $k = 3$, $n_1 = n_2 = 2$, $n_3 = 4$, $d = 0$,
inf P(CS) = .4121 (at $p = .657$); (b) $k = 3$, $n_1 = n_2 = 2$, $n_3 = 4$, $d = .25$,
inf P(CS) = .6144 (at $p = .391$); (c) $k = 3$, $n_1 = n_2 = 2$, $n_3 = 4$, $d = .50$,
inf P(CS) = .8575 (at $p = .440$); (d) $k = 3$, $n_1 = n_2 = 2$, $n_3 = 4$, $d = .75$,
inf P(CS) = .9254 (at $p = .525$)

The Gupta–Sobel procedure for unequal sample sizes requires that the inf P(CS) be computed for $k-1$ values of d'''. Denote the (ascending) ordered observed probabilities by $\hat{p}_{[1]}, \hat{p}_{[2]}, \cdots, \hat{p}_{[k]}$. The first relevant value for d''' is $\hat{p}_{[k]} - \hat{p}_{[k-1]}$ which produces a subset containing only one population. For this particular case, $d''' = \hat{p}_{[4]} - \hat{p}_{[3]} = 0.0029$. The inf P(CS)=0.2360 (at common $p=0.674$). The inf P(CS) is then checked against the pre-stated P^*-condition to determine if a subset size of one is sufficient.

The second relevant value for d''', $\hat{p}_{[k]} - \hat{p}_{[k-2]}$, yields a subset containing only two populations. For this particular case, inf P(CS)=0.2715 (at common $p=0.500$). Using $d''' = \hat{p}_{[k]} - \hat{p}_{[k-3]}$ yields a subset containing only three populations. For this case, inf P(CS)=0.2803 (at common $p = 0.532$).

Note that the inf P(CS) is extremely low; less than what would be expected by simply selecting a population by chance! For a good procedure the true P(CS) clearly should not be less than chance selection, i.e., less than $1/k$. With this selection procedure, however, the theoretic inf P(CS) can be less than $1/k$. The lack of stochastic order that occurs with unequal sample sizes may seriously weaken the Gupta–Sobel procedure. Reporting inf P(CS) less than chance would be confusing to practitioners and renders the procedure practically unusable as it stands.

Note that the common p-value that minimizes P(CS) are all 0.5 or greater. Yet the observed $\hat{p}_i$'s are all less than 0.02. If the range on the p_i's were known in advance, then a restricted procedure could improve the OC's. For this problematic case, the improvement in the inf P(CS) was moderate under fairly strong restrictions ($0.003 < p < 0.032$). However, more typical cases have shown dramatic improvement in the inf P(CS) when the range on the p_i's was conservatively restricted.

26.7 Discussion

There are many practical situations that require the application of a binomial subset selection procedure. The Gupta–Sobel procedure is intuitively natural to users and easy to implement once the computational procedures are tabled (equal sample sizes) or programmed (unequal sample sizes).

There is no problem in guaranteeing the P^*-condition; a probability statement that addresses the least favorable configuration. Ensuring adequate OC's can be a challenge. Clearly, the procedure usually will provide a P(CS) substantively greater than the inf P(CS) guaranteed by the P^*-condition. However, working with the OC's requires that the user has some prior information about the problem. Examining the published OC tables, restricting the scope of the inference, or judging the impact of data contamination all require some prior knowledge.

The lack of stochastic order that occurs with unequal sample sizes creates additional challenges. There is no difficulty in ensuring the P^*-condition, but significant theoretical and practical difficulties arise. Generally, one should attempt to "balance" the experimental results so as to avoid these difficulties.

Acknowledgement. The authors gratefully acknowledge many helpful suggestions of the referee.

References

1. Feller, W. (1968). *An Introduction to Probability Theory and Its Applications*, New York: John Wiley & Sons.

2. Gupta, S. S., and Liao, Y. (1993). Subset selection procedures for binomial models based on a class of priors and some applications, In *Multiple Comparisons, Selection, and Applications in Biometry: A Festschrift in Honor of C. W. Dunnett* (Ed., F. M. Hoppe), pp.331-351, New York: Marcel Dekker, Inc.

3. Gupta, S. S., and McDonald, G. C. (1986). A statistical selection approach to binomial models, *Journal Quality Tech.*, 18 (No. 2), 103-115.

4. Gupta, S. S., and Sobel, M. (1960). Selecting a subset containing the best of several binomial populations, In *Contributions to Probability and Statistics* (Eds., I. Olkin, S. G. Ghurye, W. Hoeffding, W. G. Madow and H. B. Mann), pp. 219-230, Stanford: Stanford University Press.

5. McDonald, G. C. (1994a). Adjusting for data contamination in statistical inference, *Journal Qual. Tech.*, 26 (No. 2), 88-95.

6. McDonald, G. C. (1994b). Analyzing randomized response data with a binomial selection procedure, In *Proceedings of the Fifth Purdue International Symposium on Statistical Decision Theory and Related Topics* (Eds., S. S. Gupta and J. O. Berger), pp. 341-351, New York: Springer-Verlag.

27

The Use of Subset Selection in Combined-Array Experiments to Determine Optimal Product or Process Designs

Thomas Santner and Guohua Pan

Ohio State University, Columbus, OH
Oakland University, Rochester, MI

Abstract: In the quality control literature, a number of authors have advocated the use of combined-arrays in screening experiments to identify robust product designs or robust process designs [Shoemaker, Tsui and Wu (1991); Nair *et al.* (1992); Myers, Khuri and Vining (1992), among others]. This paper considers product design and process design applications in which there are one or more "control" factors that can be modified by the manufacturer, and one or more "environmental" (or "noise") factors that vary under field or manufacturing conditions. We show how Gupta's subset selection philosophy can be implemented in such a setting to identify optimal combinations of the levels of the control factors [Gupta (1956, 1965)]. By optimal, we mean those settings of the control factors that yield product designs whose performance is the most robust to variations in environmental factors. For process designs, the optimal settings of the control factors yield a fabrication method whose product quality is as nearly independent as possible to variations in the uncontrollable manufacturing factors, for example, to daily temperature fluctuations.

Keywords and phrases: Combined-array, inner array, maxmin approach, outer array, product-array, quality improvement, response model, screening, subset selection, variance reduction

27.1 Introduction

In his pioneering work on product and process improvement, Taguchi (1986) emphasizes two types of factors that effect product quality: "control factors" are those variables that can be (easily) manipulated by the manufacturer and "noise factors" are those variables that represent either different environmental conditions that affect the performance of a product in the field *or* (uncontrollable) variability in component parts or raw materials that affect the performance of an end-product. For experiments to improve product or process design, Taguchi advocates using statistical designs that are products of highly fractionated orthogonal arrays in the control and noise factors. In the case of product design, for example, the goal of such experiments is to determine conditions under which the mean product quality is independent of the noise factors. While some of Taguchi's proposals have been controversial [Box (1988)], the basic viewpoint that he advocates has been applied widely and with many successes [Taguchi and Phadke (1984)].

A number of authors have proposed statistical refinements to the Taguchi methodology [Shoemaker, Tsui, and Wu (1991); Nair *et al.* (1992); Myers, Khuri, and Vining (1992), for example]. One of these proposals is to use combined-arrays in the control and noise factors to design quality improvement experiments rather than Taguchi's product-arrays. At the expense of confounding higher-order interactions, carefully chosen combined-arrays allow the experimenter to determine interactions among the control factors and interactions among the noise factors, as well as the critical control factor by noise factor interactions that allow one to minimize the effect of noise factors in product quality. A second proposal is to apply response surface methodology to the combined-array data to identify parsimonious models for the quality characteristic(s) of interest; these models can be used to select the levels of the control factors.

This paper shows how the subset selection philosophy introduced by Gupta (1956, 1965) can be fruitfully used to screen for control factor combinations in quality improvement experiments based on data analytic models. Bechhofer, Santner and Goldsman (1995) give an overview of this field and present procedures to accomplish other important experimental goals. We focus on the case where the quality control characteristic of interest is to be maximized. Section 27.2 introduces a typical example, described by Box and Jones (1992), which is a study to improve the quality of a cake mix recipe by manipulating its ingredients, the control variables, to improve the taste of the final product when the baking is performed under a variety of time and oven temperature conditions, the noise variables. A 2^{5-1} combined-array experiment is described to study these variables and a model is developed relating the results of a taste test to the control and noise variables. Section 27.3 introduces the subset selection procedure proposed for identifying the recipe that maximizes the minimum mean taste test response where the minimum is taken over the levels of the oven temperature $\times$ baking time variables. The critical value required to implement the procedure is identified. Some generalizations and caveats are presented in the final section.

27.2 The Data

Box and Jones (1992) describe a study whose goal is to improve the taste of a cake mix that consumers bake under conditions that can vary from the directions printed on the cakebox. There are three control factors in this experiment: the amounts of Flour, Shortening, and Egg used in the cake mix which will be denoted by F, S, and E throughout. Box and Jones study two noise factors corresponding to variations of the baking temperature, denoted by T, and baking time, denoted by Z ("Zeit" is "time" in German). The response is a taste test, denoted by Y, with values ranging from 1 to 8; larger values correspond to a better tasting product.

While Box and Jones' emphasis was on the use of randomization in conducting a complete 2^5 experiment involving these factors, our goal is to illustrate the use of subset selection to screen candidates for the optimal recipe based on the data from a fractional factorial (combination-array) experiment. We use the 2_V^{5-1} design with defining contrast $I = FSETZ$ in this paper. The response and treatment combinations are listed in Table 27.1. Let

$$D \equiv \{(i, j, k, \ell, m) \text{ in the } 2_V^{5-1} \text{ design of Table 27.1}\}. \qquad (27.1)$$

Table 27.1: Results of 2_V^{5-1} cake tasting experiment with defining fraction $I = EFSTZ$

Flour (F)	Shortening (S)	Egg (E)	Temp (T)	Time (Z)	Taste (Y)
0	0	0	1	0	1.6
0	0	0	0	1	1.2
1	0	0	0	0	2.2
1	0	0	1	1	6.5
0	1	0	0	0	1.3
0	1	0	1	1	1.7
1	1	0	1	0	3.5
1	1	0	0	1	3.8
0	0	1	0	0	1.6
0	0	1	1	1	4.4
1	0	1	1	0	6.1
1	0	1	0	1	4.9
0	1	1	1	0	2.4
0	1	1	0	1	2.6
1	1	1	0	0	5.2
1	1	1	1	1	6.0

For this experimental design, the main effects are confounded with the 4-way interactions and the 2-way interactions are confounded with 3-way interactions.

The coefficients based on the model with all main effects and 2-way interactions are listed in Table 27.2 and a normal probability plot is shown in Figure 27.1. The normal probability plot suggests that only the shortening × temperature interaction, and the main effects of flour, egg, temperature, and time are important.

Table 27.2: Estimated effects for the fully saturated quadratic model based on the data in Table 27.1. The rows in boldface give, for the reduced model (27.2)–(27.3), the estimated effects and P-values for individual tests that coefficients are zero. Based on (27.2)–(27.3), the estimated standard error of each boldface coefficient is 0.1247

Term	Estimated Coefficient	P-value
(Intercept)	**3.4375**	**0.0000**
flour	**1.3375**	**0.0000**
shortening	**-0.125**	**0.3424**
egg	**0.7125**	**0.0003**
temp	**0.5875**	**0.0011**
time	**0.45**	**0.0057**
flour:shortening	-0.025	
flour:egg	0.0625	
flour:temp	0.1625	
flour:time	0.075	
shortening:egg	0.025	
shortening:temp	**-0.5**	**0.0031**
shortening:time	-0.2375	
egg:temp	-0.0125	
egg:time	-0.125	
temp:time	0.175	

We fit the reduced hierarchical model that includes the shortening × temperature interaction and the significant main effects

$$Y_{ijk\ell m} = \mu_{ijk\ell m} + \epsilon_{ijk\ell m} \quad (i,j,k,\ell,m) \in D \tag{27.2}$$

where we assume

$$\mu_{ijk\ell m} = \mu_0 + F_i + S_j + E_k + T_\ell + Z_m + (ST)_{j\ell}. \tag{27.3}$$

holds for all observed and unobserved treatment combinations i, j, k, ℓ, and m. The terms F_i, S_j, E_k, T_ℓ, and Z_m are the shortening, flour, egg, temperature, and time main effects, respectively. As usual, the sum of the parameters over any subscript is zero and $\epsilon_{ijk\ell m}$ are independent and $N(0, \sigma^2)$ distributed.

For Model 27.3, Table 27.2 lists, in boldface, the estimated coefficients and tests of the null hypotheses that individual coefficients are zero. The estimated

σ^2 is $s^2 = (0.4989)^2$ based on 9 degrees of freedom. The procedure of Section 27.3 will be applied to Model 27.3.

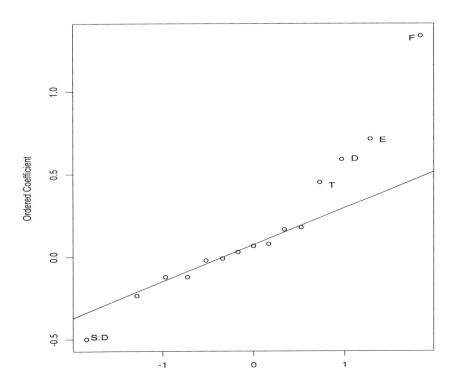

Figure 27.1: Normal probability plot of the 15 estimated coefficients in Table 27.2

27.3 The Procedure

Consider a generic setup in which there are r combinations of the levels of the control factor(s) and c combinations of the levels of the noise factor(s). For the cake tasting example introduced in Section 27.2, $r = 8 = 2^3$ and $c = 4 = 2^2$. In the following, we regard the r and c combinations as the number of levels of a single (composite) control factor and a single (composite) noise factor.

In this generic setup, suppose that the quality measure at i-th level of the control variable and j-th level of the noise variable is μ_{ij} which satisfies

$$\mu_{ij} = \boldsymbol{x}_{ij}^\top \boldsymbol{\beta} \quad (1 \le i \le r, 1 \le j \le c) \tag{27.4}$$

where $\boldsymbol{x}_{ij}$ is a vector of known covariates. The quantity $\xi_i = \min_{1 \le j \le c}\{\mu_{ij}\}$ measures the worst case (mean) performance of the ith level of the control variable against the noise variable. Let

$$\xi_{[1]} \le \cdots \le \xi_{[r]}$$

denote the ordered ξ_i values.

Suppose that we observe

$$Y_{ij} = \mu_{ij} + \epsilon_{ij} \quad (i,j) \in D^\star \tag{27.5}$$

where $D^\star$ is a subset of the direct product $\{1,\ldots,r\} \times \{1,\ldots,c\}$ and the $\{\epsilon_{ij}\}$ are independent mean zero normal random variables with common, unknown variance σ^2. We assume that the design is such that the least squares estimator of $\boldsymbol{\beta}$ based on (27.4) exists for data $\{Y_{ij}|(i,j) \in D^\star\}$. Throughout we subscript (and regard) probabilities involving the Y_{ij} as functions of $\boldsymbol{\mu}$ since our focus is selection in terms of means; we could equally well subscript them by $\boldsymbol{\beta}$.

Based on the data (27.5) from the combined-array experiment, our goal is to screen the levels of the control variable by selecting a subset of the control levels $\{1,\ldots,r\}$ so as to contain the optimum control level, $[r]$. We wish to achieve this goal subject to the following performance requirement of the procedure.

Design Requirement: Given α with $0 < \alpha < 1$, we desire that

$$P_{\boldsymbol{\mu}}\{CS\} \ge 1 - \alpha \tag{27.6}$$

for all $\boldsymbol{\mu}$ satisfying Model (27.4) where $[CS]$ denotes the event that the selected subset contains the control level associated with $\xi_{[r]}$.

To achieve the design goal (27.6), we use the LS estimator of $\boldsymbol{\beta}$ based on (27.4)–(27.5), denoted by $\widehat{\boldsymbol{\beta}}$, to form $\widehat{\mu}_{ij} = \boldsymbol{x}_{ij}^\top \widehat{\boldsymbol{\beta}}$ $(1 \le i \le r, 1 \le j \le c)$ and then estimate ξ_i by

$$\widehat{\xi}_i = \min_{1 \le j \le c}\{\widehat{\mu}_{ij}\} \quad (1 \le i \le r).$$

Let

$$\widehat{\xi}_{[1]} \le \cdots \le \widehat{\xi}_{[r]}$$

denote the ordered $\widehat{\xi}_i$. Also suppose that an estimator of σ^2 is available, denoted by S^2, for which $\nu S^2/\sigma^2$ has a chi-square distribution with known degrees of freedom ν and whose distribution is independent of $\{\widehat{\mu}_{ij}\}$. We propose the following procedure to select levels of the control factor.

Procedure $\mathcal{G}$: For given α, $0 < \alpha < 1$, select control level i if and only if

$$\widehat{\xi}_i \geq \widehat{\xi}_{[r]} - h\,S$$

where h is chosen so that

$$\min P_{\boldsymbol{\mu}}\{\widehat{\xi}_{(r)} \geq \widehat{\xi}_{[r]} - h\,S\} \geq 1 - \alpha, \tag{27.7}$$

$\widehat{\xi}_{(r)}$ denotes the estimator associated with $\xi_{[r]}$, and the minimum is taken over all $\boldsymbol{\mu} = (\mu_{ij})$ that satisfy (27.4).

In general, the choice of the constant h depends on the least favorable configuration, i.e., the $\boldsymbol{\mu}$ that minimizes the probability of correct selection in (27.7), which in turns depends on the specific model used in the experiment. As an example of the general methodology, this paper analyzes the Box and Jones cake mix data based on Model (27.2)–(27.3). As noted above, in this case the generic index i on ξ is the triple (i, j, k) with $r = 8$ values, the generic j on ξ is (ℓ, m) with $c = 4$ values, and

$$\xi_{ijk} = \min\{\mu_{ijk00}, \mu_{ijk01}, \mu_{ijk10}, \mu_{ijk11}\}.$$

For Model (27.2), it is straightforward to compute that

$$
\begin{aligned}
\xi_{ijk} &= F_i + E_k + \mu_0 + S_j + \min_{\ell,m}\{T_\ell + Z_m + (ST)_{j\ell}\} \\
&= F_i + E_k + \mu_0 + S_j - |Z_1| + \min_{\ell}\{T_\ell + (ST)_{j\ell}\} \tag{27.8}
\end{aligned}
$$

where (27.8) holds because $\min_{(a,b)\in A\times B}\{s_a + t_b\} = \min_{a\in A}\{s_a\} + \min_{b\in B}\{t_b\}$ and $\min_m\{Z_m\} = -|Z_1|$.

Some notation will be required to describe and analyze Procedure $\mathcal{G}$ for Model (27.2). Throughout, it will be convenient to let $\boldsymbol{\mu}$ denote the 8×4 matrix with (i, j, k)th row $(\mu_{ijk00}, \mu_{ijk01}, \mu_{ijk10}, \mu_{ijk11})$ and with row elements arranged so that the means corresponding to $S = 0$ are first and lexicographically according to (F, E) within each S level. Every ξ_{ijk} is the minimum of one row of $\boldsymbol{\mu}$. We regard the data to be collected using experimental design $\mathcal{D}$, denoted by $\boldsymbol{Y_D}$, as arranged conformably with $\boldsymbol{\mu}$ but with missing data in the positions where the design of Table 27.1 collects no observations. For example, using lower case notation for observed values, we have

$$
\underset{8\times4=}{\boldsymbol{y_D}}
\begin{bmatrix}
- & 1.2 & 1.6 & - \\
1.6 & - & - & 4.4 \\
2.2 & - & - & 6.5 \\
- & 4.9 & 6.1 & - \\
1.3 & - & - & 1.7 \\
- & 2.6 & 2.4 & - \\
- & 3.8 & 3.5 & - \\
5.2 & - & - & 6.0
\end{bmatrix}
\begin{matrix}
(F,S,E) \\
(0,0,0) \\
(0,0,1) \\
(1,0,0) \\
(1,0,1) \\
0,1,0) \\
(0,1,1) \\
(1,1,0) \\
(1,1,1)
\end{matrix}
$$

$$
\begin{matrix}
(0,0) & (0,1) & (1,0) & (1,1) \\
& (T,Z) &
\end{matrix}
$$

is the data in Table 27.1 arranged in this manner for D. The phrase "$\boldsymbol{Y_D}$ has mean $\boldsymbol{\mu}$" expresses the fact that the components of $\boldsymbol{Y_D}$ have means given by the corresponding element of $\boldsymbol{\mu}$.

The estimated ξ_{ijk} are

$$
\widehat{\xi}_{ijk} = \min\{\widehat{\mu}_{ijk00}, \widehat{\mu}_{ijk01}, \widehat{\mu}_{ijk10}, \widehat{\mu}_{ijk11}\} \tag{27.9}
$$

where

$$
\widehat{\mu}_{ijk\ell m} = \widehat{\mu}_0 + \widehat{F}_i + \widehat{S}_j + \widehat{E}_k + \widehat{T}_\ell + \widehat{Z}_m + \widehat{(ST)}_{j\ell}
$$

and the components are estimated by least squares based on (27.2). Explicitly, we have $\widehat{\mu}_0 = \overline{Y}_{.....}$, $\widehat{F}_i = \overline{Y}_{i....} - \overline{Y}_{.....}$, $\widehat{S}_j = \overline{Y}_{.j...} - \overline{Y}_{.....}$, $\widehat{E}_k = \overline{Y}_{..k..} - \overline{Y}_{.....}$, $\widehat{T}_\ell = \overline{Y}_{...\ell.} - \overline{Y}_{.....}$, $\widehat{Z}_m = \overline{Y}_{....m} - \overline{Y}_{.....}$, and $\widehat{(ST)}_{j\ell} = \overline{Y}_{.j.\ell.} - \overline{Y}_{.j...} - \overline{Y}_{...\ell.} + \overline{Y}_{.....}$. Here the averages are taken over the observations in the design D. The estimated $\widehat{\mu}_{ijk\ell m}$ for each control and noise factor combination are listed in Table 27.3 together with $\widehat{\xi}_{ijk}$ based on the four $T \times Z$ baking combinations.

Table 27.3: Estimated control variable means and row minimums for each (F, S, E) combination

(F,S,E)	(0,0)	(0,1)	(1,0)	(1,1)	$\widehat{\xi}_{F,S,E}$
(0,0,0)	1.3	1.2	1.6	3.1	1.2
(0,0,1)	1.6	2.3	3.5	4.4	1.6
(1,0,0)	2.2	3.2	5.5	6.5	2.2
(1,0,1)	4.1	4.9	6.1	6.3	4.1
(0,1,0)	1.3	1.5	1.2	1.7	1.2
(0,1,1)	1.9	2.6	2.4	2.2	1.9
(1,1,0)	3.7	3.8	3.5	4.2	3.5
(1,1,1)	5.2	5.5	5.8	6.0	5.2

(column group header: (T,Z) spanning (0,0), (0,1), (1,0), (1,1))

To achieve confidence level $100 \times (1 - \alpha)\%$, procedure $\mathcal{G}$ selects those (F,S,E) combinations i, j, k for which

$$\widehat{\xi}_{ijk} \geq \max_{i^\star, j^\star, k^\star} \widehat{\xi}_{i^\star j^\star k^\star} - h\,S \tag{27.10}$$

where the maximum is over $(i^\star, j^\star, k^\star) \in \{0, 1\}^3$ and h is given by Equation (27.11).

Theorem 27.3.1 *The procedure $\mathcal{G}$ attains confidence level $100 \times (1 - \alpha)\%$ for Model (27.2)–(27.3) if h is the solution of the equation*

$$P_{\boldsymbol{\mu}_{LFC}} \left\{ \widehat{\xi}_{111} \geq \max_{i,j,k} \widehat{\xi}_{ijk} - h\sqrt{W/9} \right\} = 1 - \alpha \tag{27.11}$$

where $\boldsymbol{Y_D}$ has mean $\boldsymbol{\mu}_{LFC} = \begin{pmatrix} 0 & +\infty \\ 0 & 0 \end{pmatrix} \otimes J_{4 \times 2}$ with $\otimes$ denoting Kronecker product, $J_{4 \times 2}$ is the 4×2 matrix of 1's, the variance is unity of each $Y_{ijk\ell m}$ with $(i, j, k, \ell, m) \in D$, and W has a chi-squared distribution with 9 degrees of freedom that is independent of $\boldsymbol{Y_D}$.

Before proving this result, note that $\xi_{ijk} = 0$ for all (i, j, k) under the (least favorable) configuration (LFC) of the means $\boldsymbol{\mu}_{LFC}$ described in the theorem. Hence we also have $\max_{i,j,k} \xi_{ijk} = 0$. This result is similar to many others in the subset selection literature concerning least favorable configurations—the parameter of interest is identical for the LFC.

The minimum of the probability of correct selection corresponds to the event that the sample estimator for the *last row*, $\widehat{\xi}_{111}$, is greater than the remaining seven $\widehat{\xi}_{ijk}$. This is different from many other (simple) problems in which the estimators of the parameters of interest are often independent and identically distributed; in the latter case, one computes the minimum probability over the parameter space as the probability that the estimator of any specific parameter exceeds the remaining parameter estimators by the yardstick for that problem. In this example, the marginal distributions of the $\widehat{\xi}_{ijk}$ are not all the same. The reason for focusing on $\widehat{\xi}_{111}$ will be demonstrated in the proof.

PROOF OF THEOREM 27.3.1. By Lemmas A.1 and A.2 in the Appendix, the LFC has the form—a Kronecker product of $\begin{pmatrix} 0 & v_{01} \\ 0 & v_{11} \end{pmatrix}$, $\begin{pmatrix} v_{00} & 0 \\ 0 & v_{11} \end{pmatrix}$, $\begin{pmatrix} 0 & v_{01} \\ v_{10} & 0 \end{pmatrix}$, or $\begin{pmatrix} 0 & v_{01} \\ 0 & v_{11} \end{pmatrix}$ and $J_{4 \times 2}$ where the $v_{i\ell}$ are non-negative. We consider only the first case, but the other three possibilities can be analyzed by analogous arguments and lead to infimums of the PCS that are equivalent to (27.11). Given $\boldsymbol{\mu}$, define

$$Y^\star_{ijk\ell m} \equiv \begin{cases} Y_{ijk\ell m} - v_{11} & \text{, when } (j, \ell) = (1, 1) \\ Y_{ijk\ell m} & \text{, when } (j, \ell) \neq (1, 1) \end{cases}$$

for $(i,j,k,\ell,m) \in D$. Then $\boldsymbol{Y_D}^\star_D$ has mean $\boldsymbol{\mu}^\star = \left(\begin{smallmatrix} 0 & v_{01} \\ 0 & 0 \end{smallmatrix}\right) \otimes J_{4\times2}$. Notice that all ξ_{ijk} and $\xi_{ijk}^\star$ are zero. Calculation shows that $\widehat{\xi}_{i0k}^\star = \widehat{\xi}_{i0k}$ and $\widehat{\xi}_{i1k}^\star = \widehat{\xi}_{i1k} - v_{11}$ for all (i,k). When the first element, say, of the last row is $\epsilon > 0$ then the event of correct selection is $[\widehat{\xi}_{111} \geq \widehat{\xi}_{ijk} - hS \ \forall \ (i,j,k)]$ and as $\epsilon \downarrow 0$, the PCS converges to

$$P_{\boldsymbol{\mu}}\{\widehat{\xi}_{111} \geq \widehat{\xi}_{ijk} - hS \ \forall \ (i,j,k)\}$$

$$= P_{\boldsymbol{\mu}}\{\widehat{\xi}_{111} - v_{11} \geq \widehat{\xi}_{ijk} - v_{11} \times I[j=1] - hS$$
$$- v_{11} + v_{11} \times I[j=1] \ \forall \ (i,j,k)\} \quad (27.12)$$

where I[E] is the indicator function of the event E. This gives

$$(27.12) \geq P_{\boldsymbol{\mu}}\{\widehat{\xi}_{111}^\star \geq \widehat{\xi}_{ijk} - hS \ \forall \ (i,j,k)\} \quad (27.13)$$
$$= P_{\boldsymbol{\mu}^\star}\{\widehat{\xi}_{111} \geq \widehat{\xi}_{ijk} - hS \ \forall \ (i,j,k)\}$$
$$= P_{\boldsymbol{\mu}^\star}\{CS\}$$

where (27.13) holds because $v_{11} \times I[i=1] - v_{11} \leq 0$. In a similar way it is straightforward to prove that $P_{\boldsymbol{\mu}}\{\widehat{\xi}_{111} \geq \widehat{\xi}_{ijk} - hS \ \forall \ (i,j,k)\}$ is decreasing in v_{01} when $\boldsymbol{\mu}$ has the form $\left(\begin{smallmatrix} 0 & v_{01} \\ 0 & 0 \end{smallmatrix}\right) \otimes J_{4\times2}$. At the configuration $\boldsymbol{\mu}_{LFC} = \left(\begin{smallmatrix} 0 & +\infty \\ 0 & 0 \end{smallmatrix}\right) \otimes J_{4\times2}$, after division of each $\widehat{\xi}_{ijk}$ by σ, we obtain (27.11) as the infimum of the PCS over all $\boldsymbol{\mu}$ satisfying (27.2).   ∎

The critical point h is computed as the $1-\alpha$ percentile of the distribution

$$V = \sqrt{9}(\widehat{\xi}_{111} - \max_{ijk}\widehat{\xi}_{ijk})/\sqrt{W}$$

where $W \sim \chi_9^2$ and each $\widehat{\xi}_{ijk}$ is calculated from (27.9) based on $\boldsymbol{Y_D}$ with the mean and variance structure of Theorem 27.3.1. Based on this description, we used simulation to calculate that $h = 1.344$ for our application when $\alpha = .05$. Thus the subset selection procedure selects those treatment combinations for which

$$\widehat{\xi}_{ijk} \geq 5.2 - 1.344 \times (.4989) = 5.2 - .67 = 4.53$$

which results in the *single combination* $(F,S,E) = (1,1,1)$ being selected (flour, shortening, and egg all at their high levels).

27.4 Discussion

Like many multiple comparison procedures for linear models, the confidence level of the procedure proposed in this paper is conditional based on the validity of the data analytic model determined in the first stage of the analysis. The procedure is fully justified if the model can be determined a priori by physical considerations or from a pilot data set. Its unconditional use with a single data set requires that a (complicated) two-stage stage procedure be developed that describes the probabilistic choice of data analytic model, including the subjective graphical aspects of the process, given a true model as well as the subsequent selection process for each possible model.

This paper develops a subset selection procedure for a specific model that describes the means of a factorial experiment. A general theory that will allow the determination of the least favorable configuration for arbitrary models is under development. Equally important, software that allows experimenters to easily compute critical values for such procedures will be required.

In practice, despite the fact that the combined-array fractional factorial experiments studied in this paper have relatively few runs compared to their full factorial versions, many experimenters may desire to use some form of randomization restriction in conducting the experiment. Typically, the use of randomization restriction will complicate the appropriate model by adding interaction terms. For example, Pan and Santner (1996) analyze procedures for selecting and screening best control treatments using complete factorial experiments run according to a split-plot design. It would also be possible to analyze such a procedure in this setting.

If the model used for the subset selection procedure does not permit degrees of freedom for estimating σ^2, then the experimenter has to either know σ^2 or have available an independent (chi-square) estimate from other sources. Subset selection cannot be performed unless one of these circumstances holds

This paper has restricted attention to the case where the experimenter wishes to maximize a quality characteristic. In other applications, it may be desired to move the process to a target value τ_0, say, subject to minimizing the product variability about τ_0. In the latter case it may be more appropriate to screen control treatments to find those treatments i having small values of $s_i^2 = \sum_{j=1}^{n_e} (\mu_{ij} - \tau_0)^2$ or some other measure of spread about the target. For normally distributed data, the former problem amounts to selection in terms of a non-central chi-square random variable and can be solved using techniques similar to those of Alam and Rizvi (1966).

Acknowledgements. Research of the first author was supported in part by Sonderforschungsbereich 386, Ludwig-Maximilians-Universität München, 80539

München. The authors would like to thank an anonymous referee for comments that helped improve this paper.

Appendix

Let Ω denote the set of $\boldsymbol{\mu}$ satisfying (27.2) and Ω_0 be those $\boldsymbol{\mu} \in \Omega$ for which $F_i \overset{i}{=} 0$, $E_k \overset{k}{=} 0$, and $Z_m \overset{m}{=} 0$. Thus

$$\Omega_0 \equiv \left\{ \boldsymbol{\mu} = \begin{pmatrix} v_{00} & v_{01} \\ v_{10} & v_{11} \end{pmatrix} \otimes J_{4\times 2} \,\middle|\, v_{ab} \quad \text{are real numbers,} \quad a, b = 0, 1 \right\}$$

Lemma A.1 *Suppose $\boldsymbol{Y_D}$ has mean $\boldsymbol{\mu}$ satisfying (27.2). Then Procedure $\mathcal{G}$ satisfies*

$$\inf_{\Omega} P\{CS\} = \inf_{\Omega_0} P\{CS\} \tag{A.1}$$

PROOF. Given $\boldsymbol{\mu} \in \Omega$ set

$$Y^*_{ijk\ell m} \equiv Y_{ijk\ell m} - F_i - E_k - Z_m$$

for $(i, j, k, \ell, m) \in D$. Then $Y^*_{ijk\ell m} \sim N(\mu^*_{ijk\ell m} = \mu_0 + S_j + T_\ell + (ST)_{j\ell}, \sigma^2)$ and $\boldsymbol{\mu}^* = (\mu^*_{ijk\ell m}) \in \Omega_0$. It is straightforward to show that the ξ_{ijk} based on the $\mu_{ijk\ell m}$ and the ξ^*_{ijk} based on the $\mu^*_{ijk\ell m}$ are related by $\xi_{ijk} = F_i + E_k - |Z_1| + \xi^*_{ijk}$. Similarly, the $\widehat{\xi}_{ijk}$ based on $\boldsymbol{Y_D}$ and the $\widehat{\xi}^*_{ijk}$ based on $\boldsymbol{Y^*_D}$ satisfy $\widehat{\xi}_{ijk} = \widehat{F}_i + \widehat{E}_k - |\widehat{Z}_1| + \min_\ell \{\widehat{\mu}_0 + \widehat{S}_j + \widehat{T}_\ell + (\widehat{ST})_{j\ell}\} = \widehat{\xi}^*_{ijk} + F_i + E_k + |\widehat{Z}_1 - Z_1| - |\widehat{Z}_1|$ for all i, j, k, ℓ, and m. We assume without loss of generality that $F_1 = \max_i\{F_i\}$, $E_1 = \max_k\{E_k\}$, and $\min_\ell(S_1 + (ST)_{1\ell}) = \max_j \min_\ell(S_j + (ST)_{j\ell})$ so that

$$\max_{i,j,k}\{\xi_{ijk}\} = F_1 + E_1 + Z_1 + \max_j \min_\ell\{\mu_0 + S_j + T_\ell + (S \times T)_{j\ell}\} = \xi_{111}$$

and $\xi^*_{111} \geq \xi^*_{ijk}$ for all i, j, and k. Then by definition,

$$
\begin{aligned}
P_{\boldsymbol{\mu}}\{CS\} &= P_{\boldsymbol{\mu}}\{\widehat{\xi}_{111} \geq \widehat{\xi}_{ijk} - hS \ \forall \ (i,j,k)\} \\
&= P_{\boldsymbol{\mu}}\{\widehat{\xi}_{111} - F_1 - E_1 - |\widehat{Z}_1 - Z_1| + |\widehat{Z}_1| \\
&\quad\quad \geq \widehat{\xi}_{ijk} - F_i - E_k - |\widehat{Z}_1 - Z_1| + |\widehat{Z}_1| \\
&\quad\quad\quad + F_i - F_1 + E_k - E_1 - hS \ \forall \ (i,j,k)\} \\
&\geq P_{\boldsymbol{\mu}}\{\widehat{\xi}^*_{111} \geq \widehat{\xi}^*_{ijk} - hS \ \forall \ (i,j,k)\} \tag{A.2} \\
&= P_{\boldsymbol{\mu}^*}\{\widehat{\xi}_{111} \geq \widehat{\xi}_{ijk} - hS \ \forall \ (i,j,k)\} \tag{A.3}
\end{aligned}
$$

where (A.2) holds because $F_i - F_1 \leq 0$ for $i = 0, 1$ and $E_k - E_1 \leq 0$ for $k = 0, 1$. ∎

Given $\boldsymbol{\mu} \in \Omega_0$ let $v = v(\boldsymbol{\mu})$ be the 2×2 matrix defined by $\mu_{ijk\ell m} = \mu_0 + S_j + T_\ell + (ST)_{j\ell} = v_{j\ell}$, say. Let

$$\Omega_{00} \equiv \{\boldsymbol{\mu} \in \Omega_0 | \min\{v_{00}, v_{01}\} = 0 = \min\{v_{10}, v_{11}\}\}.$$

Notice that $v_{j\ell} \geq 0$ for all $\boldsymbol{\mu} \in \Omega_{00}$, and that at least two of the $v_{j\ell}$ are exactly zero in one of the patterns

$$\begin{pmatrix} 0 & v_{01} \\ 0 & v_{11} \end{pmatrix}, \begin{pmatrix} v_{00} & 0 \\ 0 & v_{11} \end{pmatrix}, \begin{pmatrix} 0 & v_{01} \\ v_{10} & 0 \end{pmatrix}, \text{ or } \begin{pmatrix} 0 & v_{01} \\ 0 & v_{11} \end{pmatrix}.$$

Lemma A.2 *Suppose* $\boldsymbol{Y_D}$ *has mean* $\boldsymbol{\mu}$ *satisfying (27.2). Then Procedure* $\mathcal{G}$ *satisfies*

$$\inf_{\Omega_0} P\{CS\} = \inf_{\Omega_{00}} P\{CS\} \tag{A.4}$$

PROOF. Given $\boldsymbol{\mu} \in \Omega_0$, let $v_j = \min\{v_{j0}, v_{j1}\}$ for $j = 0, 1$. Then $\xi_{ijk} = v_j$ for all i, j, k. Without loss of generality assume that $v_1 = \max\{v_0, v_1\}$. Thus $\xi_{010} = \xi_{011} = \xi_{110} = \xi_{111} = v_1 \geq \xi_{000} = \xi_{001} = \xi_{100} = \xi_{101} = v_0$. Set $Y^*_{ijk\ell m} \equiv Y_{ijk\ell m} - v_j$ for $(i, j, k, \ell, m) \in D$. We have $Y^*_{ijk\ell m} \sim N(\mu^*_{ijk\ell m}, \sigma^2)$ where

$$\mu^*_{ijk\ell m} = v_{j\ell} - v_j = \begin{cases} = 0 & \text{if } v_{j\ell} = v_j \\ \geq 0 & \text{if } v_{j\ell} \geq v_j \end{cases}$$

so that $\boldsymbol{\mu}^* = (\mu^*_{ijk\ell m}) \in \Omega_{00}$. It can be calculated that $\xi^*_{ijk} = \xi_{ijk} - v_j$ and $\widehat{\xi}^*_{ijk} = \widehat{\xi}_{ijk} - v_j$ for all (i, j, k). Arguing as in Lemma A.1, we obtain that the probability of correct selection satisfies

$$\begin{aligned} P_{\boldsymbol{\mu}}\{CS\} &= P\{\widehat{\xi}_{111} \geq \widehat{\xi}_{ijk} - hS \ \forall \ (i,j,k)\} \\ &= P\{\widehat{\xi}_{111} - v_1 \geq \widehat{\xi}_{ijk} - v_j + v_j - v_1 - hS \ \forall \ (i,j,k)\} \\ &\geq P\{\widehat{\xi}_{111} - v_1 \geq \widehat{\xi}_{ijk} - v_j - hS \ \forall \ (i,j,k)\} \tag{A.5} \\ &= P_{\boldsymbol{\mu}^*}\{\widehat{\xi}_{111} \geq \widehat{\xi}_{ijk} - hS \ \forall \ (i,j,k)\} \tag{A.6} \end{aligned}$$

where (A.5) holds because $v_1 \geq v_j$. ∎

References

1. Alam, K. and Rizvi, M. H. (1966). Selection from multivariate normal populations, *Annals of the Institute of Statistical Mathematics*, **18**, 307–318.

2. Bechhofer, R. E., Santner, T. J. and Goldsman, D. M. (1995). *Designing Experiments for Statistical Selection, Screening, and Multiple Comparisons*, New York: John Wiley & Sons.

3. Box, G. E. P. (1988). Signal-to-Noise ratios, performance criteria, and transformations, *Technometrics*, **30**, 1–17.

4. Box, G. E. P. and Jones, S. (1992). Split-plot designs for robust product experimentation, *Journal of Applied Statistics*, **19**, 3–26.

5. Gupta, S. S. (1956). *On A Decision Rule for a Problem in Ranking Mean*, Ph.D. Dissertation, Institute of Statistics, University of North Carolina, Chapel Hill, NC.

6. Gupta, S. S. (1965). On some multiple decision (selection and ranking) rules, *Technometrics*, **7**, 225–245.

7. Myers, R. H., Khuri, A. I., and Vining, G. (1992). Response surface alternatives to the Taguchi robust parameter design approach, *The American Statistician*, **46**, 131–139.

8. Nair, V. J. *et al.* (1992). Taguchi's parameter design: a panel discussion, *Technometrics*, **34**, 127–161.

9. Pan, G. and Santner, T. J. (1996). Selection and screening procedures to determine optimal product designs (submitted for publication).

10. Shoemaker, A. C., Tsui, K., and Wu, C.-F. J. (1991). Economical experimental methods for robust design, *Technometrics*, **33**, 415–427.

11. Taguchi, G. (1984). *Introduction to Quality Engineering: Designing Quality into Products and Processes*, Tokyo: Asian Productivity Organization.

12. Taguchi, G. and Phadke, M. S. (1984). Quality engineering through design optimization, Atlanta: Conference Record, *GLOBECOM84 Meeting, IEEE Communications Society*, 1106–1113.

28

Large–Sample Approximations to the Best Linear Unbiased Estimation and Best Linear Unbiased Prediction Based on Progressively Censored Samples and Some Applications

N. Balakrishnan and C. R. Rao

McMaster University, Hamilton, Canada
Pennsylvania State University, University Park, PA

Abstract: In this paper, we consider the situation where a life-testing experiment yields a Type-II progressively censored sample. We then develop large-sample approximations to the best linear unbiased estimators for the scale-parameter as well as for the location-scale parameter families of distributions. Large-sample expressions are also derived for the variances and covariance of these estimators. These results are used further to develop large-sample approximations to the best linear unbiased predictors of future failures. Finally, we present two examples in order to illustrate the methods of inference developed in this paper.

Keywords and phrases: Progressive censoring, order statistics, life-testing, best linear unbiased estimation, best linear unbiased prediction, exponential distribution, extreme value distribution, Weibull distribution, Uniform distribution

28.1 Introduction

Progressive Type-II censoring occurs when some live units are removed at the times of failure of a few units. Such a progressive Type-II censored sampling is certainly an economical way of securing data from a life-testing experiment, as compared to the cost of obtaining a complete sample. It also enables the observation of some extreme life-times while a conventional Type-II right censored sampling will prohibit the observation of extreme life-times. Thus, progressive Type-II censored sampling provides a nice compromise between the experiment completion time and the observation of some extreme life-times. Some properties of order statistics arising from a progressive Type-II censored sample, termed hereafter as *progressive Type-II censored order statistics*, have been discussed recently by Balakrishnan and Sandhu (1995) and Aggarwala and Balakrishnan (1997). Inference procedures based on progressive Type-II censored samples have also been developed for a number of distributions by several authors including Cohen (1963, 1966, 1975, 1976, 1991), Cohen and Norgaard (1977), Mann (1971), Thomas and Wilson (1972), Nelson (1982), Cohen and Whitten (1988), Balakrishnan and Cohen (1991), Viveros and Balakrishnan (1994), and Balakrishnan and Sandhu (1996).

Let us now consider the following progressive Type-II censoring scheme: Suppose n identical units with life-time distribution $F(\cdot)$ are placed on a life-testing experiment; at the time of the first failure, R_1 surviving units (out of $n-1$) are withdrawn from the experiment randomly; the life-test continues on with the remaining units and at the time of the second failure, R_2 surviving units (out of the $n-2-R_1$) are withdrawn from the experiment randomly; finally, at the time of the m-th failure, all the remaining R_m surviving units are withdrawn from the experiment. Clearly, $n = m + \sum_{i=1}^{m} R_i$, where m is the number of failed units and $\sum_{i=1}^{m} R_i$ is the number of progressively censored units. Let us denote these progressive Type-II censored order statistics by $X_{1:m:n}^{(R_1,\ldots,R_m)}, X_{2:m:n}^{(R_1,\ldots,R_m)}, \ldots, X_{m:m:n}^{(R_1,\ldots,R_m)}$. Then, if the life-times are from a continuous population with cumulative distribution function $F(x)$ and probability density function $f(x)$, the joint density function of the above m progressive Type-II censored order statistics is given by

$$f(x_1, x_2, \ldots, x_m) = C \prod_{i=1}^{m} f(x_i)\{1 - F(x_i)\}^{R_i}, \ x_1 < x_2 < \cdots < x_m, \quad (28.1)$$

where

$$C = n(n-1-R_1)(n-2-R_1-R_2)\cdots(n-m+1-R_1-\cdots-R_{m-1}).$$
$$(28.2)$$

From the joint density in (28.1), if one determines the means, variances and covariances of progressive Type-II censored order statistics, then these quantities can be used to determine the best linear unbiased estimators (BLUEs) of the scale or the location and scale parameters of the underlying life-time distribution. Unfortunately, the explicit derivation of the means, variances and covariances of progressive Type-II censored order statistics is possible only in very few cases such as the uniform, exponential and Pareto distributions. In all other cases, these quantities have to be determined with great difficulty by means of numerical integration techniques; even if this was done, there are simply too many possible configurations for $(m, R_1, \ldots, R_m)$ thus making the numerical tabulation of BLUEs almost an impossible task for most practical values of n.

In this chapter, we therefore develop large-sample approximations to the BLUEs for the scale-parameter as well as for the location-scale parameter families of distributions. Large-sample approximations are also developed for the variances and covariance of these estimators. These results are used further to develop large-sample approximations to the best linear unbiased predictors (BLUPs) of future failures. Finally, we present two examples in order to illustrate the methods of inference developed in this paper.

28.2 Approximations to Means, Variances and Covariances

Suppose the progressive Type-II censored order statistics have come from the Uniform(0,1) distribution. For convenience in notation, let us denote them by $U_{1:m:n}^{(R_1,\ldots,R_m)}, U_{2:m:n}^{(R_1,\ldots,R_m)}, \ldots, U_{m:m:n}^{(R_1,\ldots,R_m)}$. Then, from (28.1), we readily have their joint density function to be

$$f(u_1, u_2, \ldots, u_m) = C \prod_{i=1}^{m} (1 - u_i)^{R_i}, \quad 0 < u_1 < \cdots < u_m < 1, \qquad (28.3)$$

where C is as given in (28.2). From (28.3), Balakrishnan and Sandhu (1995) have established that the random variables

$$V_i = \frac{1 - U_{m-i+1:m:n}^{(R_1,\ldots,R_m)}}{1 - U_{m-i:m:n}^{(R_1,\ldots,R_m)}}, \quad i = 1, \ldots, m-1, \text{ and } V_m = 1 - U_{1:m:n}^{(R_1,\ldots,R_m)} \qquad (28.4)$$

are all mutually independent, and further that

$$W_i = V_i^{i+R_m+R_{m-1}+\cdots+R_{m-i+1}}, \quad i = 1, 2, \ldots, m \qquad (28.5)$$

are all independently and identically distributed as Uniform(0,1).

From (28.4), we can readily write

$$U_{i:m:n}^{(R_1,\ldots,R_m)} \stackrel{d}{=} 1 - \prod_{j=m-i+1}^{m} V_j, \quad i = 1, 2, \ldots, m, \qquad (28.6)$$

where V_j's are independently distributed as $\text{Beta}(j + \sum_{k=m-j+1}^{m} R_k, 1)$. Using this result, one can readily derive from (28.5) explicit expressions for means, variances and covariances of progressive Type-II censored order statistics from Uniform(0,1) distribution as follows:

$$E\left(U_{i:m:n}^{(R_1,\ldots,R_m)}\right) = \Pi_i = 1 - b_i, \quad i = 1, 2, \ldots, m, \qquad (28.7)$$

$$\text{Var}\left(U_{i:m:n}^{(R_1,\ldots,R_m)}\right) = a_i b_i, \quad i = 1, 2, \ldots, m, \qquad (28.8)$$

and

$$\text{Cov}\left(U_{i:m:n}^{(R_1,\dots,R_m)}, U_{j:m:n}^{(R_1,\dots,R_m)}\right) = a_i b_j, \qquad 1 \le i < j \le m, \qquad (28.9)$$

where, for $i = 1, 2, \dots, m$,

$$\begin{aligned}
a_i &= \prod_{k=1}^{i}\left\{\frac{m-k+2+R_k+R_{k+1}+\dots+R_m}{m-k+3+R_k+R_{k+1}+\dots+R_m}\right\} \\
&\quad - \prod_{k=1}^{i}\left\{\frac{m-k+1+R_k+R_{k+1}+\dots+R_m}{m-k+2+R_k+R_{k+1}+\dots+R_m}\right\}
\end{aligned} \qquad (28.10)$$

and

$$b_i = \prod_{k=1}^{i}\left\{\frac{m-k+1+R_k+R_{k+1}+\dots+R_m}{m-k+2+R_k+R_{k+1}+\dots+R_m}\right\}. \qquad (28.11)$$

Observe that, if we set $R_1 = \dots = R_m = 0$ so that there is no progressive censoring, then the above expressions deduce to $E(U_{i:n}) = \frac{i}{n+1}$, $\text{Var}(U_{i:n}) = \frac{i(n-i+1)}{(n+1)^2(n+2)}$ and $\text{Cov}(U_{i:n}, U_{j:n}) = \frac{i(n-j+1)}{(n+1)^2(n+2)}$ which are well-known results on order statistics from the Uniform$(0,1)$ distribution; see, for example, David (1981, pp. 35–36) and Arnold, Balakrishnan and Nagaraja (1992, pp. 19–20).

We shall now use these expressions to get large-sample approximations to the means, variances and covariances of progressive Type-II censored order statistics from an arbitrary continuous distribution $F(\cdot)$. From the inverse probability integral transformation, we readily have the relationship

$$Y_{i:m:n}^{(R_1,\dots,R_m)} \overset{d}{=} F^{-1}\left(U_{i:m:n}^{(R_1,\dots,R_m)}\right), \qquad (28.12)$$

where $F^{-1}(\cdot)$ is the inverse cumulative distribution function of the life-time distribution from which the progressive censored sample has come. Expanding the function on the right hand side of (28.12) in a Taylor series around $E(U_{i:m:n}^{(R_1,\dots,R_m)}) = \Pi_i$ and then taking expectations and retaining only the first term, we obtain the approximation

$$E\left(Y_{i:m:n}^{(R_1,\dots,R_m)}\right) \simeq F^{-1}(\Pi_i), \quad i = 1, 2, \dots, m, \qquad (28.13)$$

where Π_i is as given in (28.7). Proceeding similarly, we obtain the approximations

$$\text{Var}\left(Y_{i:m:n}^{(R_1,\dots,R_m)}\right) \simeq \left\{F^{-1^{(1)}}(\Pi_i)\right\}^2 a_i b_i, \quad i = 1, 2, \dots, m, \qquad (28.14)$$

and

$$\text{Cov}\left(Y_{i:m:n}^{(R_1,\dots,R_m)}, Y_{j:m:n}^{(R_1,\dots,R_m)}\right) \simeq F^{-1^{(1)}}(\Pi_i)F^{-1^{(1)}}(\Pi_j)a_i b_j, \ i < j, \quad (28.15)$$

where a_i and b_i are as given in (28.10) and (28.11), respectively, and $F^{-1^{(1)}}(u) = \frac{d}{du} F^{-1}(u)$. This type of a Taylor series approximation for the usual order statistics was given by David and Johnson (1954). As a matter of fact, if we set $R_1 = \cdots R_m = 0$, the above approximations reduce to the first terms in the corresponding formulas given by David and Johnson (1954). If more precision is needed in the approximations, then more terms could be retained in the Taylor series expansion.

28.3 Approximation to the BLUE for Scale-Parameter Family

Let us assume that the progressive Type-II censored sample $X_{1:m:n}^{(R_1,\ldots,R_m)}$, $X_{2:m:n}^{(R_1,\ldots,R_m)}, \ldots, X_{m:m:n}^{(R_1,\ldots,R_m)}$ has arisen from a life-time distribution belonging to the scale-parameter family, that is, with density function $f(x;\sigma) = \frac{1}{\sigma} f(\frac{x}{\sigma})$. In this case, either from the generalized least-squares theory [see Lloyd (1952)] or through Lagrangian multiplier method [see Balakrishnan and Rao (1997a,b)], one can get the BLUE of σ as

$$\sigma^* = \left(\frac{\boldsymbol{\alpha}^T \boldsymbol{\Sigma}^{-1}}{\boldsymbol{\alpha}^T \boldsymbol{\Sigma}^{-1} \boldsymbol{\alpha}} \right) \boldsymbol{X} \tag{28.16}$$

and its variance as

$$\text{Var}(\sigma^*) = \sigma^2 / (\boldsymbol{\alpha}^T \boldsymbol{\Sigma}^{-1} \boldsymbol{\alpha}). \tag{28.17}$$

In the above formulas, $\boldsymbol{X}$ denotes the observed vector of progressive Type-II censored order statistics $(X_{1:m:n}^{(R_1,\ldots,R_m)}, \ldots, X_{m:m:n}^{(R_1,\ldots,R_m)})^T$, $\boldsymbol{Y} = \boldsymbol{X}/\sigma$ denotes the vector of corresponding standardized order statistics, $\boldsymbol{\alpha} = E(\boldsymbol{Y})$, and $\boldsymbol{\Sigma} = \text{Var}(\boldsymbol{Y})$. The usage of the BLUE in (28.16) becomes somewhat restrictive due to the complications in determining the variance-covariance matrix $\boldsymbol{\Sigma}$ for any given life-time distribution F and also in getting $\boldsymbol{\Sigma}^{-1}$ for large m.

We, therefore, seek to derive a large-sample approximation to the BLUE of σ in (28.16) by making use of the approximate expressions of $\boldsymbol{\alpha}$ and $\boldsymbol{\Sigma}$ presented in the last section. To this end, we first note from (28.14) and (28.15) that the large-sample approximation to the variance-covariance matrix $\boldsymbol{\Sigma} = ((\Sigma_{i,j}))$ has the special form

$$\Sigma_{i,j} = \beta_i \gamma_j, \qquad 1 \le i \le j \le m, \tag{28.18}$$

where

$$\beta_i = F^{-1^{(1)}}(\Pi_i) a_i \quad \text{and} \quad \gamma_j = F^{-1^{(1)}}(\Pi_j) b_j. \tag{28.19}$$

Then, $\boldsymbol{\Sigma}^{-1} = ((\Sigma^{i,j}))$ is a symmetric tri-diagonal matrix given by [see, for example, Graybill (1983, p. 198) and Arnold, Balakrishnan and Nagaraja (1992, pp. 174–175)]

$$\Sigma^{i,j} = \begin{cases} \dfrac{a_2}{\{F^{-1^{(1)}}(\Pi_1)\}^2 \, a_1(a_2b_1 - a_1b_2)} & , \quad i = j = 1 \\[2ex] \dfrac{a_{i+1}b_{i-1} - a_{i-1}b_{i+1}}{\{F^{-1^{(1)}}(\Pi_i)\}^2 \, (a_ib_{i-1} - a_{i-1}b_i)(a_{i+1}b_i - a_ib_{i+1})} & , \quad i = j = 2, \ldots, m-1 \\[2ex] \dfrac{b_{m-1}}{\{F^{-1^{(1)}}(\Pi_m)\}^2 \, b_m(a_mb_{m-1} - a_{m-1}b_m)} & , \quad i = j = m \\[2ex] \dfrac{-1}{F^{-1^{(1)}}(\Pi_i)F^{-1^{(1)}}(\Pi_{i+1})(a_{i+1}b_i - a_ib_{i+1})} & , \quad i = j - 1 = 1, \ldots, m-1 \\[2ex] 0 & , \quad \text{otherwise.} \end{cases}$$

$$(28.20)$$

Upon using the expressions in (28.13) and (28.20), we obtain the large-sample approximation to the BLUE of σ in (28.16) as

$$\begin{aligned} \sigma^* = \frac{1}{\Delta}\Bigg[&\left\{ \frac{F^{-1}(\Pi_1)a_2}{\{F^{-1^{(1)}}(\Pi_1)\}^2 \, a_1(a_2b_1 - a_1b_2)} \right. \\ &\left. - \frac{F^{-1}(\Pi_2)}{F^{-1^{(1)}}(\Pi_1)F^{-1^{(1)}}(\Pi_2) \, (a_2b_1 - a_1b_2)} \right\} X_{1:m:n}^{(R_1,\ldots,R_m)} \\ &+ \sum_{i=2}^{m-1} \left\{ \frac{-F^{-1}(\Pi_{i-1})}{F^{-1^{(1)}}(\Pi_{i-1})F^{-1^{(1)}}(\Pi_i)(a_ib_{i-1} - a_{i-1}b_i)} \right. \\ &\quad - \frac{F^{-1}(\Pi_{i+1})}{F^{-1^{(1)}}(\Pi_i)F^{-1^{(1)}}(\Pi_{i+1})(a_{i+1}b_i - a_ib_{i+1})} \\ &\quad \left. + \frac{F^{-1}(\Pi_i)(a_{i+1}b_{i-1} - a_{i-1}b_{i+1})}{\{F^{-1^{(1)}}(\Pi_i)\}^2 \, (a_ib_{i-1} - a_{i-1}b_i)(a_{i+1}b_i - a_ib_{i+1})} \right\} X_{i:m:n}^{(R_1,\ldots,R_m)} \\ &+ \left\{ \frac{-F^{-1}(\Pi_{m-1})}{F^{-1^{(1)}}(\Pi_{m-1})F^{-1^{(1)}}(\Pi_m)(a_mb_{m-1} - a_{m-1}b_m)} \right. \\ &\quad \left. + \frac{F^{-1}(\Pi_m)b_{m-1}}{\{F^{-1^{(1)}}(\Pi_m)\}^2 \, b_m(a_mb_{m-1} - a_{m-1}b_m)} \right\} X_{m:m:n}^{(R_1,\ldots,R_m)} \Bigg], \end{aligned}$$

$$(28.21)$$

where

$$\Delta = \frac{\{F^{-1}(\Pi_1)\}^2 \, a_2}{\{F^{-1^{(1)}}(\Pi_1)\}^2 \, a_1(a_2b_1 - a_1b_2)}$$

$$+ \sum_{i=2}^{m-1} \frac{\{F^{-1}(\Pi_i)\}^2 \, (a_{i+1}b_{i-1} - a_{i-1}b_{i+1})}{\{F^{-1(1)}(\Pi_i)\}^2 \, (a_i b_{i-1} - a_{i-1}b_i)(a_{i+1}b_i - a_i b_{i+1})}$$

$$+ \frac{\{F^{-1}(\Pi_m)\}^2 \, b_{m-1}}{\{F^{-1(1)}(\Pi_m)\}^2 \, b_m(a_m b_{m-1} - a_{m-1}b_m)}$$

$$- 2 \sum_{i=1}^{m-1} \frac{F^{-1}(\Pi_i) F^{-1}(\Pi_{i+1})}{F^{-1(1)}(\Pi_i) F^{-1(1)}(\Pi_{i+1})(a_{i+1}b_i - a_i b_{i+1})} . \qquad (28.22)$$

Furthermore, the large-sample approximation to the variance of the BLUE of σ in (28.17) is given by

$$\mathrm{Var}(\sigma^*) = \sigma^2/\Delta, \qquad (28.23)$$

where Δ is as given in (28.22).

Example 28.3.1 Nelson (1982, p. 228, Table 6.1) reported data on times to breakdown of an insulating fluid in an accelerated test conducted at various test voltages. For illustrating the method of estimation developed in this section, let us consider the following progressive Type-II censored sample of size $m = 8$ generated from the $n = 19$ observations recorded at 34 kilovolts in Nelson's (1982) Table 6.1, as given by Viveros and Balakrishnan (1994):

Table 28.1: Progressive censored sample generated from the times to breakdown data on insulating fluid tested at 34 kilovolts by Nelson (1982)

i	1	2	3	4	5	6	7	8
$x_{i:8:19}$	0.19	0.78	0.96	1.31	2.78	4.85	6.50	7.35
R_i	0	0	3	0	3	0	0	5

In this case, let us assume a scale-parameter exponential distribution with density function

$$f(x;\sigma) = \frac{1}{\sigma} \, e^{-x/\sigma}, \qquad x \geq 0, \ \sigma > 0 \qquad (28.24)$$

as the time-to-breakdown distribution. Then, we note that the standardized variable $Y = X/\sigma$ has a standard exponential distribution in which case

$$F(y) = 1 - e^{-y}, \ F^{-1}(u) = -\ln(1-u) \text{ and } F^{-1(1)}(u) = 1/(1-u).$$

From (28.21), we now determine the BLUE of σ to be

$$\begin{aligned}
\sigma^* &= (0.12305 \times 0.19) + (0.12305 \times 0.78) + (0.47255 \times 0.96) \\
&\quad + (0.12458 \times 1.31) + (0.47153 \times 2.78) + (0.12798 \times 4.85) \\
&\quad + (0.12808 \times 6.50) + (0.82641 \times 7.35) \\
&= 9.57.
\end{aligned}$$

Similarly, from (28.22) we determine

$$\text{Var}(\sigma^*) = 0.11713\,\sigma^2$$

so that we obtain the standard error of the estimate σ^* to be

$$\text{SE}(\sigma^*) = \sigma^*(0.11713)^{1/2} = 3.28.$$

It is of interest to mention here that in this case the exact BLUE of σ (which is also the MLE) is given by $\frac{1}{m}\sum_{i=1}^{m}(R_i + 1)X_{i:m:n}^{(R_1,\ldots,R_m)}$ and its exact variance is σ^2/m; see, for example, Viveros and Balakrishnan (1994) and Balakrishnan and Sandhu (1996). These results yield the exact BLUE of σ to be 9.09 and its standard error to be 3.21. Observe the closeness of these values to the corresponding values obtained above from the large-sample approximations even though we have n and m to be small, viz., $n = 19$ and $m = 8$.

28.4 Approximations to the BLUEs for Location-Scale Parameter Family

Let us assume that the progressive Type-II censored sample $X_{1:m:n}^{(R_1,\ldots,R_m)}$, $X_{2:m:n}^{(R_1,\ldots,R_m)}, \ldots, X_{m:m:n}^{(R_1,\ldots,R_m)}$ has come from a life-time distribution belonging to the location-scale parameter family, that is, with density function $f(x;\mu,\sigma) = \frac{1}{\sigma}\,f(\frac{x-\mu}{\sigma})$. In this case, either from the generalized least-squares theory [see Lloyd (1952)] or through Lagrangian multiplier method [see Balakrishnan and Rao (1997a,b)], one can derive the BLUEs of μ and σ as

$$\mu^* = -\frac{1}{\Delta}\,\boldsymbol{\alpha}^T\boldsymbol{\Gamma}\boldsymbol{X} \quad \text{and} \quad \sigma^* = \frac{1}{\Delta}\,\mathbf{1}^T\boldsymbol{\Gamma}\boldsymbol{X} \tag{28.25}$$

and their variances and covariance as

$$\begin{aligned}
\text{Var}(\mu^*) &= \sigma^2(\boldsymbol{\alpha}^T\boldsymbol{\Sigma}^{-1}\boldsymbol{\alpha})/\Delta, \quad \text{Var}(\sigma^*) = \sigma^2(\mathbf{1}^T\boldsymbol{\Sigma}^{-1}\mathbf{1})/\Delta, \\
\text{Cov}(\mu^*,\sigma^*) &= -\sigma^2(\boldsymbol{\alpha}^T\boldsymbol{\Sigma}^{-1}\mathbf{1})/\Delta.
\end{aligned} \tag{28.26}$$

In the above formulas, $\boldsymbol{X}$ denotes the observed vector of progressive Type-II censored order statistics $(X_{1:m:n}^{(R_1,\ldots,R_m)}, \ldots, X_{m:m:n}^{(R_1,\ldots,R_m)})^T$, $\boldsymbol{Y} = (\boldsymbol{X} - \mu\mathbf{1})/\sigma$ denotes the vector of corresponding standardized order statistics, $\mathbf{1}$ is a $m \times 1$ column vector of 1's, $\boldsymbol{\alpha} = E(\boldsymbol{Y})$, $\boldsymbol{\Sigma} = \text{Var}(\boldsymbol{Y})$, $\boldsymbol{\Gamma}$ is a skew-symmetric matrix given by

$$\boldsymbol{\Gamma} = \boldsymbol{\Sigma}^{-1}(\mathbf{1}\boldsymbol{\alpha}^T - \boldsymbol{\alpha}\mathbf{1}^T)\boldsymbol{\Sigma}^{-1}, \tag{28.27}$$

and

$$\Delta = (\boldsymbol{\alpha}^T\boldsymbol{\Sigma}^{-1}\boldsymbol{\alpha})(\mathbf{1}^T\boldsymbol{\Sigma}^{-1}\mathbf{1}) - (\boldsymbol{\alpha}^T\boldsymbol{\Sigma}^{-1}\mathbf{1})^2. \tag{28.28}$$

As in the last section, we seek to derive here large-sample approximations to the BLUEs of μ and σ in (28.25) by making use of the approximate expressions of $\boldsymbol{\alpha}$ and $\boldsymbol{\Sigma}^{-1}$ presented in (28.13) and (28.20). For this purpose, let us denote

$$
\begin{aligned}
\Delta_1 = & \frac{\{F^{-1}(\Pi_1)\}^2 \, a_2}{\{F^{-1^{(1)}}(\Pi_1)\}^2 \, a_1(a_2 b_1 - a_1 b_2)} \\
& + \sum_{i=2}^{m-1} \frac{\{F^{-1}(\Pi_i)\}^2 \, (a_{i+1}b_{i-1} - a_{i-1}b_{i+1})}{\{F^{-1^{(1)}}(\Pi_i)\}^2 \, (a_i b_{i-1} - a_{i-1}b_i)(a_{i+1}b_i - a_i b_{i+1})} \\
& + \frac{\{F^{-1}(\Pi_m)\}^2 \, b_{m-1}}{\{F^{-1^{(1)}}(\Pi_m)\}^2 \, b_m(a_m b_{m-1} - a_{m-1}b_m)} \\
& - 2 \sum_{i=1}^{m-1} \frac{F^{-1}(\Pi_i) F^{-1}(\Pi_{i+1})}{F^{-1^{(1)}}(\Pi_i) F^{-1^{(1)}}(\Pi_{i+1})(a_{i+1}b_i - a_i b_{i+1})} ,
\end{aligned}
\tag{28.29}
$$

$$
\begin{aligned}
\Delta_2 = & \frac{a_2}{\{F^{-1^{(1)}}(\Pi_1)\}^2 \, a_1(a_2 b_1 - a_1 b_2)} \\
& + \sum_{i=2}^{m-1} \frac{a_{i+1}b_{i-1} - a_{i-1}b_{i+1}}{\{F^{-1^{(1)}}(\Pi_i)\}^2 \, (a_i b_{i-1} - a_{i-1}b_i)(a_{i+1}b_i - a_i b_{i+1})} \\
& + \frac{b_{m-1}}{\{F^{-1^{(1)}}(\Pi_m)\}^2 \, b_m(a_m b_{m-1} - a_{m-1}b_m)} \\
& - 2 \sum_{i=1}^{m-1} \frac{1}{F^{-1^{(1)}}(\Pi_i) F^{-1^{(1)}}(\Pi_{i+1})(a_{i+1}b_i - a_i b_{i+1})} ,
\end{aligned}
\tag{28.30}
$$

$$
\begin{aligned}
\Delta_3 = & \frac{F^{-1}(\Pi_1) a_2}{\{F^{-1^{(1)}}(\Pi_1)\}^2 \, a_1(a_2 b_1 - a_1 b_2)} \\
& + \sum_{i=2}^{m-1} \frac{F^{-1}(\Pi_i)(a_{i+1}b_{i-1} - a_{i-1}b_{i+1})}{\{F^{-1^{(1)}}(\Pi_i)\}^2 \, (a_i b_{i-1} - a_{i-1}b_i)(a_{i+1}b_i - a_i b_{i+1})} \\
& + \frac{F^{-1}(\Pi_m) b_{m-1}}{\{F^{-1^{(1)}}(\Pi_m)\}^2 \, b_m(a_m b_{m-1} - a_{m-1}b_m)} \\
& - \sum_{i=1}^{m-1} \frac{F^{-1}(\Pi_i) + F^{-1}(\Pi_{i+1})}{F^{-1^{(1)}}(\Pi_i) F^{-1^{(1)}}(\Pi_{i+1})(a_{i+1}b_i - a_i b_{i+1})} ,
\end{aligned}
\tag{28.31}
$$

and

$$
\Delta = \Delta_1 \Delta_2 - \Delta_3^2.
\tag{28.32}
$$

Then, from (28.26) we have the variances and covariance of the BLUEs to be

$$
\text{Var}(\mu^*) = \sigma^2 \Delta_1 / \Delta, \quad \text{Var}(\sigma^*) = \sigma^2 \Delta_2 / \Delta, \quad \text{Cov}(\mu^*, \sigma^*) = -\sigma^2 \Delta_3 / \Delta.
\tag{28.33}
$$

Example 28.4.1 Consider the log-times to breakdown of an insulating fluid in an accelerated test conducted at various test voltages reported by Nelson (1982, p. 228, Table 6.1). In this case, we have the following progressive Type-II censored sample of size $m = 8$ generated from the $n = 19$ observations recorded at 34 kilovolts in Nelson's (1982) Table 6.1, as presented by Viveros and Balakrishnan (1994):

Table 28.2: Progressive censored sample generated from the log-times to breakdown data on insulating fluid tested at 34 kilovolts by Nelson (1982)

i	1	2	3	4	5	6	7	8
$x_{i:8:19}$	-1.6608	-0.2485	-0.0409	0.2700	1.0224	1.5789	1.8718	1.9947
R_i	0	0	3	0	3	0	0	5

Now, let us assume an extreme value distribution with density function

$$f(x; \mu, \sigma) = \frac{1}{\sigma} e^{(x-\mu)/\sigma} e^{-e^{(x-\mu)/\sigma}}, \qquad -\infty < x < \infty, \qquad (28.34)$$

for the distribution of log-times to breakdown. Realize here that this assumption means that we are assuming a two-parameter Weibull distribution for the lifelengths of the units in the original time scale. Then, we have the standardized variable $Y = (X - \mu)/\sigma$ to have the standard extreme value distribution in which case

$$F(y) = 1 - e^{-e^y}, \quad F^{-1}(u) = \ln(-\ln(1-u)) \text{ and } F^{-1(1)}(u) = \frac{-1}{(1-u)\ln(1-u)}.$$

From (28.25), we determine the BLUEs of μ and σ to be

$$\begin{aligned}
\mu^* &= (-0.09888 \times -1.6608) + (-0.06737 \times -0.2485) + (-0.00296 \times -0.0409) \\
&\quad + (-0.04081 \times 0.2700) + (0.12238 \times 1.0224) + (0.00760 \times 1.5789) \\
&\quad + (0.04516 \times 1.8718) + (1.03488 \times 1.9947) \\
&= 2.456
\end{aligned}$$

and

$$\begin{aligned}
\sigma^* &= (-0.15392 \times -1.6608) + (-0.11755 \times -0.2485) + (-0.11670 \times -0.0409) \\
&\quad + (-0.10285 \times 0.2700) + (-0.03942 \times 1.0224) + (-0.07023 \times 1.5789) \\
&\quad + (0.04037 \times 1.8718) + (0.64104 \times 1.9947) \\
&= 1.31377.
\end{aligned}$$

From (28.33), we determine

$$\text{Var}(\mu^*) = 0.16442\sigma^2, \ \text{Var}(\sigma^*) = 0.10125\sigma^2, \ \text{Cov}(\mu^*, \sigma^*) = 0.06920\sigma^2$$

so that we obtain the standard errors of the estimates μ^* and σ^* to be

$$\text{SE}(\mu^*) = \sigma^*(0.16442)^{1/2} = 0.53272 \quad \text{and} \quad \text{SE}(\sigma^*) = \sigma^*(0.10125)^{1/2} = 0.41804.$$

It may be noted here that, by employing a numerical maximization method, Viveros and Balakrishnan (1994) determined in this case the maximum likelihood estimates of μ and σ to be 2.222 and 1.026, respectively. Keep in mind that these maximum likelihood estimates are both biased estimates.

28.5   Approximations to the BLUPs

Suppose a progressive Type-II censored sample has been observed from a life-testing experiment with R_m (> 0) units having been censored at the time of the last failure, viz., $X_{m:m:n}^{(R_1,\ldots,R_m)}$. Then, the experimenter may be interested in predicting the life-times of the last R_m units withdrawn from the life-test. For example, the experimenter may be interested in estimating the time of the next failure had the life-test experiment continued. Note in this case that we are simply interested in predicting $X_{m+1:m+1:n}^{(R_1,\ldots,R_{m-1},0,R_m-1)}$.

As Doganaksoy and Balakrishnan (1997) have shown recently, the best linear unbiased predictor of $X_{m+1:m+1:n}^{(R_1,\ldots,R_{m-1},0,R_m-1)}$ is that value which, when assumed to be the observed value of $X_{m+1:m+1:n}^{(R_1,\ldots,R_{m-1},0,R_m-1)}$, will make the BLUE of the parameter (either the scale or location) based on this progressive Type-II censored sample of size $m+1$ to be exactly equal to the corresponding BLUE based on the already observed progressive Type-II censored sample of size m. Consequently, the large-sample approximations to the BLUEs derived in Sections 28.3 and 28.4 can be used to determine the BLUPs of failure times corresponding to units censored after the last observed failure.

For illustration, let us consider the large-sample approximation to the BLUE σ^* presented in (28.21) based on the progressive Type-II censored sample $X_{1:m:n}^{(R_1,\ldots,R_m)},\ldots,X_{m:m:n}^{(R_1,\ldots,R_m)}$, and denote it by σ_m^* for convenience. Next, let us assume that we have the progressive Type-II censored sample $X_{1:m+1:n}^{(R_1,\ldots,R_{m-1},0,R_m-1)},\ldots,X_{m:m+1:n}^{(R_1,\ldots,R_{m-1},0,R_m-1)},X_{m+1:m+1:n}^{(R_1,\ldots,R_{m-1},0,R_m-1)}$; let us denote the BLUE σ^* in (28.21) computed based on this sample by σ_{m+1}^*. Note that $X_{i:m+1:n}^{(R_1,\ldots,R_{m-1},0,R_m-1)} = X_{i:m:n}^{(R_1,\ldots,R_m)}$ for $i = 1, 2, \ldots, m$. Then, the BLUP of $X_{m+1:m+1:n}^{(R_1,\ldots,R_{m-1},0,R_m-1)}$ can be determined simply from the equation $\sigma_{m+1}^* = \sigma_m^*$.

Let us consider Example 28.3.1 once again. The eighth failure was observed at 7.35 at which time the remaining five surviving units were withdrawn from the life-test. Instead, if the experiment had continued, at what time would the next failure have occurred? To answer this question, we proceed as follows. First of all, we have here $n = 19$, $m = 8$ and $\sigma_8^* = 9.57$. Next, from (28.21) we have the BLUE σ_9^* to be

$$
\begin{aligned}
\sigma_9^* &= (0.10915 \times 0.19) + (0.10916 \times 0.78) + (0.41919 \times 0.96) \\
&\quad + (0.11051 \times 1.31) + (0.41829 \times 2.78) + (0.11353 \times 4.85) \\
&\quad + (0.11362 \times 6.50) + (0.11376 \times 7.35) \\
&\quad + \left(0.61933 \times X_{9:9:19}^{(0,\dots,0,4)} \right) \\
&= 3.9412 + \left(0.61933 \times X_{9:9:19}^{(0,\dots,0,4)} \right).
\end{aligned}
$$

Upon equating this to $\sigma_8^* = 9.57$, we simply obtain the BLUP of $X_{9:9:19}^{(0,\dots,0,4)}$ to be 9.0885, which provides the answer to the question posed earlier.

Next, let us consider again Example 28.4.1. The eighth log-time to breakdown was observed as 1.9947 when the remaining five surviving units were withdrawn from the life-test. If we want to predict the log-time of the next breakdown if the life-test experiment had continued, we proceed as follows. First of all, we have here $n = 19$, $m = 8$ and $\mu_8^* = 2.456$. Next, from (28.25) we have the BLUE μ_9^* to be

$$
\begin{aligned}
\mu_9^* &= (-0.06042 \times -1.6608) + (-0.03942 \times -0.2485) + (0.01365 \times -0.0409) \\
&\quad + (-0.01952 \times 0.2700) + (0.10649 \times 1.0224) + (0.01596 \times 1.5789) \\
&\quad + (0.04286 \times 1.8718) + (0.07576 \times 1.9947) \\
&\quad + \left(0.86464 \times X_{9:9:19}^{(0,\dots,0,4)} \right) \\
&= 0.46973 + \left(0.86464 \times X_{9:9:19}^{(0,\dots,0,4)} \right).
\end{aligned}
$$

Upon equating this to $\mu_8^* = 2.456$, we simply obtain the BLUP of $X_{9:9:19}^{(0,\dots,0,4)}$ to be 2.2972. It is important to note here that we would have come up with the same prediction had we used the BLUE of σ instead. In order to see this, let us consider from (28.25) the BLUE σ_9^* given by

$$
\begin{aligned}
\sigma_9^* &= (-0.12772 \times -1.6608) + (-0.09850 \times -0.2485) + (-0.10539 \times -0.0409) \\
&\quad + (-0.08834 \times 0.2700) + (-0.05025 \times 1.0224) + (-0.06454 \times 1.5789) \\
&\quad + (-0.04194 \times 1.8718) + (-0.01236 \times 1.9947) \\
&\quad + \left(0.58903 \times X_{9:9:19}^{(0,\dots,0,4)} \right) \\
&= -0.039382 + \left(0.58903 \times X_{9:9:19}^{(0,\dots,0,4)} \right).
\end{aligned}
$$

Upon equating this to $\sigma_8^* = 1.31377$, we simply obtain the BLUP of $X_{9:9:19}^{(0,\dots,0,4)}$ to be 2.2972, which is exactly the same as the predicted value obtained earlier through the BLUE of μ. Furthermore, we may note that the predicted value of the log-time of breakdown obtained from the exponential model (which, incidentally, is not the BLUP) turns out to be $ln(9.0885) = 2.2070$.

The best linear unbiased prediction of the failure times corresponding to the remaining surviving units can be developed similarly by repeating this process and treating the predicted values as the observed values of the corresponding units.

References

1. Aggarwala, R. and Balakrishnan, N. (1997). Some properties of progressive censored order statistics from arbitrary and uniform distributions, with applications to inference and simulation, *Journal of Statistical Planning and Inference* (to appear).

2. Arnold, B. C., Balakrishnan, N. and Nagaraja, H. N. (1992). *A First Course in Order Statistics*, New York: John Wiley & Sons.

3. Balakrishnan, N. and Cohen, A. C. (1991). *Order Statistics and Inference: Estimation Methods*, San Diego: Academic Press.

4. Balakrishnan, N. and Rao, C. R. (1997a). A note on the best linear unbiased estimation based on order statistics, *The American Statistician* (to appear).

5. Balakrishnan, N. and Rao, C. R. (1997b). On the efficiency properties of BLUEs, *Journal of Statistical Planning and Inference* (to appear).

6. Balakrishnan, N. and Sandhu, R. A. (1995). A simple simulational algorithm for generating progressive Type-II censored samples, *The American Statistician*, **49**, 229–230.

7. Balakrishnan, N. and Sandhu, R. A. (1996). Best linear unbiased and maximum likelihood estimation for exponential distributions under general progressive Type-II censored samples, *Sankhyā, Series B*, **58**, 1–9.

8. Cohen, A. C. (1963). Progressively censored samples in life testing, *Technometrics*, **5**, 327–329.

9. Cohen, A. C. (1966). Life testing and early failure, *Technometrics*, **8**, 539–549.

10. Cohen, A. C. (1975). Multi-censored sampling in the three parameter log-normal distribution, *Technometrics*, **17**, 347–351.

11. Cohen, A. C. (1976). Progressively censored sampling in the three parameter log-normal distribution, *Technometrics*, **18**, 99–103.

12. Cohen, A. C. (1991). *Truncated and Censored Samples*, New York: Marcel Dekker.

13. Cohen, A. C. and Norgaard, N. J. (1977). Progressively censored sampling in the three parameter gamma distribution, *Technometrics*, **19**, 333–340.

14. Cohen, A. C. and Whitten, B. J. (1988). *Parameter Estimation in Reliability and Life Span Models*, New York: Marcel Dekker.

15. David, F. N. and Johnson, N. L. (1954). Statistical treatment of censored data. I. Fundamental formulae, *Biometrika*, **41**, 228–240.

16. David, H. A. (1981). *Order Statistics*, Second edition, New York: John Wiley & Sons.

17. Doganaksoy, N. and Balakrishnan, N. (1997). A useful property of best linear unbiased predictors with applications to life-testing, *The American Statistician* (to appear).

18. Graybill, F. A. (1983). *Matrices with Applications in Statistics*, Second edition, Belmont, California: Wadsworth.

19. Lloyd, E. H. (1952). Least-squares estimation of location and scale parameters using order statistics, *Biometrika*, **39**, 88-95.

20. Mann, N. R. (1971). Best linear invariant estimation for Weibull parameters under progressive censoring, *Technometrics*, **13**, 521–533.

21. Nelson, W. (1982). *Applied Life Data Analysis*, New York: John Wiley & Sons.

22. Thomas, D. R. and Wilson, W. M. (1972). Linear order statistic estimation for the two-parameter Weibull and extreme-value distributions from Type-II progressively censored samples, *Technometrics*, **14**, 679–691.

23. Viveros, R. and Balakrishnan, N. (1994). Interval estimation of life characteristics from progressively censored data, *Technometrics*, **36**, 84–91.

Subject Index